# Handbook of Mycology

# Handbook of Mycology

Edited by Thomas Carrey

MURPHY & MOORE
www.murphy-moorepublishing.com

Published by Murphy & Moore Publishing,
1 Rockefeller Plaza,
New York City, NY 10020, USA
www.murphy-moorepublishing.com

**Handbook of Mycology**
Edited by Thomas Carrey



International Standard Book Number: 978-1-63987-289-3 (Hardback)



**Cataloging-in-Publication Data**

Handbook of mycology / edited by Thomas Carrey.
p. cm.
Includes bibliographical references and index.
ISBN 978-1-63987-289-3
1. Mycology. 2. Microbiology. 3. Fungi. I. Carrey, Thomas.
QK603 .H36 2022
579.5--dc23

# TABLE OF CONTENTS

# PREFACE

This book has been a concerted effort by a group of academicians, researchers and scientists, who have contributed their research works for the realization of the book. This book has materialized in the wake of emerging advancements and innovations in this field. Therefore, the need of the hour was to compile all the required researches and disseminate the knowledge to a broad spectrum of people comprising of students, researchers and specialists of the field.

The branch of biology concerned with the study of the genetic and biochemical properties of fungi, their taxonomy, and their application as a source for food, tinder and medicine is known as mycology. Fungi are crucial to life on Earth due to their role as symbionts. Further, fungi can break complex biomolecules and pollutants and therefore are critical to the global carbon cycle. They are also the source of many drugs, such as penicillin, psilocybin, ciclosporin, etc. Mycology is closely related to the discipline of phytopathology, which studies plant disease. This book traces the progress of mycology and highlights some of its key applications. It will also provide interesting topics for research which interested readers can take up. Scientists and students actively engaged in this field will find this book full of crucial and unexplored concepts.

At the end of the preface, I would like to thank the authors for their brilliant chapters and the publisher for guiding us all-through the making of the book till its final stage. Also, I would like to thank my family for providing the support and encouragement throughout my academic career and research projects.

**Editor**

# 1

# A flavoprotein supports cell wall properties in the necrotrophic fungus *Alternaria brassicicola*

Sandrine Pigné[1], Agata Zykwinska[2,3], Etienne Janod[2], Stéphane Cuenot[2], Mohammed Kerkoud[1], Roxane Raulo[1], Nelly Bataillé-Simoneau[1], Muriel Marchi[1], Anthony Kwasiborski[1], Guillaume N'Guyen[1], Guillaume Mabilleau[4], Philippe Simoneau[1] and Thomas Guillemette[1*]

**Abstract**

**Background:** Flavin-dependent monooxygenases are involved in key biological processes as they catalyze a wide variety of chemo-, regio- and enantioselective oxygenation reactions. Flavoprotein monooxygenases are frequently encountered in micro-organisms, most of which require further functional and biocatalytic assessment. Here we investigated the function of the *AbMak1* gene, which encodes a group A flavin monooxygenase in the plant pathogenic fungus *Alternaria brassicicola*, by generating a deficient mutant and examining its phenotype.

**Results:** Functional analysis indicates that the AbMak1 protein is involved in cell wall biogenesis and influences the melanization process. We documented a significant decrease in melanin content in the Δ*abmak1* strain compared to the wild-type and complemented strains. We investigated the cell wall morphology and physical properties in the wild-type and transformants using electron and atomic force microscopy. These approaches confirmed the aberrant morphology of the conidial wall structure in the Δ*abmak1* strain which had an impact on hydrophilic adhesion and conidial surface stiffness. However, there was no significant impairment in growth, conidia formation, pathogenicity or susceptibility to various environmental stresses in the Δ*abmak1* strain.

**Conclusion:** This study sheds new light on the function of a fungal flavin-dependent monooxygenase, which plays an important role in melanization.

**Keywords:** Monooxygenase, Cell wall, Melanin, Fungus, Flavoprotein

## Background

Flavin-dependent monooxygenases are involved in a wide variety of biological processes, such as biosynthesis, catabolism and detoxification of various natural compounds and xenobiotics, in both prokaryotes and eukaryotes. They catalyze the incorporation of one atom of molecular oxygen into the substrate and these oxygenation reactions include, for instance, hydroxylation, epoxidation, Baeyer–Villiger oxidation or sulfoxidation (for reviews, see [1–3]). Specific monooxygenase-driven transformations are usually hard to achieve without using these enzymatic catalysts, which is why such enzymes (particularly flavin-dependent monooxygenases and cytochrome P450 monooxygenases) are of great interest for synthetic purposes. An analysis of genome sequences revealed that flavoprotein monooxygenases are frequently encountered in micro-organisms, most of which require functional and biocatalytic assessments [2, 3]. Eight groups of flavin monooxygenases can be distinguished on the basis of their structural features and functions [2]. Group A flavin monooxygenases are single-component enzymes that contain typical FAD binding regions and rely on NAD(P)H as their external electron donor. Typical class A substrates are aromatic compounds containing an hydroxyl or amino group. Only

*Correspondence: thomas.guillemette@univ-angers.fr
[1] IRHS, Agrocampus-Ouest, INRA, Université d'Angers, SFR 4207 QuaSaV, 49071 Beaucouzé, France
Full list of author information is available at the end of the article

about 70 group A monooxygenase members are currently known, many of which do not yet have an Enzyme Commission (EC) number.

In phytopathogenic fungi, one of the best known group A flavin monooxygenases is the MAK1 protein from *Nectria haematococca*. MAK1 specifically hydroxylates medicarpin and maackiain, converting them into less fungitoxic derivatives [4]. Medicarpin and maackiain are antifungal phytoalexins produced by many legumes, and are thought to be important components of the defense response of these legumes to certain fungal pathogens. In a previous study, in *Alternaria brassicicola*, we identified a gene encoding a class A flavin monooxygenase which was found to be upregulated by camalexin, the major phytoalexin produced by *Arabidopsis thaliana* [5]. *A. brassicicola* causes black spot disease in a wide range of Brassicaceae plants and is routinely used as a model necrotrophic pathogen in studies with *A. thaliana*. In the present study, we investigated the function of the *AbMak1* gene by generating a knockout mutant and examining its phenotype. Unexpectedly, our functional analyses showed that this protein is involved in cell wall biogenesis and influences the melanization process. Like other filamentous fungi, *Alternaria* species synthesize melanin via a 1,8-dihydroxynaphthalene (DHN) intermediate [6]. Melanins constitute a group of related pigments that are polymers of phenolic compounds, although the exact arrangement of these phenolic subunits is generally unclear [7]. These ubiquitous pigments are known to provide protection against damaging effects of environmental stresses such as ultraviolet (UV) irradiation, enzymatic lysis, extreme temperatures, oxidizing agents and ionizing radiation [8]. In addition, they play a role in the pathogenesis of some human and plant pathogenic fungi [9, 10].

## Results

### AbMak1 encodes a class A flavoprotein monooxygenase

The camalexin-induced sequence P1B3 (GenBank accession No. DY543080) was previously identified in *A. brassicicola* as an EST encoding a predicted protein having matches with the flavin-containing monooxygenase MAK1 described in *N. haematococca* [5]. The corresponding protein, here referred to as AbMak1 and annotated AB02358.1 through the interactive JGI fungal portal MycoCosm (http://genome.jgi.doe.gov/Altbr1/Altbr1.home.html), displayed strong sequence similarities with hypothetical proteins from other Dothideomycetes species. For instance, the resulting AbMak1 protein had 95% identity or more to the corresponding proteins described in *Pyrenophora teres* or in various *Cochliobolus* species. AbMak1 also displayed high similarities with homologs from other lineages, such as *Podospora anserina* (76%) and *Aspergillus niger* (40%). A lower degree of identity (26%) was obtained by alignment with the *N. haematococca* MAK1 protein. Only one copy of the gene was present in the draft genome sequence of *A. brassicicola*. Automatic annotation at the locus encoding AB02358.1 predicted four introns and an encoded 528 amino acid protein. Nevertheless, examination of the transcribed sequence showed that automated prediction of the fourth intron was wrong (data not shown), resulting in a protein containing 497 amino acids (Fig. 1; Additional file 1: Figure S1). This protein contains typical motifs of flavin monooxygenases belonging to group A. The first fingerprint sequence is the Rossmann fold, or $\beta\alpha\beta$-fold (containing the GXGXXG sequence), which is crucial for binding the ADP moiety of FAD [11]. The second FAD binding motif contains the GD sequence which contacts the riboflavin moiety of FAD [12]. The last fingerprint contains a highly conserved DG motif, which is involved in binding the pyrophosphate moieties of both FAD and NADPH [13].

### Generation of the AbMak1 disruption mutant and major morphological traits

Disruption of *AbMak1* in *A. brassicicola* was accomplished by replacing a part of the *AbMak1* ORF with a hygromycin B (Hyg B) resistance cassette (Fig. 2). One Hyg B-resistant transformant was generated (Δ*abmak1*) after transformation of protoplasts of the wild-type strain, regeneration and purification by single-spore isolation. PCR screening, carried out using primers homologous to the hygromycin resistance cassette and genomic sequence outside of the flanking regions, confirmed that integration of the replacement construct occurred by homologous recombination at the targeted loci for this mutant (Fig. 2). Disruption of the coding region in the transformant was further confirmed using internal primers. One complemented strain (Δ*abmak1-c)* was obtained by reintroducing the wild-type copy of *AbMak1* into the Δ*abMak1* mutant. In this case, the replacement cassette included a 629 bp 5′ region of the gene *AbMak1*, the *Nat* gene cassette conferring resistance to nourseothricin, and the wild-type allele of the *AbMak1* gene with its native promoter. The homologous recombination event occured at the previously transformed locus in the Δ*abMak1* mutant (as shown in Fig. 2) and led to the replacement of the the *Hph* gene cassette by the *Nat* gene cassette and to the intoduction of a wild-type copy of the *AbMak1* gene.

Microscopic observations revealed that Δ*abmak1* and Δ*abmak1-c* mutants displayed a normal hyphal morphology, growth rate and conidiation rate on standard potato dextrose agar (PDA) medium. The most obvious morphological change was the color of Δ*abmak1* conidia

**a**

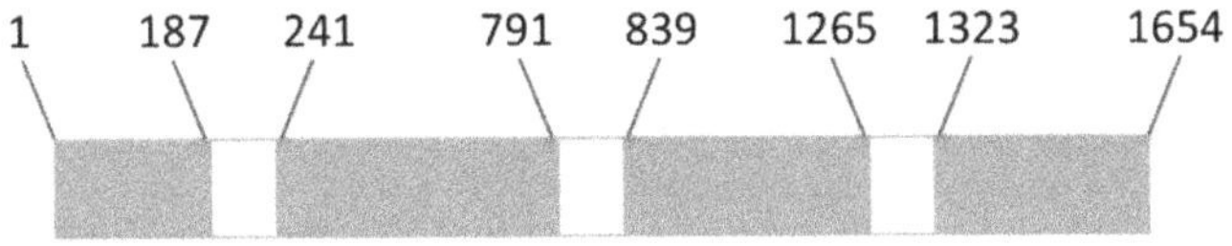


**b**

MPHALTSRDSTMESKPLGSYPNTGIDVLIV**G**T**G**LA**G**LVAALECTRKGHNVRVLERNADINTAGDMYFMGLS
ATRFLKHWPELLKEYKKISLHNAWIETFKHSGEVVIGPKKVADRLRAQGLDPDTPPGEFQMRPLVYKMFVSAV
ENLGVSVEFNRRVVDYFEDEQTGKGGCTTDDGKRYEADVVIAA**DG**VGSKSQKLVGGQVRARPSGRAMWRAAF
PREALDKDPEVKEFFKMMPGNEPIVRTWLGPSTYALTLSREDVMVWIMNHDVTGTEKESWNNTIEKEEVLDGM
DNMPGMETIKWAPIFKKLVETTPPNTIVNFELLWRNPQPSWTSPAARVIQI**GD**AAHSYLPASGNGATQAIED
AISIASCLQVGGKDNIPQSVRVHIRFRFIRNSCAQKLGFSNAELLQDTDWDKVKLDPRRAAPKLPKWVWSHDP
EAYVYENYELSAQSLKEGIDMKDEDKFEPNYPKGYKYEPWSIEKIMDDMRAGKPIDLGAGDWE

**Fig. 1** Sequence features of the *AbMak1* gene and corresponding protein. **a** Schematic *AbMak1* gene structure. The predicted intronic sequences are indicated in *white*. **b** Amino acid sequence of AbMak1. Typical motifs of flavin monooxygenases belonging to group A are indicated in *bold*

that appeared lighter than that of conidia produced by the reference strain or the Δ*abmak1-c* strain (Fig. 3). In order to validate a potential pigmentation alteration in Δ*abmak1* conidia, a semi-quantitative assay of melanin, based on spectrophotometry readings at 459 nm on melanin extracted from either isolated conidia or mycelium, was first used, as described by Babitskaia et al. [14]. The results showed a significant decrease in the melanin content of the Δ*abmak1* sample compared to the wild-type sample and partial recovery of the melanin level in the Δ*abmak1-c* strain (Fig. 4).

Another method, based on calculation of the integral intensity of electron paramagnetic resonance (EPR) spectra, was used and confirmed this significant melanin loss in *Δabmak1* cell walls. Different kinds of melanin biopolymers, such as eumelanin, pheomelanin and neuromelanin, exist and they have some interesting features such as redox properties [15]. These properties are due to delocalization of an electron between orthoquinone and catecholic moieties, giving rise to semiquinone free radicals. These radicals offer the opportunity for melanins to be involved in one- and two-electron redox reactions. The semiquinone free radicals trapped in melanin are responsible for the paramagnetic properties of melanin detectable by EPR. More precisely, melanin contains both o-semiquinone free radicals upon which unpaired electrons are localized on oxygen atoms with spin of ½ and biradicals with spin of 1 [15, 16]. EPR spectroscopy detects the absorption of energy relative to the transition of unpaired electrons from a low to a higher energy level. EPR spectra provide information about the concentration of paramagnetic centers, their type and distribution (homogeneous or non-homogeneous) in the samples [16].

Figure 5a shows three spectra recorded in the linear regime, with microwave powers lower than typical $P_{1/2}$ values of 1 mW. Such low $P_{1/2}$ values are in good agreement with those previously reported by Sarna and Hyde [17] for different melanin samples. These spectra were normalized with respect to both the experimental conditions and sample mass. For all samples, the signal shape was in general slightly asymmetric, resembling that expected from a powder pattern arising from a π–electron radical with an approximately axial g-tensor [17]. The same $S = ½$ free-radical signal was observed for the three samples, with a g-factor of 2.0035 for the wild-type and *Δabmak1-c* and of 2.0030 for *Δabmak1*, respectively. These very close values—consistent with those reported in the literature for o-semiquinone free radicals—indicated that the same type of melanin was present in the three samples [15, 16]. Indeed, the weak difference of 5e−4 in the g-factor is much smaller than the dispersion of g-values published for melanin, i.e. in the 2.0030–2.0060 range [15, 16]. The integral intensity of EPR spectra was calculated to estimate the relative contents of free radicals in the three samples. Indeed, the free radical concentration is directly proportional to this value [16]. The integral intensity calculation was done without any assumption on the spectra shape and the area under the absorption curves was obtained by double integrations. Figure 5b presents the integral intensities normalized with respect to that of the wild-type. These results clearly showed that the relative free radical

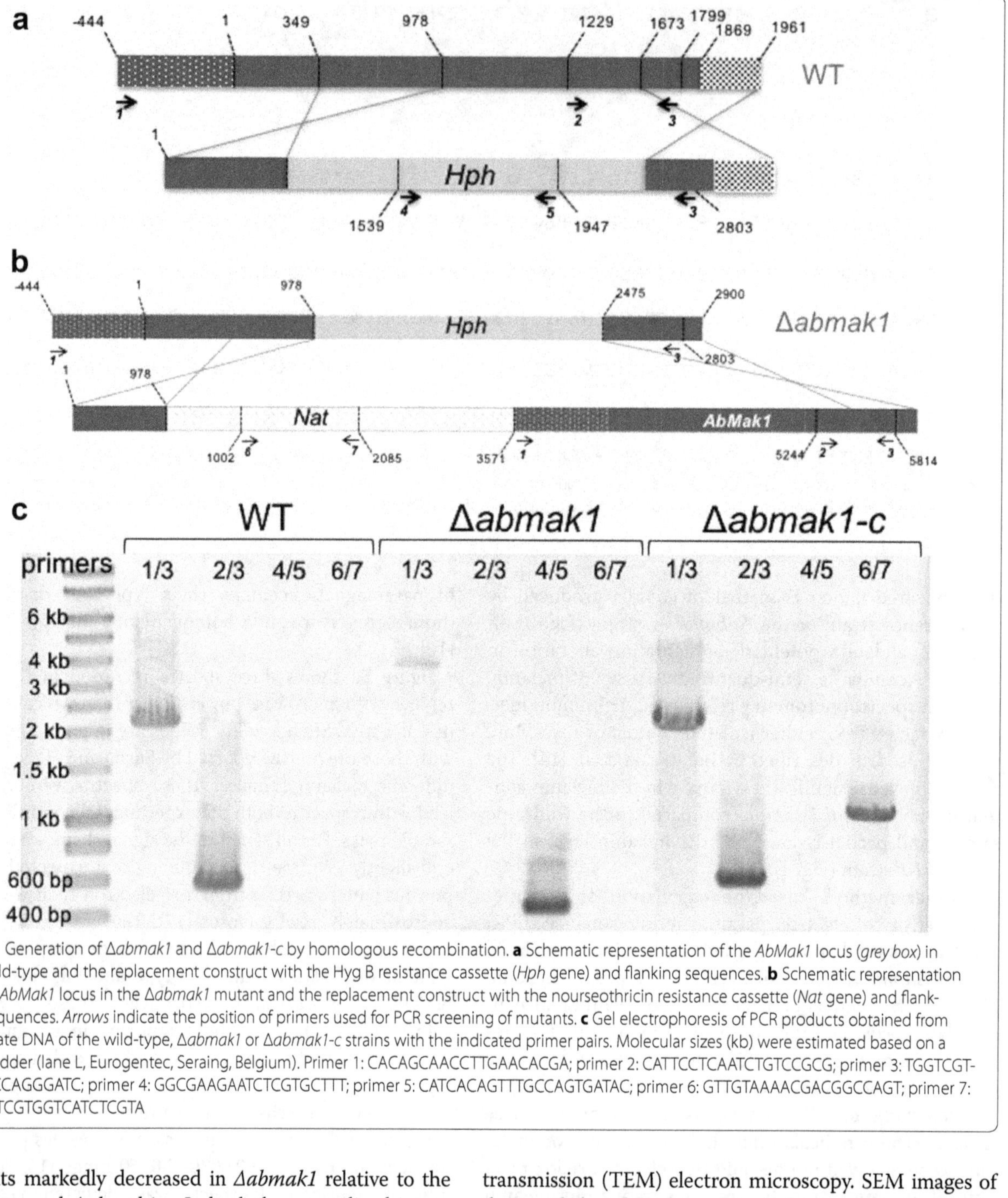


**Fig. 2** Generation of *Δabmak1* and *Δabmak1-c* by homologous recombination. **a** Schematic representation of the *AbMak1* locus (*grey box*) in the wild-type and the replacement construct with the Hyg B resistance cassette (*Hph* gene) and flanking sequences. **b** Schematic representation of the *AbMak1* locus in the *Δabmak1* mutant and the replacement construct with the nourseothricin resistance cassette (*Nat* gene) and flanking sequences. *Arrows* indicate the position of primers used for PCR screening of mutants. **c** Gel electrophoresis of PCR products obtained from template DNA of the wild-type, *Δabmak1* or *Δabmak1-c* strains with the indicated primer pairs. Molecular sizes (kb) were estimated based on a 1 kb ladder (lane L, Eurogentec, Seraing, Belgium). Primer 1: CACAGCAACCTTGAACACGA; primer 2: CATTCCTCAATCTGTCCGCG; primer 3: TGGTCGTTACACCAGGGATC; primer 4: GGCGAAGAATCTCGTGCTTT; primer 5: CATCACAGTTTGCCAGTGATAC; primer 6: GTTGTAAAACGACGGCCAGT; primer 7: GGCTTCGTGGTCATCTCGTA

contents markedly decreased in *Δabmak1* relative to the wild-type and *Δabmak1-c*. Indeed, the normalized intensities were about 7 ± 2 and 70 ± 10% for *Δabmak1* and *Δabmak1-c*, respectively. Hence, these results suggest a substantial loss of melanin in *Δabmak1* and partial recovery of melanin in *Δabmak1-c*.

**Conidial wall ultrastructure imaged by electron microscopy**

We investigated the morphology of the wild-type and transformant conidia using scanning (SEM) and transmission (TEM) electron microscopy. SEM images of the conidial surface showed typical ornamentation for the wild-type and *Δabmak1-c* strains (Fig. 6a). However, the *Δabmak1* mutant displayed a highly altered conidial surface. Ornamentation was not found to be as regular as in the wild strain and surface subsidences were clearly visible.

The aberrant morphology of the mutant conidia was confirmed on TEM images of the conidial wall ultrastructure (Fig. 6b). Typical conidial walls are composed of several superimposed layers, with a thick electron transparent

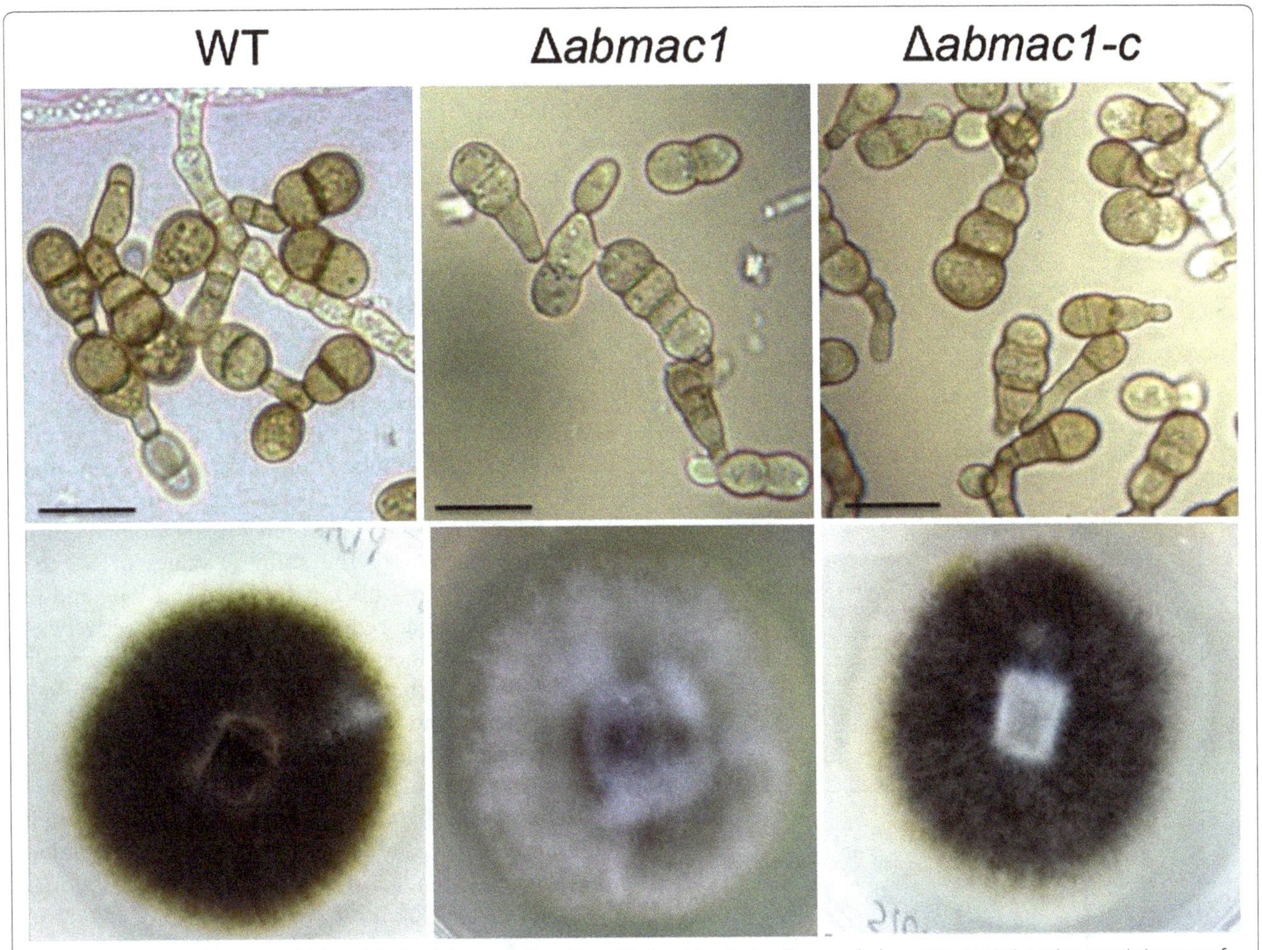


**Fig. 3** Morphological features of *A. brassicicola* wild-type, *Δabmak1 and Δabmak1-c* strains. *Top panels* show representative microscopic images of conidia. *Scale bars* indicate 20 μm. *Bottom panels* show representative macroscopic images of a colony

inner layer, a middle cell wall layer and a thin electron-dense outermost layer. TEM showed that, in the conidial cell wall of *Δabmak1*, the contours of the outermost layer appeared irregular and that the separation between primary and secondary walls was no longer visible. It is also apparent that the thickness of *Δabmak1* walls (511 nm ± 120) is greatly reduced compared to the thickness of *Δabmak1-c* (687 nm ± 113) and wild-type walls (809 nm ± 128) (Student's *t* test, $P < 0.01$). The *Δabmak1-c* cell wall ultrastucture was found to be much closer to that of wild-type cell walls.

### Investigation of the conidial surface by atomic force microscopy (AFM): imaging and force spectroscopy measurements

The AFM images of *A. brassicicola* conidia presented in the Additional file 2: Figure S2A revealed the presence of ornamentation on the wild-type cell wall surface, in agreement with the SEM observations (Fig. 6). In contrast, the *Δabmak1* conidial surface was significantly affected in the *Δabmak1* mutant, which led to less regular ornamentation and the presence of large smooth zones on the conidial surface. The cell wall surface morphology of *Δabmak1-c* conidia showed more similarities to that of the wild-type than to that of *Δabmak1*. However, neither the wild-type nor its mutants presented specific nanoscale structures on the cell wall surface, which remained perfectly smooth (Additional file 2: Figure S2B).

To investigate the chemical nature of the cell wall surface and probe any difference between the cell wall composition of Abra43 and that of *Δabmak1*, non-specific force-curves were measured with OH-modified tips (Fig. 7). Chemical force spectroscopy measurements using OH-modified probes revealed the presence of hydrophilic components, such as polysaccharides, OH groups of melanin and glycoproteins, on the outer conidial cell wall surface [18]. Force-curves recorded on the surface of wild-type conidia with OH tips showed large adhesion forces of 1.15 ± 0.3 nN, whereas a three-fold lower value of 0.35 ± 0.15 nN was obtained on the

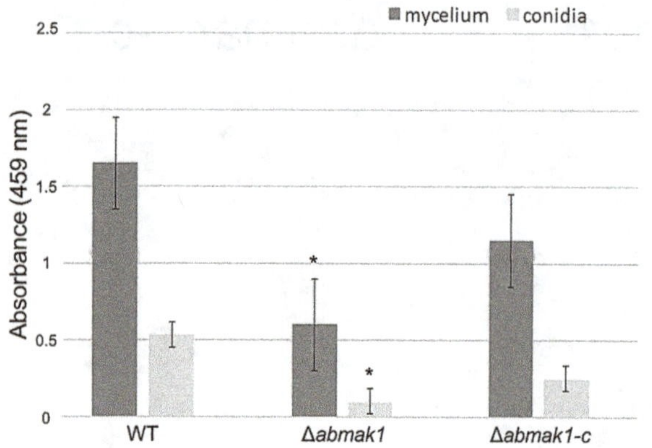


**Fig. 4** Semi-quantitative assay of melanin based on spectrophotometry readings at 459 nm on extracted melanin from either isolated conidia or mycelium. Mycelia or isolated conidia from 7-day-old cultures were extracted by an alkali–acid method to obtain melanin. The amount of melanin was extrapolated from the photometry absorbance results at 459 nm, as previously reported by Babitskaia et al. [14] and Alviano et al. [48]. *Asterisk* indicate a significant difference between the mutant and the parental isolate (Student's *t* test, P < 0.01)

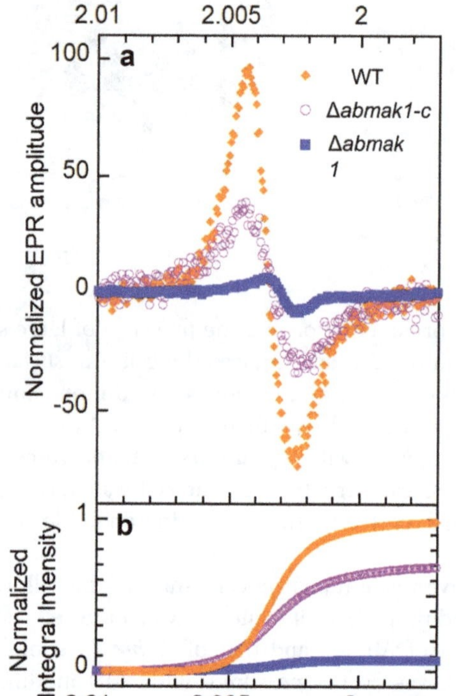


**Fig. 5** Calculation of the integral intensity of electron paramagnetic resonance (EPR) spectra. **a** Normalized X-band EPR spectra of *A. brassicicola* wild-type, *Δabmak1 Δabmak1-c* samples represented as a function of the g-factor. **b** Normalized integral intensities with respect to the maximum intensity of the wild-type

*Δabmak1* conidial surface. Hydrophilic OH/OH adhesion forces of 0.9 ± 0.3 nN measured on the *Δabmak1-c* conidial surface were very close to those obtained for the wild-type. The uncertainties, which correspond to the experimental dispersion, are determined from the Gaussian fits performed on each histogram (Fig. 7). A decrease in OH/OH adhesion measured on the *Δabmak1* conidial surface suggested that some cell wall components bearing OH groups were lacking on its surface. On the contrary, similar OH/OH adhesion values obtained on the wild-type and the *Δabmak1-c* cell wall surfaces indicated that their composition was similar.

Measurement of the conidial surface stiffness of different samples seemed to confirm that the conidial cell wall surface composition differed between them (Fig. 7). Indeed, twofold lower cell wall stiffness of 1.5 ± 0.5 N $m^{-1}$ was measured on the *Δabmak1* surface in comparison to the stiffness values of 3.2 ± 1 and 2.5 ± 0.7 N $m^{-1}$ obtained on the wild-type and the *Δabmak1-c* conidial cell wall surfaces, respectively. As for hydrophilic adhesion measurements, the uncertainties are obtained from the Gaussian fits of the stiffness measurements, and they reflect the experimental dispersion. The significantly lower elasticity measured on the *Δabmak1* surface highlighted that the mutation strongly impacted the cell wall composition. The hydrophilic adhesion and conidial surface stiffness results obtained for *Δabmak1-c* tended to be closer to those of the wild-type than to those of *Δabmak1*.

**Susceptibility of the Δabmak1 mutant to stress conditions**

Monitoring growth in solid PDA medium or in liquid PDB medium did not reveal any significant effect of *AbMak1* disruption, compared to the parental and complemented strains, on the mycelium growth rate, conidia germination and initial hyphal growth (data not shown). The results of analyses of growth curves in liquid medium supplemented with $H_2O_2$, allyl-isothiocyanate (Al-ITC) and camalexin or in solid PDA medium supplemented with Congo red (CR) and Calcofluor white (CFW) were used to assess the susceptibility of the monooxygenase mutant to oxidative stress, plant defense metabolites and cell wall stress. As shown in Table 1, none of the strains (*Δabmak1* and *Δabmak1-c)* showed increased susceptibility to any of these stresses compared to the wild-type. As PDA is a relatively rich medium, we also monitored fungal growth in a minimal medium (10 g $L^{-1}$ glucose, 1.65 g $L^{-1}$ $(NH_4)2SO_4$, 15 g $L^{-1}$ agar, 1 g $L^{-1}$ $KH_2PO_4$, 0.5 g $L^{-1}$ KCl, 0.5 g $L^{-1}$ $MgSO_4$, $7H_2O$, 0.01 g $L^{-1}$ $FeSO_4$, $7H_2O$). We did not noticed any significative susceptibility to plant defense metabolites and cell wall stress in this particular growth condition (data not shown).

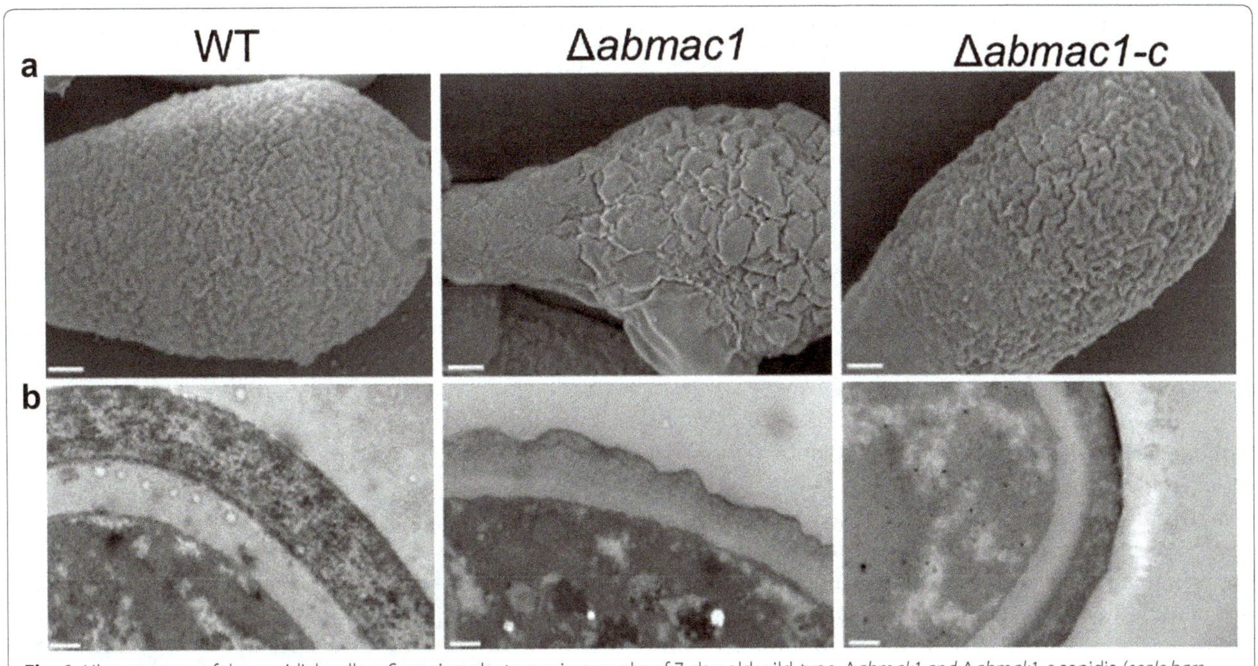


**Fig. 6** Ultrastructure of the conidial walls. **a** Scanning electron micrographs of 7-day-old wild-type, *Δabmak1 and Δabmak1-c* conidia (*scale bars* 1 μm). **b** Transmission electron micrographs of 7-day-old wild-type, *Δabmak1* and *Δabmak1-c* conidia (*scale bars* 0.2 μm)

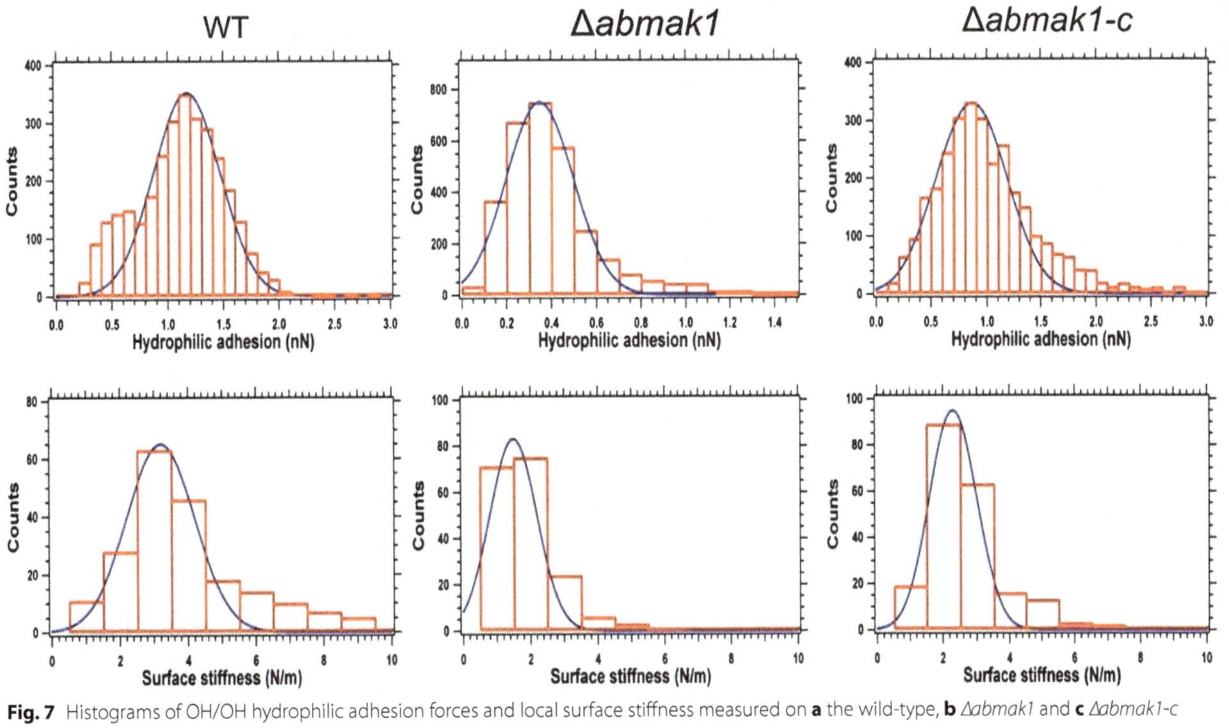


**Fig. 7** Histograms of OH/OH hydrophilic adhesion forces and local surface stiffness measured on **a** the wild-type, **b** *Δabmak1* and **c** *Δabmak1-c* strains

**Table 1 Susceptibility of *A. brassicicola* wild-type, *Δabmak1* and *Δabmak1-c* strains to different stress conditions**

| | $H_2O_2$ | | ITC | Camalexin | CR | | CFW | |
|---|---|---|---|---|---|---|---|---|
| | 5 mM | 10 mM | 2.5 mM | 25 μM | 200 mg $L^{-1}$ | 400 mg $L^{-1}$ | 200 mg $L^{-1}$ | 400 mg $L^{-1}$ |
| WT | 24 ± 4 | 84 ± 14 | 18 ± 5 | 51 ± 4 | 42 ± 7 | 55 ± 5 | 62 ± 5 | 73 ± 5 |
| *Δabmak1* | 33 ± 5 | 97 ± 1 | 20 ± 5 | 45.5 ± 2 | 46 ± 2 | 59 ± 5 | 61 ± 5 | 76 ± 3 |
| *Δabmak1-c* | 42 ± 5 | 89.5 ± 8 | 27 ± 8 | 51.5 ± 8 | 42 ± 2 | 56 ± 7 | 58 ± 5 | 81 ± 8 |

The results are expressed as the percentage of inhibition in treated samples compared to the control without additive. Conidia of each genotype were used to inoculate microplate wells containing standard PDB medium supplemented with the appropriate compound. Nephelometric growth was automatically recorded for 33 h at 24 °C. Each condition was tested in triplicate and the experiments were repeated twice. The areas under the curves were used to calculate the percentages of inhibition for each treatment compared to the control growth curves. Values are means of three biological repetitions and represent the percentage growth inhibition under stress conditions compared with standard growth conditions

### Pathogenic behavior of replacement mutants on vegetative organs

*Brassica oleracea* leaves were inoculated with drops of conidia suspension ($10^5$, $10^4$ or $10^3$ conidia/mL) to test the effects of targeted *AbMak1* gene knockout on pathogenicity (Additional file 3: Figure S3). The wild-type, *Δabmak1* and *Δabmak1-c* were all able to produce typical symptoms and, as determined from the lesion sizes at various inoculum loads, no significant decreases in aggressiveness were recorded for the mutants. Regardless of the inoculated strain, small necrotic symptoms were already observed at 3 days post-inoculation (dpi) and they continued to expand into large typical necrotic areas surrounded by chlorotic halos at 6 dpi. During late stages of infection, necrotic spots exhibited a dense conidia formation on the surface.

Inoculation experiments were also performed on leaves of two *A. thaliana* ecotypes: Landsberg erecta (*Ler*) and Columbia (Col-0). The wild-type and mutants were all able to produce typical symptoms and visual evaluation of the necrotic lesion area at 6 dpi showed no alteration in mutant aggressiveness (Fig. 8).

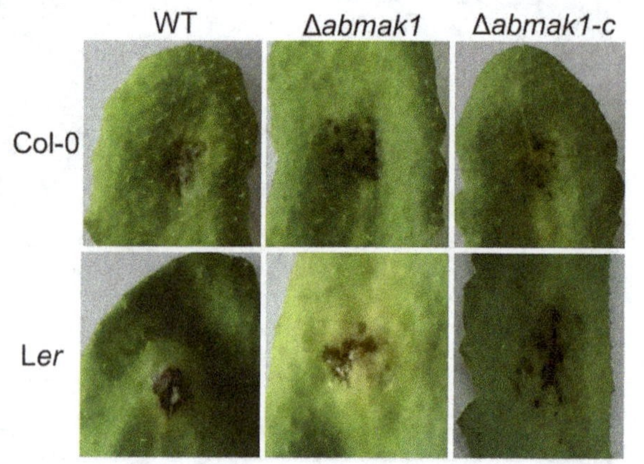


**Fig. 8** Representative symptoms obtained by inoculation of wild-type, *Δabmak1* and *Δabmak1-c* on Columbia (Col-0) and Landsberg erecta (*Ler*) *A. thaliana* ecotypes

## Discussion

We investigated the role of AbMak1, a group A flavin monooxygenase, in the plant pathogenic fungus *A. brassicicola* by generating a disruption mutant for the corresponding gene. Group A flavin monooxygenases comprise single-component enzymes that combine flavin reduction and monooxygenation in one polypeptide chain. They use FAD as a prosthetic group and mainly NADPH as an electron donor [19]. Well studied examples of enzymes belonging to subclass A are p-hydroxybenzoate hydroxylase, which is involved in degradation of aromatic compounds, and squalene monooxygenase, which catalyzes the first oxygenation step in sterol biosynthesis [20, 21]. Based on sequence homology and the gene expression profile, we initially hypothesized that AbMak1 could be involved in the metabolization of plant phytoalexins, as previously reported for the flavoprotein protein MAK1 from the filamentous fungus *Nectria haematococca*. Indeed, MAK1 is known to specifically hydroxylate the legume phytoalexins medicarpin and maackiain, converting them to less fungitoxic derivatives [4]. Moreover, this hypothesis was consistent with the fact that *AbMak1* was overexpressed upon exposure to camalexin, the major phytoalexin in *A. thaliana* [5]. However, a major contradictory result is that the mutant was not affected in its susceptibility to camalexin (Table 1) or to other cruciferous phytoalexins (data not shown). More generally, there was no significant impairment in growth, conidia formation or pathogenicity of the *Δabmak1* mutant. These results showed that AbMak1 was not a MAK1 ortholog in *A. brassicicola*.

As an unexpected phenotype, we observed that the loss of function of AbMak1 altered the melanin content. Like other dematiaceous fungi, *Alternaria* species produce 1,8-dihydroxynaphthalene (1,8-DHN) melanin that accumulates mainly in conidia cell walls. This ubiquitous pigment protects them from the damaging effects of environmental stress, contributes to the ability of fungi to survive in harsh environments and allows overwintering or dormancy of fungal propagules [22, 23], while also playing a role in fungal pathogenesis. In human pathogens, fungal melanin can modulate the host immune response by interfering with the normal function of phagocytic cells or altering the cytokine levels [24, 25]. Melanin is also critical to host invasion in some plant pathogens, such as *Magnaporthe oryzae*, by providing mechanical strength to the appressoria and allowing the organisms to penetrate plant tissues [26]. *A. brassicicola* does not use this mechanical strategy to penetrate host tissues, which may explain why the pathogenicity of the *Δabmak1* strain is not affected. As an *AbMak1* homologous gene exists in *M. oryzae*, it would be interesting to determine the extent to which these genes are involved in its pathogenicity. The lack of effect of melanin deficiency on the pathogenicity has also been reported in other plant-pathogenic fungi [27, 28]. Cho et al. [6] reported the functions of Amr1, a transcription factor that regulates melanin biosynthesis in *A. brassicicola*. These authors determined that *Δamr1* mutants were melanin-deficient but, unexpectedly, more virulent than the wild-type, suggesting that loss of the gene was beneficial to pathogenesis. RNA-seq analysis of interactions during late-stage pathogenesis revealed that the expression of a subset of genes involved in the melanin biosynthesis pathway was regulated by Amr1. AB02358.1, that encodes AbMak1, belongs to this subset. In contrast, many hydrolytic enzyme-coding genes were expressed at higher levels in Amr1 mutants than in the wild-type during pathogenesis, indicating that this subset of genes was negatively regulated by the transcription factor during this infection process. The authors speculated that this transcription factor promotes long-term survival due to its role in melanin biosynthesis, at the expense of virulence, thus contributing to the success of *A. brassicicola* as a competitive saprophyte and plant parasite. Interestingly, four other genes located in the vicinity of *AbMak1* were also found to be regulated by Amr1 (Additional file 4: Figure S4) suggesting that this region represents a gene cluster involved in melanin biosynthesis. One gene (AB02359.1) encode a cytochrome P450 and the other three (AB02355.1, AB02356.1 and AB02357.1) encode hypothetical proteins. Deletion of these genes should be considered and could lead to obtain more marked phenotypes.

The physical conidia surface properties were also investigated in the different *A. brassicicola* strains by atomic force microscopy. Chemical force spectroscopy measurements using OH-modified probes revealed a potential increase in *Δabmak1* conidial surface hydrophobicity since some cell wall components bearing OH groups are lacking on its surface. It should be noted that typical hydrophobic rodlet layers, which have been visualised by AFM examination of the *A. fumigatus* conidial surface [29], have not been observed in the *A. brassicicola* wild-type or in *A. brassicicola* mutants. These rodlet layers are formed by the self-assembly of particular amphiphatic proteins called hydrophobins, that are able to coat hydrophobic or hydrophilic surfaces and reverse their hydropathy character [30]. According to our observations, it therefore seems that these proteins do not participate in the hydropathy profile of *A. brassicicola* conidia and they are not altered by *AbMak1* disruption. The *Δabmak1* surface also exhibited a significantly lower elasticity compared to the wild-type or complemented strain. This character is probably linked to the decreased thickness of *Δabmak1* walls we observed on TEM images. This result highlighted the fact that the mutation not only impacted the surface hydropathy character but also, more generally, the cell wall structure and properties. However, this alteration did not seem to contribute to the marked loss of conidia adherence properties since the pathogenicity on leaves was not impaired. Contrary to our results, the complete loss of melanin in *A. fumigatus* led to the lack of some hydrophobic components on the conidial surface and modified the conidial adhesion to laminin and fibronectin [29]. Another mechanism by which pigment may contribute to virulence concerns its ability to confer some resistance to reactive oxygen species (ROS), a major host antimicrobial effector system. Jahn et al. [31] reported that an *A. fumigatus* isolate lacking conidial pigmentation displayed higher susceptibility to oxidative attack in vitro. In this study, the *AbMak1* mutant did not show increased susceptibility to hydrogen peroxyde or allyl-ITC, which are known to induce intracellular ROS accumulation [32]. More generally, it should be noted that the marked phenotypes, such as the increased susceptibility to oxidants or the loss of virulence that were reported in other studies on fungal melanins, were obtained from nonpigmented isolates. As shown in Fig. 4, the *A. brassicicola AbMak1* mutant did not exhibit a complete melanin defect. This partial melanin deficiency observed in *Δabmak1* could therefore explain the lack of increased susceptibility to the applied stresses.

At present, it is hard to determine how the monooxygenase AbMak1 specifically acts on the melanin structure. Melanin is an amorphous polymer of phenolic compounds that is both hydrophobic and negatively charged [33]. Nevertheless, the precise physicochemical nature of melanin is not yet fully understood [34], mainly because melanin is insoluble, so many traditional biochemical techniques are unsuitable for studying this pigment [7, 9]. However, microscopic studies have revealed that melanins form granular particles localized in cell walls, where they are likely crosslinked to polysaccharides [35–37]. The wide variety of pathogenic dematiceous fungi synthesize their melanins from 1,8-dihydroxynaphthalene (DHN)-melanin from the precursor molecules acetyl coA or malonyl coA. The first step is the synthesis of 1,3,6,8-tetrahydroxynaphthalene (1,3,6,8-THN), which is catalyzed by a polyketide synthase. Then, a series of reduction and dehydration reactions produce the intermediates scytalone, 1,3,8-trihydroxynaphthalene, vermelone, and finally 1,8-dihydroxynaphthalene (DHN), whose polymerization leads to melanin formation [7, 9]. We do not believe that the AbMak1 protein is involved in the early stages of synthesis since these steps have been relatively well studied in fungi [9, 38], and also because the application of pyroquilon, an inhibitor of the hydroxynaphtalene reductase, did not modify the *Δabmak1* phenotype compared to the wild-type (data not shown). Most of the later reaction steps involved in melanin synthesis would require further investigation, and we favor the hypothesis that the protein is involved in melanin polymerization or in crosslinking to cell wall components, such as chitin [39], via a hydroxylation step.

Detailed insight into the DHN-melanin synthesis process in fungi is important primarily because this pigment contributes to virulence in several human and plant pathogenic fungi. Enzymes involved in the DHN-melanin biosynthetic pathway are thus emerging targets for the development of selective fungicides since this pigment is not synthesized in host organisms [40]. Biological control strategies have also been envisaged to limit the accumulation and persistence of plant pathogens by degrading their melanin content or inhibiting its production [41].

## Conclusions

In this study, we identified a fungal flavin-dependent monooxygenase that plays a role in DHN-melanization. Mutation of the gene encoding AbMak1 resulted in major alteration of the cell wall structure, including a decrease in the melanin content and a probable modification of its chemical structure. The physical and chemical properties of the conidia surface were also altered but not enough to impact the pathogenicity and susceptibility of the fungus to various stress conditions. As homologous genes are present in other Ascomycota genomes, this enzyme likely has a major role throughout this phylum. As the definition of the melanin structure is beyond our current technological capability, the exact impact of this flavin-dependent monooxygenase on the melanin chemical structure remains unclear.

## Methods

### Strains and growth conditions

The *A. brassicicola* wild-type strain Abra43 used in this study has previously been described [42]. For routine cultures, fungi were maintained at 24 °C by transferring hyphal plugs on 3.9% (w/v) PDA (Difco) or on agar-solidified Vogel's medium N supplemented with 1.2% (wt/vol) sucrose. For radial growth assays, agar disks were cut from the margin of a 7-day-old colony growing on PDA and were transferred onto the centre of PDA medium supplemented with the compounds under investigation (at concentrations specified in the Results) and incubated at 24 °C. Colony diameters were measured daily and used for calculation of radial growth (mm day$^{-1}$). To study hyphal growth in liquid media, conidial suspensions ($10^5$ spores mL$^{-1}$, final concentration) were inoculated into microplate wells containing the appropriate test substances in PDB in a total volume of 300 µL. Microplates were placed in a laser-based microplate nephelometer (NEPHELOstar, BMG Labtech) and growth was monitored automatically over a 30 h period, as described by Joubert et al. [43]. Data were exported from Nephelostar Galaxy software in ASCII format and further processed in Microsoft Excel.

### Generation of the targeted gene disruption mutant

The gene disruption cassettes were generated using the double-joint PCR procedure described by Yu et al. [44]. The selectable marker inserted in the PCR constructs corresponded to the *Hph* gene cassette (1436 bp) from pCB1636 [45] or the *Nat* gene cassette (2150 bp) from pNR [46] conferring resistance to hygromycin B and nourseothricin, respectively. The final products of each disruption construct consisted of the chosen selective marker with 0.5–1.0 kb 5′ and 3′ spart of the targeted gene as illustrated in Fig. 2. These products were purified and used to transform *A. brassicicola* protoplasts as described in [47]. The Hyg B resistant mutants were selected and prescreened by PCR with relevant primer combinations to confirm integration of the replacement cassette at the targeted locus. The gene replacement mutants were further purified by three rounds of single-spore isolation. The *A. brassicicola* wild-type Abra43 was used to obtain the single hygromycin resistant transformant strain *Δabmak1*. The *Δabmak1* genotype was used to obtain the complemented nourseothricin resistant *Δabmak1-c* strain.

### Melanin extraction

After 7 days of culture on PDA medium, the mycelium (entire colony) or only the conidia were harvested and used for melanin pigment extraction. Melanin was extracted as previously reported by Babitskaia et al. [14] and Alviano et al. [48]. Briefly, samples were lyophilized and, for each genotype, the same amounts of powder were incubated in 2% NaOH (dilution coefficient 1:10) at 100 °C for 2 h in a water bath. The extract was cooled and acidified with concentrated HCl to pH 2.0. The coagulated pigment was separated by centrifugation at 6000*g* for 15 min and dissolved in 0.1 M HCl. Finally it was dialysed against distilled water and lyophilized. The amount of melanin was determined from the photometry absorbance results at 459 nm.

### Infection assays

*Arabidopsis thaliana* plants were grown to the 8- to 12-leaf stage in controlled environment rooms (21–19 °C day and night temperature respectively) and a 8 h light photoperiod. *Brassica oleracea* plants were grown in a greenhouse for 5 weeks. For inoculations, 5 μL drops of *A. brassicicola* conidia suspension ($10^5$, $10^4$ or $10^3$ conidia/mL in water) were deposited on intact leaves from 5 week-old plants. Drops of sterile water were applied on control plants. The plants were then maintained under saturating humidity (100% relative humidity) in a plastic box. Symptoms were observed at 6 dpi.

### Electron microscopy

The conidial wall ultrastructure was investigated by TEM and SEM using conidial suspensions obtained from 7-day-old cultures on PDA. Concerning the TEM sample preparation, successive steps of fixation, post-fixation, dehydratation and embedding in Epon were carried out as previously described [49]. Thin sections were contrasted with uranyle acetate and lead citrate and examined under a JEM-2010 transmission electron microscope (Jeol, Paris, France). SEM samples were prepared as described in [49]. After drying by the critical-point method, specimens were then sputtercoated with a thin carbon layer and examined under a JEOL JSM 6301-F scanning electron microscope (Jeol, Paris, France).

### AFM imaging and surface property measurement

The surface of *A. brassicicola* conidia was imaged using a NanoWizard® atomic force microscope (JPK Instruments AG, Berlin, Germany) operating in intermittent contact mode under ambient conditions. For imaging, a standard rectangular cantilever (Nanosensors NCL-W) was used at a free resonance frequency of 165 kHz and a typical spring constant of about 40 N $m^{-1}$. The tip radius curvature was ~10 nm. For adhesion measurements, gold-coated cantilevers (Olympus, Hambourg, Germany) with a spring constant of 0.01 N $m^{-1}$ were functionalized by immersion in 1 mM 11-mercapto-1-undecanol (Sigma-Aldrich) solution in ethanol for 14 h before rinsing with ethanol. Using these functionalized cantilevers, hydrophilic adhesion force measurements were performed on the conidial surface in ultrapure water [18]. A detailed analysis of the force-distance curves was performed using JPK Data Processing software (JPK Instruments AG). From these curves (2048 measurements), the mean hydrophilic adhesion was extracted from Gaussian fits performed on the histograms. For elasticity measurements, silicon nitride cantilevers having a calibrated spring constant of 0.05 N $m^{-1}$ (Cantilevers MSCT, Veeco) were used in contact mode under ambient conditions. From the force-distance curve measurements, JPK Data Processing software was used to fit the linear part of the approach curves and then to estimate the local surface stiffness [50].

### EPR measurement

Melanin samples obtained from an entire 7 day-old colony were examined at room temperature with a Bruker-Elexsys X-band (9.78 GHz) electron paramagnetic resonance (EPR) spectrometer using a magnetic modulation field at 100 kHz. For the EPR measurements, the three samples were located in thin walled glass tubes with an external diameter of 3 mm. Masses of all samples were determined. The EPR spectra were recorded at different microwave powers in the 0.04–40 mW range. The $P_{1/2}$ experimental parameter, which is the incident microwave power at which the signal is half as great as it would be in the absence of microwave power saturation, was estimated to avoid microwave saturation of the spectral line.

## Additional files

**Additional file 1: Figure S1.** Nucleotide sequence of *AbMak1*. The predicted intronic sequences are indicated in bold.

**Additional file 2: Figure S2.** AFM amplitude images of *A. brassicicola* strains. (**A**) Conidia of the wild-type (6 μm x 5 μm), *Δabmak1* (8 μm x 6 μm) and *Δabmak1-c* (10 μm x 12 μm) and (**B**) conidial surface (1.5 μm x 1.5 μm).

**Additional file 3: Figure S3.** Pathogenic behaviour of *A. brassicicola* wild-type, *Δabmak1* and *Δabmak1-c* strains. *B. oleracea* leaves were inoculated with 5 μL drops of conidia suspensions ($10^5$, $10^4$ or $10^3$ conidia/mL in water). Transformants were inoculated on the right part of the central vein and compared with the parental strain inoculated on the left part of the same leaf. Symptoms were measured at 6 dpi. Values are means of three biological repetitions.

**Additional file 4: Figure S4.** Schematic map of the genomic region including *AbMak1* and flanking genes. This map was generated and modified from the Alternaria Genomes Database (http://alternaria.vbi.vt.edu/index.html). Genes are indicated as red boxes and positions of introns by white bars.

**Authors' contributions**

MK, NBS, MM, AK, GN and RR were involved in the generation of mutants and in their phenotyping. AZ, EJ and SC performed the AFM imaging, surface properties and EPR measurements. SEM and TEM samples were prepared and examined by GM. SP, SC, AZ, PS and TG designed the experiments and were involved in writing the manuscript. All authors read and approved the final manuscript.

**Author details**

[1] IRHS, Agrocampus-Ouest, INRA, Université d'Angers, SFR 4207 QuaSaV, 49071 Beaucouzé, France. [2] UMR 6502, Institut des Matériaux Jean Rouxel,

2, Rue de la Houssinière, BP 32229, 44322 Nantes Cedex 3, France. [3] Present Address: Laboratoire Ecosystèmes Microbiens et Molécules Marines pour les Biotechnologies, IFREMER, Rue de l'île d'Yeu, BP 21105, 44311 Nantes Cedex 3, France. [4] Plateforme SCIAM, Institut de Biologie en Santé, CHU, Université d'Angers, 4, Rue Larrey, 49933 Angers Cedex, France.

**Acknowledgements**

The "Region des Pays de la Loire" is acknowledged for financial support in the framework of the "Myco-AFM" research programme. The authors thank David Manley for correcting the English version of the manuscript.

**Competing interests**

The authors declare that they have no competing interests.

**Funding**

Part of the study was funded by the "Region des Pays de la Loire" in the framework of the "Myco-AFM" research programme.

**References**

1 Cochrane RV, Vederas JC. Highly selective but multifunctional oxygenases in secondary metabolism. Acc Chem Res. 2014;47:3148–61.
2 Huijbers MM, Montersino S, Westphal AH, Tischler D, van Berkel WJ. Flavin dependent monooxygenases. Arch Biochem Biophys. 2014;544:2–17.
3 van Berkel WJ, Kamerbeek NM, Fraaije MW. Flavoprotein monooxygenases, a diverse class of oxidative biocatalysts. J Biotechnol. 2006;124:670–89.
4 Covert SF, Enkerli J, Miao VP, VanEtten HD. A gene for maackiain detoxification from a dispensable chromosome of *Nectria haematococca*. Mol Gen Genet. 1996;251:397–406.
5 Sellam A, Dongo A, Guillemette T, Hudhomme P, Simoneau P. Transcriptional responses to exposure to the brassicaceous defence metabolites camalexin and allyl-isothiocyanate in the necrotrophic fungus *Alternaria brassicicola*. Mol Plant Pathol. 2007;8:195–208.
6 Cho Y, Srivastava A, Ohm RA, Lawrence CB, Wang KH, Grigoriev IV, Marahatta SP. Transcription factor Amr1 induces melanin biosynthesis and suppresses virulence in *Alternaria brassicicola*. PLoS Pathog. 2012;8:e1002974.
7 Eisenman HC, Casadevall A. Synthesis and assembly of fungal melanin. Appl Microbiol Biotechnol. 2012;93:931–40.
8 Gomez BL, Nosanchuk JD. Melanin and fungi. Curr Opin Infect Dis. 2003;16:91–6.
9 Langfelder K, Streibel M, Jahn B, Haase G, Brakhage AA. Biosynthesis of fungal melanins and their importance for human pathogenic fungi. Fungal Genet Biol. 2003;38:143–58.
10 Ludwig N, Lohrer M, Hempel M, Mathea S, Schliebner I, Menzel M, Kiesow A, Schaffrath U, Deising HB, Horbach R. Melanin is not required for turgor generation but enhances cell-wall rigidity in appressoria of the corn pathogen *Colletotrichum graminicola*. Mol Plant Microbe Interact. 2014;27:315–27.
11 Wierenga RK, Terpstra P, Hol WG. Prediction of the occurrence of the ADP-binding beta alpha beta-fold in proteins, using an amino acid sequence fingerprint. J Mol Biol. 1986;187:101–7.
12 Eggink G, Engel H, Vriend G, Terpstra P, Witholt B. Rubredoxin reductase of *Pseudomonas oleovorans*. Structural relationship to other flavoprotein oxidoreductases based on one NAD and two FAD fingerprints. J Mol Biol. 1990;212:135–42.
13 Eppink MH, Schreuder HA, Van Berkel WJ. Identification of a novel conserved sequence motif in flavoprotein hydroxylases with a putative dual function in FAD/NAD(P)H binding. Protein Sci. 1997;6:2454–8.
14 Babitskaia VG, Shcherba VV, Filimonova TV, Grigorchuk EZ. Melanin pigments of the fungi *Paecilomyces variotii* and *Aspergillus carbonarius*. Prikl Biokhim Mikrobiol. 2000;36:153–9.
15 Godechal Q, Gallez B. The contribution of electron paramagnetic resonance to melanoma research. J Skin Cancer. 2011;2011:273280.
16 Chodurek E, Zdybel M, Pilawa B, Dzierzewicz Z. Examination by EPR spectroscopy of free radicals in melanins isolated from A-375 cells exposed on valproic acid and cisplatin. Acta Pol Pharm. 2012;69:1334–41.
17 Sarna T, Hyde JS. Electron spin-lattice relaxation times of melanin. J Chem Phys. 1978;69:1945–8.
18 Dufrene YF. Direct characterization of the physicochemical properties of fungal spores using functionalized AFM probes. Biophys J. 2000;78:3286–91.
19 Torres Pazmino DE, Winkler M, Glieder A, Fraaije MW. Monooxygenases as biocatalysts: classification, mechanistic aspects and biotechnological applications. J Biotechnol. 2010;146:9–24.
20 Entsch B. Hydroxybenzoate hydroxylase. Methods Enzymol. 1990;188:138–47.
21 Laden BP, Tang Y, Porter TD. Cloning, heterologous expression, and enzymological characterization of human squalene monooxygenase. Arch Biochem Biophys. 2000;374:381–8.
22 Rosa LH, Almeida Vieira Mde L, Santiago IF, Rosa CA. Endophytic fungi community associated with the dicotyledonous plant *Colobanthus quitensis* (Kunth) Bartl. (Caryophyllaceae) in Antarctica. FEMS Microbiol Ecol. 2010;73:178–89.
23 Zhdanova NN, Zakharchenko VA, Vember VV, Nakonechnaya LT. Fungi from Chernobyl: mycobiota of the inner regions of the containment structures of the damaged nuclear reactor. Mycol Res. 2000;104:1421–6.
24 Chai LY, Netea MG, Sugui J, Vonk AG, van de Sande WW, Warris A, Kwon-Chung KJ, Kullberg BJ. *Aspergillus fumigatus* conidial melanin modulates host cytokine response. Immunobiology. 2010;215:915–20.
25 Volling K, Thywissen A, Brakhage AA, Saluz HP. Phagocytosis of melanized *Aspergillus conidia* by macrophages exerts cytoprotective effects by sustained PI3 K/Akt signalling. Cell Microbiol. 2011;13:1130–48.
26 Howard RJ, Valent B. Breaking and entering: host penetration by the fungal rice blast pathogen *Magnaporthe grisea*. Annu Rev Microbiol. 1996;50:491–512.
27 Kawamura C, Tsujimoto T, Tsuge T. Targeted disruption of a melanin biosynthesis gene affects conidial development and UV tolerance in the Japanese pear pathotype of *Alternaria alternata*. Mol Plant Microbe Inter MPMI. 1999;12:59–63.
28 Moriwaki A, Kihara J, Kobayashi T, Tokunaga T, Arase S, Honda Y. Insertional mutagenesis and characterization of a polyketide synthase gene (PKS1) required for melanin biosynthesis in *Bipolaris oryzae*. FEMS Microbiol Lett. 2004;238:1–8.
29 Pihet M, Vandeputte P, Tronchin G, Renier G, Saulnier P, Georgeault S, Mallet R, Chabasse D, Symoens F, Bouchara JP. Melanin is an essential component for the integrity of the cell wall of *Aspergillus fumigatus* conidia. BMC Microbiol. 2009;9:177.
30 Zykwinska A, Guillemette T, Bouchara JP, Cuenot S. Spontaneous self-assembly of SC3 hydrophobins into nanorods in aqueous solution. Biochim Biophys Acta. 2014;1844:1231–7.
31 Jahn B, Koch A, Schmidt A, Wanner G, Gehringer H, Bhakdi S, Brakhage AA. Isolation and characterization of a pigmentless-conidium mutant of *Aspergillus fumigatus* with altered conidial surface and reduced virulence. Infect Immun. 1997;65:5110–7.
32 Calmes B, N'Guyen G, Dumur J, Brisach CA, Campion C, Iacomi B, Pigne S, Dias E, Macherel D, Guillemette T, Simoneau P. Glucosinolate-derived isothiocyanates impact mitochondrial function in fungal cells and elicit an oxidative stress response necessary for growth recovery. Front Plant Sci. 2015;6:414.
33 Nosanchuk JD, Casadevall A. The contribution of melanin to microbial pathogenesis. Cell Microbiol. 2003;5:203–23.
34 Wakamatsu K, Ito S. Advanced chemical methods in melanin determination. Pigment Cell Res. 2002;15:174–83.
35 Bernard M, Latge JP. *Aspergillus fumigatus* cell wall: composition and biosynthesis. Med Mycol. 2001;39(Suppl 1):9–17.
36 Eisenman HC, Nosanchuk JD, Webber JB, Emerson RJ, Camesano TA, Casadevall A. Microstructure of cell wall-associated melanin in the human pathogenic fungus *Cryptococcus neoformans*. Biochemistry. 2005;44:3683–93.
37 Youngchim S, Morris-Jones R, Hay RJ, Hamilton AJ. Production of melanin by *Aspergillus fumigatus*. J Med Microbiol. 2004;53:175–81.

38 Schumacher J. DHN melanin biosynthesis in the plant pathogenic fungus *Botrytis cinerea* is based on two developmentally regulated key enzyme (PKS)-encoding genes. Mol Microbiol. 2016;99:729–48.

39 Wang Z, Zheng L, Hauser M, Becker JM, Szaniszlo PJ. WdChs4p, a homolog of chitin synthase 3 in Saccharomyces cerevisiae, alone cannot support growth of Wangiella (Exophiala) dermatitidis at the temperature of infection. Infect Immun. 1999;67:6619–30.

40 Brunskole Svegelj M, Turk S, Brus B, Lanisnik Rizner T, Stojan J, Gobec S. Novel inhibitors of trihydroxynaphthalene reductase with antifungal activity identified by ligand-based and structure-based virtual screening. J Chem Inf Model. 2011;51:1716–24.

41 Butler MJ, Gardiner RB, Day AW. Degradation of melanin or inhibition of its synthesis: are these a significant approach as a biological control of phytopathogenic fungi? Biol Control. 2005;32:326–36.

42 Sellam A, Iacomi-Vasilescu B, Hudhomme P, Simoneau P. In vitro antifungal activity of brassinin, camalexin and two isothiocyanates against the crucifer pathogens *Alternaria brassicicola* and *Alternaria brassicae*. Plant Pathol. 2007;56:296–301.

43 Joubert A, Calmes B, Berruyer R, Pihet M, Bouchara JP, Simoneau P, Guillemette T. Laser nephelometry applied in an automated microplate system to study filamentous fungus growth. Biotechniques. 2010;48:399–404.

44 Yu JH, Hamari Z, Han KH, Seo JA, Reyes-Dominguez Y, Scazzocchio C. Double-joint PCR: a PCR-based molecular tool for gene manipulations in filamentous fungi. Fungal Genet Biol. 2004;41:973–81.

45 Sweigard JA, Carroll AM, Kang S, Farrall L, Chumley FG, Valent B. Identification, cloning, and characterization of PWL2, a gene for host species specificity in the rice blast fungus. Plant Cell. 1995;7:1221–33.

46 Malonek S, Rojas MC, Hedden P, Gaskin P, Hopkins P, Tudzynski B. The NADPH-cytochrome P450 reductase gene from *Gibberella fujikuroi* is essential for gibberellin biosynthesis. J Biol Chem. 2004;279:25075–84.

47 Cho Y, Davis JW, Kim KH, Wang J, Sun QH, Cramer RA Jr, Lawrence CB. A high throughput targeted gene disruption method for *Alternaria brassicicola* functional genomics using linear minimal element (LME) constructs. Mol Plant Microbe Interact. 2006;19:7–15.

48 Alviano CS, Farbiarz SR, De Souza W, Angluster J, Travassos LR. Characterization of *Fonsecaea pedrosoi* melanin. J Gen Microbiol. 1991;137:837–44.

49 Joubert A, Simoneau P, Campion C, Bataille-Simoneau N, Iacomi-Vasilescu B, Poupard P, Francois JM, Georgeault S, Sellier E, Guillemette T. Impact of the unfolded protein response on the pathogenicity of the necrotrophic fungus *Alternaria brassicicola*. Mol Microbiol. 2011;79:1305–24.

50 Touhami A, Nysten B, Dufrene YF. Nanoscale mapping of the elasticity of microbial cells by atomic force microscopy. Langmuir ACS J Surf Colloids. 2003;19:4539–43.

# 2

# Promoters from the itaconate cluster of *Ustilago maydis* are induced by nitrogen depletion

Thiemo Zambanini[1†], Sandra K. Hartmann[1,2†], Lisa M. Schmitz[1], Linda Büttner[3], Hamed Hosseinpour Tehrani[1], Elena Geiser[1], Melanie Beudels[1], Dominik Venc[1], Georg Wandrey[4], Jochen Büchs[4], Markus Schwarzländer[2,5,6], Lars M. Blank[1] and Nick Wierckx[1*]

**Abstract**

**Background:** *Ustilago maydis* is known for its natural potential to produce a broad range of valuable chemicals, such as itaconate, from both industrial carbon waste streams and renewable biomass. Production of itaconate, and many other secondary metabolites, is induced by nitrogen limitation in *U. maydis*. The clustered genes responsible for itaconate production have recently been identified, enabling the development of new expression tools that are compatible with biotechnological processes.

**Results:** Here we report on the investigation of two of the native promoters, $P_{tad1}$ and $P_{mtt1}$, from the itaconate cluster of *U. maydis* MB215. For both promoters the specific activation upon nitrogen limitation, which is known to be the trigger for itaconate production in *Ustilago*, could be demonstrated by *gfp* expression. The promoters cover a broad range of expression levels, especially when combined with the possibility to create single- and multicopy construct integration events. In addition, these reporter constructs enable a functional characterization of gene induction patterns associated with itaconate production.

**Conclusions:** The promoters are well suited to induce gene expression in response to nitrogen limitation, coupled to the itaconate production phase, which contributes towards the further improvement of organic acid production with *Ustilago*.

**Keywords:** *Ustilago maydis*, Itaconate cluster, Promoter characterization, GFP

## Background

The family of Ustilaginaceae has sparked great interest as promising industrial production organisms in recent years. The growing biotechnological attention results from their native ability to utilize a range of bio-based substrates [10, 18, 20, 62] and to produce a broad variety of value-added chemicals, such as glycolipids, polyols, and organic acids [12, 19, 33]. Considerable efforts have been undertaken to optimize fermentation and process conditions in order to increase the yield, titer and rate of glycolipids [42], erythritol [28], malate [61], itaconate [16, 41] and 2-hydroxyparaconate [15, 22]. The biochemical pathways and associated gene clusters of several secondary metabolites have been characterized and engineered, including those of cellobiose lipids [51], mannosyl erythritol lipids [23, 27], malate [60], itaconate [17, 59], and 2-hydroxyparaconate [16]. These efforts have accelerated recently largely thanks to a growing suite of efficient genetic engineering tools, including marker recycling through the FLP/FRT system [32], Golden Gate Cloning [53] and Cas9-based genome engineering [48]. Although originally developed for the model organism *U. maydis*, these tools can also be adapted for use in other

*Correspondence: nick.wierckx@rwth-aachen.de
†Thiemo Zambanini and Sandra K. Hartmann have contributed equally to the manuscript
[1] Institute of Applied Microbiology – iAMB, Aachen Biology and Biotechnology – ABBt, RWTH Aachen University, Worringerweg 1, Aachen 52074, Germany
Full list of author information is available at the end of the article

Ustilaginaceae [59, 60]. However, critical limitations remain that are characteristic for fungal biotechnology, including the availability of suitable promoters. Specific sets of promoters have been developed for the production of proteins and chemicals of high industrial relevance in different organisms, such as *Aspergillus niger* [55, 56], *Escherichia coli* [37, 38], and *Penicillium chrysogenum* [44]. Modern online analysis systems have become an attractive means to investigate promoter properties, enabling high-resolution characterization of the specific time of induction, the induction trigger, or the promoter strength, for different conditions or promoter variants [64]. One particularly important trigger for the production of organic acids and glycolipids in Ustilaginaceae is nitrogen limitation. Such a limitation causes a strict temporal separation of the very different cellular objectives associated with biomass growth and product formation [14, 26, 52]. This characteristic of fungal secondary metabolite production may be exploited for metabolic engineering. Constitutive promoters, such as $P_{otef}$ and $P_{oma}$ [13, 46, 49], are mainly active in the growth phase, while inducible promoters, such as *Pcrg1*, *Pnar1* or the tet-system, come with other drawbacks [4, 6, 63]. *Pcrg1* relies on arabinose and is efficiently repressed by glucose and xylose, two of the main carbon sources for industrial production processes [4]. *Pnar1* is induced by nitrate and repressed by ammonium [6] and is therefore not suitable for nitrogen-limited production. Other inducible promoters are associated with specific phases of the *Ustilago* life cycle, making them equally problematic for engineering of secondary metabolite production [2]. Hence new promoters, which are specifically activated by nitrogen limitation hold much promise for metabolic engineering.

Recently, the gene cluster responsible for itaconate production in *U. maydis* was identified [17] and its promoters are promising candidates to overcome current limitations. From the different functions of the genes in the cluster, including catalytic, transport and transcriptional regulation activities, it may be deduced that the individual promoters should possess different inherent properties with respect to induction and strength. This was confirmed by qRT–PCR [17]. Two- to ninefold elevated activity for all seven genes in this cluster, except for *rdo1*, was shown during the itaconate production phase compared to non-induced conditions in rich medium [17]. Further, the regulation of most genes in the cluster is strongly dependent on the putative transcriptional regulator *ria1*. By deletion of *ria1*, the transcription level for all genes within the cluster was lowered by up to ninefold and overexpression triggered expression of most cluster genes [17]. Therefore these promoters show much promise for biotechnological usage as expression tools.

Here we investigate a set of promoters from the genes responsible for itaconate production in *U. maydis* that are induced during the non-growing production phase as initiated by nitrogen limitation.

## Methods

### Strains and cultivation conditions

All strains used and constructed within this work are listed in Table 1.

As host for cloning experiments, *E. coli* DH5α (DSM 6897) was used. All *E. coli* strains were grown at 37 °C shaking at 200 rpm (shaking diameter 25 mm) in lysogeny broth (LB) medium. For recombinant strains 100 mg $L^{-1}$ ampicillin were added to the medium.

**Table 1 Strains used and constructed within this work**

| Name | Genotype | Origin |
|---|---|---|
| *Wildtype strains* | | |
| *U. maydis* MB215 | Wildtype | DSM 17144 |
| *E. coli* DH5α | Wildtype | DSM 6897 |
| *Recombinant U. maydis strains* | | |
| *U. maydis* MB215 $P_{otef}$-*gfp* (s) | Single integration of pUMa43-otef-gfp-nos cbx | This work |
| *U. maydis* MB215 $P_{otef}$-*gfp* (m) | Multiple integration of pUMa43-otef-gfp-nos cbx | This work |
| *U. maydis* MB215 $P_{tad1}$-*gfp* (s) | Single integration of pUMa43-$P_{tad1}$-gfp cbx | This work |
| *U. maydis* MB215 $P_{tad1}$-*gfp* (m) | Multiple integration of pUMa43-$P_{tad1}$-gfp cbx | This work |
| *U. maydis* MB215 $P_{itp1}$-*gfp* (s) | Single integration of pUMa43-$P_{itp1}$-gfp cbx | This work |
| *U. maydis* MB215 $P_{itp1}$-*gfp*(m) | Multiple integration of pUMa43-$P_{itp1}$-gfp cbx | This work |
| *U. maydis* MB215 $P_{adi1}$-*gfp* (s) | Single integration of pUMa43-$P_{adi1}$-gfp cbx | This work |
| *U. maydis* MB215 $P_{adi1}$-*gfp* (m) | Multiple integration of pUMa43-$P_{adi1}$-gfp cbx | This work |
| *U. maydis* MB215 $P_{mtt1}$-*gfp* (s) | Single integration of pUMa43-$P_{mtt1}$-gfp cbx | This work |
| *U. maydis* MB215 $P_{mtt1}$-gfp (m) | Multiple integration of pUMa43-$P_{mtt1}$-gfp cbx | This work |

*Ustilago maydis* cultures were cultivated in YEPS light medium containing 20 g $L^{-1}$ D–sucrose, 5 g $L^{-1}$ yeast extract, and 10 g $L^{-1}$ peptone at 30 °C shaking at 200 rpm (shaking diameter 25 mm).

For physiological experiments 96 flat-bottom round-well plates or 48-well flower plates were used as described below.

As cultivation medium MTM was used containing 50 g $L^{-1}$ glucose, 0.2 g $L^{-1}$ $MgSO_4 \cdot 7\ H_2O$, 0.01 g $L^{-1}$ $FeSO_4 \cdot 7\ H_2O$, 0.5 g $L^{-1}$ $KH_2PO_4$, 1 mL $L^{-1}$ vitamin solution, 10 mL $L^{-1}$ trace element solution, 19.5 g $L^{-1}$ 2-(N-morpholino)ethanesulfonic acid (MES) as buffer, and differing $NH_4Cl$ concentrations. MES buffer was used as stock solution with pH adjusted to 6.5 with NaOH. The vitamin solution contained (per liter) 0.05 g D-biotin, 1 g D-calcium panthotenate, 1 g nicotinic acid, 25 g myo-inositol, 1 g thiamine hydrochloride, 1 g pyridoxol hydrochloride, and 0.2 g para–aminobenzoic acid. The trace element solution contained (per liter) 1.5 g EDTA, 0.45 g $ZnSO_4 \cdot 7\ H_2O$, 0.10 g $MnCl_2 \cdot 4\ H_2O$, 0.03 g $CoCl_2 \cdot 6\ H_2O$, 0.03 g $CuSO_4 \cdot 5\ H_2O$, 0.04 g $Na_2MoO_4 \cdot 2\ H_2O$, 0.45 g $CaCl_2 \cdot 2\ H_2O$, 0.3 g $FeSO_4 \cdot 7\ H_2O$, 0.10 g $H_3BO_3$, and 0.01 g KI [19]. For phosphate limitation 3.2 g $L^{-1}$ $NH_4Cl$ and 0.1 g $L^{-1}$ $KH_2PO_4$ were used.

Precultures for analytical experiments were cultivated for 24 h in 500 mL shake flasks without baffles with 50 mL MTM containing 4 g $L^{-1}$ $NH_4Cl$, to ensure that no nitrogen limitation occurs prior to inoculation of the main culture.

**Analytical methods**

All experiments were performed in triplicates, unless stated otherwise. The arithmetic mean of the biological replicates is shown. Error bars and ± values indicate standard error of the mean.

HPLC analysis was performed as described previously [62]. Centrifuged samples (13,000*g*, 5 min) were filtered through cellulose acetate filters (diameter 0.2 µm, VWR, Germany) and subsequently diluted 1:10 with distilled water. For analysis a Dionex Ultimate 3000 HPLC (Dionex, USA) with an Organic Acid Resin column (CS-Chromatographie, Germany) kept at 75 °C, with a constant flow rate of 0.8 ml $min^{-1}$ of 5 mM sulfuric acid as eluent was used. For detection, a Shodex RI101 detector at 35 °C and a variable wavelength UV detector (Dionex, USA) at 210 nm were used.

Ammonium concentration was determined by a colorimetric assay according to Willis [57]. For this 50 µL of the sample (maximal 50 mg $L^{-1-}$ $NH_4^+$) were mixed with 1 mL reagent solution and afterwards with 0.25 mL hypochlorite solution. The mixture was incubated for at least 12 min at room temperature before measuring the absorbance at 685 nm. The reagent solution contained 32 g sodium salicylate (anhydrous), 40 g trisodium phosphate (TSP) and 0.5 g sodium nitrosylpentacyanoferrate(III) (sodium nitroprusside) dissolved in 1 L of water. For the hypochlorite solution, 50 mL of commercially available bleach (Clorox) containing 5–5.25% sodium hypochlorite were diluted with water to 1 L with a final concentration of ~ 0.25% hypochlorite.

Initial promoter characterization experiments were performed in 96-well plates using a BioLector (m2p-labs, Baesweiler, Germany) at 30 °C shaking at 1000 rpm (shaking diameter: 3 mm). GFP fluorescence intensity was determined at 488/520 nm (excitation/emission) with the gain set to 70. For biomass determination backscatter intensity in the far red range 620/620 nm with the gain set to 10 was used. Specific promoter activities are expressed as GFP fluorescence over biomass backscatter. Since both these signals are measured in arbitrary units, this specific activity is expressed without dimension. Values are normalized (point-to-point) for the arithmetic mean of the corresponding wildtype. Induction ratios were determined by dividing the maximum specific promoter activity by the average activity from 2–10 h.

Induction profiles of *U. maydis* MB215 $P_{tad1}$-*gfp* under different cultivation conditions were analyzed in 48-well flower plates (M2P-labs, Baesweiler, Germany) with an in-house constructed screening system based on the established BioLector setup [45, 54] with a fluorospectrometer, which features excitation and emission monochromators for free wavelength selection in the UV/Vis range (Fluoromax-4, HORIBA Jobin–Yvon). GFP fluorescence intensity was determined at 495/507 nm (band-pass: 8 nm, integration time: 600 ms) and biomass was measured via backscatter intensity at 650/650 nm (band-pass: 4 nm, integration time: 1200 ms).

**Cloning procedures**

The genome of *U. maydis* 521 was taken as reference sequence [29].

Plasmids $P_{tad1}$-*gfp*, $P_{itp1}$-*gfp*, $P_{adi1}$-*gfp*, and $P_{mtt1}$-*gfp* were cloned by amplifying the promoter regions of genes *UMAG_05076* (*tad1*), *UMAG_05077* (*itp1*), *UMAG_05078* (*adi1*), and *UMAG_05079* (*mtt1*) with the oligonucleotide primers listed in Table 2 and exchanging $P_{otef}$ on the vector pUMa43 [34]. For *UMAG_05077* (*itp1*), *UMAG_05078* (*adi1*), and *UMAG_05079* (*mtt1*) 2000 bp in front of the respective start codon were used and for *UMAG_05076* (*tad1*) 1500 bp were used (Fig. 1).

The resulting plasmids were linearized with the restriction enzyme *SspI* prior to *U. maydis* transformation and constructs were integrated into the *ip*-locus of *U. maydis* strain MB215 by homologous recombination [40]. Transformation of *U. maydis* was performed using standard

**Table 2 Oligonucleotide primers with restriction sites**

| Plasmid | Primers/specifications | Restriction enzymes |
|---|---|---|
| pUMa43 | $P_{otef}$-*gfp*-$T_{nos}$; ori ColE1; ampR; *U. maydis ip*$^R$-locus | |
| $P_{tad1}$-*gfp* | fwd:<br>CTCGGTACCTTCGACTTGGTGGATACTGCGGCTGTTG<br>rev:<br>GTCGGTCTTGACTCGACCCAGCTCACCGGATCCGTA | *KpnI/BamHI* |
| $P_{itp1}$-*gfp* | fwd:<br>TCGAAATTCGAGCTCGGTACCGACCTGACAGAAGAGATAG<br>rev:<br>AGCTCCTCGCCCTTGCTCACCATGGTTCGACTTGGTGGATACTG | *KpnI/NcoI* |
| $P_{adi1}$-*gfp* | fwd:<br>TCGAAATTCGAGCTCGGTACCTCAGCCGATAGGTTTCAC<br>rev:<br>AGCTCCTCGCCCTTGCTCACCATGGTGAGCTGGGTCGAGTC | *KpnI/NcoI* |
| $P_{mtt1}$-*gfp* | fwd:<br>TCGAAATTCGAGCTCGGTACCGTTTACCGCACGCTGTA<br>rev:<br>AGCTCCTCGCCCTTGCTCACCATGGTGGATGACGAATCTCAAG | *KpnI/NcoI* |

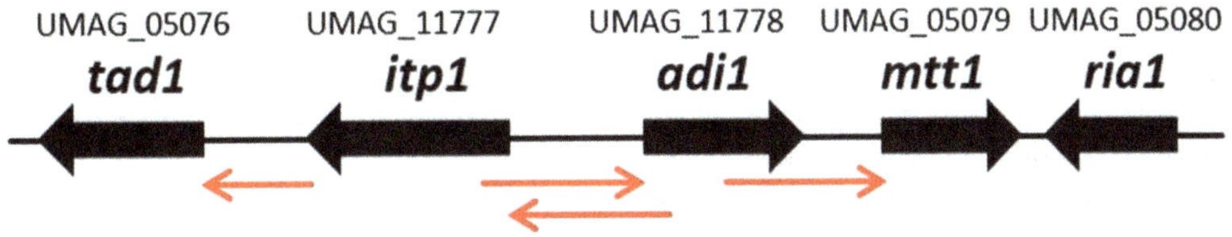


**Fig. 1** Clustered itaconate genes in *U. maydis* including *trans*-aconitate decarboxylase (*tad1*), a major facilitator superfamily transporter (*itp1*), an aconitate-Δ-isomerase (*adi1*), a mitochondrial tricarboxylate transporter (*mtt1*), and a transcriptional regulator (*ria1*) [17]. Red arrows indicate the promoter regions investigated in this study

protocols [47]. For selection of transformants, PDA plates with 2 µg ml$^{-1}$ carboxin were used. Correct integration of constructs and number of integration events was verified by Southern blot analysis using enzyme *EcoRV* for genome restriction and the Cbx-cassette from vector pMF1-C [5] as a probe for the detection of specific DNA fragments.

## Results and discussion

### Expression strengths of promoters from the itaconate cluster of *U. maydis*

The recently characterized genes encoding core catalytic enzymes and transporters, which are required for itaconate production (*tad1*, *itp1*, *adi1*, *mtt1*) were chosen as primary targets for the investigation of promoter activities. From previous studies it was known that their native activity is strongly influenced by the activity of the transcriptional regulator Ria1 which is itself also encoded in the itaconate gene cluster [17].

For analysis of induction conditions and promoter strength we fused each of the four promoters to *gfp*. GFP has been shown to be a suitable tool for expression analysis in several organisms including *Escherichia coli* [11, 43], *Saccharomyces cerevisiae* [1] and *Pseudomonas putida* [64]. Also in *U. maydis* GFP was used before to investigate host–pathogen interactions [49], to identify and localize proteins, for example a motor protein [39], and to investigate promoter activities [3, 49].

The plasmids containing the promoter-*gfp* fusions were integrated into the *ip*-locus of *U. maydis* MB215 [9, 31]. The resulting mutants were screened for *ip*–locus and single (s) or multiple (m) integration events by PCR and Southern Blotting. This targeted in-locus integration avoids undesired polar effects, such as gene disruption, and ensures optimal comparability between different strains by avoiding locus- and copy number-dependent expression differences [49]. For all promoters, one clone with single construct integration was chosen for further characterization (Fig. 2).

Fluorescence increased over time for all strains (data not shown), however, with great differences in the absolute intensity range. While $P_{itp1}$ and $P_{adi1}$ showed very low fluorescence intensities after 24 h of cultivation, about two-to-three orders of magnitude below that of the strong $P_{otef}$, $P_{tad1}$ showed 48-fold and $P_{mtt1}$ fivefold lower intensities. These data indicate pronounced differences

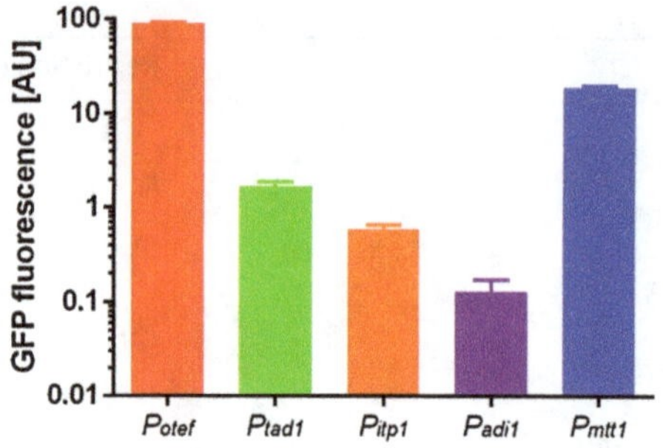


**Fig. 2** Fluorescence signals for *U. maydis* single integration mutants cultivated in MTM containing 0.8 g $L^{-1}$ $NH_4Cl$ and 50 g $L^{-1}$ glucose expressing *gfp* under the control of four promoters from the itaconate gene cluster. GFP fluorescence intensities (normalized against the wildtype) are shown for single integration mutants under control of $P_{otef}$ (red), $P_{tad1}$ (green), $P_{itp1}$ (orange), $P_{adi1}$ (purple), and $P_{mtt1}$ (blue) after 24 h of cultivation. Error bars indicate standard error of the mean ($n = 3$)

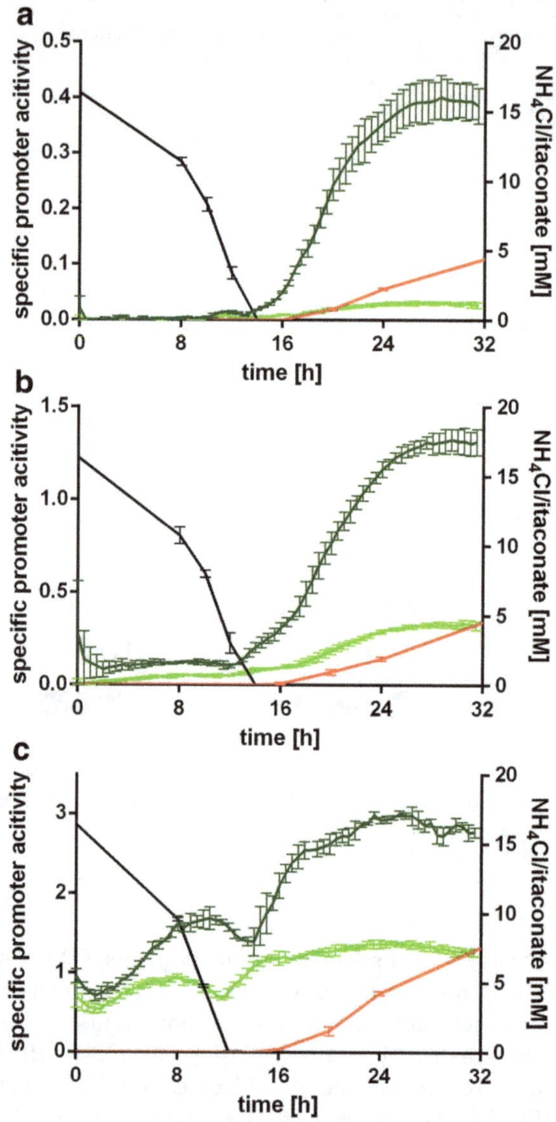


**Fig. 3** Induction profiles of *gfp* under control of $P_{tad1}$ (**a**), $P_{mtt1}$ (**b**), and $P_{otef}$ (**c**). Itaconate concentration (red) and $NH_4Cl$ concentration (black) correlated to the specific promoter activity (GFP over biomass, normalized against the WT) for *U. maydis* strains containing single (light green) and multiple (dark green) insertions. Error bars indicate standard error of the mean ($n = 3$)

of promoter activity within the itaconate cluster from *U. maydis*, with $P_{adi1}$ and $P_{itp1}$ as relatively weak promoters and $P_{mtt1}$ as a strong promoter. The differences are in line with the physiological roles of the regulated genes. The mitochondrial transporter Mtt1 is the limiting step for itaconate biosynthesis in *U. maydis* [17], requiring relatively high expression. In contrast, high expression of Adi1 would result in a surplus of *trans*–aconitate, which is a potent inhibitor for vital metabolic reactions, such as the conversion of citrate to isocitrate by aconitase in the TCA–cycle [21]. The *in-trans* expression of $P_{adi1}$ and $P_{itp1}$ outside of their original genomic context might influence their activity, since the orientation in the genome suggests a possible bidirectional promoter (Fig. 1), which might be influenced by additional upstream elements in their native context. Based on those results, $P_{tad1}$ and $P_{mtt1}$ were chosen as the two strongest promoters for further characterization and for the development as expression tools.

#### Activation of the $P_{tad1}$ and $P_{mtt1}$ occurs in response to nitrogen limitation

For expression of genes coupled to itaconate production the induction time can be critical, depending on the expressed product, since some products are toxic during the growth phase, and the production of itaconate poses a drain on the primary metabolite *cis*-aconitate. Consequently we investigated the induction time and conditions (Fig. 3) for $P_{tad1}$ and $P_{mtt1}$ using $P_{otef}$ as a control. Additionally, we investigated the differences resulting from clones with single construct integration compared to multiple integration, hypothesizing that multiple insertions may allow for a wider range of activities. It has to be noted, however, that the copy number for multiple integration events could not be quantified.

Promoter activities are expressed as GFP fluorescence intensity over biomass (measured as scattered light). The activity of both $P_{tad1}$ and $P_{mtt1}$ increased strongly after 14 h of cultivation, corresponding to the time point where ammonium was depleted from the culture medium (Fig. 3). Production of itaconate started approximately two hours later. This correlates well with the known induction of itaconate production for *U. maydis* [14]. In contrast, the strong constitutive $P_{otef}$ promoter gave a high and relatively stable signal during the growth- and production phases. The low fluorescence intensity of the $P_{tad1}$ construct during the initial growth phase hints at a weak, but tightly controlled activity with an induction ratio of 72 (single integration) and 122 (multiple integrations). $P_{mtt1}$ in contrast is a strong, but leaky promoter, with a basal activity during the growth phase and a lower induction ratio of 13 (single integration) and 11 (multiple integrations). The basal activity of $P_{mtt1}$ in the growth phase may result from the generally high promoter activity, or from an activation due to expression out of the original chromosomal context.

We next investigated the impact of multiple integration on the expression activity. With values of 0.03 ($P_{tad1}$, single, 23–32 h), 0.40 ($P_{tad1}$, multi, 28–29 h), 0.33 ($P_{mtt1}$, single, 27–32 h), 1.32 ($P_{mtt1}$, multi, 29–32 h), 1.38 ($P_{otef}$, single, 25–26 h), and 3.00 ($P_{otef}$, multi, 25–26 h), the maximum expression level increased by about twofold ($P_{otef}$), fourfold ($P_{mtt1}$), and 13-fold ($P_{tad1}$) for multiple construct integrations compared to single integration. Taking into consideration the promoter activities of single-copy $P_{tad1}$ and $P_{mtt1}$ constructs, and multiple integration, expression levels covering a nearly 50-fold dynamic range with high resolution appears realistic. This value may even be increased by selecting higher integration numbers for $P_{mtt1}$.

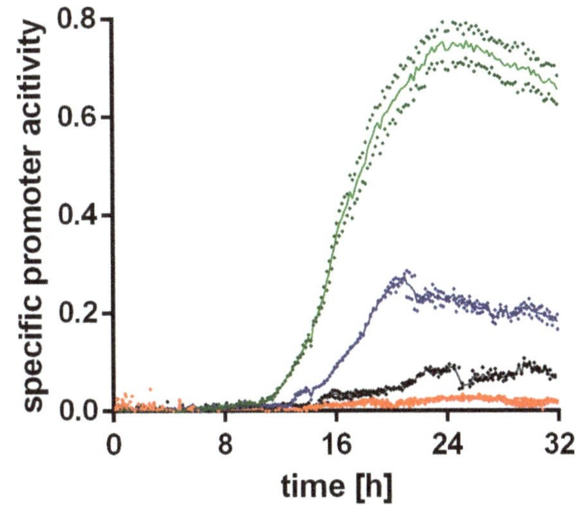


**Fig. 4** Impact of growth limiting nutrients on $P_{tad1}$ activity. Specific promoter activity (GFP over biomass, normalized against the wildtype) for *U. maydis* $P_{tad1}$-*gfp* cultivated in MTM containing 0.5 g $L^{-1}$ $KH_2PO_4$ with 0.8 g $L^{-1}$ (green), 1.6 g $L^{-1}$ (blue), and 3.2 g $L^{-1}$ (black) $NH_4Cl$, and 0.1 g $L^{-1}$ $KH_2PO_4$ with 3.2 g $L^{-1}$ $NH_4Cl$ (red). Single values are shown for two biological replicates ($n = 2$; diamonds) and the mean value (line)

### The impact of growth limiting nutrients on $P_{tad1}$

Itaconate production by *U. maydis* is generally induced upon nitrogen limitation. However, the amount of nitrogen source as a growth-limiting nutrient has a strong impact on the efficiency of itaconate production [16, 41]. In other itaconate producers, such as *Aspergillus terreus*, phosphate-limiting media are generally used [36, 58], although a P-limitation is not strictly needed for efficient induction [25, 35]. In order to further investigate the effect of different growth limitations on the activation of $P_{tad1}$, we cultivated *U. maydis* $P_{tad1}$-*gfp* with different combinations of $NH_4Cl$ and $KH_2PO_4$. Keeping the initial concentration of $KH_2PO_4$ at 0.5 g $L^{-1}$ we increased the initial $NH_4Cl$ concentrations from 0.8 to 1.6 and 3.2 g $L^{-1}$, resulting in higher biomass concentrations and nitrogen limitation at a later time-point during cultivation. To ensure a phosphate-limited culture, we lowered the initial concentration of $KH_2PO_4$ to 0.1 g $L^{-1}$ combined with an initial $NH_4Cl$ concentration of 3.2 g $L^{-1}$ (Fig. 4).

Higher $NH_4Cl$ concentrations resulted in a delay of $P_{tad1}$ induction, correlated with a later depletion of nitrogen. However, specific $P_{tad1}$ promoter activity was about 2.5-fold (1.6 g $L^{-1}$) and 7.5-fold (3.2 g $L^{-1}$) lower for the cultures containing more $NH_4Cl$, compared to the culture containing 0.8 g $L^{-1}$ $NH_4Cl$, and for the phosphate-limited culture the specific promoter activity was negligible (Fig. 4). These data further support the role of nitrogen limitation as major trigger for the production of itaconate, as opposed to general growth limitation. They also explain previous observations, where the use of 4 and 0.8 g $L^{-1}$ $NH_4Cl$ resulted in similar volumetric production rates [16], even though much more biomass was formed at 4 g $L^{-1}$ $NH_4Cl$. Apparently, lower nitrogen concentrations result in a stronger induction of itaconate production genes, leading to more efficient production. This may be related to pH in the employed system of batch cultures with a soluble 100 mM MES buffer starting at pH 6.5. Higher nitrogen concentrations may lead to a stronger pH drop during growth, and itaconate formation might be triggered at a sub-optimal pH level with preference over other secondary metabolites. Indeed, *U. maydis* favors the production of glycolipids over organic acids at a low pH [24, 27, 50], and the regulatory interplay between different secondary metabolites is complex [7, 8, 30]. A strong relationship between pH and induction of itaconate production has also been observed for *A. terreus* [35].

In summary, we were able to show that the activation of the itaconate cluster in *U. maydis* is induced specifically in response to nitrogen limitation, and the level of induction was strongly dependent on the initial nitrogen concentration. Although the depletion of nitrogen triggers the activation, different other factors, such as pH and the concentration of other nutrients, can be optimized for fine tuning of the promoters activities.

## Conclusions

The potential of Ustilaginaceae as production organisms for different industrially-relevant compounds has been convincingly demonstrated in several instances. The presented investigation of the promoters from *Ustilago*'s itaconate cluster provides a new set of genetic tools that will enable heterologous gene expression under nitrogen limitation. Activity of these promoters is clearly coupled to the production phase, with a broad range of activities that reach up to the level of the commonly used $P_{otef}$. The investigation of these promoters opens new doors for future metabolic engineering strategies. These strategies aim for an improved match with the different cellular objectives during growth- and production phases compared to $P_{otef}$ and $P_{oma}$, which are mainly active during growth phase or e.g. *Pcrg1* and *Pnar1*, which are activated or repressed under conditions incompatible with *Ustilago*'s production phase. In addition, the GFP fusions enable further detailed investigations into the mechanism of induction of secondary metabolite production in *U. maydis*, and specifically give further insight into the regulation of the itaconate cluster of *Ustilago* and of the itaconate production pathway.

#### Authors' contributions

NW, MS, LMB, and TZ conceived and designed the project. NW, TZ, SKH, GW, JB, and LMB designed experiments and analyzed results. SKH, LMS, LB, HHT, EG, MB, DV, GW, and TZ constructed the strains and performed screening experiments. TZ and NW wrote the manuscript with the help of LMB. All authors read and approved the final manuscript.


#### Author details

[1] Institute of Applied Microbiology – iAMB, Aachen Biology and Biotechnology – ABBt, RWTH Aachen University, Worringerweg 1, Aachen 52074, Germany. [2] BioSC, c/o Forschungszentrum Jülich, 52425 Jülich, Germany. [3] Department of Biology, Philipps-University Marburg, Karl-von-Frisch-Straße 8, 35032 Marburg, Germany. [4] AVT-Aachener Verfahrenstechnik, Biochemical Engineering, RWTH Aachen University, Forckenbeckstraße 51, 52074 Aachen, Germany. [5] Institute of Crop Science and Resource Conservation (INRES), Rheinische Friedrich-Wilhelms-Universität Bonn, 53113 Bonn, Germany. [6] Institute of Plant Biology and Biotechnology (IBBP), Westfälische Wilhelms-Universität Münster, 48143 Münster, Germany.



#### Acknowledgements

We gratefully acknowledge Dr. Kerstin Schipper and Prof. Michael Feldbrügge (Heinrich Heine University Düsseldorf) and Prof. Michael Bölker (Philipps University of Marburg) for providing *Ustilago* vectors and advice.


#### Competing interests

The authors declare that they have no competing interests.


#### Funding

This work was performed as a part of the Cluster of Excellence "Tailor-Made Fuels from Biomass", which is funded by the Excellence Initiative of the German federal and state governments to promote science and research at German universities. The scientific activities of the Bioeconomy Science Center were supported financially by the Ministry of Innovation, Science and Research within the framework of the NRW Strategieprojekt BioSC (No. 578 313/323-400-002 13).


#### References


1. Alper H, Fischer C, Nevoigt E, Stephanopoulos G. Tuning genetic control through promoter engineering. Proc Natl Acad Sci USA. 2005;102:12678–83.
2. Basse CW. Dissecting defense-related and developmental transcriptional responses of maize during *Ustilago maydis* infection and subsequent tumor formation. Plant Physiol. 2005;138:1774–84.
3. Basse CW, Stumpferl S, Kahmann R. Characterization of a *Ustilago maydis* gene specifically induced during the biotrophic phase: evidence for negative as well as positive regulation. Mol Cell Biol. 2000;20:329–39.
4. Bottin A, Kamper J, Kahmann R. Isolation of a carbon source-regulated gene from *Ustilago maydis*. Mol Gen Genet. 1996;253:342–52.
5. Brachmann A, König J, Julius C, Feldbrügge M. A reverse genetic approach for generating gene replacement mutants in *Ustilago maydis*. Mol Genet Genom. 2004;272:216–26.
6. Brachmann A, Weinzierl G, Kamper J, Kahmann R. Identification of genes in the bW/bE regulatory cascade in *Ustilago maydis*. Mol Microbiol. 2001;42:1047–63.
7. Brakhage AA. Regulation of fungal secondary metabolism. Nat Rev Microbiol. 2013;11:21–32.
8. Brakhage AA, Schroeckh V. Fungal secondary metabolites—strategies to activate silent gene clusters. Fungal Genet Biol. 2011;48:15–22.
9. Broomfield PL, Hargreaves JA. A single amino-acid change in the iron-sulphur protein subunit of succinate dehydrogenase confers resistance to carboxin in *Ustilago maydis*. Curr Genet. 1992;22:117–21.
10. Couturier M, Navarro D, Olive C, Chevret D, Haon M, Favel A, Lesage-Meessen L, Henrissat B, Coutinho PM, Berrin JG. Post-genomic analyses of fungal lignocellulosic biomass degradation reveal the unexpected potential of the plant pathogen *Ustilago maydis*. BMC Genom. 2012;13:57.
11. De Mey M, Maertens J, Lequeux GJ, Soetaert WK, Vandamme EJ. Construction and model-based analysis of a promoter library for *E. coli*: an indispensable tool for metabolic engineering. BMC Biotechnol. 2007;7:34.
12. Feldbrügge M, Kellner R, Schipper K. The biotechnological use and potential of plant pathogenic smut fungi. Appl Microbiol Biotechnol. 2013;97:3253–65.
13. Flor-Parra I, Vranes M, Kamper J, Perez-Martin J. Biz1, a zinc finger protein required for plant invasion by *Ustilago maydis*, regulates the levels of a mitotic cyclin. Plant Cell. 2006;18:2369–87.
14. Geiser E. Itaconic acid production by *Ustilago maydis*. Aachen: Apprimus; 2015.
15. Geiser E, Przybilla SK, Engel M, Kleineberg W, Buttner L, Sarikaya E, Hartog TD, Klankermayer J, Leitner W, Bolker M. Genetic and biochemical insights into the itaconate pathway of *Ustilago maydis* enable enhanced production. Metab Eng. 2016;38:427–35.
16. Geiser E, Przybilla SK, Engel M, Kleineberg W, Buttner L, Sarikaya E, Hartog TD, Klankermayer J, Leitner W, Bolker M, Blank LM, Wierckx N. Genetic and biochemical insights into the itaconate pathway of *Ustilago maydis* enable enhanced production. Metab Eng. 2016;38:427–35.
17. Geiser E, Przybilla SK, Friedrich A, Buckel W, Wierckx N, Blank LM, Bölker M. *Ustilago maydis* produces itaconic acid via the unusual intermediate *trans*-aconitate. Microb Biotechnol. 2016;9:116–26.
18. Geiser E, Reindl M, Blank LM, Feldbrugge M, Wierckx N, Schipper K. Activating intrinsic carbohydrate-active enzymes of the smut fungus *Ustilago maydis* for the degradation of plant cell wall components. Appl Environ Microbiol. 2016;82:5174–85.
19. Geiser E, Wiebach V, Wierckx N, Blank LM. Prospecting the biodiversity of the fungal family Ustilaginaceae for the production of value-added chemicals. BMC Fungal Biol Biotechnol. 2014;1:2.

20. Geiser E, Wierckx N, Zimmermann M, Blank LM. Identification of an endo-1,4-beta-xylanase of *Ustilago maydis*. BMC Biotechnol. 2013;13:59.
21. Glusker J. Aconitase. In: Boyer P, editor. The enzymes. 3rd ed. New York: Academic Press; 1971. p. 413–39.
22. Guevarra ED, Tabuchi T. Production of 2-hydroxyparaconic and itatartaric acids by *Ustilago cynodontis* and simple recovery process of the acids. Agric Biol Chem. 1990;54:2359–65.
23. Gunther M, Grumaz C, Lorenz S, Stevens P, Lindemann E, Hirth T, Sohn K, Zibek S, Rupp S. The transcriptomic profile of *Pseudozyma aphidis* during production of mannosylerythritol lipids. Appl Microbiol Biotechnol. 2015;99:1375–88.
24. Günther M, Hirth T, Zibek S, Rupp S. Produktion von Biotensiden mit Pseudozyma-Stämmen. BIOspektrum. 2013;19:813–5.
25. Hevekerl A, Kuenz A, Vorlop KD. Filamentous fungi in microtiter plates-an easy way to optimize itaconic acid production with *Aspergillus terreus*. Appl Microbiol Biotechnol. 2014;98:6983–9.
26. Hewald S, Josephs K, Bölker M. Genetic analysis of biosurfactant production in *Ustilago maydis*. Appl Environ Microbiol. 2005;71:3033–40.
27. Hewald S, Linne U, Scherer M, Marahiel MA, Kämper J, Bölker M. Identification of a gene cluster for biosynthesis of mannosylerythritol lipids in the basidiomycetous fungus *Ustilago maydis*. Appl Environ Microbiol. 2006;72:5469–77.
28. Jeya M, Lee KM, Tiwari MK, Kim JS, Gunasekaran P, Kim SY, Kim IW, Lee JK. Isolation of a novel high erythritol-producing *Pseudozyma tsukubaensis* and scale-up of erythritol fermentation to industrial level. Appl Microbiol Biotechnol. 2009;83:225–31.
29. Kämper J, Kahmann R, Bölker M, Ma LJ, Brefort T, Saville BJ, Banuett F, Kronstad JW, Gold SE, Muller O, Perlin MH, Wosten HAB, de Vries R, Ruiz-Herrera J, Reynaga-Pena CG, Snetselaar K, McCann M, Perez-Martin J, Feldbrügge M, Basse CW, Steinberg G, Ibeas JI, Holloman W, Guzman P, Farman M, Stajich JE, Sentandreu R, Gonzalez-Prieto JM, Kennell JC, Molina L, Schirawski J, Mendoza-Mendoza A, Greilinger D, Munch K, Rossel N, Scherer M, Vranes M, Ladendorf O, Vincon V, Fuchs U, Sandrock B, Meng S, Ho ECH, Cahill MJ, Boyce KJ, Klose J, Klosterman SJ, Deelstra HJ, Ortiz-Castellanos L, Li WX, Sanchez-Alonso P, Schreier PH, Hauser-Hahn I, Vaupel M, Koopmann E, Friedrich G, Voss H, Schluter T, Margolis J, Platt D, Swimmer C, Gnirke A, Chen F, Vysotskaia V, Mannhaupt G, Guldener U, Munsterkotter M, Haase D, Oesterheld M, Mewes HW, Mauceli EW, DeCaprio D, Wade CM, Butler J, Young S, Jaffe DB, Calvo S, Nusbaum C, Galagan J, Birren BW. Insights from the genome of the biotrophic fungal plant pathogen *Ustilago maydis*. Nature. 2006;444:97–101.
30. Keller NP, Turner G, Bennett JW. Fungal secondary metabolism—from biochemistry to genomics. Nat Rev Microbiol. 2005;3:937–47.
31. Keon JP, White GA, Hargreaves JA. Isolation, characterization and sequence of a gene conferring resistance to the systemic fungicide carboxin from the maize smut pathogen, *Ustilago maydis*. Curr Genet. 1991;19:475–81.
32. Khrunyk Y, Munch K, Schipper K, Lupas AN, Kahmann R. The use of FLP-mediated recombination for the functional analysis of an effector gene family in the biotrophic smut fungus *Ustilago maydis*. New Phytol. 2010;187:957–68.
33. Klement T, Milker S, Jäger G, Grande PM, de Maria PD, Büchs J. Biomass pretreatment affects *Ustilago maydis* in producing itaconic acid. Microb Cell Fact. 2012;11:43.
34. König J. Die Identifikation von Ziel-Transkripten des RNA bindenden Proteins Rrm4 aus *Ustilago maydis*. Germany: Philipps-Universität Marburg; 2008.
35. Krull S, Hevekerl A, Kuenz A, Prusse U. Process development of itaconic acid production by a natural wild type strain of *Aspergillus terreus* to reach industrially relevant final titers. Appl Microbiol Biotechnol. 2017;101:4063–72.
36. Kubicek CP, Punt P, Visser J. Production of organic acids by filamentous fungi. In: Hofrichter M, editor. Industrial applications. Berlin: Springer; 2011. p. 215–34.
37. Ladner T, Mühlmann M, Schulte A, Wandrey G, Büchs J. Prediction of *Escherichia coli* expression performance in microtiter plates by analyzing only the temporal development of scattered light during culture. Biotechnol J. 2017;11:20.
38. Lara AR, Jaén KE, Sigala J-C, Mühlmann M, Regestein L, Büchs J. Characterization of endogenous and reduced promoters for oxygen-limited processes using *Escherichia coli*. ACS Synth Biol. 2017;6:344–56.
39. Lehmler C, Steinberg G, Snetselaar KM, Schliwa M, Kahmann R, Bölker M. Identification of a motor protein required for filamentous growth in *Ustilago maydis*. EMBO J. 1997;16:3464–73.
40. Loubradou G, Brachmann A, Feldbrugge M, Kahmann R. A homologue of the transcriptional repressor Ssn6p antagonizes cAMP signalling in *Ustilago maydis*. Mol Microbiol. 2001;40:719–30.
41. Maassen N, Panakova M, Wierckx N, Geiser E, Zimmermann M, Bölker M, Klinner U, Blank LM. Influence of carbon and nitrogen concentration on itaconic acid production by the smut fungus *Ustilago maydis*. Eng Life Sci. 2013;14:129–34.
42. Morita T, Fukuoka T, Imura T, Kitamoto D. Production of glycolipid biosurfactants by basidiomycetous yeasts. Biotechnol Appl Biochem. 2009;53:39–49.
43. Mutalik VK, Guimaraes JC, Cambray G, Lam C, Christoffersen MJ, Mai QA, Tran AB, Paull M, Keasling JD, Arkin AP, Endy D. Precise and reliable gene expression via standard transcription and translation initiation elements. Nat Methods. 2013;10:354–60.
44. Polli F, Meijrink B, Bovenberg RAL, Driessen AJM. New promoters for strain engineering of *Penicillium chrysogenum*. Fungal Genet Biol. 2016;89:62–71.
45. Samorski M, Müller-Newen G, Büchs J. Quasi-continuous combined scattered light and fluorescence measurements: a novel measurement technique for shaken microtiter plates. Biotechnol Bioeng. 2005;92:61–8.
46. Sarkari P, Reindl M, Stock J, Muller O, Kahmann R, Feldbrugge M, Schipper K. Improved expression of single-chain antibodies in *Ustilago maydis*. J Biotechnol. 2014;191:165–75.
47. Schulz B, Banuett F, Dahl M, Schlesinger R, Schafer W, Martin T, Herskowitz I, Kahmann R. The b alleles of *U. maydis*, whose combinations program pathogenic development, code for polypeptides containing a homeodomain-related motif. Cell. 1990;60:295–306.
48. Schuster M, Schweizer G, Reissmann S, Kahmann R. Genome editing in *Ustilago maydis* using the CRISPR-Cas system. Fungal Genet Biol. 2016;89:3–9.
49. Spellig T, Bottin A, Kahmann R. Green fluorescent protein (GFP) as a new vital marker in the phytopathogenic fungus *Ustilago maydis*. Mol Gen Genet. 1996;252:503–9.
50. Teichmann B. The cellobiose lipid ustilagic acid from *Ustilago maydis*: biosynthesis and transcriptional regulation. Marburg: Philipps University Marburg; 2009.
51. Teichmann B, Labbe C, Lefebvre F, Bolker M, Linne U, Belanger RR. Identification of a biosynthesis gene cluster for flocculosin a cellobiose lipid produced by the biocontrol agent *Pseudozyma flocculosa*. Mol Microbiol. 2011;79:1483–95.
52. Teichmann B, Liu LD, Schink KO, Bölker M. Activation of the ustilagic acid biosynthesis gene cluster in *Ustilago maydis* by the $C_2H_2$ zinc finger transcription factor Rua1. Appl Environ Microbiol. 2010;76:2633–40.
53. Terfrüchte M, Joehnk B, Fajardo-Somera R, Braus GH, Riquelme M, Schipper K, Feldbrügge M. Establishing a versatile Golden Gate cloning system for genetic engineering in fungi. Fungal Genet Biol. 2014;62:1–10.
54. Wandrey G, Bier C, Binder D, Hoffmann K, Jaeger K-E, Pietruszka J, Drepper T, Büchs J. Light-induced gene expression with photocaged IPTG for induction profiling in a high-throughput screening system. Microb Cell Fact. 2016;15:63.
55. Wanka F, Arentshorst M, Cairns TC, Jorgensen T, Ram AF, Meyer V. Highly active promoters and native secretion signals for protein production during extremely low growth rates in *Aspergillus niger*. Microb Cell Fact. 2016;15:145.
56. Wanka F, Cairns T, Boecker S, Berens C, Happel A, Zheng X, Sun J, Krappmann S, Meyer V. Tet-on, or Tet-off, that is the question: Advanced conditional gene expression in *Aspergillus*. Fungal Genet Biol. 2016;89:72–83.
57. Willis RB, Montgomery ME, Allen PR. Improved method for manual, colorimetry determination of total Kjeldahl nitrogen using salicylate. J Agric Food Chem. 1996;44:1804–7.
58. Willke T, Vorlop KD. Biotechnological production of itaconic acid. Appl Microbiol Biotechnol. 2001;56:289–95.
59. Zambanini T, Hosseinpour Tehrani H, Geiser E, Merker D, Schleese S, Krabbe J, Buescher JM, Meurer G, Wierckx N, Blank LM. Efficient itaconic

acid production from glycerol with *Ustilago vetiveriae* TZ1. Biotechnol Biofuels. 2017;10:131.

60. Zambanini T, Hosseinpour Tehrani H, Geiser E, Sonntag CK, Buescher JM, Meurer G, Wierckx N, Blank LM. Metabolic engineering of *Ustilago trichophora* TZ1 for improved malic acid production. Metab Eng Commun. 2017;4:12–21.
61. Zambanini T, Kleineberg W, Sarikaya E, Buescher JM, Meurer G, Wierckx N, Blank LM. Enhanced malic acid production from glycerol with high-cell-density *Ustilago trichophora* TZ1 cultivations. Biotechnol Biofuels. 2016;9:135.
62. Zambanini T, Sarikaya E, Kleineberg W, Buescher JM, Meurer G, Wierckx N, Blank LM. Efficient malic acid production from glycerol with *Ustilago trichophora* TZ1. Biotechnol Biofuels. 2016;9:67.
63. Zarnack K, Maurer S, Kaffarnik F, Ladendorf O, Brachmann A, Kämper J, Feldbrügge M. Tetracycline-regulated gene expression in the pathogen *Ustilago maydis*. Fungal Genet Biol. 2006;43:727–38.
64. Zobel S, Benedetti I, Eisenbach L, de Lorenzo V, Wierckx N, Blank LM. Tn7-based device for calibrated heterologous gene expression in *Pseudomonas putida*. ACS Synth Biol. 2015;4:1341–51.

# 3

# Vita activa in biotechnology: what we do with fungi and what fungi do with us

Martin Weinhold[1], Edeltraud Mast-Gerlach[2] and Vera Meyer[2*]

**Abstract**

Filamentous fungi are fascinating microorganisms. One of the reasons why it is so worthwhile to take a closer look at them is their capacity to produce secondary metabolites. Some of these substances have the potential to be of great use for mankind, such as it was the case with penicillin and its discovery in 1928. Almost a century later, the situation in healthcare could possibly turn back to the state before the development of the first antibiotics. Due to an overuse of antibiotics we are facing a surge of multiresistant bacteria that are not inhibited by any of the currently known drugs. That was part of the background why a European research project was launched in October 2013, titled "Quantitative Biology for Fungal Secondary Metabolite Producers", or "QuantFung". Fifteen young scientists embarked on a new phase in their career, moving to new work environments within Europe and dedicating their work lives intensively to the quest for useful secondary metabolites. After 4 years, the QuantFung project concluded in October this year. In this commentary, we aim to convey what it means to work in this field of fungal biotechnology and how important it is to improve the efficiency of the research therein. We introduce five out of the fifteen fellows at length and let them have their say about the adventure of science, euphoric moments, prospects and doubts. We also raise questions about the current state of research in academia, something the QuantFung fellows experienced first-hand. Being a scientist often goes beyond earning money to make one's living. This is why we also reflect on aspects of the meaning of work in our western society, where production for profit's sake is a main driver. For that we refer to one of the most distinguished thinkers of the twentieth century, to Hannah Arendt.

**Keywords:** Innovative Training Network (ITN), Marie Curie, Hannah Arendt, Fungal biotechnology, QuantFung, Natural product, Secondary metabolite, Systems biology, Synthetic biology, EUROFUNG

## Background

In 1960, political theorist Hannah Arendt chose "Vita activa" as the title for the German edition of her groundbreaking book "The Human Condition", a highly complex text elaborating on what we actually do when we work [1]. "Vita activa" as a philosophical term means to participate socially in the complex structures humans create to organize themselves. Arendt was highly in favour of using the precious good of a given lifetime for the real capacities that humans possess. Applied to today's world of work, it would be about capacities that go way beyond the daily toil which is necessary to make ends meet. In a job-focused society, where income often matters more than input, it may sound demanding to ask for a truly meaningful work life. Though closely observing recent phenomena in the realms of science or industry, it shows that we are limiting our perspectives when narrowing them down to profit margins. In one example, as it was more profitable, the German car industry was investing in new software to deceive testing methods for exhaust fumes instead of developing new technologies that would make the era of internal combustion engines a historic one altogether. Turning towards the antibiotics market: since the rewards are too low, the number of pharmaceutical companies worldwide went down from 18 in 1990 to 4 in 2010 [2]. Having the current crisis of antibiotic resistance in mind, that is threatening news.

*Correspondence: vera.meyer@tu-berlin.de
[2] Department of Applied and Molecular Microbiology, Institute of Biotechnology, Technische Universität Berlin, Gustav-Meyer-Allee 25, 13355 Berlin, Germany
Full list of author information is available at the end of the article

In October 2017, a European research project concluded that produced a network for 15 young researchers working in five different European countries. The project started in October 2013 as a 4-year multi-partner Innovative Training Network (ITN), as part of the Marie-Curie Actions in the Seventh Framework Programme 'People' (FP7 people) and was provided with €3.9 million (EU Grant number 607332). The focus of the "QuantFung" project was to identify and isolate novel bioactive molecules from filamentous fungi. One of the triggers for launching this 4-year project was the crisis of antibiotic resistance. QuantFung was offering what should be a matter of course for such an important field of science: an efficient, result-based network of researchers with diverse backgrounds and expertise, where communication can flow freely and no efforts are doubled. Even though the work of these 15 highly committed researchers took into account that finally their new knowledge may result in new products for the market, it was not its driving force. Motivations to join QuantFung were manifold. For all participants the chance to gain new knowledge in a fascinating field of research was compelling—"unlocking (a) secret door", as Min Jin Kwon, one of the researchers put it [3]. Nonetheless, the idea of producing something that society can really benefit from, is of major importance for them. Referring back to Hannah Arends idea of a meaningful life, in which humans can unlock their own capacities and talents, working for the QuantFung project truly qualifies for that. In this paper, we take stock of what was achieved within the framework of this exciting project, and also shed light on life models in (fungal) biotechnology. Kindly, a number of QuantFung fellows agreed to give a glimpse of what their life was like during the last 4 years and what they think can be learnt from their experience for future research projects.

## The origins of QuantFung

The quest for novel bioactive fungal products goes far beyond the problem of the antibiotics crisis, as it includes the need for novel drugs for multiple human health problems ranging from different cancers to increasing neurodegenerative diseases, which are especially prominent in aging societies. The necessity for a multidisciplinary training network that can unlock the genomic potential of filamentous fungi to produce novel bioactive compounds for human welfare was identified at a meeting of the EUROFUNG consortium in November 2010. EUROFUNG is a virtual centre of excellence and includes 35 academic members from 13 European countries and an associated industrial platform of 9 small, medium-sized and large European biotechnological and pharmaceutical companies [4]. A cross sector group of EUROFUNG active in the field of fungal bioactive compounds developed the initial idea for the QuantFung project (Technische Universität Berlin, Universität Göttingen, Danmarks Tekniske Universitet, DSM). QuantFung stands for "Quantitative Biology for Fungal Secondary Metabolite Producers". The concept was further developed at the ESF-EMBO symposium "Synthetic biology of antibiotic production" in October 2011 [5], where additional groups joined (Rijksuniversiteit Groningen, Leibniz-Institut für Naturstoff-Forschung und Infektionsbiologie Jena—Hans-Knöll-Institut, Chalmers tekniska högskola, Debreceni Egyetem, Christian-Albrechts-Universität zu Kiel, Bioviotica Naturstoffe, HiTeXacoat, Planton GmbH, Codexis Laboratories Hungary LTD). The intended mission of the ITN was to combine internationally renowned research experience of the EUROFUNG members with new groups and their expertise in new technologies for a training network exploring the potential of fungi for innovative, bioactive products for biotechnology and, at the same time, training high quality researchers. This intention resulted in a first proposal submitted in January 2012 that scored 93.0 points and was placed on the EU reserve list for funding. A revised second proposal had been improved by addressing the helpful and constructive comments from the reviewers of the first evaluation. It also benefited from new ideas and refinements that came from a two-day workshop in October 2012 in Berlin. The submission of the subsequent proposal happened in November 2012 and it was approved in April 2013 with a score of 93.2.

After negotiations involving all academic and industrial partners, the QuantFung project eventually started in October 2013 as the first European training platform for the production of novel bioactive compounds based on fungal systems and synthetic biology [6]. Six months were devoted to selecting and recruiting the best fellows among the applicants. The eventual start-up meeting for all research projects within QuantFung was in March 2014, which was nearly 3.5 years after the initial idea for this EU project was launched.

## Combining excellence in training and research

The driving force of QuantFung was the collaboration of 8 academic and 5 industrial partners to expedite the application of new secondary metabolites in areas such as health care, nutrition or agriculture. The main objectives of the consortium were to find novel bioactive molecules by exploiting the wealth of fungal biodiversity and to translate these into useful products. This required the (re-)design and engineering of fungal organisms with new characteristics, using highly sophisticated synthetic biology tools. The educational idea of QuantFung was to use this research context to train 11 Ph.D. students and 4 Post-Docs as new problem-solving, creative European scientists in interdisciplinary and intersectorial biotechnological research. The fields varied from modelling and network analysis to systems biology (e.g. genomics, transcriptomics, proteomics and metabolomics), from molecular biology, like fungal genetics and biochemistry, to synthetic biology methods. Different work packages were designed which focused on discovery of secondary metabolite gene clusters, targeted activation of gene clusters, quantification of secondary metabolites in industrial hosts, and bioactivity testing to identify their mode of action. In order to bridge the gap between academia and industry, the industrial partners offered secondments and training modules to share their experience with the trainees. As these multidisciplinary projects required physical and intellectual flexibility, the training program for the young fellows included defined work periods in different QuantFung laboratories, as well as local and networked training events for researchers. The vision of the QuantFung partners was that such an approach would promote the development of a new generation of fungal biotechnologist with experience in both the academic and the industrial work culture, comprising significant translational and entrepreneurial skills.

## What has been achieved?

We decided to address this question on two different levels: On the one hand, we gathered all relevant data and information regarding the scientific output of QuantFung (e.g. publications, patent applications, poster presentations, talks at conferences). On the other hand, we were using interviews to discuss with members of the consortium the impact QuantFung had on their personal life and personal development.

Regarding the evaluation of "hard facts", we took the project's official end as an effective end date (October 1st, 2017), even with the knowledge that much of QuantFung's output cannot yet be taken into account as many manuscripts are still in preparation or have just been submitted and are currently under review. Therefore, the assessment of the endeavor's effectiveness and outcome can currently only be a partial one. The overall impact will be most likely revealed in 1 or 2 years' time from now. However, the fellows' accomplishments are already very impressive at this early stage: On average, each fellow presented his/her work at four (inter)national conferences and has been co-author of 1.9 publications (Table 1). Three of these publications have receive particular attention by the research community: (1) the establishment of a CRISPR/Cas9 based genome editing tool for *Penicillium chrysogenum* [7], (2) the sequencing of nine different *Penicillium* genomes and the identification of 1317 putative secondary metabolite gene clusters hidden there [8], (3) the establishment of polycistronic gene expression in the cell factory *Aspergillus niger* as a tool for high level production of secondary metabolites [9]. As a team, the QuantFung participants jointly organised one Mini-symposium in 2014 during the annual meeting of the German Society of General and Applied Microbiology (VAAM) and a session during the highly prestigious Conference on the Physiology of Yeast and Filamentous Fungi 2016 in Lisbon (PYFF6). The fellows used the opportunity to publicize QuantFung to the broader audience of biotechnologists and the interested public, and collectively co-wrote a commentary piece in 2015 [3], which resulted in over 2590 downloads during the first 2 years after its publication.

On paper, these years of research on filamentous fungi were very successful. But what impact did this time have on the fellows? We focused on five fellows who provided a personal perspective—Min Jin Kwon, Sietske Grijseels, Yvonne Nygård, Carsten Pohl and Jens Christian Frøslev Nielsen.

### The lab away from home

An essential part of the QuantFung ITN project was its mobility rule: the fellows had to move to a European country where they had not lived during the past 3 years.

For Yvonne, one of the experienced researchers in the group, it meant going from Finland to the Netherlands. As she is used to being much further away for work assignments (e.g. being for some time in Berkeley, USA, as a visiting researcher), the distance between her home country and the Netherlands was not a major issue. Going for visits back to Finland on an average of every 2 months proved to be sufficient, and convenient due to non-stop flights from Amsterdam to Helsinki. In Yvonne's perception the main cultural difference was the very direct way of Dutch communication, although in the lab Dutch people actually were a minority, as she says, there it was way more international.

Min Jin, another of the experienced researchers, moved from the Netherlands to Berlin to participate in the project. It was a familiar place for her, as she had come to the German capital before for her master's degree after graduating in South Korea where she was born. Min Jin had lived and worked for 5 years straight in the Netherlands before QuantFung, therefore she was qualifying for the mobility rule.

For Sietske, Carsten and Jens their respective change of country kept them culturally fairly close to what they knew from home. There were no greater "shocks" to cope with than maybe sweeter pastry for Sietske in Denmark or for Jens the Swedish consensus culture which was quite different from what he was used to in Denmark, he explains in the interview. Carsten even said about his time in Groningen (Netherlands) that he sometimes felt

**Table 1 Dissemination of QuantFung results (referring only to publications, patents or presentations within the 4-years funding period)**

| | Co-author of peer-reviewed publications[b] | Co-author of patent applications | Presentations at conferences (poster or talk) | Future perspectives/remark |
|---|---|---|---|---|
| Ph.D.[a] | | | | |
| 1 | 2 | – | 8 | Successfully defended Ph.D. and continues research career as Post-Doc in academia working on fungi |
| 2 | 1 | – | 8 | Ph.D. defence date is envisaged, currently in job interviews |
| 3 | 3 | 2 | 4 | Ph.D. project continues for a 4th year (not funded by QuantFung) |
| 4 | 1 | – | 4 | Ph.D. project continues for a 4th year (not funded by QuantFung) |
| 5 | 3 | – | 5 | Ph.D. defence date is envisaged, currently in job interviews with academia |
| 6 | 2 | – | 4 | Ph.D. project continues for a 4th year (not funded by QuantFung). Already obtained a permanent position in a University to work on fungi as a fermentation engineer |
| 7 | 1 | – | 2 | Ph.D. project continues, which is however not funded by QuantFung |
| 8 | 2 | – | 4 | Ph.D. defence date is envisaged, currently in job interviews with industry |
| 9 | 4 | – | 4 | Ph.D. project continues (not funded by QuantFung) |
| 10 | 1 | – | – | Ph.D. project continues for a 4th year (not funded by QuantFung) |
| 11 | 1 | – | 3 | Ph.D. project continues for a 4th year (not funded by QuantFung) |
| Post-Doc[a] | | | | |
| 1 | 1 | – | 7 | Continues career in a private-public cooperation project at an University working on fungi |
| 2 | 1 | – | 3 | Continues career in industry |
| 3 | 3 | 2 | 3 | Obtained an Assistant Professorship position |
| 4 | 2 | – | 2 | Obtained a permanent position at a research institution |

[a] Note that projects funded by QuantFung for Ph.D. students were for 36 months, whereas post-doctoral fellows were funded for 24 months only

[b] Note that one publication was a joined publication of all fellows [2]

as if he were just in another German city. "But where I really felt like being in Europe was in my lab. There was the world at home." he continues. Counting all the nationalities he was working with in the lab actually takes him a while, "...that was like a world community, you can surely say that". The international aspect of their research environment—and its positive effect—is a shared perception among the QuantFung fellows. As Jens mentions about his group in Gothenburg: "The main nationality here is not Swedish, it's Chinese. Then you have Indian, quite a few Germans and so on. So it's not only the new Swedish culture I learnt about, it's about all other cultures from just anywhere in the world."

Nonetheless there are challenges when working abroad. For instance, figuring out things like tax authorities and insurances, which at the same time trains your ability for trouble-shooting and asking colleagues who have been in the same situation. Then there were the long commutes when you wanted to see family and friends during a weekend, using up most of the Friday for travelling. Even seemingly close distances within Europe take a toll and, as it was also about getting to know the other country and the people there, travel back home did not occur every weekend. For Jens it is one of the reasons why he will look for employment in his home country Denmark after the QuantFung project. His overall balance with the project is absolutely positive but he wants to be back where family and friends are. "This kind of work life-style (like during an ITN project) does come at a cost" he says. Whereas for Sietske going back to the Netherlands would now feel a bit like going back to a previous life. After the project she will move to Tromsø in Norway instead, for living there with somebody she got to know during her time in Denmark. "In that way you could say it's an outcome of the QuantFung project." she says with a laugh.

**The balance between work and private life**

According to Hannah Arendt's *Vita activa* (*The Human Condition*), excellence can only be achieved in the public realm, as this is the place where we can prove ourselves as being one of the best, as the place where we can stand out because there is comparison and because there are people who will notice our achievements. Nothing humans ever do in private can achieve this potential. Nonetheless there is the need to have a retreat where we are not seen and heard all the time and can recover [1].

**Fig. 1** Sietske during a summer weekend trip to Denmark's largest lake Arresø (© by Aaron Andersen)

**Fig. 2** Testing the European mobility rule: Carsten racing to the interview appointment in Berlin, 2017

The QuantFung fellows certainly moved onto a public stage when entering the international research community during the project. How did they perceive the balance between their work and the private life?

Sietske (Fig. 1) is looking back very happily with what she experienced in Denmark at DTU. There was spare time, there were weekends for going out by bike for camping tours, which she loves doing. There was also a life besides fungi. However, considering future work options for herself she tends towards favoring a job in industry, despite the fact that she really likes her field of research and enjoys being at a university. There is the difference between her work life during QuantFung, where it was totally okay to go home after 5 pm, and then there is the other side of academia that she observed: "When I see the professors around here at DTU, many of them have a family but they work day and night to do the research and to apply for grants at the same time." Especially for starting a family it could be very hard, she thinks "You find a post-doc for one or two years and during that period you already have to start writing grant applications or applying for new positions. ... It's really insecure and it's not sure whether you can stay in the same country. I think it could be pretty hard to start a family in such a position."

In the project's beginning, Carsten (Fig. 2) was involved in sporting activities outside work, like fencing classes, "...but then there was such a dynamics that literally sucked me in. One thing led to another and it became more and more exciting and then you were checking the clock and it was 9 pm and you said to yourself, okay now it really is too late for exercise..." For a tangible break he needs a change of location, as work issues remain too preoccupying for as long as he is close to the lab: "I don't go home in Groningen, close my door and then I don't think about my work anymore, that's nothing I am capable of doing. ... There was an intense phase during the project in which I actually didn't see the people at home, in my flat-sharing community, at all." Carsten sees it also as a way to find a personal limit with work. He says it is mainly your own responsibility to decide how much you let yourself being consumed personally through work. "For me it was voluntary, I really enjoyed that and it was not done under pressure and force at all. ... But it's not a life model that works as a permanent set-up, certainly not." During longer vacations that brought him away from Groningen and the lab, Carsten participated for instance twice in a huge bicycle race at the Nürburgring (Rad am Ring), something that required proper preparation and therefore was a real change of subject.

Min Jin (Fig. 3) was living with her family in Berlin and therefore combined being a mother of two kids and pursuing the project's ambitious research goals. Within the project's run she took a 10-months maternity leave after giving birth to her second child. That is why she says without hesitation: "Within the QuantFung project there definitely was a good balance for me between private

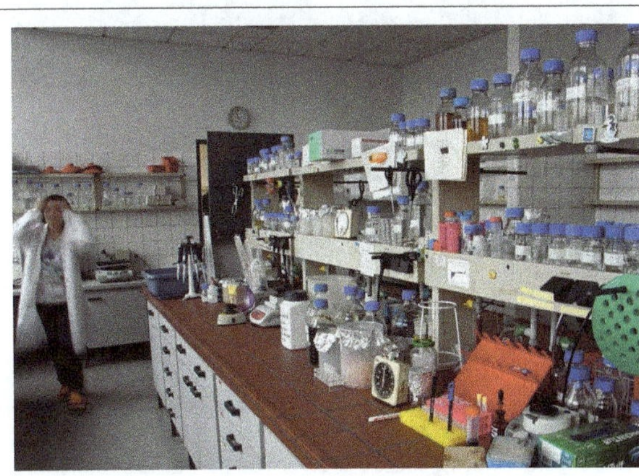

**Fig. 3** Min Jin entering one of the labs of the TU Berlin, 2015

and work life." The favourite thing besides the kingdom of fungi for her is playing with her children. But during the 6 months in which she had to come to the lab late at night for fostering demanding fungal bioreactor cultivations, she discovered Harry Potter novels as a diverting read. "Whenever I was going by public transportation at night or in the early morning, I appreciated J.K. Rowling, as she made me forget all the tiredness."

Jens (Fig. 4) had a very intense first year with the group in Gothenburg that he had chosen for its combination of biology with informatics. It was the year before his girlfriend moved to Gothenburg. Being by himself in the beginning meant that he "could just give the project all I had". The fantastic research environment he found in Sweden was exciting and working intensely was therefore nothing he had to force himself to. Nonetheless he would not like to continue that way in the future: "Even though I really liked this intense time, I think it's just important to also have a balance." And having a balance could mean for him something as simple as watching a Champion's League game together with some childhood friends on a Wednesday evening. Jens mentions another important aspect about the intensity of research work: "the fear of failing, as you have this limited time". The need to publish papers, as otherwise you cannot finish the Ph.D., is quite a common fear in the back of the mind of Ph.D. students, he says. Therefore, the motivation to work hard comes from within. "Hopefully it's more based on your excitement to the work, which is definitely a major driver, but also to some degree that you don't want to fail."

**Fig. 4** Jens during a day trip to the forests of the greater Gothenburg area—apparently being successful at picking edible mushrooms, 2016

Even if there are intense work periods, Yvonne (Fig. 5) tries to make enough time also for leisure activities and meeting friends and family. "I like to think that I work very systematically and organized. And all my spare time and all my sports activities are also in my calendar." She agrees that sports are the best thing you can do to rejuvenate your energy, a common view among all the interviewed QuantFung fellows. Another favourite pastime for Yvonne is being in nature. Asking her about what she did in nature in the Netherlands she is just laughing, "That is what I missed most in the Netherlands - nature - there is not much nature around." When the subject of the blurred border between work and private life comes up she quickly replies "There is no border for me." and continues: "Of course it's problematic because it means that a lot of weekends I spend working, but I do it because I enjoy my work and want to push forward."

**Results and revelations**

The outcome of an intense project time usually is a diverse one. With respect to QuantFung, for one there are measurable, scientific results—as everybody was hoping for, even though all the fellows were very much aware of the nature of research, i.e. "not all the reasonable approaches give the results as expected" (Min Jin). Then there is also the personal experience gained as a researcher when exposed to a new work environment and to a new lab with different opportunities to pursue the respective project goals. We asked the QuantFung fellows about results and revelations, about personal changes they noticed, as well as for their perspective on fungi after being immersed in this field of research for a substantial amount of time.

Sietske describes a measurable outcome that was very satisfying for her, and is an example of why and how important it is to work together with researchers with different expertise. "There was one fungus that produces a secondary metabolite, calbistrin, that was found to have an anti-cancer activity ... We wanted to identify the genes encoding the biosynthetic pathway for formation of calbistrin. For this we decided to sequence the genome of the fungus in which Jens [from the QuantFung group in Goteborg] had the leading role. I analysed the structure of the compound to predict which enzymatic activities would likely be involved in the formation of the secondary metabolite and using these predictions I compared the genome sequences of three species that all produce

**Fig. 5** Yvonne in the Madurodam themepark in The Hague, picture taken by her mother (© Marianne Nygård) during a visit of Yvonne's parents in 2016

this secondary metabolite, to predict which genes could be essential for its formation. Then we had an idea of what the gene cluster could be and Carsten [from the QuantFung group in Groningen] deleted several of the genes with the CRISPR/Cas9 based method that they had developed in the QuantFung team in Groningen. Fortunately, I saw that deletion of the genes resulted in mutant strains that did not produce calbistrin anymore, showing the power of combining all these different techniques The knowledge gained in this project could lead to the development of metabolic engineering strategies for improving calbistrin yields in future cell factories." Sietske is convinced that "doing research changes your personality because it is so unpredictable with respect to the outcome. There are so many things not working and every time you have to tell yourself: Next time it might work, okay, there we go again" and that she "can deal with that better, after these three years. If things don't work out - it doesn't matter, we will try to find a different way." Was her perception of the fungal world changed throughout the years? Yes. "It is really fascinating how they produce these secondary metabolites. ... That they have these clusters of genes in the genome that all encode for enzymes to work together for making these secondary metabolites. They are so interesting because they're all so different and we don't actually know what these secondary metabolites are doing in nature, why they are there."

One of Carsten's focal points with his QuantFung group in Groningen was to develop new synthetic biology tools that can be applied to fungi. The group was also involved in quantitative biology. That was something he enjoyed and was a point of pride, as this work went beyond descriptive research to generate usable facts. Another measurable outcome was the first paper he published, which Carsten describes as follows: "We established the CRISPR-Cas9 tool in two different ways in our fungus. There were papers about this tool working in fungi before, but we showed a new way where we brought the protein into the cell directly, without it being expressed by a plasmid. It's a method that can accelerate the process. Additionally, I was able to show that it works the same with one of the other QuantFung fellows' fungi [Sietske at DTU]. It was very satisfying to see that it's not just a one-time-wonder I discovered but a method that also works reliably with a different fungus. On top, it was a method helpful for my own research experiments, as it saved a lot of time." What did this kind of collaboration do with Carsten? "I learnt how much it matters on a social level to really find a way of collaboration. I think in the beginning I was very naïve, I thought it would work automatically and we would get along just because of the great and common goal. Then I was surprised to see how much the human factor mattered, despite the common goal. The individual aspect must not be underestimated also when specialists are expected to work together. It's something I intend to improve for myself, to be better at team-building." And: "I still underestimated the diversity of fungi. I was specifically working with one organism. However, I think one really should review the kingdom of fungi as broad as possible to seek after useful molecules."

Jens' field is bioinformatics. Therefore, you won't find him in the lab working with fungi, as he sits in front of the computer most of his work day. For the QuantFung project he was looking at genome data from various fungi. Studying these masses of genetic information led to an outcome he was very happy about: "I ended up publishing a paper in a journal called Nature Microbiology [8]. As I am just a Ph.D. student, I don't have a huge amount of papers yet but I ended up getting something quite impactful out of the work we did, out of all these work hours spent. That was super nice. That particular paper may also help me in the future. As for the biggest revelation - how conserved everything is in life. What I find fascinating about this biology is the similarity between all species. You can take any genome of any species - if it's a human, a bacterium or if it's a filamentous fungus - and they actually look quite similar from my point of view. I think this "conservedness" of biology is what I find most fantastic and what I realized during my Ph.D. studies. That's, of course, a little bit "nerdy" but I think that holds true and is really fascinating to me."

In Berlin, Min Jin was looking for global transcription factors that regulate secondary metabolite gene expression in *A. niger.* From meta-analyses of transcriptomic data she predicted regulators in silico and verified their role in secondary metabolism by generating mutant strains. Even though she is still finalizing the data and preparing the paper, this discovery already makes Min Jin very happy as an outcome, as she describes: "I was afraid that two selected transcription factors maybe not be true regulators of secondary metabolism, so I was still a bit doubtful about our candidate genes. But when I observed colour changes of the cultivation media of the mutant strains and also sclerotia formation induced by overexpression of one transcription factor, it was the moment that I was very happy. A colour change suggests that some secondary metabolites were produced and also sclerotia formation is not a very common phenomenon with *Aspergillus niger*. Of course, I was very excited to see that." And her perspective on fungi? "At the beginning I thought, oh, *Aspergillus niger* is very interesting, whereas now I realize that many filamentous fungi are attractive to be studied."

Yvonne was working for QuantFung in part at the University of Groningen and for half the time she had a secondment with DSM, where she was employed by one of the project's industrial partners. In her QuantFung project they were developing CRISPR gene editing tools and also expression systems. The results are partly published [7], which is a measurable outcome. New methodologies developed during the QuantFung project are now routinely used at the University of Groningen. With respect to the importance of comparable data sets for research, she says QuantFung could be a good starting point, as the fellows got to know the different procedures and methods in the labs where they were located. She carries on: "So you could take that to initiate some more standardization but it would need to be more long-term and also require commitment." When we asked her about her relationship with fungi, she had to laugh: "Well, I think it's a funny way of looking at it. But yes, that happens, you get to personalize your microorganism that you work with; I don't know whether it's always all that healthy...".

**Better is the enemy of good**

"I think it was a very successful project, in general it went well." This is as brief as Yvonne puts her answer when asked for an overall balance of QuantFung. That it went very well is a common assessment among the fellows who participated in the interviews. All of them see the time with QuantFung as an important and substantial period for making progress in their respective fields of research. Without a shadow of a doubt it was a true gain for their future career path. However, even with a successful project there is always space for improvement. Like Carsten, one of the Ph.D. students, puts it "... that's nothing that can be preconceived theoretically. ... Such a project is basically like one of my experiments—of course you try to make as much risk assessment as possible but in the end you have to do it and see what it does in real life." Which is why we asked the fellows, now in hindsight, about valued advantages with an ITN, about the things that went well in their view and about those aspects that were not perfect, so that one can learn from their experiences in order to make the next project an even better one.

The complementing effect of working on the same project, that is on the same fungus, with different tools yet the same goal, and communicating about it, is the first thing Sietske mentions with respect to advantages of this ITN project. During QuantFung she fully realized "... how valuable it is to work with researchers that have slightly different backgrounds and work on the same project. ... You have the biology and the chemistry and the bioinformatics and try to combine all of that somehow. That is something I really enjoyed - to understand that all parts are needed to make one story out of that. It is also very nice to be able to talk with each other about the research - and truly understand each other. When you are doing your Ph.D. you get so specific about things that even with people who studied the same field, or even work in the same lab, it can be quite difficult to talk in depth about what you are doing and what your struggles are. But when you're working on the same project and writing papers together, then you can discuss in depth about 'where do we want to go' and 'what do we need'. ... And the third thing for me: It is also just fun and nice to work together."

Carsten "can recommend such an ITN to any young scientist, I certainly would do that again, absolutely. It really broadens the mind when you can go somewhere else and don't stay where you are at home. That's an important aspect. It was a great opportunity for networking, as I don't know whether I would have been able to organize such an access to other colleagues. You may be able to meet somebody occasionally during a conference but it is not that kind of access that was provided through the meetings we had with QuantFung." Where would be potential for improvement? "It would be good to start and extend the networking earlier. It's just so important to be capable of teamwork in science."

Jens found the overall balance in general a good one. "I would recommend everybody joining an ITN. Being connected with other research groups within Europe has been a real advantage. I really enjoyed that I basically could come to all the meetings and talk to some of the greatest experts in the field. All the issues I had since the last meeting, I can take up with some people who are the best to talk to about that. This connection for sharing knowledge was great." What was good and helpful within QuantFung for your project? "What has been working very well for me was that I had a very strong collaboration since the beginning of the project with partners at DTU in Denmark [Sietske's QuantFung group], where we've been working really closely together. I was doing a lot of the computational analysis and they were doing the experimental work. With Sietske, you can say both of our entire Ph.D. projects were totally intertwined, which has been very nice. My research group [in Gothenburg] has not been working that much with the filamentous fungi but with the bioinformatics techniques. ... so I had that place in Sweden where I could discuss the methodology and so on, while in QuantFung I had another group of people where I could share and discuss more biological problems. And that kind of interplay has been working very well for me."

For Min Jin the main advantage of the ITN was "to meet the different people regularly. The network an ITN provides the "human infrastructure" for discussing ideas and questions with other researchers. So that you are not working isolated by yourself all the time but regularly have an exchange - and having this possibility always available. ... I think it worked very well for me and also for my family. I sure would join such a project again." And how about potentials for improvement? "Maybe to be better organized from the very beginning. When starting as a Ph.D. or a postdoc, it's always good to have plan, even though we don't know many things in the beginning...".

For Yvonne, "the network and the collaboration is the most valuable aspect for me. It was good for building up a network for your future research career. Also, it was very valuable to spend time both in the industry and at the university. For me the collaboration worked very well but for future projects there also should be some flexibility. Because it doesn't fit everybody. It was somewhat difficult for me to move to the other part of the Netherlands [for the DSM secondment]. It's all these practical issues that are quite challenging."

**Continuation and prospects**

Carsten will be returning to Germany after the QuantFung project. He wants to apply for a job in biotechnology in the vicinity of his hometown Potsdam. Besides being reunited with his girl-friend, what is he looking forward to when he is back home? "My grandmother has a garden in Potsdam; as a pastime I actually would like to do again some 'farming', to watch how things grow that I planted." With respect to his future work, Carsten would like to continue to work with fungi. In general, the wasteful use of resources in our affluent society makes him very concerned, which is why the ecological niche for biotechnology is a lot on his mind, like fungal-based materials. "I am truly impressed by start-up companies such as Ecovative and Mycoworks who turn by-products like leftovers from cereal processing into useful, biodegradable products such as packaging materials and animal-free leather." Though, he also stresses that "Biotechnology of the 21st century is going to be very radical when fully applied to replace an existing product. When I think about the future of synthetic biology and its possibilities, I am at first very excited about it. But then you should also realize that one need to pay price for change: it could become a competition to established industries and their products—and in the end, this poses a direct competition for human labour itself.".

Sietske is hoping to find a research position in Tromsø so that she can continue with her work after the move from Denmark to Norway. In Tromsø there is a focus on secondary metabolites and enzymes from marine organisms, as marine organisms often produce bioactive secondary metabolites and enzymes that are active under cold conditions. She would like to join one of the research groups working on that and apply the skills she learnt from working with fungi. In every case she wants to continue with the topic of industrial microbiology. And another thing matters to her: "What I really would like to see in the next project is working with other researchers and their different perspectives. I like approaching research in an interdisciplinary way, as I did with QuantFung where I occupied myself with a lot of different aspects. ...I could see me working in a project like that again."

Min Jin is going to stay at the TU Berlin, where she was working during QuantFung. She already started a private-public collaboration project working with another filamentous fungus. She says: "It is exciting to work with different fungi to see some similarity and also differences." Min Jin is preparing a grant proposal for a follow-up project based on the QuantFung results. Obviously, her passion for that field of research continues. As for her private life, she thinks that it is comparatively easier having a family within the university than in the industry, even though it is never going to be the fixed work day from nine to five. Asked about a working hours scheme that she would imagine in a perfect world, she says the four-days work week would be her favorite: "...for having a longer weekend and especially for having one day just for myself."

Jens will return to Denmark, but this comes with the decision of what to do after his Ph.D.. In his view, it is basically about two options: either following an academic track or a career in industry. "I am somewhat inclined to go the industrial track where I have a bit more defined work hours." he says. Jens has a lot of positive to say about academia and a lot of ideas for new projects to do after his Ph.D.. But, "With academia it's just so hard because the thing about working as a professor is that it is so competitive; so if you are just going home early and you don't get your papers and you don't get your grants then you don't have any Ph.D. students—and it's just this vicious circle." On the other hand he thinks it's tempting to be in an academic environment where you are not as predefined in your research as you may encounter in the industry where the need to produce a specific product is prevailing. With respect to weighing the options he continues: "I have former colleagues who went straight to industry after their Ph.D. and then they thought: 'that's too strict'. Then they went to academia but realized that's also tough. It can be hard to choose what is exactly good for you. Maybe you should end up being your own boss by starting some company based on secondary metabolites and filamentous fungi; that could also be a good idea. ... That's something I dream about. Let's see. It could be amazing, I think."

Yvonne already moved from the Netherlands to Sweden for a position as assistant professor at the Chalmers University of Technology in Gothenburg (one of the QuantFung partner universities). The focus of her research has now shifted from the quest for secondary metabolites to produce antibiotics to the production of fuels and chemicals, where for example the substitution of fossil fuels is one of the goals. She says both topics are really interesting and that it really matters to her that the research she does potentially has an application. In that respect her view is a very pragmatic one. "If I would have happened to study solar energy then probably I would be all into developing solar panels because that's also important. I want to do meaningful work and at the moment this is the meaningful work that I do. ... I think being a scientist is a privilege. I am interested in how things work and I daily get to figure out how things work. I think it's a lot of fun."

## The QuantFung legacy

The interdisciplinary research and training programme of QuantFung ensured that the fellows became well qualified with a broad spectrum of expertise in the field of fungal systems and synthetic biology. They are now equipped with skills and knowledge to awaken the natural product reservoir of filamentous fungi, to overproduce compounds of interest and to publish their work in peer-reviewed high impact journals. The collaboration of different academic and industrial groups led to researchers with experience of public and private work cultures and an understanding of the wider and commercial potential of their work.

For the life sciences and especially biotechnology it is essential—like a *conditio sine qua non*—to work together with each other. And this is what the QuantFung fellows have learnt. They never hesitated to ask each other because they were peers; there were no obstacles by hierarchy, they all communicated on eye level with each other. The fellows are thus much more self-confident after these years with QuantFung, as well as they are more articulate profound in scientific debates. They are not intimidated by questions anymore. Instead of evading questions they now come up with ideas of how to handle an issue, which is an important outcome of the constant communication process among the QuantFung fellows. They are also way more active asking colleagues during presentations, as they now have the courage and the experience to do that. These young scientists think along with their peers.

Another remarkable progress is the way these young scientists handle criticism. Criticism now is perceived as something positive among the fellows that underwent the QuantFung years. They understand that critical questions are nothing negative but see these questions as pieces of advice to think about for further improvement of their work. The QuantFung fellows thus became more mature scientists who grasp critical questions as part of the scientific discourse. Their next step would be to explain and picture the research work for the general public, something which they hopefully will learn during their next career steps. All things considered, ITNs such as QuantFung are ideal training networks for the new generation of (fungal) scientists. Here they can learn how to share scientific know-how and research infrastructures and how to turn research results into new technologies and products. Furthermore, they experience hands-on the importance of the human factor and of cultural exchange for unlocking their own capacities and talents. The essential assets of ITNs are indeed a very good way to advance science and enable young scientists to find their paths into a truly meaningful work life.

## Everything perfect then?

No, unfortunately not. As indicated by some of the fellows in their interviews, the job of a scientist and the career path in academia is not as attractive as it used to be. Not because one can make more money in other jobs—this is often not as relevant for students and established scientists alike, as they are intrinsically motivated and gain satisfaction from their daily work. It is because the journey into this professional future is such an unsecure one. The employment situation in academia is particularly vulnerable and one has to face the continuous challenge to get funding for new projects. This is why professors are constantly "swamped and work like mad", to the effect that they are no longer perceived as role models for their students. Recently, the head of the Department of Experimental Neurology at the Charité Berlin, Ulrich Dirnagl, gave a glimpse of what a regular Saturday is like for him: Having to read nine project proposals of 50 pages each, plus working on a grant proposal himself. He also points out that even under the best of circumstances the chance of getting a proposal granted is 50%; in the recent past that number is closer to a mere 10%. If turned down, all the resources that went in the application process are basically wasted [10]. When looking at the success rate in ITN applications, that number even dropped to 7% (Table 2). A substantial part of scientist's resources is thus burnt: All the work invested in a complex ITN application that can easily take up 2–3 months, including the creation of a network, the communication with potential partners and the actual grant writing which comes easily close to 50–60 pages. All of this is wasted capacity when the proposal is not successful in the end. In fact, it can be considered true economic damage, as so many work hours rendered by highly qualified experts are simply lost.

How can we make science and research more efficient? One suggestion for a new modus operandi to finance research in academia would be granting the funds in retrospect. This is a complete change of the current models where proposals are evaluated in foresight, proposals that are consequently rather promises. How could retrospect funding work? Here's a possible scenario. University departments receive a particular budget enabling them to do teaching *and* research. After a period of 3–4 years there will be the evaluation of what is achieved with the provided funds—e.g. how many (under)graduate students defended there thesis (BSc MSc, Ph.D.) and how many publications and patents were gained. This assessment then is the base for funding for the next 3–4 years. That means the core funding may vary, according to the performance and achievements, and scientists would not thus need to spend most of their time on grant writing with insecure outcome but on doing research. Even years with a somewhat lower funding could be used for the proper preparation of new projects. Another advantage of an assessment in retrospect would be that institutions are not getting generous funds only because they have by chance somebody in their employ who is brilliant at writing grants. Funding then would rather be given to the ones demonstrated to bring the best of results.

Another suggestion, particularly for ITNs, would be the implementation of so-called two-stage proposals. Applicants write at first a project sketch that may consist of 2–3 pages only. Here they briefly outline the intended application's concept, mention potential partners and the coordinator's CV is added, for proving there is the experience to realize such a project. In 2017, about 1400 complete ITN applications were submitted; each of them containing 50–60 pages. If these applications were reduced to 2–3 pages project sketches and that would be taken as base for the reviewer panel to decide who is being invited to submit the full application—it clearly would make the application/funding process and thus science more efficient. Conceivable is a rate of 300 of the original 1400 submissions being invited to write full proposals with a chance to get one granted being about 30%. In that case it also would give the reviewer panel sufficient time to evaluate full proposals thoroughly. This system is in place in Germany, e.g. for calls from the Federal Ministry of Education and Research (BMBF).

**Table 2 Chances of success for getting an ITN proposal granted**

| Year of evaluation | Budget (Mio €) | No. of proposals submitted | No. of proposals granted | Success rate (%) |
|---|---|---|---|---|
| 2010 | 243 | 900 | 92 | 10.4 |
| 2011 | 318 | 863 | 63 | 7.4 |
| 2012 | 423 | 1022 | 128 | 12.6 |
| 2013 | 470 | 1175 | 150 | 12.9 |
| 2014 | 405 | 1161 | 121 | 10.5 |
| 2015 | 370 | 1566 | 106 | 6.8 |
| 2016 | 370 | 1611 | 109 | 7.0 |
| 2017 | 370 | 1437 | 98 | 6.8 |

Science needs its path! True innovation is often only possible at the margins of different scientific disciplines. But they must, at first, come together and get to know each other; they must comprehend how the other ones are working, what their methods are. Out of the 4 years with QuantFung, the real time for project work was 3 years for Ph.D. students and only 2 years for the post-docs. After these 2–3 years, the QuantFung fellows now have a much better idea of what needs to be considered when one intends to find and overproduce new secondary metabolites from fungi, how important the merge of different disciplines is and have achieved their first promising results. However, QuantFung as a consortium is still far away from actually having produced new classes of antibiotics, although industrial partners for the full pipeline to develop such products were within that network. Clearly, 2–3 years of research are not enough to really accomplish such a challenging mission. It is worth remembering that it was well over a decade between the discovery of penicillin and effective production [11]. This is why we propose that the EU consider applications for follow-up ITN projects. They should be limited and awarded in case of outstanding scientific achievements of an ITN consortium.

**Fig. 6** Champi(gn)ons, V. meer, 2017. Parasol mushroom, iron stand, shellac, rust (Reproduced with permission from [13])

## Conclusions

The QuantFung participants were so immersed in the field of fungi, got so intrigued, that actually a number of fellows will continue doing research work with these organisms. Another QuantFung post-doc, Danielle Troppens, became a passionate blogger on the world of fungi. She regularly reports on findings that are curious and fascinating [12]. And as for QuantFung's coordinator, Vera Meyer, another effect came along with her work for the QuantFung project (Fig. 6). "Fungi turned out to be not only objects of scientific study for me. I also see and use them as art objects. It is an opportunity to convey the work that we do for the general public, to get them interested. I don't see these fields separated, science and art. Especially art for me is a means of communication. Since I have seen fungi so many times under the microscope, they became also aesthetically fascinating creatures for me. Fungi are champions for me, and it really intrigues me how to use art to give our science, the science around fungi, a face."

ITNs represent one of the best funding instruments the EU is offering. They provide an ideal framework for international exchange and for building up a vital and innovative research environment where knowledge, state-of-the-art technologies and the most modern equipment can be shared. All of which is what is needed to train our next generation of high-calibre scientists, to foster their creativity and to jointly come to new scientific breakthroughs. How beneficial it is to have such a network was strikingly proven by the success that QuantFung had on many levels. There are the measurable results and there is the fact that 15 young researchers are now exquisitely prepared for a future career in biotechnology. The fellows are well qualified with a broad portfolio of skills and practical experience of cross sector working; this will make them strong candidates for future employers and underpin their career progression. Importantly, taking responsibility for their own development has instilled in them at an early stage the importance of continuous professional development. Such high quality and highly competitive candidates are needed in both the public and private sectors. Although academia may lose great talents, they have excellent chances to take up industry posts.

"Antibiotic resistance is one of the biggest threats to global health, food security, and development today." [14]. This is the first sentence of the World Health Organization's internet fact sheet addressing the crisis of more and more frequently failing antibiotic medication. Which leaves nobody in doubt about the problem's dimension. There are ambitious researchers in academia who eagerly want to contribute to the problem's solution. But they should be given the opportunity to pursue their respective work lives in a way that leaves them space to breathe, instead of keeping them in the uncertainty of perpetual-temporary contracts and in the frenzy of constant grant writing. A "Vita activa" in Hannah Arendt's view means a life spent reaching up to higher goals. Advancing the deeper understanding of fascinating organisms like filamentous fungi for the sake of human health is certainly such a goal.

**Authors' contributions**

VM conceived the idea and drafted the manuscript. MW directed the interviews. VM, MW and EMG were involved in writing the manuscript. All authors read and approved the final manuscript.

**Authors' information**

Martin Weinhold is a communications professional by trade and a documentarian by passion; he has explored occupational life models for more than a decade in an epic project called WorkSpace Canada. In this project the young nation Canada is serving as a template for the present-day world of work. Producing a multi-faceted overall map of its variety is Weinhold's ambition. Since 2006 he portrays in his photographs people and their work landscapes, gathers sound samples from potash mines, or interviews the last remaining fur traders [15]. When approached with the request to contribute to this paper, he was struck by the same curiosity to learn about QuantFung that led him already to more than 200 occupational fields in his own project work. Edeltraud Mast-Gerlach is a chemist by training and holds Ph.D. in biotechnology with a focus on yeast fermentation. She also holds a Master of Science degree in communication and marketing. She worked for 8 years in an industrial biotech company before resuming an academic career at the Technische Universität Berlin, Germany. She is the managing director of the Innovation Centre Technologies for Health and Food at the TU Berlin and was involved in the QuantFung project as management assistant. Vera Meyer studied biotechnology and worked as scientist in Berlin, London and Leiden. Currently, she works as a professor at the Technische Universität Berlin, Germany and runs the Department for Applied and Molecular Microbiology. She was the driving force behind the QuantFung project and acted as its coordinator. Prof. Meyer is interested in decoding nature's genetic principles underlying growth and metabolism of fungal microorganisms. By bridging the gap between systems and synthetic biology, together with her team she develops new fungal cell factories for the sustainable production of pharmaceuticals and proteins. As a self-taught artist, she does not necessarily follow explicit project purposes but rather goes where improvisation and intuition will take her. Her work is inspired by a very conscious perception of nature phenomena and everyday objects. Likewise, ideas are sparked by her work with fungi or when travelling the world [13].

**Author details**

[1] Menschenfotograf, Am Goldmannpark 21, 12587 Berlin, Germany. [2] Department of Applied and Molecular Microbiology, Institute of Biotechnology, Technische Universität Berlin, Gustav-Meyer-Allee 25, 13355 Berlin, Germany.

**Acknowledgements**

This project was funded by the Marie Curie Innovative Training Network QuantFung (FP7-People-2013-ITN, Grant 607332).

**Competing interests**

The authors declare that they have no competing interests.

**References**

1. Arendt H. Vita activa oder Vom tätigen Leben. 3rd ed. Munich: Piper; 2005.
2. Cooper MA, Shlaes D. Fix the antibiotics pipeline. Nature. 2011;472:32.
3. Büttel Z, Díaz R, Dirnberger B, Flak M, Grijseels S, Kwon MJ, Nielsen JCF, Nygård Y, Phule P, Pohl C, Prigent S, Randelovic M, Schütze T, Troppens D, Viggiano A. Unlocking the potential of fungi: the QuantFung project. Fungal Biol Biotechnol. 2015;2:6.
4. http://mikrobiologie.eurofung.tu-berlin.de.
5. Takano E, Bovenberg RA, Breitling R. A turning point for natural product discovery: ESF-EMBO research conference: synthetic biology of antibiotic production. Mol Microbiol. 2012;83:884–93.
6. http://intern.mikrobiologie.tu-berlin.de.
7. Pohl C, Kiel JA, Driessen AJ, Bovenberg RA, Nygård Y. CRISPR/Cas9 Based Genome Editing of *Penicillium chrysogenum*. ACS Synth Biol. 2016;5:754–64.
8. Nielsen JC, Grijseels S, Prigent S, Ji B, Dainat J, Nielsen KF, Frisvad JC, Workman M, Nielsen J. Global analysis of biosynthetic gene clusters reveals vast potential of secondary metabolite production in *Penicillium* species. Nat Microbiol. 2017;2:17044.
9. Meyer V, Schütze T. Polycistronic gene expression in *Aspergillus niger*. Microb Cell Factories. 2017;16:162.
10. Ulrich Dirnagl. Werden Sie Forschungsförderer. In: Laborjournal 2017-06, pp 22–23.
11. https://www.pfizer.com/about/history/all.
12. www.microbe-scope.com.
13. www.v-meer.de.
14. www.who.int/mediacentre/factsheets/antibiotic-resistance/en/.
15. www.menschenfotograf.de/en/workspace-canada-2.

4

# Spontaneous and CRISPR/Cas9-induced mutation of the osmosensor histidine kinase of the canola pathogen *Leptosphaeria maculans*

Alexander Idnurm[1*], Andrew S. Urquhart[1], Dinesh R. Vummadi[1], Steven Chang[2], Angela P. Van de Wouw[1] and Francisco J. López-Ruiz[2]

**Abstract**

**Background:** The dicarboximide fungicide iprodione has been used to combat blackleg disease of canola (*Brassica napus*), caused by the fungus *Leptosphaeria maculans*. For example, in Australia the fungicide was used in the late 1990s but is no longer registered for use against blackleg disease, and therefore the impact of iprodione on *L. maculans* has not been investigated.

**Results:** Resistance to iprodione emerged spontaneously under in vitro conditions at high frequency. A basis for this resistance was mutations in the *hos1* gene that encodes a predicted osmosensing histidine kinase. While loss of the homologous histidine kinase in some fungi has deleterious effects on growth and pathogenicity, the *L. maculans* strains with the *hos1* gene mutated had reduced growth under high salt conditions, but were still capable of causing lesions on *B. napus*. The relative ease to isolate mutants with resistance to iprodione provided a method to develop and then optimize a CRISPR/Cas9 system for gene disruptions in *L. maculans*, a species that until now has been particularly difficult to manipulate by targeted gene disruptions.

**Conclusions:** While iprodione is initially effective against *L. maculans* in vitro, resistance emerges easily and these strains are able to cause lesions on canola. This may explain the limited efficacy of iprodione in field conditions. Iprodione resistance, such as through mutations of genes like *hos1*, provides an effective direction for the optimization of gene disruption techniques.

**Keywords:** *Agrobacterium*-mediated transformation, Canola, Gene editing, HOG pathway

## Background

Canola (*Brassica napus*) is a major crop worldwide, and is also grown as part of the crop rotation systems with cereals [1]. Control of the main disease of canola, blackleg, is through farming practices that minimize exposure to the infectious spores, sowing cultivars that carry resistance genes, and more recently relying on fungicides. Blackleg disease is caused by a species complex in the genus *Leptosphaeria* (Dothideomycetes; Pleosporales) [2–4], with most crop losses due to *L. maculans*. Although a number of molecular biology resources are available for *L. maculans*, including a genome sequence [5], investigations of gene functions in the fungus has been hampered by the low rates of homologous integration of constructs used in generating gene deletion strains, with just nine gene knock outs reported in the literature [6–11]. Hence, this aspect of the fungus has limited the ability to test how specific genes may impact the ability of this fungus to cause disease on canola.

The application of fungicides has provided large yield increases to crops by reducing the symptoms caused by fungal diseases. In the case of blackleg disease, different fungicides have been and continue to be employed [1]. In Australia, currently these are in the triazole class [Fungicide Resistance Action Committee (FRAC) group 3], and used as seed dressings, combined with fertilizer, or as a foliar

*Correspondence: alexander.idnurm@unimelb.edu.au
[1] School of BioSciences, University of Melbourne, Building 122, Parkville, VIC 3010, Australia
Full list of author information is available at the end of the article

spray. In another class, the dicarboximide iprodione (FRAC group 2; trade name Rovral® produced by Bayer CropScience), was approved for use against blackleg disease and used for about 5 years before being unregistered at the end of the 1990s. Iprodione currently can still be used for treatment of Sclerotinia stem rot of canola, which is a disease that has an overlapping distribution as blackleg.

Resistance to dicarboximide fungicides, like iprodione, can occur if mutations arise in the high osmolarity glycerol response (HOG) pathway [12]. The HOG pathway was first characterized in *Saccharomyces cerevisiae* for its role in enabling growth under hyper- and hypo-osmotic conditions [13–15], and subsequent research in numerous fungal species has defined multiple phenotypes of strains with mutations in the signaling genes. The HOG pathway features a sensing histidine kinase that transfers environmental information into a cascade of three mitogen activated protein kinases [12]. Mutations often occur in the homologs of the histidine kinase in other fungi to confer resistance to the dicarboximide fungicides [16].

Exposure to fungicides is linked to the emergence of fungicide resistance, thereby rendering specific fungicides or entire classes ineffective. We recently initiated an investigation into the levels of resistance to triazole fungicides in *L. maculans* populations in Australia [17]. As a control, isolates were tested for their responses against the unrelated chemical iprodione. Spontaneous resistance to this molecule was commonly observed, leading to the investigation into its basis and the impact of those mutations on pathogenicity. Subsequently, the *hos1* gene that was mutated in these resistant strains was used as a tool for the development of the clustered regularly interspaced short palindromic repeats (CRISPR)-Cas9 system to make targeted mutations in *L. maculans*.

## Methods

### Fungal strains and culturing

Routine culturing of *L. maculans* was on 10% V8 juice with 2% agar. The wild type strains or those isolated during this study are listed in Table 1. Because there is often limited contrast between the color of the V8 juice and fungal hyphae, to increase the contrast in the figures the strains were cultured on potato dextrose agar. Iprodione was dissolved in dimethyl sulfoxide, and added to agar media at final concentrations of 5 or 10 μg/ml.

**Table 1 Strains of *Leptosphaeria* spp. used in this study**

| Strain name(s) | Genotype | Origin |
|---|---|---|
| *L. biglobosa* 06J154 | Wild type | Burren Junction, NSW, Australia, 2006 [19] |
| D5 (IBCN18; M1) | Wild type | Penshurst, VIC, Australia, 1988 |
| D5-Ip$^R$ | *hos1*$^-$ | Selection of D5 on iprodione |
| D5-Ip$^R$ + *hos1* | *hos1*$^-$ + *hos1* | Transformation of D5-Ip$^R$ with wild type *hos1* |
| D2 (IBCN15) | Wild type | Streatham, VIC, Australia, 1988 [2] |
| D2-Ip$^R$ | *hos1*$^-$ | Selection of D2 on iprodione |
| D2-Ip$^R$ + *hos1* | *hos1*$^-$ + *hos1* | Transformation of D2-Ip$^R$ with wild type *hos1* |
| D3 (IBCN16) | Wild type | Mt Barker, WA, Australia, 1988 [2] |
| D3-Ip$^R$ | *hos1*$^-$ | Selection of D3 on iprodione |
| D3-Ip$^R$ + *hos1* | *hos1*$^-$ + *hos1* | Transformation of D3-Ip$^R$ with wild type *hos1* |
| 14P290 | Wild type | Katanning, WA, Australia, 2014 [17] |
| 14P290-Ip$^R$ | *hos1*$^-$ | Selection of 14P290 on iprodione |
| 14P290-Ip$^R$ + *hos1* | *hos1*$^-$ + *hos1* | Transformation of 14P290-Ip$^R$ with wild type *hos1* |
| D13 (09SMW024) | Wild type | Cummins, SA, Australia, 2009 [52] |
| DV1 | *hos1 guide* | D5 |
| DV2 | *hos1 guide; cas9* | DV1 |
| DV3 | *hos1 guide; cas9; hos1*$^-$ | DV2 |
| D13-CoT | *cas9; hos1 guide* | D13 cotranformed with both guide RNA (hyg) and Cas9 (G418) |
| D13-Ip$^R$1 | *cas9; hos1 guide; hos1*$^-$ | Selection of D13-CoT on iprodione |
| D13-Ip$^R$2 | *cas9; hos1 guide; hos1*$^-$ | Selection of D13-CoT on iprodione |
| v23.1.3 | Wild type | *In vitro* crosses, France [5] |
| JN3-Cas9 | *cas9* | v23.1.3 transformed with pMAI23 |
| JN3-avrLm1-1 | *cas9; AvrLm1 guide; avrLm1*$^-$ | JN3-Cas9 |
| JN3-avrLm1-2 | *cas9; AvrLm1 guide; avrLm1*$^-$ | JN3-Cas9 |

All strains are *L. maculans*, with the exception of one *L. biglobosa* strain used as a source of DNA for constructs. The numerous *hos1* mutants isolated from CRISPR-Cas9 sources and the 28 progeny from the D3-Ip$^R$ × D13 cross are not listed. IBCN indicates a strain in the International Blackleg of Crucifers Network collection

For genetic segregation analysis, crosses were set up between strain D3-Ip$^R$ and strain D13, on 20% V8 juice and $CaCO_3$ medium. After 1 week of growth, the plates were overlaid with water agar. Cultures were incubated at 14 °C with alternating 12 h dark-blacklight cycles for 6 weeks. At this point, the plates were examined for the formation of pseudothecia. Asci were released by placing the pseudothecia in sterile water whereby ascospores were discharged naturally. Individual ascospores were then collected and allowed to germinate on 2% water agar plates before being hyphal-tip subcultured to create individual strains. A total of 28 progeny was collected and analyzed.

**DNA isolation of *L. maculans*, PCR and sequencing**

*Leptosphaeria maculans* mycelia were cultured in 10% cleared V8 juice medium (pH 6). Mycelia were freeze-dried, broken with 2 mm glass beads, and DNA extracted in a CTAB buffer and incubation at 65 °C, followed by one chloroform extraction, and precipitation with an equal volume of isopropanol [18].

For sequencing, *hos1* was amplified with different primer combinations to cover different regions of this large gene. Amplicons used to identify spontaneous mutations were MAI0218-MAI0223 and MAI0220-MAI0224. Primer sequences used in this study are found in Additional file 1: Table S1. Primers used to amplify and then identify mutations in *hos1* induced by CRISPR-Cas9 were MAI0220-MAI0224.

To resolve which *hos1* allele is present in progeny of the D3-Ip$^R$ × D13 cross, the region was amplified with primers MAI0220-MAI0376. The DNA was precipitated and then digested with AgeI restriction enzyme, which cuts the amplicon of the wild type copy but not the iprodione resistance allele.

Primers used to amplify the *AvrLm1* gene to identify mutations induced by CRISPR-Cas9 were MAI0353-MAI0354.

**Construction of plasmids for transformation of *L. maculans***

Plasmids were constructed for the introduction of T-DNA molecules into *L. maculans* using *Agrobacterium tumefaciens* mediated transformation.

Two plasmids conferring resistance to G418 or hygromycin were made, in which gene expression was from the promoter and terminator of the actin gene of *L. biglobosa* strain 06J154 [19]. G418 resistance has not been used previously in *L. maculans* transformation. For the G418 construct, the promoter region was amplified with primers MAI0014-MAI0015, and terminator region with primers MAI0016-MAI0017 from genomic DNA of strain 06J154, isolated as for *L. maculans*. The open reading for the aminoglycoside phosphotransferase that confers resistance to G418 was amplified with primers ALID0835-ALID0836 from plasmid pPZP-NEO1 [20]. The three pieces were joined by overlap PCR using primers MAI0014-MAI0017, and cloned into the TOPO pCR2.1 plasmid (Invitrogen). To ensure the expression system worked for resistance to G418, the equivalent plasmid was made to confer resistance to hygromycin. The promoter was amplified with primers MAI0018-MAI0024 and terminator with MAI0020-MAI0021. The hygromycin phosphotransferase gene was amplified with primers MAI0022-MAI0023 from plasmid pPZPHyg*Hin*dX [21]. The three pieces were joined together using primers MAI0018-MAI0021, and cloned into the TOPO 4.0 plasmid (Invitrogen). All PCRs used Platinum® *Pfx* DNA polymerase (Invitrogen). Plasmids containing clones without PCR-derived errors were identified, by sequencing the inserted fragments. The G418 resistance construct was excised with EcoRI and cloned into the EcoRI site of plasmid pPZP-201BK, which is able to replicate in *Agrobacterium tumefaciens* [22], to form pMAI2. The hygromycin resistance construct was excised with KpnI-SpeI and cloned into the KpnI-XbaI site of pPZP-201BK to form plasmid pMAI6.

Plasmid pLAU2 was constructed by cloning two fragments, the *L. maculans* actin (*act1*) promoter and the *trp3* terminator, into pPZPHyg*Hin*dX [21] digested with AscI and PacI using Gibson Assembly (New England Biolabs). Primers used to amplify the actin promoter, with Platinum® *Pfx* DNA Polymerase (Invitrogen), were AU1 and AU2 and for the *trp3* terminator were AU5 and AU6. Primers were designed with additional nucleotides such that a BglII site was introduced between the promoter and terminator. Green fluorescent protein (GFP) was amplified with primers AU28 and AU31, and cloned into the BglII site of pLAU2 to form plasmid pLAU17. The *act1* promoter and *trp3* terminator combination was excised from plasmid pLAU2 using restriction enzymes SpeI and NheI and cloned into the XbaI site of plasmid pMAI2, to form pLAU53.

For complementation with the wild type copy of *hos1*, the gene was amplified with primers MAI0206 and MAI0207 using Q5 DNA polymerase (New England Biolabs) from genomic DNA of wild type isolate D5, and cloned using Gibson assembly (New England Biolabs) into plasmid pMAI2 that had been linearized with EcoRV and XhoI.

Constructs were made to express either the Cas9 endonuclease or the CRISPR guide RNAs using the actin promoter of *L. maculans*. Cas9 was amplified with primers MAI0225 and MAI0226 from plasmid pHSN401 [23] and cloned into the BglII sites of pLAU2 and pLAU53 by Gibson assembly to form plasmids pMAI22 and pMAI23, respectively. The DNA fragment to target the endonuclease to *hos1*, guide RNA and two ribozymes were synthesized by Thermo Fisher Scientific (sequence in Additional file 2) and provided as a cloned product. The fragment was amplified using primers MAI0228 and MAI0229 and then inserted into the BglII sites of both plasmids pLAU2 and pLAU53 using Gibson assembly.

To streamline the production of the guide RNAs, two additional plasmids were made that incorporate the hepatitis delta virus (HDV) ribozyme, such that a single oligonucleotide of about 100 nt can be used for cloning the targeting RNA, rather than a synthesized and cloned DNA fragment. A XhoI restriction enzyme site was included to facilitate subsequent cloning of the gene-specific fragments. A DNA molecule (Additional file 2) was synthesized by Thermo Fisher Scientific, and amplified with primers MAI0228-MAI0229 and cloned into the BglII sites of both plasmids pLAU2 and pLAU53 using Gibson assembly to form plasmids pMAI75 and pMAI97, respectively.

A construct to produce a guide RNA to target mutations to the *AvrLm1* gene was generated, by amplification off oligonucleotide MAI0336 with primers MAI0309-MAI0310, and cloning the amplicon into plasmid pMAI75 linearized with XhoI.

In all cases of plasmid construction that used amplification of DNA and subsequent cloning, the inserted DNA molecules in the plasmids were sequenced to either confirm that no PCR-induced errors occurred or to identify error-free clones.

#### Transformation of *L. maculans* with *Agrobacterium tumefaciens*

The plasmids were transformed into *A. tumefaciens* strain EHA105 using electroporation, and selected on LB medium + kanamycin (50 µg/ml). The *Agrobacterium* strains were then used to transform strains of *L. maculans*, with selection of fungal transformants using either G418 (100 µg/ml) or hygromycin (50 µg/ml), and cefotaxime (100–150 µg/ml) to inhibit *Agrobacterium* growth. The transformation of *L. maculans* was as follows. Overnight cultures of *Agrobacterium* in LB + kanamycin were diluted in sterile water and plated onto *Agrobacterium* induction medium [24] solidified with 2% agar (25 ml in 15 cm diameter Petri dishes) with pycnidiospores harvested in sterilized water from the *L. maculans* strains. Bacterial and fungal cells were spread across the plate. After 3 days incubation at 22 °C in darkness, 25 ml of cleared V8 juice 1.5% agar media containing selective antibiotics were overlaid. Transformed colonies emerged through the overlay agar 10–18 days later. The transformed strains were subcultured at least once onto V8 juice agar supplemented with the antibiotic suitable to select for fungal transformation and cefotaxime to inhibit *Agrobacterium* growth.

#### Quantification of antifungal drug resistance

Levels of resistance to two dicarboximide fungicides and a triazole fungicide were quantified in a radial growth assay, as in Ref. [25] with some modifications. In brief, for each strain 4 mm diameter mycelium plugs were inoculated into the center of 9-cm PDA petri dishes amended with a range of concentrations of technical grade iprodione (0.195–50 µg/ml), procymidone (0.195–50 µg/ml) and tebuconazole (0.0782–5 µg/ml) dissolved in dimethyl sulfoxide (DMSO). The colony diameter was measured in two perpendicular directions and values recorded in millimeters. $EC_{50}$ values were calculated as described previously [16]. $EC_{50}$ values of wild type and complemented strains were analyzed by the Mann–Whitney *U*-test.

#### Microscopy

Spores of wild type and a GFP-expressing strain were germinated in cleared V8 juice (10%) and examined 3 days later using a Leica DM6000 microscope with an attached digital camera.

#### Plant inoculations and pathogenicity testing

*Brassica napus* cultivar Westar was grown in soil in growth cabinets. Two weeks after sowing the seed, pycnidiospore suspensions ($10^6$ spores/ml) from the *L. maculans* strains were placed as 10 µl drops onto wounded cotyledons. Lesions were scored on a 0–9 scale, as previously described [26], 11–14 days later.

## Results

#### Development of a positive selection system in *L. maculans* based on resistance to iprodione due to mutations in the *hos1* gene

Wild type strains of *L. maculans* developed resistance to iprodione readily when mycelial plugs were inoculated on V8 juice agar medium supplemented with iprodione at concentrations up to 10 µg/ml. The fungicide at first inhibited growth, and then after several days a section of the mycelial plug initiated growth. These sectors were cultured and purified as single spore isolates.

A single homolog, named *hos1*, of the osmosensing histidine kinase is present in *L. maculans* as assessed by BLAST analysis of the genome sequence [5]. Amplification of the *hos1* gene and sequencing revealed mutations in the gene in iprodione resistant mutants, all of which are predicted to cause a loss-of-function (Fig. 1a). Two mutants, one arising from strain D2 (referred to as D2-$Ip^R$) and the other from strain 14P290 (P290-$Ip^R$), cause frame shifts in the reading frame, leading to the introduction of premature stop codons. The mutations that occurred in the mutants arising from strains D5 (D5-$Ip^R$) and D3 (D3-$Ip^R$) cause amino acid substitutions (R923K and G929R, respectively). These two amino acid residues are highly

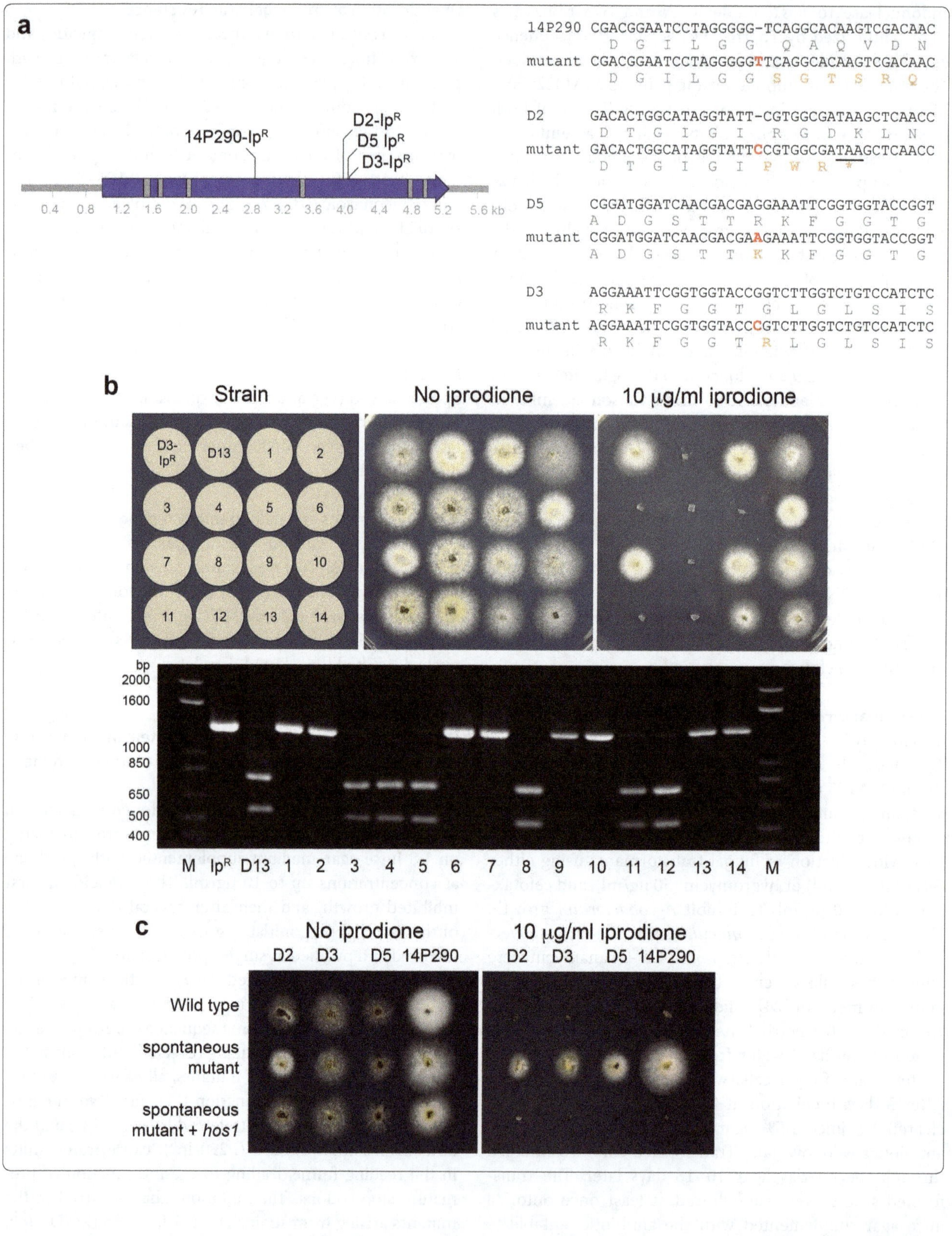
a
14P290-Ip^R
D2-Ip^R
D5 Ip^R
D3-Ip^R
0.4 0.8 1.2 1.6 2.0 2.4 2.8 3.2 3.6 4.0 4.4 4.8 5.2 5.6 kb
14P290 CGACGGAATCCTAGGGGG-TCAGGCACAAGTCGACAAC
D G I L G G Q A Q V D N
mutant CGACGGAATCCTAGGGGGTTCAGGCACAAGTCGACAAC
D G I L G G S G T S R Q
D2 GACACTGGCATAGGTATT-CGTGGCGATAAGCTCAACC
D T G I G I R G D K L N
mutant GACACTGGCATAGGTATTCCGTGGCGATAAGCTCAACC
D T G I G I P W R *
D5 CGGATGGATCAACGACGAGGAAATTCGGTGGTACCGGT
A D G S T T R K F G G T G
mutant CGGATGGATCAACGACGAAGAAATTCGGTGGTACCGGT
A D G S T T K K F G G T G
D3 AGGAAATTCGGTGGTACCGGTCTTGGTCTGTCCATCTC
R K F G G T G L G L S I S
mutant AGGAAATTCGGTGGTACCCGTCTTGGTCTGTCCATCTC
R K F G G T R L G L S I S
b
Strain
No iprodione
10 μg/ml iprodione
D3-Ip^R D13 1 2
3 4 5 6
7 8 9 10
11 12 13 14
bp
2000
1600
1000
850
650
500
400
M Ip^R D13 1 2 3 4 5 6 7 8 9 10 11 12 13 14 M
c
No iprodione
10 μg/ml iprodione
D2 D3 D5 14P290
D2 D3 D5 14P290
Wild type
spontaneous mutant
spontaneous mutant + hos1

(See figure on previous page.)
**Fig. 1** Spontaneous iprodione resistance occurs through mutation of the *hos1* gene. **a** Positions and nature of spontaneous mutations in iprodione resistant strains from four wild type strains relative to the exon (blue) and intron (grey) structure of the *hos1* gene. The sequence alignments are of the wild type and mutant strains, with the nucleotide mutations that occurred in the four strains in red bold, and the predicted amino acid sequences underneath. The mutations in strains 14P290 and D2 cause frame shifts (the stop codon in the D2 mutant is underlined, in 14P290 the new stop codon is beyond the sequence shown). The mutations in M1 and D3 cause amino acid substitutions in residues that are invariant across *hos1* homologs in the fungi. **b** A mutation in *hos1* co-segregates with iprodione resistance. Growth of two parents and 14 progeny (from 28 total) of a D13 × D3-Ip$^R$ cross between the parents on PDA with or without iprodione. The alleles of *hos1* were assessed by PCR from genomic DNA of the two parents and 14 progeny from a cross between them, and subsequent digestion with AgeI restriction enzyme. M is the Invitrogen 1 kb + ladder. **c** Complementation of iprodione resistance back to sensitivity by the wild type *hos1* gene. Mycelial plugs were inoculated onto PDA medium with or without iprodione (10 µg/ml) and cultured 4 days. The strains are four wild type strains, four spontaneous mutants derived from them, and the four strains whereby the wild type copy of *hos1* was transformed into the mutants. The wild type copy of *hos1* returned the strains to the wild type sensitive phenotype

conserved in homologs of the histidine kinase, because BLAST analysis of the fungal genomes available through the MycoCosm Portal of the Joint Genome Institute [27] revealed that both residues are invariant in all fungal species, including those of the early diverging lineages commonly termed the chytrids and zygomycetes.

Two methods were used to confirm that mutations in the *hos1* gene caused the resistance to iprodione. The first approach was to analyze the segregation of traits and genotypes in progeny from a cross. Resistant strain D3-Ip$^R$ was crossed with sensitive strain D13, and 28 progeny obtained. The strains were scored for growth on medium containing iprodione and genotyped for the *hos1* allele by PCR–RFLP. The AgeI restriction enzyme recognition site (ACC<u>G</u>GT) of the wild type strain is lost in strain D3-Ip$^R$ due to a base pair substitution (ACC<u>C</u>GT), as underlined. Nine progeny were sensitive to iprodione and 19 were resistant, which is not statistically different from the expected 1:1 ratio based on a $\chi^2$-test. The AgeI cut site polymorphism in *hos1* co-segregated with the iprodione sensitive or resistant phenotype (Fig. 1b). The second approach was complementation of the mutant phenotype. A wild type copy of the *hos1* gene was amplified and cloned into a plasmid that confers resistance to G418 for selection when the T-DNA is transformed into *L. maculans*. This construct was transformed into four strains that were iprodione resistant (i.e. D2-Ip$^R$, D3-Ip$^R$, D5-Ip$^R$ and 14P290-Ip$^R$). Transformants were tested on medium containing iprodione. Reintroducing the wild type copy of *hos1* into the strains caused them to become sensitive once again to iprodione, indicating that the mutations identified in *hos1* were the cause of the resistance to this antifungal agent (Fig. 1c).

A quantitative assay was used to measure the level of resistance to iprodione, a second dicarboximide chemical, procymidone, and an azole, tebuconazole, in four sets of strains (Table 2). The minimum inhibitory concentration was between 1 and 2 µg/ml for the wild type strains

**Table 2 Half maximal effective concentration (EC$_{50}$) measurements of fungicide action for strains of *L. maculans***

| Strain | Iprodione EC$_{50}$ (µg/ml) | | Procymidone EC$_{50}$ (µg/ml) | | Tebuconazole EC$_{50}$ (µg/ml) | |
|---|---|---|---|---|---|---|
| | Average | SD | Average | SD | Average | SD |
| P290 | 1.25 | 0.09 | 1.75 | 0.04 | 0.29 | 0.01 |
| P290-Ip$^R$ | > 50 | | > 50 | | 0.53 | 0.01 |
| P290-Ip$^R$ + *hos1* | 1.32 | 0.09 | 1.70 | 0.01 | 1.12 | 0.34 |
| D2 | 1.10 | 0.22 | 1.94 | 0.12 | 0.40 | 0.14 |
| D2-Ip$^R$ | > 50 | | > 50 | | 0.39 | 0.02 |
| D2-Ip$^R$ + *hos1* | 0.94 | 0.11 | 2.66 | 0.16 | 0.43 | 0.00 |
| D3 | 1.21 | 0.06 | 1.84 | 0.13 | 0.91 | 0.01 |
| D3-Ip$^R$ | > 50 | | > 50 | | 0.51 | 0.02 |
| D3-Ip$^R$ + *hos1* | 1.07 | 0.10 | 1.73 | 0.06 | 0.77 | 0.01 |
| D5 | 0.43 | 0.27 | 1.57 | 0.20 | 0.50 | 0.03 |
| D5-Ip$^R$ | > 50 | | > 50 | | 0.10 | 0.05 |
| D5-Ip$^R$ + *hos1* | 0.48 | 0.28 | 1.25 | 0.23 | 0.14 | 0.00 |

*SD* standard deviation

for the dicarboximides. The strains derived from selection on iprodione were not inhibited with these chemicals at concentrations up to 50 μg/ml. Complementation of the strains with the wild type copy of *hos1* restored the $EC_{50}$ values to close to those seen in the wild type parents. In contrast, resistance to the unrelated molecule, tebuconazole (FRAC group 3), was not altered. Analysis of the results revealed no significant differences between wild type and restored strains on iprodione ($p = 0.200$–$1.00$), procymidone ($p = 0.057$–$0.686$) and tebuconazole ($p = 0.333$–$1.00$).

Mutation of the HOG pathway components is also responsible for other phenotypes in fungi, the best known being changes in growth in the presence of the phenylpyrrole fungicide fludioxonil (FRAC group 12) and salt [12]. Using growth on plates, a wild type D5, *hos1*$^{-}$ mutant (D5-Ip$^{R}$) and a + *hos1* complemented strain (D5-Ip$^{R}$ + *hos1*) were examined for these properties. As has been reported for other fungi, mutation of *hos1* resulted in increased resistance to fludioxonil and an increased sensitivity to sodium chloride (Fig. 2).

Growth *in planta* can be considered an environment of high stress to plant pathogenic fungi. Three *L. maculans* strains were inoculated onto wounded *B. napus* cotyledons, and lesion formation was examined over time. No difference in pathogenicity was observed between the three strains (average disease scores for wild type 6.14, *hos1*$^{-}$ mutant 6.03 and + *hos1* complementated 6.73; Fig. 3), indicating that *hos1* is not required for the ability of *L. maculans* to cause disease on canola.

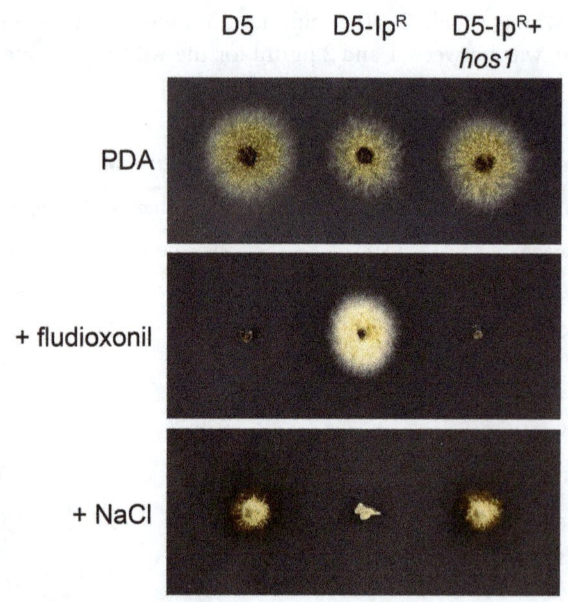


**Fig. 2** Mutation of *hos1* causes *L. maculans* to become resistant to the fungicide fludioxonil and more sensitive to NaCl. Strains D5, D5-Ip$^{R}$ and D5-Ip$^{R}$ + *hos1* and were cultured of PDA, and with PDA supplemented with fludioxonil (1 μg/ml) or NaCl (0.5 M)

**Development of a CRISPR/Cas9 gene disruption system for *L. maculans***

Identification of gene functions in *L. maculans* through targeted gene replacements has been inefficient due to low rates of homologous integration of constructs. The potential to use a positive selection system, i.e. growth on iprodione when *hos1* is mutated, was an impetus to use the *hos1* gene for the development of gene targeting methods, specifically through CRISPR-Cas9.

A versatile pair of plasmids (pLAU2 and pLAU53) was created for strong constitutive transcription of DNA sequences that are cloned into them. Both plasmids feature the 1001 bp prior to the start codon (the promoter) of the *act1* gene, encoding an actin subunit, and the terminator of *trp3* encoding anthranilate synthase of *L. maculans*. Actin is considered to be constitutively expressed and is commonly used as the reference gene in quantitative reverse transcriptase PCR experiments [21, 28, 29]. The plasmid includes a BglII site between the promoter and terminator into which genes or other DNA fragments can be cloned. To test if this promoter and terminator combination was able to drive protein production, the open reading frame for GFP was cloned into plasmid pLAU2, and the T-DNA transformed into wild type *L. maculans*. Fluorescence was abundant in spores and hyphae of transformants, indicating that the construct induces gene expression and yields high and stable protein synthesis (Additional file 3: Fig. S1).

The open reading frame of the Cas9 endonuclease was amplified and cloned into both the pLAU2 or pLAU53 constructs. Similarly, a *hos1* RNA guide construct was cloned into both pLAU2 and pLAU53. As a consequence, different options were available for the order of transformation and selection of transformants. The T-DNAs were sequentially introduced into wild type strain D5, either as the *hos1* RNA guide first and Cas9 second (e.g. strains DV1 and DV2), or in the other order, and the transformants were plated onto media containing iprodione (5 μg/ml) to isolate resistant strains (Fig. 4a).

Iprodione resistant strains were cultured as mycelia, genomic DNA isolated, and the region spanning the site for CRISPR-Cas9 induced mutation in the *hos1* gene amplified. Amplicons were digested with KpnI restriction enzyme and/or sequenced. All isolates derived from strains expressing both Cas9 and the *hos1* RNA guide had mutations in this region (Fig. 4b). Most mutations were either the insertion of an additional nucleotide, or the deletion of one or several nucleotides (Fig. 4b). One strain, which is not illustrated in Fig. 4b due to the size of the DNA sequence, had a tandem duplication of 69 bp

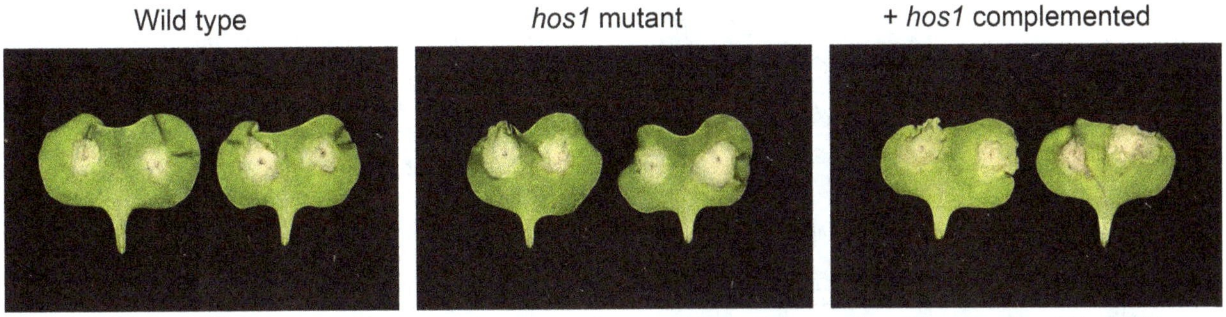


**Fig. 3** *L. maculans hos1* mutants are pathogenic on canola. Lesions on *B. napus* cv. Westar 11 days post inoculation caused by wild type isolate 14P290, a spontaneous mutation in *hos1* isolated on iprodione medium, and that strain complemented with a wild type copy of *hos1*

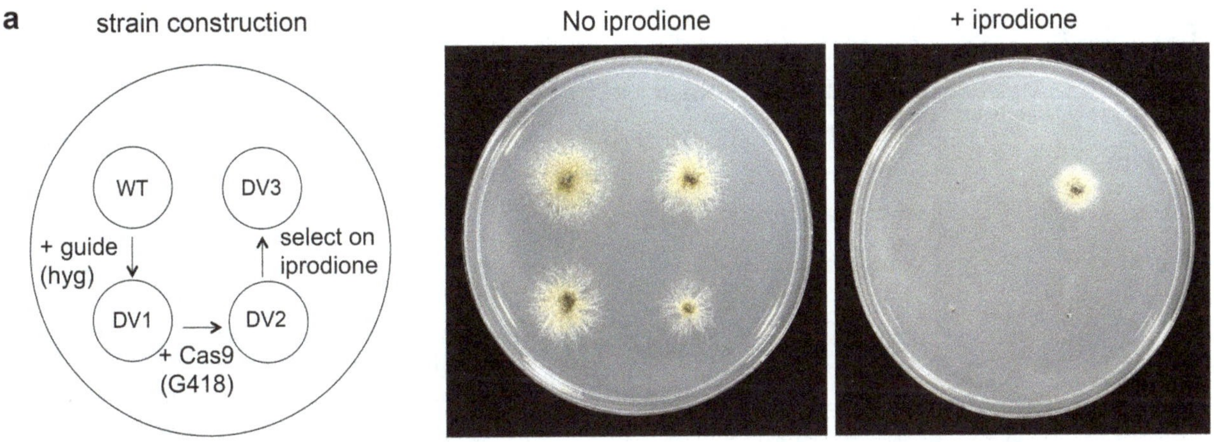


**b**

```
Wild type  ACTTGGCGGACAGGCTACAGTACATGGTGTCGAGGGT-ACCTGGAAAGACCTCACGGAAAACGTGAATGGTATGGCCATGAATCTCACAACCCAGGTGCGAGAAATCGCAGAAGTC
AI0676     ACTTGGCGGACAGGCTACAGTACATGGTGTCGAGGGTTACCTGGAAAGACCTCACGGAAAACGTGAATGGTATGGCCATGAATCTCACAACCCAGGTGCGAGAAATCGCAGAAGTC
TC1        ACTTGGCGGACAGGCTACAGTACATGGTGTCGAGGGTTACCTGGAAAGACCTCACGGAAAACGTGAATGGTATGGCCATGAATCTCACAACCCAGGTGCGAAAAATCGCAGAAGTC
AI0671     ACTTGGCGGACAGGCTACGGTAC-----------------CTGGAAAGACCTCACGGAAAACGTGAATGGTATGGCCATGAATCTCACAACCCAGGTGCGAGAAATCGCAGAAGTC
AI0668     ACTTGGCGGACAGGCTACAGTACAT-------------------------------------------GGTATGGCCATGAATCTCACAACCCAGGTGCGAGAAATCGCAGAAGTC
AI0675     ACTTGG----------------------------------------------------------------TATGGCCATGAATCTCACAACCCAGGTGCGAGAAATCGCAGAAGTC
AI0669     ACTTGGCGGACAGGCTACAGTACAT--------------------AAGACCTCACGGAAAACGTGAATGGTATGGCCATGAATCTCACAACCCAGGTGCGAGAAATCGCAGAAGTC
AI0677     ACTTGGCGGACAGGCTACAGTACATGGTGTC----------------GACCTCACGGAAAACGTGAATGGTATGGCCATGAATCTCACAACCCAGGTGCGAGAAATCGCAGAAGTC
TC6        ACTTGGCGGACAG----------------------------------------------------------------------------------------------------GTC
AI0678     ACTTGGCGGACAGGCTACAG------------------ACCTGGAAAGACCTCACGGAAAACGTGAATGGTATGGCCATGAATCTCACAACCCAGGTGCGAGAAATCGCAGAAGTC
TC2        ACTTGGCGGACAGGCTACAGT-----------------ACCTGGAAAGACCTCACGGAAAACGTGAATGGTATGGCCATGAATCTCACAACCCACGTGCGAGAAATCGCAAAGGTC
TC3        ACTTGGCGGACAGGCTACAGTACAT----------------------GACGTCAAGGAAAACGTGAATGGTATGGCCATGAATCTCACAACCCACGTGCGAGAAATCACATAAGTC
```

**Fig. 4** Development of the CRISPR/Cas9 system for targeted gene disruption in *L. maculans*. **a** Phenotype of transformants on plates with or without iprodione. Three strains derive from sequential modification of the wild type (WT) strain D5, first by transformation of the *hos1* guide RNA construct (strain DV1), then transformation of the Cas9 construct (strain DV2), and lastly by selection on iprodione (strain DV3). **b** Alignment of sequences of *hos1* from the wild type and 11 independently-created iprodione resistant mutants. On the wild type sequence the protospacer adjacent motif (PAM) is in blue highlight, region incorporated in the guide RNA in green highlight, and the KpnI restriction enzyme site used for screening is underlined. Changes in the sequence in the mutants are in red text

(ACCTGGAAAGACCTCACGGAAAACGTGAATGGTATGGCCATGAATCTCACAACCCAGGTGCGAGAAATC). In all cases the mutations in *hos1* were near the region of the genome where the guide RNA would target Cas9, indicating that they were derived from inaccurate repair of DNA damage by the endonuclease. The types of mutations also often differed from those found in the spontaneous mutants as many featured large deletions, compared to single nucleotide substitutions or insertions.

The pathogenicity of strains from one set of gene manipulations were tested by inoculating *B. napus* cotyledons. The three strains derived from the wild type produced lesions like the wild type isolate (Additional file 4: Fig. S2). This indicates that the introduction of Cas9

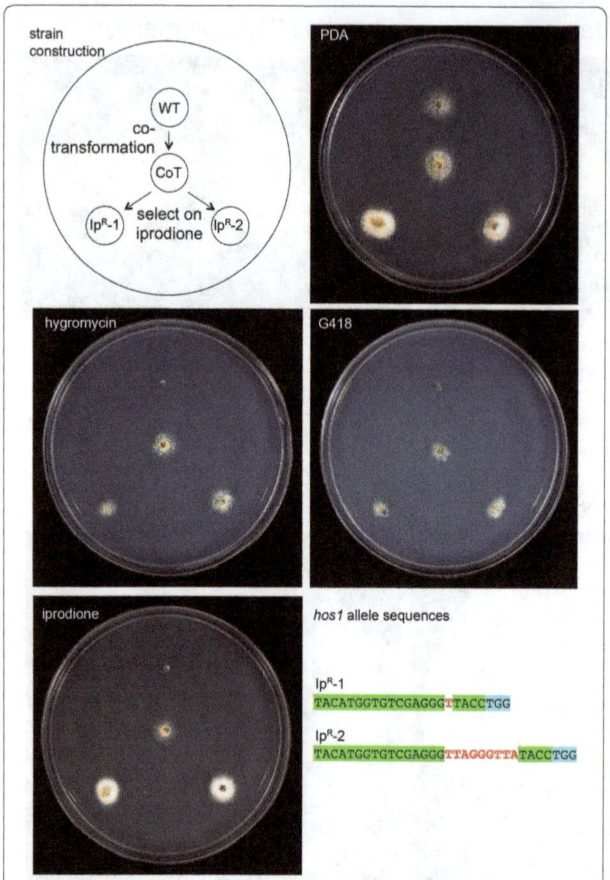


**Fig. 5** Co-transformation of Cas9 and the CRISPR guide RNA constructs into *L. maculans* by *Agrobacterium*-mediated transformation. Wild type strain D13 was co-transformed with T-DNAs from both constructs, and selected on media containing both hygromycin and G418. One transformant (D13-CoT) was cultured on medium containing iprodione. Two independent resistant strains were obtained, their DNA isolated, and the *hos1* mutations determined by amplification and sequencing. The region incorporated into the guide RNA is highlighted in green, and adjacent PAM site in blue highlight. The two iprodione resistant strains have additional nucleotides (red font) within the *hos1* gene

or a guide RNA into *L. maculans* does not impact its pathogenicity.

### Improvements to mutation by CRISPR/Cas9

One disadvantage of the method to induce mutations by the CRISPR/Cas9 system developed here was the need to perform two rounds of transformation, and hence co-transformation was therefore tested. The wild type strain D13 was co-transformed with both the *hos1* RNA guide and Cas9 constructs, with simultaneous co-selection on media containing G418 and hygromycin. One double-drug resistant transformant (strain D13-CoT) was then cultured on medium containing iprodione, and two iprodione resistant isolates (D13-Ip$^R$1 and D13-Ip$^R$2) characterized by sequencing the *hos1* region (Fig. 5). Both strains have mutations caused by additional base pairs that can be attributed to the CRISPR-Cas9 system. While iprodione is generally considered non-mutagenic, the proportion of iprodione resistant spores were compared between the wild type D13 and the D13-CoT strains by culturing these in the absence of iprodione and then plating onto media with or without the fungicide. While the proportion of spores resistant to iprodione was less than 1 in 10,000 for the wild type, 54% of spores were resistant from the strain carrying the Cas9 endonuclease and *hos1*-guide construct.

To eliminate the requirement to order synthesized DNA fragments, which have to be cloned due to the complexity in secondary structure (Thermo Fisher Scientific), two plasmids were created that have the regions for the HDV ribozyme and Cas9-binding RNA. This new plasmid system was then used to target the first *Avr* gene identified in *L. maculans, AvrLm1*. The hammerhead ribozyme and a region to target the *AvrLm1* gene was synthesized as an 101 nucleotide oligonucleotide, amplified by PCR and cloned into the XhoI site of plasmid pMAI75. Cas9 was transformed into wild type strain v23.1.3, and then the guide RNA construct to mutate *AvrLm1* was transformed into this strain. Genomic DNA isolated from the resulting transformants was used as the templates to amplify a fragment of *AvrLm1*, and amplicons then cut with NlaIII restriction enzyme. Two strains without the NlaIII site were obtained and *AvrLm1* was amplified and sequenced from them. The mutant alleles had a single additional base pair or a 27 bp deletion (Additional file 5: Fig. S3). Thus, CRISPR-Cas9 can be used to isolate strains with mutations in genes, without a strong selection system as used for *hos1*.

## Discussion

Resistance to fungicides is a major problem in many areas of disease management, especially in agriculture with the high levels of antifungals applied to ensure maximum yield returns. For example, currently in Australia canola growers may apply triazole fungicides three times during a season, including as a seed dressing, in combination with fertilizer and as a foliar spray [1]. Stubble is retained in the field after harvest, and the subsequent crop in the same field may receive fungicide treatments, thereby inadvertently causing additional exposure to the *L. maculans* populations while growing as a saprophyte in the canola stubble.

Resistance to iprodione and other dicarboximide fungicides and the underlying mechanisms have been well characterized in other plant pathogenic fungi, such as *Botrytis cinerea* [30–32]. The impact of mutation of *hos1* in plant pathogens varies, in some cases impacting pathogenicity and in others having no effect [16]. Hallmarks of impairing the HOG pathway, i.e. increased resistance to fludioxonil and sensitivity to salt, occur upon mutation of the *hos1* gene in *L. maculans*, but the important ability to cause disease on plants is not. These results indicate that while iprodione is initially effective in vitro, resistance emerges easily, and these strains are still pathogenic on canola, potentially in part explaining the limited efficacy of iprodione in field conditions. Studies in other plant pathogenic fungi show differing results in terms of the contribution of the histidine kinase to pathogenicity. Similar to *L. maculans*, disruption of the gene does not impair pathogenicity in *Alternaria alternata*, *Parastagonospora nodorum* and *Pyricularia oryzae* [16, 33, 34]. In contrast, the homolog is required for pathogenicity in *Botrytis cinerea*, *Fusarium oxysporum*, *Monilinia fructicola*, *Sclerotinia sclerotiorum* and *Ustilago maydis* [35–39]. Deletion strains in *Alternaria longipes* form larger lesions on *Nicotiana tabacum* than the wild type [40]. There is ambiguity about the role of the gene in *A. brassicicola* with isolates with point mutations having wild type pathogenicity, while a deletion allele, albeit analyzed in a large scale study, being less pathogenic [41, 42]. The histidine kinase contributes to the virulence of the human pathogens *Candida albicans* and *Cryptococcus neoformans* [43, 44].

*Leptosphaeria maculans* continues to be a recurrent disease of oilseed Brassicas around the world because the factors that this fungus produces to cause disease are mostly unknown. One challenge with finding new ways to combat *L. maculans* is that identifying gene functions has been technically challenging. Gene disruption in *L. maculans* is inefficient, with only nine genes disrupted as reported in the literature [6–11]. Constructs require large amounts of DNA for targeting by homologous recombination and even then the proportion of gene deletion events versus ectopic integration of the constructs is low, e.g. an efficiency of just one knock out from > 450 transformants screened [6]. Advances in improving the proportion of targeted gene replacements versus ectopic integrations have been made, including the use of a counter selection system against ectopic insertion events [9] or using the selectable marker split into two pieces [11]. However, isolating the large DNA fragments needed and/or the cloning into suitable vectors imposes limitations to the efficiency of created targeted mutations. For this reason, alternative methods to disrupt genes are needed for *L. maculans*, and the method employing CRISPR-Cas9 was explored.

CRISPR/Cas is a combination of an endonuclease that is guided to a specific site in a genome using an RNA molecule, found in Bacteria and Archaea for recognition of parasitic DNA elements and their specific cleavage. Modified for use in other organisms, its ability to make specific double-stranded breaks in DNA that are then inaccurately repaired to induce small mutations, such as within genes, is on the cusp of revolutionizing methods for gene functional studies, including in fungal species that have until now been difficult to manipulate genetically.

Here we used iprodione resistance due to mutation of the *hos1* gene as an easy screening tool to develop CRISPR/Cas9 for *L. maculans*. Iprodione resistant strains derived from strains expressing Cas9 and a guide RNA targeting *hos1* all had mutations at the place within *hos1* where the endonuclease would cut the DNA. As proof-of-function that the method could work on other genes, the first avirulence gene that was identified in this fungus, *AvrLm1*, was disrupted [45]. Targeting these effectors, all found to date to lie within distinctive large regions of AT-rich and highly repetitive DNA [5, 46], has not been possible using homologous recombination. Curiously, the avirulence profile of these strains did not change as predicted (data not shown), and will require additional experiments to understand what is emerging in *L. maculans* as complex multigene sets of interactions between fungal avirulence genes and plant resistance genes [46, 47].

In the first iteration of CRISPR/Cas9 for *L. maculans*, two rounds of transformation were used to separately introduce the guide RNA and Cas9 expression constructs into the fungus. After seeking a suitable promoter for regulation by RNA polymerase III in *L. maculans* without success, the dual ribozyme system to process the guide RNA when expressed from an RNA polymerase II promoter was used. This dual ribozyme approach was developed for plant transformation [48], and has recently also been employed in *Aspergillus* spp. [49], the basidiomycete human pathogen *Cryptococcus neoformans* [50] and the ascomycete plant pathogen *Alternaria alternata* [51]. The disadvantage of using ribozymes for processing the guide RNA is that they add size to the constructs. To alleviate this issue, we created a vector such that just one of the ribozymes and the RNA fragment to target Cas9 to the gene to be mutated are synthesized: the hammerhead ribozyme requires folding with part of the target RNA and hence there is a requirement for long ~ 100 nucleotide oligonucleotides. The current method, although involving two transformation steps or co-transformation of both constructs, is suitable for making targeted mutations in genes, and has been tested in multiple wild type isolates. We have mutated more than 24 other genes in *L. maculans* to date (unpublished data). Potential refinements to the method in the near future will likely make it even more effective as a mutational tool to discover gene functions in *L. maculans*.

## Additional files

**Additional file 1: Table S1.** Oligonucleotide primers used in this study.

**Additional file 2.** Sequence of the guide RNA constructs that were synthesized. Colors of the nucleotides infer different purposes. Blue, primer binding sites for amplification and cloning into plasmids; grey, ribozymes; red, 20 nucleotides specific to *hos1*; purple, guide RNA; black, stop codon for *trp3*. Underlined nucleotides will base pair in the hammerhead ribozyme. Bold is the XhoI restriction enzyme site.

**Additional file 3: Fig. S1.** Constitutive expression of genes using the actin regulatory sequences. The *L. maculans* actin (*act1*) promoter and 5′ UTR were cloned to allow expression of adjacent genes. In this case, GFP was fused to this region, and the construct transformed into *L. maculans*. Spores were germinated for 3 days in 10% cleared V8 juice media. Bar = 50 μm.

**Additional file 4: Fig. S2.** Expression of the CRISPR components or mutation of *hos1* does not impair pathogenicity on plants. Cotyledons of *B. napus* cv. Westar were inoculated with four strains, and lesions measured 14 days later. The wild type isolate D5 was sequentially transformed with the *hos1* guide RNA construct (to create strain DV1) and the Cas9 construct (DV2), and then plated on iprodione to isolate a resistant strain (DV3).

**Additional file 5: Fig. S3.** Targeted mutation of the *AvrLm1* gene in *L. maculans* by CRISPR-Cas9. Alignment of the coding region of *AvrLm1* from wild type and two mutant alleles. On the wild type sequence the PAM is in blue highlight, region incorporated in the guide RNA in green, and the NlaIII restriction enzyme site used for screening is underlined. The differences in sequence in the two mutants are in red, as an extra A nucleotide or a deletion of 27 nucleotides.

### Authors' contributions

AI, DRV, and FJL-R conceived and designed the project. AI, ASU, DRV, SC, APVdW, and FJL-R designed and performed the experiments, and analyzed the results. AI wrote the manuscript. All authors read and approved the final manuscript.

### Author details

[1] School of BioSciences, University of Melbourne, Building 122, Parkville, VIC 3010, Australia. [2] Department of Environment and Agriculture, Centre for Crop and Disease Management, Curtin University, Bentley, WA 6102, Australia.

### Acknowledgements

We thank Edwin Lampugnani for providing the plasmid containing the Cas9 open reading frame.

### Competing interests

The authors declare that they have no competing interests.

### Funding

The research was supported by the Australian Grains Research and Development Corporation (Projects CUR00023 and UM00050, and Scholarship GRS11006) and the Australian Research Council (Grant FT130100146).

### References


1. Van de Wouw AP, Marcroft SJ, Howlett BJ. Blackleg disease of canola in Australia. Crop Pasture Sci. 2016;67:273–83.
2. Mendes-Pereira E, Balesdent M-H, Brun H, Rouxel T. Molecular phylogeny of the *Leptosphaeria maculans-L. biglobosa* species complex. Mycol Res. 2003;107:1287–304.
3. Voigt K, Cozijnsen AJ, Kroymann J, Pöggeler S, Howlett BJ. Phylogenetic relationships between members of the crucifer pathogenic *Leptosphaeria maculans* species complex as shown by mating type (*MAT1-2*), actin, and β-tubulin sequences. Mol Phylogenet Evol. 2005;37:541–57.
4. de Gruyter J, Woudenberg JHC, Aveskamp MM, Verkley GJM, Groenewald JZ, Crous PW. Redisposition of *Phoma*-like anamorphs in *Pleosporales*. Stud Mycol. 2013;75:1–36.
5. Rouxel T, Grandaubert J, Hane JK, Hoede C, van de Wouw AP, Couloux A, Dominguez V, Anthouard V, Bally P, Bourras S, et al. Effector diversification within compartments of the *Leptosphaeria maculans* genome affected by repeat-induced point mutations. Nat Commun. 2011;2:202.
6. Idnurm A, Taylor JL, Pedras MSC, Howlett BJ. Small scale functional genomics of the blackleg fungus, *Leptosphaeria maculans*: analysis of a 38 kb region. Australas Plant Pathol. 2003;32:511–9.
7. Idnurm A, Warnecke D, Heinz E, Howlett BJ. Characterisation of neutral trehalase and UDP-glucose: sterol glucosyltransferase genes from the plant pathogenic fungus *Leptosphaeria maculans*. Physiol Mol Plant Pathol. 2003;62:305–13.
8. Wilson LM, Idnurm A, Howlett BJ. Characterization of a gene (*sp1*) encoding a secreted protein from *Leptosphaeria maculans*, the blackleg pathogen of *Brassica napus*. Mol Plant Pathol. 2002;3:487–93.
9. Gardiner DM, Howlett BJ. Negative selection using thymidine kinase increases the efficiency of recovery of transformants with targeted genes in the filamentous fungus *Leptosphaeria maculans*. Curr Genet. 2004;45:249–55.
10. Gardiner DM, Cozijnsen AJ, Wilson LM, Pedras MSC, Howlett BJ. The sirodesmin biosynthetic gene cluster of the plant pathogenic fungus *Leptosphaeria maculans*. Mol Microbiol. 2004;53:1307–18.
11. Feng J, Zhang H, Strelkov SE, Hwang S-F. The *LmSNF1* gene is required for pathogenicity in the canola blackleg pathogen *Leptosphaeria maculans*. PLoS ONE. 2014;9:e92503.
12. Bahn Y-S. Master and commander in fungal pathogens: the two-component system and the HOG signaling pathway. Eukaryot Cell. 2008;7:2017–36.
13. Maeda T, Takekawa M, Saito H. Activation of yeast PBS2 MAPKK by MAPKKKs or by binding of an SH3-containing osmosensor. Science. 1995;269:554–8.
14. Posas F, Wurgler-Murphy SM, Maeda T, Witten EA, Thai TC, Saito H. Yeast HOG1 MAP kinase cascade is regulated by a multistep phosphorelay mechanism in the SLN1-YPD1-SSK1 "two-component" osmosensor. Cell. 1996;86:865–75.
15. Brewster JL, de Valoir T, Dwyer ND, Winter E, Gustin MC. An osmosensing signal transduction pathway in yeast. Science. 1993;259:1760–3.
16. John E, Lopez-Ruiz F, Rybak K, Mousley CJ, Oliver RP, Tan K-C. Dissecting the role of histidine kinase and HOG1 mitogen-activated protein kinase signalling in stress tolerance and pathogenicity of *Parastagonospora nodorum* on wheat. Microbiology. 2016;162:1023–36.
17. Van de Wouw AP, Elliott VL, Chang S, López-Ruiz FJ, Marcroft SJ, Idnurm A. Identification of isolates of the plant pathogen *Leptosphaeria maculans* with resistance to the triazole fungicide fluquinconazole using a novel *in planta* assay. PLoS ONE. 2017;12:e0188106.
18. Pitkin JW, Panaccione DG, Walton JD. A putative cyclic peptide efflux pump encoded by the *TOXA* gene of the plant-pathogenic fungus *Cochliobolus carbonum*. Microbiology. 1996;142:1557–65.
19. Van de Wouw AP, Thomas VL, Cozijnsen AJ, Marcroft SJ, Salisbury PA, Howlett BJ. Identification of *Leptosphaeria biglobosa* 'canadensis' on *Brassica juncea* stubble from northern New South Wales, Australia. Aust Plant Dis Notes. 2008;3:124–8.
20. Walton FJ, Idnurm A, Heitman J. Novel gene functions required for melanization of the human pathogen *Cryptococcus neoformans*. Mol Microbiol. 2005;57:1381–96.
21. Elliott CE, Howlett BJ. Overexpression of a 3-ketoacyl-CoA thiolase in *Leptosphaeria maculans* causes reduced pathogenicity on *Brassica napus*. Mol Plant Microbe Interact. 2006;19:588–96.
22. Covert SF, Kapoor P, Lee M-H, Briley A, Nairn CJ. *Agrobacterium tumefaciens*-mediated transformation of *Fusarium circinatum*. Mycol Res. 2001;105:259–64.
23. Xing H-L, Dong L, Wang Z-P, Zhang H-Y, Han C-Y, Liu B, Wang X-C, Chen Q-J. A CRISPR/Cas9 toolkit for multiplex genome editing in plants. BMC Plant Biol. 2014;14:327.

24. Bundock P, den Dulk-Ras A, Beijersbergen A, Hooykaas PJJ. Trans-kingdom T-DNA transfer from *Agrobacterium tumefaciens* to *Saccharomyces cerevisiae*. EMBO J. 1995;14:3206–14.
25. Eckert MR, Rossall S, Selley A, Fitt BDL. Effects of fungicides on in vitro spore germination and mycelial growth of the phytopathogens *Leptosphaeria maculans* and *L. biglobosa* (phoma stem canker of oilseed rape). Pest Manag Sci. 2010;66:396–405.
26. Elliott VL, Marcroft SJ, Howlett BJ, Van de Wouw AP. Gene-for-gene resistance is expressed in cotyledons, leaves and pods, but not during late stages of stem colonisation in the *Leptosphaeria maculans–Brassica napus* pathosystem. Plant Breeding. 2016;135:200–7.
27. Grigoriev IV, Nikitin R, Haridas S, Kuo A, Ohm R, Otillar R, Riley R, Salamov A, Zhao X, Korzeniewski F, et al. MycoCosm portal: gearing up for 1000 fungal genomes. Nucleic Acids Res. 2014;42:D699–704.
28. Lowe RGT, Cassin A, Grandaubert J, Clark BL, Van de Wouw AP, Rouxel T, Howlett BJ. Genomes and transcriptomes of partners in plant-fungal-interactions between canola (*Brassica napus*) and two *Leptosphaeria* species. PLoS ONE. 2014;9:e103098.
29. Plissonneau C, Daverdin G, Ollivier B, Blaise F, Degrave A, Fudal I, Rouxel T, Balesdent M-H. A game of hide and seek between avirulence genes *AvrLm4-7* and *AvrLm3* in *Leptosphaeria maculans*. New Phytol. 2016;209:1613–24.
30. Grabke A, Fernández-Ortuño D, Amiri A, Li X, Peres NA, Smith P, Schnabel G. Characterization of iprodione resistance in *Botrytis cinerea* from strawberry and blackberry. Phytopathology. 2014;104:396–402.
31. Liu W, Leroux P, Fillinger S. The HOG1-like MAP kinase Sak1 of *Botrytis cinerea* is negatively regulated by the upstream histidine kinase Bos1 and is not involved in dicarboximide- and phenylpyrrole-resistance. Fungal Genet Biol. 2008;45:1062–74.
32. Cui W, Beever RE, Parkes SL, Weeds PL, Templeton MD. An osmosensing histidine kinase mediates dicarboximide fungicide resistance in *Botryotinia fuckeliana* (*Botrytis cinerea*). Fungal Genet Biol. 2002;36:187–98.
33. Lin C-H, Chung K-R. Specialized and shared functions of the histidine kinase- and HOG1 MAP kinase-mediated signaling pathways in *Alternaria alternata*, a filamentous fungal pathogen of citrus. Fungal Genet Biol. 2010;47:818–27.
34. Motoyama T, Kadokura K, Ohira T, Ichiishi A, Fujimura M, Yamaguchi I, Kudo T. A two-component histidine kinase of the rice blast fungus is involved in osmotic stress response and fungicide action. Fungal Genet Biol. 2005;42:200–12.
35. Viaud M, Fillinger S, Liu W, Polepalli JS, Le Pêcheur P, Kunduru AR, Leroux P, Legendre L. A class III histidine kinase acts as a novel virulence factor in *Botrytis cinerea*. Mol Plant Microbe Interact. 2006;19:1042–50.
36. Rispail N, Di Pietro A. The two-component histidine kinase Fhk1 controls stress adaptation and virulence of *Fusarium oxysporum*. Mol Plant Pathol. 2010;11:395–407.
37. Ma Z, Luo Y, Michailides T. Molecular characterization of the two-component histidine kinase gene from *Monilinia fructicola*. Pest Manag Sci. 2006;62:991–8.
38. Duan Y, Ge C, Liu S, Wang J, Zhou M. A two-component histidine kinase *Shk1* controls stress response, sclerotial formation and fungicide resistance in *Sclerotinia sclerotiorum*. Mol Plant Pathol. 2013;14:708–18.
39. Yun YH, Oh MH, Kim JY, Kim SH. *UmTco1*, a hybrid histidine kinase gene, is essential for the sexual development and virulence of *Ustilago maydis*. J Microbiol Biotechnol. 2017;27:1010–22.
40. Luo YY, Yang JK, Zhu ML, Liu CJ, Li HY, Lu ZB, Pan WZ, Zhang ZH, Bi W, Zhang KQ. The group III two-component histidine kinase AlHK1 is involved in fungicides resistance, osmosensitivity, spore production and impacts negatively pathogenicity in *Alternaria longipes*. Curr Microbiol. 2012;64:449–56.
41. Cho Y, Kim KH, La Rota M, Scott D, Santopietro G, Callihan M, Mitchell TK, Lawrence CB. Identification of novel virulence factors associated with signal transduction pathways in *Alternaria brassicicola*. Mol Microbiol. 2009;72:1316–33.
42. Iacomi-Vasilescu B, Bataille-Simoneau N, Campion C, Dongo A, Laurent E, Serandat I, Hamon B, Simoneau P. Effect of null mutations in the *AbNIK1* gene on saprophytic and parasitic fitness of *Alternaria brassicicola* isolates highly resistant to dicarboximide fungicides. Plant Pathol. 2008;57:937–47.
43. Bahn Y-S, Kojima K, Cox GM, Heitman J. A unique fungal two-component system regulates stress responses, drug sensitivity, sexual development, and virulence of *Cryptococcus neoformans*. Mol Biol Cell. 2006;17:3122–35.
44. Yamada-Okabe T, Mio T, Ono N, Kashima Y, Matsui M, Arisawa M, Yamada-Okabe H. Roles of three histidine kinase genes in hyphal development and virulence of the pathogenic fungus *Candida albicans*. J Bacteriol. 1999;181:7243–7.
45. Gout L, Fudal I, Kuhn M-L, Blaise F, Eckert M, Cattolico L, Balesdent M-H, Rouxel T. Lost in the middle of nowhere: the *AvrLm1* avirulence gene of the Dothideomycete *Leptosphaeria maculans*. Mol Microbiol. 2006;60:67–80.
46. Rouxel T, Balesdent M-H. Life, death and rebirth of avirulence effectors in a fungal pathogen of Brassica crops, *Leptosphaeria maculans*. New Phytol. 2017;214:526–32.
47. Petit-Houdenot Y, Fudal I. Complex interactions between fungal avirulence genes and their corresponding plant resistance genes and consequences for disease resistance management. Front Plant Sci. 1072;2017:8.
48. Gao Y, Zhao Y. Self-processing of ribozyme-flanked RNAs into guide RNAs in vitro and in vivo for CRISPR-mediated genome editing. J Integr Plant Biol. 2014;56:343–9.
49. Nødvig CS, Nielsen JB, Kogle ME, Mortensen UH. A CRISPR-Cas9 system for genetic engineering of filamentous fungi. PLoS ONE. 2015;10:e0133085.
50. Arras SDM, Chua SMH, Wizrah MSI, Faint JA, Yap AS, Fraser JA. Targeted genome editing via CRISPR in the pathogen *Cryptococcus neoformans*. PLoS ONE. 2016;11:e0164322.
51. Wenderoth M, Pinecker C, Voß B, Fischer R. Establishment of CRISPR/Cas9 in *Alternaria alternata*. Fungal Genet Biol. 2017;101:55–60.
52. Van de Wouw AP, Marcroft SJ, Ware A, Lindbeck K, Khangura R, Howlett BJ. Breakdown of resistance to the fungal disease, blackleg, is averted in commercial canola (*Brassica napus*) crops in Australia. Field Crops Res. 2014;166:144–51.

# 5

# *Cinnamomum zeylanicum* bark essential oil induces cell wall remodelling and spindle defects in *Candida albicans*

Zinnat Shahina[1], Amira M. El-Ganiny[2], Jessica Minion[3], Malcolm Whiteway[4], Taranum Sultana[1*†] and Tanya E. S. Dahms[1,3*†]

**Abstract**

**Background:** Cinnamon (*Cinnamomum zeylanicum*) bark extract exhibits potent inhibitory activity against *Candida albicans* but the antifungal mechanisms of this essential oil remain largely unexplored.

**Results:** We analyzed the impact of cinnamon bark oil on *C. albicans* RSY150, and clinical strains isolated from patients with candidemia and candidiasis. The viability of RSY150 was significantly compromised in a dose dependent manner when exposed to cinnamon bark oil, with extensive cell surface remodelling at sub inhibitory levels (62.5 µg/mL). Atomic force microscopy revealed cell surface exfoliation, altered ultrastructure and reduced cell wall integrity for both RSY150 and clinical isolates exposed to cinnamon bark oil. Cell wall damage induced by cinnamon bark oil was confirmed by exposure to stressors and the sensitivity of cell wall mutants involved in cell wall organization, biogenesis, and morphogenesis. The essential oil triggered cell cycle arrest by disrupting beta tubulin distribution, which led to mitotic spindle defects, ultimately compromising the cell membrane and allowing leakage of cellular components. The multiple targets of cinnamon bark oil can be attributed to its components, including cinnamaldehyde (74%), and minor components (< 6%) such as linalool (3.9%), cinamyl acetate (3.8%), α-caryophyllene (5.3%) and limonene (2%). Complete inhibition of the mitotic spindle assembly was observed in *C. albicans* treated with cinnamaldehyde at MIC (112 µg/mL).

**Conclusions:** Since cinnamaldehyde disrupts both the cell wall and tubulin polymerization, it may serve as an effective antifungal, either by chemical modification to improve its specificity and efficacy or in combination with other antifungal drugs.

**Keywords:** *Cinnamomum zeylanicum*, *Candida albicans*, Essential oil, Cell wall remodelling, Spindle defects

## Background

*Candida albicans*, a commensal fungi, develops into a resilient pathogen under low host immunity such as that for immunocompromised individuals with HIV/AIDS, and patients undergoing cancer chemotherapy [1–3]. A limited number of antifungals are available for treating such infections, and the use of these antifungal classes, including azoles, echinocandins, polyenes and allylamines, can be accompanied by side effects. Poor bioavailability requires higher doses, which can ultimately result in resistance and ineffectiveness [4]. The continued high morbidity following systemic fungal infection and emerging resistance to antifungal agents underscore a clear need for alternatives [5]. In this context, essential oils (EOs) are gaining popularity due to their strong antimicrobial and antibiofilm activity [6, 7]. EO combinations with other essential oils or existing antifungal agents could vastly reduce the probability of multi-drug resistance [6–11]. However, comprehensive studies are required to fully assess their independent

*Correspondence: tanya.dahms@uregina.ca; taranum.sultana@uregina.ca
†Taranum Sultana and Tanya E. S. Dahms contributed equally to the design of experiments, student supervision and manuscript writing
[1] Department of Chemistry and Biochemistry, University of Regina, 3737 Wascana Parkway, Regina, SK, Canada
Full list of author information is available at the end of the article

pharmacological properties and potential side effects prior to consideration for clinical use as antifungal agents.

Cinnamon oil is an aromatic liquid obtained from the twigs, bark and leaves of *Cinnamomum zeylanicum* [12]. Extracts of cinnamon bark (CNB) and leaves (CNL) have been used extensively as therapeutics in many cultures since antiquity. The anti-candida activity of CNB oil against planktonic and biofilm culture of *C. albicans* and *non-albicans* spp. has been documented [7, 13–15]. The main constituents of CNB oil include trans-cinnamaldehyde, and minor components such as eugenyl acetate, linalool, and benzyl benzoate, each having antifungal activity [16–20]. CNB oil has been shown to alter cell membrane permeability and fluidity, and inhibit biofilm formation [7, 13, 15, 21], but the mechanisms of toxicity remain unknown. On the other hand, each component has been extensively studied, showing effects at various cellular sites, including the cell membrane and cytosol. For example, cinnamaldehyde, the major constituent of CNB oil, targets the membrane and causes increased cell wall thickness in *C. albicans* [16], attributed to β-1-3-glucan synthase inhibition as observed in *Saccharomyces cerevisiae* [22]. The increase in bud scar formation upon cinnamaldehyde exposure also suggests an impact on cell division, resulting in decreased viability [16, 23]. Benzyl benzoate and linalool affect membrane fluidity and induce cell cycle arrest at the G2-M and G1 phases, respectively [20] at concentrations greater than the minimum inhibitory concentration (MIC) [7, 16, 17, 23]. We hypothesized that the cell wall and membrane are primary targets of CNB oil, which in turn disrupt intracellular processes vital to *Candida* survival.

Here, we report a detailed characterization of the anticandidal effects of CNB oil using atomic force microscopy (AFM), laser scanning confocal microscopy (LSCM) and traditional biochemical assays. AFM quantitative imaging (QI™) is a powerful tool for assessing the impact of antifungals [24–28], nutrient stress [29], oxidative stress [30] and characterizing yeast genetic mutants [31], while LSCM imaging of fluorescent markers can delineate defects in intracellular processes. AFM was used to quantify the morphological, ultrastructural and biophysical properties of RSY150 and a clinical isolate exposed to CNB oil. The RSY150 strain of *C. albicans* with RFP tagged histone protein B (Htb-RFP) and GFP tagged β-tubulin (Tub2-GFP) was used to track cell cycle defects in response to CNB oil exposure. Finally biochemical assays were used to verify physiological changes identified by imaging. We report for the first time that CNB oil causes β-tubulin depolymerisation and cell cycle arrest, which we attribute to its major constituent cinnamaldehyde.

## Methods

### Chemicals and media

The cinnamon bark essential oil (Chemical Abstract Service (CAS), registry number 8015-91-6) was a steam distilled extract from the dried inner bark (Now foods, USA) of *Cinnamon zeylanicum* (cinnamon). Bacto™ agar, yeast extract and peptone were obtained from Difco (BD Biosciences, NJ, USA), and all other chemicals were purchased from Sigma Chemical Co. (St. Louis, MO, USA).

### CNB essential oil analysis

CNB oil was analyzed using gas chromatography-flame ionization detection (GC-FID) and gas chromatography–mass spectrometry (GC–MS) with an Agilent 7890A GC according to previously reported methods [32, 33]. Briefly, 1 µl of diluted CNB oil (1:10 in ethanol) was injected onto a HP-5 column (30 mm × 0.32 mm × 0.25 µm) and separated using a carrier gas of helium (constant flow of 1.1 mL/min) with the following temperature gradient: 60 °C for 5 min, increased to 210 °C at 3 °C/min and to 260 °C at 10 °C/min. The GC-FID injector and detector temperatures were 250 °C and the split ratio of injection was 20:1. For GC–MS using similar conditions, 0.2 µl was injected with a split ratio of 200:1 and separated on a HP-5MS column (30 mm × 0.25 mm × 0.25 um).

Retention indices (RI) were determined using a C8-20 standard (~ 40 mg/L each, in hexanes) for both the GC-FID and GC–MS methods. RI values were averaged between the methods for each component of the oil and identified using the NIST14 database and reported literature [32, 34].

### Strains and culture conditions

*Candida albicans* strains and clinical isolates used in this study are described in Table 1. The RSY150 and RSY35 strains were a kind gift of Dr. Richard J. Bennett. RSY150 expresses Tub2-GFP and Htb-RFP and RSY35 is a Kar3 deletion mutant that express only Tub2-GFP [35]. Kar3 is a bifunctional protein having a kinesin-like motor domain joined to a distinct microtubule binding domain that is essential for yeast nuclear fusion during mating [36]. Strains were stored as 50% glycerol stocks at − 80 °C and were freshly revived on yeast-extract peptone dextrose agar (YPDA) containing 1% Bacto-yeast extract, 2% Bacto-peptone, 2% glucose and 2% Bacto-agar prior to each experiment. All strains were grown with continuous shaking (200 rpm) at 30 °C in YPD broth and the mutant strains supplemented with 80 mg/L uridine (YPDU). For the yeast to hyphal transition and leakage assays, cells were grown to mid log phase before exposure to CNB oil. For hyphal induction

**Table 1 *Candida albicans* strains used in this study**

| Strain | Genotype | References |
|---|---|---|
| RSY150 | *TUB2-GFP-SAT1/TUB2_ HTB1-RFP-ARG4_/HTB1_ arg4_/_* | [35] |
| RSY35 | *leu2::hisG/leu2::hisG his1::hisG/his1::hisG arg4::hisG/arg4::hisG kar3::LEU2/kar3::HIS1 TUB2/TUB2-GFP::SAT1* | [35] |
| CASS1 | *his3::hisG/his3::hisGleu2::tetRGAL4AD-URA3/LEU2* | [37] |
| ATCC 64548 | Reference strain | Cedarlane Labs, Ontario, Canada |
| ATCC 10231 | Reference strain for clinical isolates | RQHR, Regina, |
| SK, Canada | | |
| *C. albicans* (1-4) | Clinical isolates; 2 blood, 2 genital | RQHR, Regina, SK, Canada |
| | *C. albicans* knockout mutants | |
| VPS28 | *his3::hisG/his3::hisGleu2::tetRGAL4AD-URA3/LEU2-VPS28* | [37] |
| CRH11 | *his3::hisG/his3::hisGleu2::tetRGAL4AD-URA3/LEU2-Crh11* | [37] |
| SSU81 | *his3::hisG/his3::hisGleu2::tetRGAL4AD-URA3/LEU2-SSU81* | [37] |
| DFG5 | *his3::hisG/his3::hisGleu2::tetRGAL4AD-URA3/LEU2-DGF5* | [37] |

experiments, cells were grown in YPD broth with 10% fetal bovine serum and 2% glucose.

#### Minimum inhibitory concentration (MIC)

The MIC of CNB oil was determined for all strains listed in Table 1 following the guidelines of the Clinical and Laboratory Standards Institute [CLSI 2014] [38] and previously reported method [15], with slight modifications. Briefly, 100 μl of CNB oil (stock concentration 1000 μg/mL) was serially diluted in triplicate in the wells of flat-bottom polystyrene 96-well microtiter plates (Sarstedt, Nümbrecht, Germany). A suspension ($OD_{600} = 0.001$) of *C. albicans* ($2.2 \times 10^5$ cells/mL) in YPD was added, with appropriate positive (amphotericin B) and negative (*Candida* only in media) controls and a blank (CNB oil in media) included. The microtiter plates were sealed with parafilm prior to incubation at 30 °C to avoid oil evaporation, and the $OD_{600}$ recorded after 24 h (Biotek Epoch; Northern Vermont, USA). The endpoint was defined as the lowest concentration of the compound resulting in total inhibition (MIC 100%) of growth, compared to the growth in negative control wells. All experiments were performed in triplicate. The MICs for cinnamaldehyde (stock concentration 450 μg/mL) and linalool (stock concentration 9.85 mg/mL) were also tested for RSY150 and RSY35 [16, 39, 40].

#### Growth and viability of *C. albicans* with CNB oil exposure

##### *Growth curves*

To investigate the effect of CNB oil on growth kinetics of *C. albicans*, an overnight culture of *C. albicans* RSY150 diluted to contain $2.2 \times 10^5$ cells/mL ($OD_{600}$ of 0.001) was treated with CNB oil (62.5, 31.25 and 15.1 μg/mL corresponding to 1/2 MIC, 1/4 MIC and 1/8 MIC) in separate wells of a microtiter plate in triplicate and incubated at 30 °C [41, 42]. The $OD_{600}$ was measured at 30 min intervals for 24 h using a BioTek Synergy HTX multi-mode microplate reader (Northern Vermont, USA). Media with CNB oil was used to determine background absorbance for three independent experiments.

##### *Cell viability*

The cell viability after CNB oil exposure was determined using the methylene blue dye exclusion assay, as reported previously [43]. Briefly, mid logarithmic phase *C. albicans* RSY150 cultures ($2.2 \times 10^5$ cells/mL) were treated with various concentrations (62.5, 31.25 and 15.1 μg/mL corresponding to 1/2 MIC, 1/4 MIC and 1/8 MIC) of CNB oil at 30 °C for 24 h. Cells fixed with formaldehyde were used as positive controls for staining. The cells were washed and resuspended in phosphate buffered saline (PBS;0.01 M pH 7.4) to ~ $10^7$ CFU/mL, then 100 μL of treated and control cell suspensions mixed with 100 μL methylene blue (0.1 mg/mL stock solution in 2% sodium citrate) and incubated for 5 min at room temperature. Cells were examined using an Eclipse 80i microscope (Nikon) at 40× magnification. A minimum of 100 cells in consecutive visual fields were examined and the percentage of stained cells were calculated using the Nikon SPOT software [44].

To determine colony forming units (CFU), overnight cultures of CNB oil-treated *C. albicans* RSY150 (1/2 MIC, 1/4 MIC, and 1/8 MIC) along with controls were serially diluted ($10^3$, $10^4$, $10^5$ cells/mL) and plated in duplicate on YPD agar [45]. Viable colonies were counted and recorded at each specific concentration of CNB oil prepared in triplicate.

### Morphological analysis

Calcofluor white (CFW) at a concentration of 0.01 μg/mL was used as a chitin specific dye [35] to highlight the gross morphology and chitin distribution in CNB oil treated RSY150 along with blood and genital clinical isolates. Cell suspensions at two growth phases each, mid logarithmic ($1 \times 10^7$ cells/mL) and stationary ($2.2 \times 10^5$ cells/mL), were treated with MIC and 1/2 MIC of CNB oil and imaged on an Axio Observer Z1 inverted epifluorescence microscope (Oberkochen, Germany) at 63× magnification ($\lambda_{ex} = 365$ nm; $\lambda_{em} = 435$ nm).

### Hyphal induction

Hyphal induction in *C. albicans* RSY150 was performed according to the literature [46]. Briefly, a yeast suspension ($1 \times 10^7$ CFU/mL) was prepared from mid logarithmic phase cells in pre-warmed YPD with 10% fetal bovine serum (FBS) and deposited into a 12 well plate with the appropriate amounts of CNB oil to achieve MIC and 1/2 MIC. Control cultures lacked CNB oil. The cells were stained with CFW (0.01 μg/mL) to highlight hyphae and pseudo hyphae after 4 h incubation at 37 °C, and images captured on an AxioObserver Z1 inverted epifluorescence microscope (Oberkochen, Germany). Germ tubes were identified when the cell projection was equal to the size of the blastospore. Results from three independent experiments were reported as average ± standard deviation.

### Ultrastructural and mechanical analysis

The cell surface biophysical properties of CNB oil treated *C. albicans* RSY150 and a clinical isolate from blood were analyzed by Quantitative Imaging (QI™) using a Nanowizard III AFM (JPK Instruments, Berlin, Germany) as described previously [47, 48]. Briefly, cell suspensions ($2.2 \times 10^5$ cells/mL) from both strains treated either with YPD only (control) or YPD with 1/2 MIC CNB oil for 24 h were deposited onto poly-L-Lysine coated cover slips for 1 h, fixed with formalin and air dried prior to AFM imaging. Cells treated with the fungal wall degrading enzyme, glucanase, served as a positive control. Samples were imaged with silicon nitride cantilevers (HYDRA6R-200NG; Nanosensors, Neuchatel, Switzerland) having calibrated spring constants ranging from 0.03 to 0.062 N/m.

QI force curves (JPK software) obtained at each pixel of a 128 × 128 raster scan were collected using a Z-length of 7 μm and a raster scan of 100 μm/s for QI™. Adhesion and Young's modulus calculations were made from approximately 16,536 force curves collected from each of five biological replicates. All force curves within a 200 × 200 nm square in the center of the cell were batch processed and histogram data exported from the JPK software using Excel. Adhesion was determined using the distance between the lowest point and baseline of the retract curve and Young's moduli determined using the Hertz model (JPK software), an estimate of cell envelope elasticity. Surface roughness was measured at the midpoint of the cell using the QI™ height images [48] and cellular volume was calculated using "Ellipsoid" (http://planetcalc.com/) [49] for at least 20 different cells from three different samples.

### Membrane integrity

To assess the integrity of the cell membrane following CNB oil exposure, the cellular content leakage assay [43] was carried out with slight modification. Briefly, a mid logarithmic phase culture of *C. albicans* RSY150 was washed three times and resuspended to ~$10^7$ CFU/mL in PBS. Cell suspensions were transferred to a 24 well plate containing CNB oil at MIC, 1/2, 1/4 and 1/8 MIC in PBS, and incubated at 30 °C for 6 h with shaking (200 rpm). CNB oil in PBS served as a blank, untreated cells in PBS as negative controls, and amphotericin B at MIC served as a positive control. Following incubation, cell leakage was analyzed from supernatant diluted 1:10 with PBS based on absorbance at 260 nm (Varian Cary 100 BIO, UV–VIS spectrophotometer; Midland, ON, Canada) against a blank lacking CNB oil. Nucleotides represent one class of leakage components that absorb at 260 nm, for which uracil containing compounds have the highest absorbance. Mean ratios for each treatment from three independent experiments were calculated and compared to that of corresponding untreated samples.

### Cell wall stress

The sensitivity of *C albicans* to cell wall perturbing agents (Congo red (CR) and CFW) after exposure to CNB oil was tested following a previously reported method [6], with minor modifications. Briefly, overnight stationary phase cultures of *C. albicans* RSY150 ($2.2 \times 10^5$ cell/mL) in YNB complete media, unexposed or exposed to CNB oil at 1/2 MIC and 1/4 MIC, were incubated for 24 h with shaking (200 rpm) at 30 °C in a microtiter plate. Cells were washed with PBS to eliminate oil residues and any carryover effect of the oil to different wells. An aliquot (5 μL) of each was diluted to final densities of $10^5$, $10^4$, $10^3$ cells/mL and were spotted on YNB plates supplemented with one of the following cell wall disrupting agents: CFW (15 μg/mL) or CR (15 μg/mL). The plates were incubated at 30 °C and monitored for *Candida* growth over 3 days. Each experiment was performed in triplicate and colonies counted for statistical analysis and graphical representation.

### Confocal microscopy

The effects of CNB oil treatment on the cell cycle was determined by laser scanning confocal microscopy (LSCM) [35]. RSY150 cells expressing Tub2-GFP and Htb-RFP [35] at mid logarithmic phase ($10^7$ cells/mL) were treated with CNB oil, cinnamaldehyde and linalool at MIC and 1/2 MIC for 4 h. Treated and control cells were transferred to glass slides sealed with clean coverslips and imaged at 63 × magnification on a Zeiss LSM 780 system (Carl Zeiss Micro imaging, Oberkochen, Germany) using an argon laser ($\lambda_{ex} = 488$ nm; $\lambda_{em} = 512$ nm) for Tub2-GFP and HeNe laser ($\lambda_{ex} = 543$ nm; $\lambda_{em} = 605$ nm) for Htb-RFP. Cells were identified and enumerated in each cycle phase category. Statistical significance was calculated from three independent experiments.

### Statistical analysis

All experiments were performed in triplicate, unless otherwise stated, and GraphPad Prism7 used for statistical analysis. The results were reported as mean ± standard deviation (SD), differences assessed using a two-tailed unpaired *t* test with Welch's correction at a 95% confidence interval, for which $p < 0.05$ was considered statistically significant.

## Results

### CNB oil composition

CNB oil analyzed by GC-FID and GC–MS identified several known constituents. The peaks identified by GC-FID were further confirmed with GC–MS and previously reported RI [32, 34]. The composition of our CNB is shown in Additional file 1: Table S1 (see Additional file 2: Figure S1), for which compounds are listed in the order of elution on the HP-5 column. The major compounds (concentrations > 2.0 as relative % peak area from GC-FID and GC–MS) were E-cinnamaldehyde (74%), α-caryophyllene (5.3%), linalool (3.9%) and E-cinnamyl acetate (3.8%). Minor components with known function included limonene (2%), eugenyl acetate (0.6%), α-pinene (0.3%) and benzyl benzoate (0.6%). Another minor constituent was p-cymene (1.4%).

### RSY150 and clinical isolates of *C. albicans* respond differently to CNB oil

The relative susceptibilities of the RSY150 strain and clinical isolates to CNB oil were determined using a broth microdilution method. The planktonic growth of *C. albicans* lab strains (ATCC 64548 and RSY150) was effectively inhibited by 125 µg/mL CNB oil, whereas the clinical strain isolated from blood and the ATCC 10231 strain were more resistant and required higher concentrations (250 µg/mL) of CNB oil to produce an inhibitory effect. This clinical strain isolated from blood was selected for detailed analysis and further experiments. Clinical isolates from blood and genital infections displayed varying susceptibility to CNB oil. In general, a blood isolate was resistant to CNB oil with a MIC of 500 µg/mL, whereas one genital strain was more resistant to CNB oil with a MIC at 1000 µg/mL and the other more susceptible with a MIC of 125 µg/mL. The data is summarized in Table 2. RSY35, a mutant strain for microtubule motor protein Kar3, showed growth inhibition at 125 µg/mL of CNB oil. Similarly, the growth of RSY150 and RSY35 was inhibited by cinnamaldehyde with a MIC of 112.5 µg/mL. Both strains showed identical response to linalool with MICs at 2.46 mg/mL.

### Dose dependence of *C. albicans* RSY150 growth, viability and membrane integrity

To determine the dose dependence of CNB oil over a 24 h period, growth curves were constructed at different concentrations of CNB oil (15.625–125 µg/mL) in YPD media. Growth inhibition was dose dependent with complete inhibition of growth at MIC (125 µg/mL) and increased growth at progressively lower sublethal concentrations: 62.5 µg/mL (1/2 MIC), 31.25 µg/mL (1/4 MIC) and 15.625 µg/mL (1/8 MIC) (Fig. 1a), used for subsequent experiments.

The viability of cells cultured to logarithmic phase after treatment for 4 h with different concentrations of CNB oil (125–62.5 µg/mL) were assessed using the vital dye, methylene blue (Fig. 1b), which enters both live and dead cells. Since live cells are able to reduce the dye [50, 51] they become colorless, whereas dead cells remain blue. The cells were non-viable following CNB oil treatment, confirmed

**Table 2 Number of reference and clinical strains at various MIC values**

| CNB oil (µg/ml) | Blood | Genital | RSY150 | RSY35 | ATCC 10231 | ATCC 64548 |
|---|---|---|---|---|---|---|
| 1000 | | 1 | | | | |
| 500 | 1 | | | | | |
| 250 | *1 | | | | | |
| 125 | 3 | **1 | 1 | 1 | 1 | 1 |

*Used as representative clinical strain in all further experiments

**Tested for chitin staining

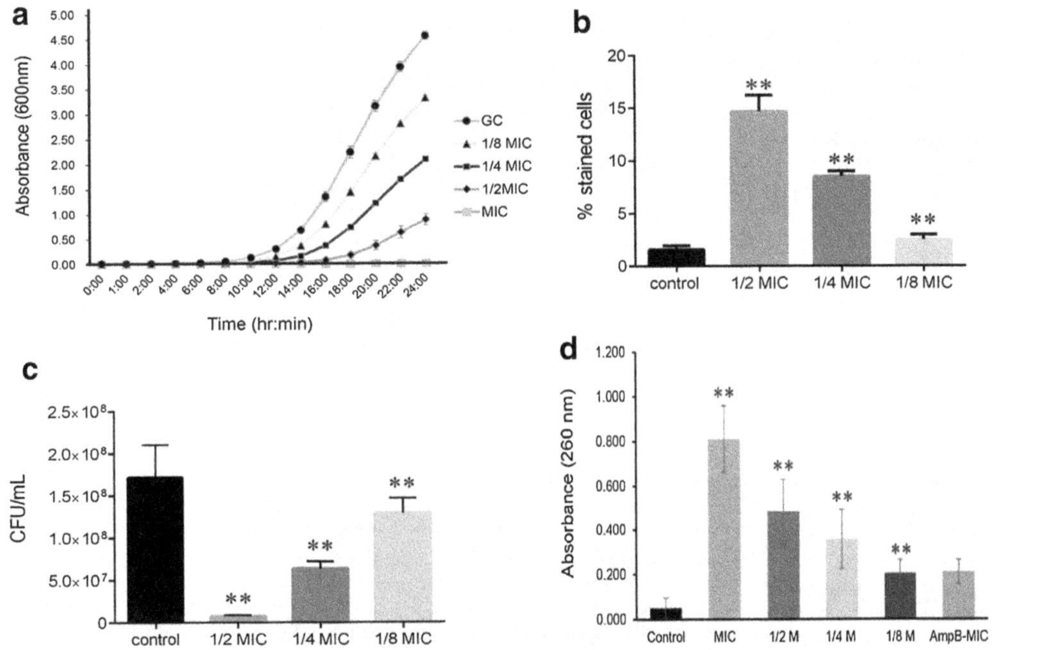


**Fig. 1** CNB oil decreased cell viability and compromised cell membrane integrity in RSY150. **a** Growth curves of RSY150 exposed to 125, 62.5, 31.25 and 15.125 µg/mL of CNB oil, with absorbance measured at 30 min intervals to show the antifungal time course of the CNB oil. **b** Methylene blue staining of cells exposed to CNB concentrations as in (**a**) showed decreased cell viability. **c** Corresponding CFU counts of viable CNB exposed cells as compared to control. **d** Leakage of cellular content as a function of absorbance at 260 nm following exposure to CNB oil at concentrations as in (**a**) confirm a dose dependent compromise of the cell membrane in (**a**). GC indicates growth controls of cells incubated without CNB oil. Double asterisks represent statistical significance ($p < 0.05$)

by CFU for each CNB oil treatment (1/2, 1/4 and 1/8 MIC and control). There was a significant reduction in the number of colonies treated with 1/2 MIC CNB oil in comparison to cells treated at 1/4, 1/8 MIC and control (Fig. 1c).

Membrane disruption in CNB oil treated RSY150 cells was assessed as leakage of cellular contents ($OD_{600} = 260$ nm) in the supernatant [43]. The amount of leaked cellular content increased as a function of CNB oil concentration, from 1/8 MIC to MIC (Fig. 1d), showing that compromised cell membrane integrity is dose dependent. The positive control amphotericin B showed leakage at MIC.

## Sublethal doses of CNB oil altered chitin distribution in *C. albicans*

To determine morphological changes, we analyzed bright field images of *C. albicans* in stationary and logarithmic phases exposed to sublethal concentrations of CNB oil for 24 h. Stationary phase *C. albicans* RSY150 cells exposed to sublethal CNB were oval and swollen, with distinct changes in chitin distribution, but with normal budding compared to control cells (Fig. 2a, d). Epifluorescence images of stationary phase control RSY150 show an even distribution of chitin around the cell wall and septal region (Fig. 2a, d). In contrast, cells treated with CNB oil at 62.5 µg/mL (1/2 MIC) had an increase in chitin content in the lateral cell wall and more intense staining at septal regions (Fig. 2a, d). This effect was absent at lower concentrations (31.625 µg/mL; 1/4 MIC) of CNB oil.

Similarly, the effects of CNB oil on metabolically active cells were distinct. Logarithmic phase cells exposed to CNB oil for 4 h at MIC and 1/2 MIC had intense chitin staining in the bud neck region and uneven lateral cell wall staining (Fig. 2b, e) compared to control cells that stained evenly at the bud and lateral wall. We also observed an increased frequency of elongated cells during exposure to CNB oil at MIC, which became less prominent at 1/2 MIC (Additional file 3: Figure S2a, b). The clinical isolate from blood, on the other hand, showed overall stronger staining than RSY150, with increased intensity at the bud neck upon exposure to CNB oil at MIC and 1/2 MIC (Additional file 4: Figure S3a). In contrast, the genital clinical isolate having a similar MIC to RSY150 showed a normal chitin distribution (Additional file 4: Figure S3b).

Based on the altered phenotype during CNB exposure at sublethal levels, we studied the impact of CNB on hyphal growth. During hyphal induction (10% serum) in the presence of CNB oil, growing hyphae had intense chitin staining at their growing tips. The intensity of chitin staining was higher in cells treated at MIC and decreased progressively as

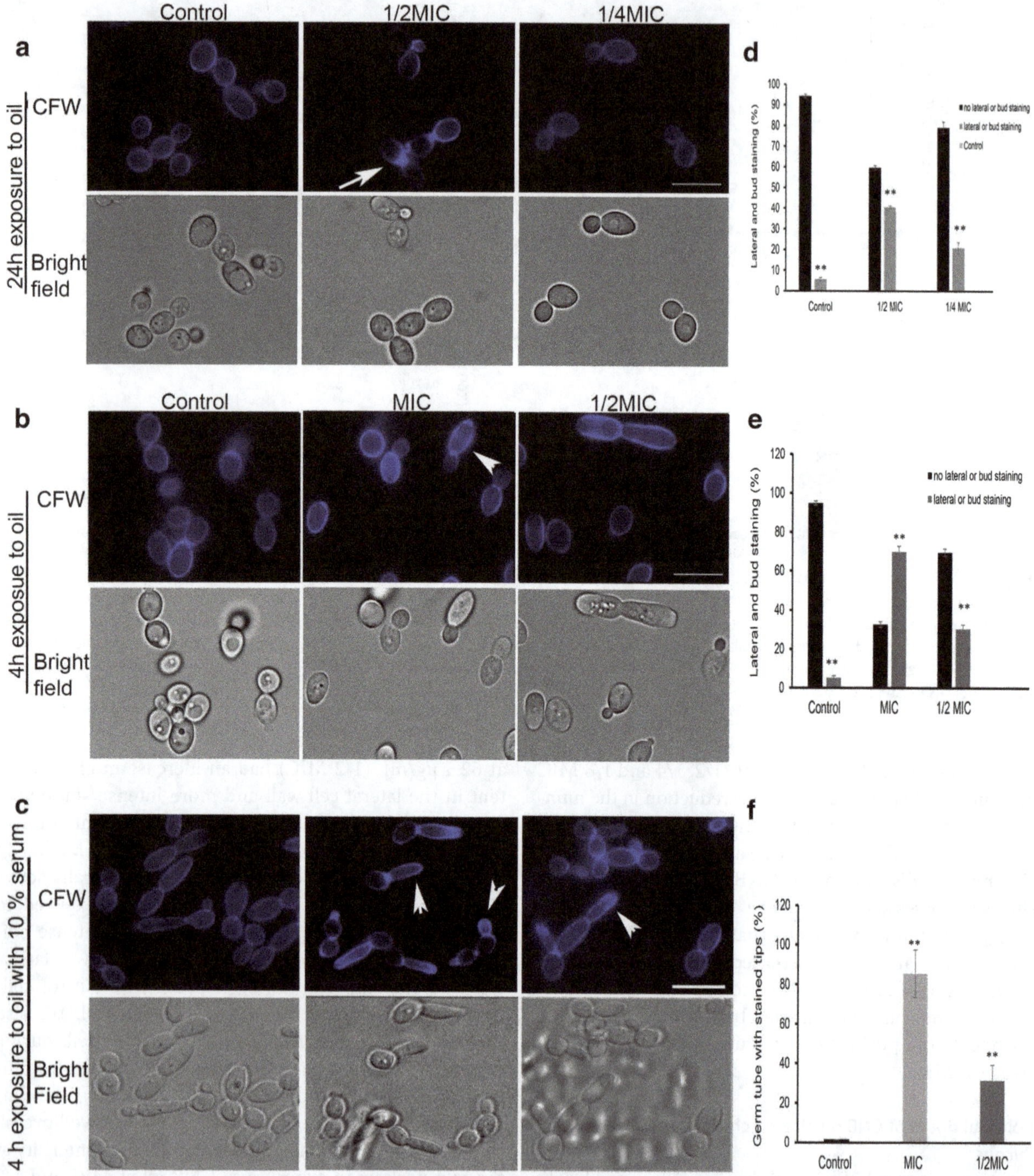


**Fig. 2** CNB oil exposure induces a cell wall stress response with altered chitin distribution in RSY150. Epifluorescence (top panel) and bright field (bottom panel) images of CFW stained cells. Cells were grown in YPD in (**a**) and (**b**) while cells in (**c**) were grown in YPD plus 10% serum for hyphal induction. **a** RSY150 cells were incubated for 24 h with and without 1/2 and 1/4 MIC CNB oil in YPD media and stained with CFW. **b** Mid log phase RSY150 were exposed to MIC and 1/2 MIC for 4 h stained with CFW and imaged. **c** Cells grown in YPD with 10% serum were exposed to the same concentration as in (**b**) for 4 h, stained with CFW and imaged. (**d**, **e**, **f**) are bar graphs showing enumeration of cells in (**a**, **b** and **c**) respectively, where double asterisks indicate $p < 0.05$. Bars = 5 μm

a function of CNB oil concentration, at 1/2 MIC (Fig. 2c, f). At 1/4 MIC CNB the intensity of CFW staining was comparable to control, however hyphal width remained the same.

### Sublethal CNB oil induces cell wall remodelling in *C. albicans*

AFM is a powerful tool for imaging ultrastructural changes at the cell surface and probing the associated biophysical properties. We chose QI™ mode, which records a large number of force curves (n = 16,536; 128 pixel$^2$ resolution) for a detailed analysis. Representative QI™ images of RSY150 and a clinical isolate of *C. albicans* exposed to sub lethal concentrations of CNB oil, along with comparable controls, are shown in Fig. 3a–d. Changes in RSY150 and the clinical isolate after CNB treatment are summarized in Table 3. There was a clear difference between the CNB treated RSY150 and the clinical isolate, with 50% of the RSY150 cells showing leakage and exfoliation while the clinical isolate showed leakage with increased surface roughness. The exfoliation in these cells was usually limited to a small area on the cell

**Table 3 Various changes to *C. albicans* cells with CNB oil exposure revealed by AFM**

| *C. albicans* treated with CNB oil (total # of cells) | Leakage only | Exfoliation only | Leakage and exfoliation | Surface roughness only | Leakage and surface roughness |
|---|---|---|---|---|---|
| RSY150 (n = 24) | 9 | 3 | 12 | – | – |
| Clinical strains (n = 21) | 1 | 1 | 1 | 5 | 13 |

Control cells had no visual changes

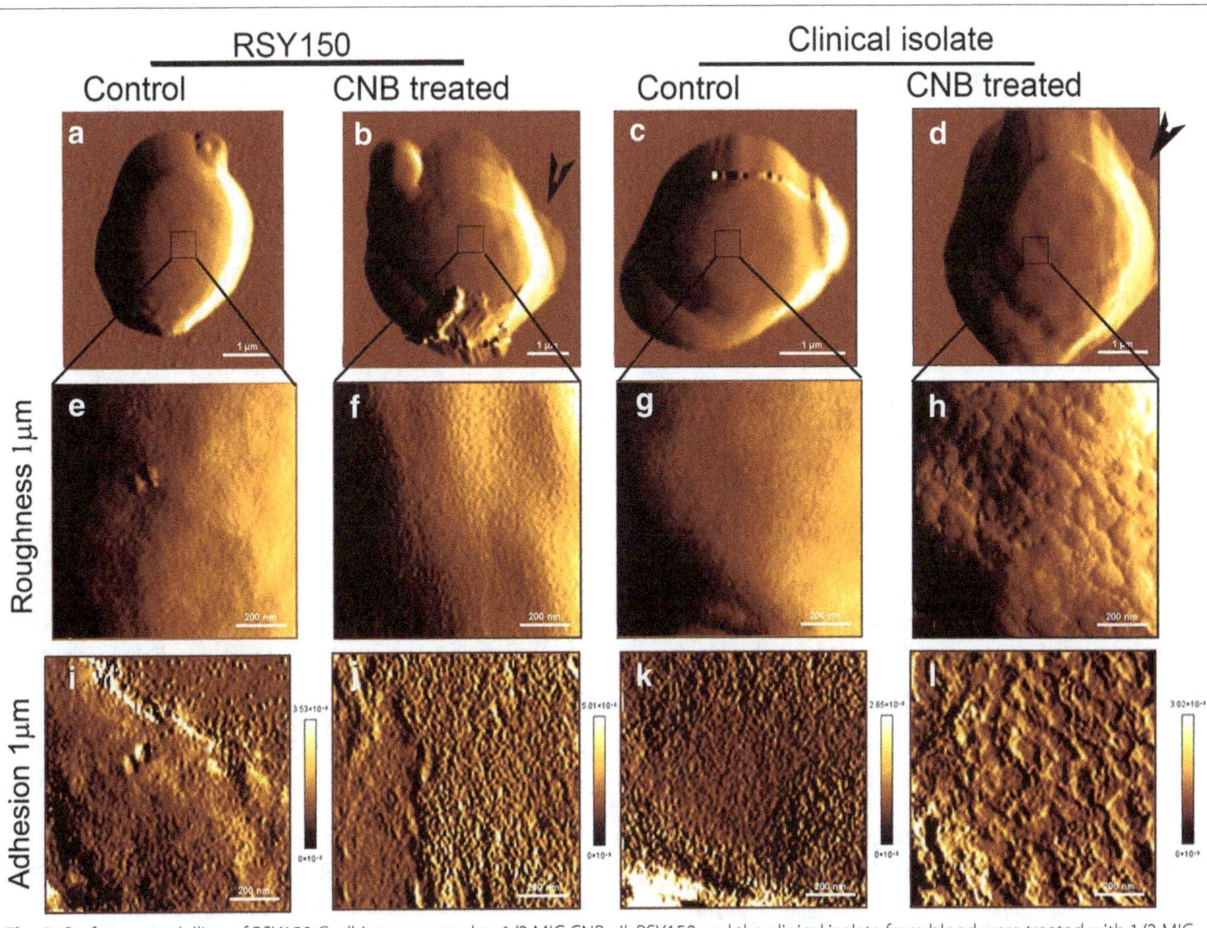


**Fig. 3** Surface remodelling of RSY150 *C. albicans* exposed to 1/2 MIC CNB oil. RSY150 and the clinical isolate from blood were treated with 1/2 MIC (62.5 μg/mL) CNB oil for 24 h, deposited and fixed on PLL-coated cover slips for AFM imaging. Images (1 μm; 128 px$^2$) of control and CNB oil treated cells are at low resolution for control (**a, c**) and treated (**b, d**) *C. albicans* showing leakage (arrow heads) in treated cells. High resolution (1 μm) height images of the cell surface (boxed area) for control (**e, g**) and CNB treated (**f, h**) *C. albicans* reveal increased surface roughness in the treated clinical isolate as compared to the treated RSY150 strain. (**i, j, k, l**) High resolution (1 μm) adhesion maps from corresponding height images. Bar a–d = 1 μm; Bar e–l = 200 nm

in the form of an aberration, while the rest of the surface showed little change. The clinical isolate exposed to CNB oil had rough cell surfaces, with or without leakage.

*Candida albicans* RSY150 and the clinical isolate displayed cell swelling and substantial cell wall structural and mechanical changes in response to CNB oil (Fig. 3). QI™ AFM of RSY150 ($n = 24$) and a clinical isolate from blood ($n = 21$) exposed to 1/2 MIC of CNB oil showed obvious changes in cell volume, roughness, surface adhesion and cell wall elasticity.

Images of untreated RSY150 (Fig. 3a, c) revealed a normal oval cell with a height of 1.65 ± 0.19 μm and cell volume of 9.41 ± 1.7 $\mu m^3$, which significantly ($p < 0.0001$) increased to a height of 2.18 ± 0.31 μm and volume of 18.4 ± 4.7 $\mu m^3$, respectively, following CNB oil treatment. The change in height (2.07 ± 0.29 μm vs. control 1.83 ± 0.27 μm) ($p = 0.01$) and volume (volume: treated 12.7 ± 5.9 $\mu m^3$ vs. control 16.2 ± 4.7 $\mu m^3$) ($p = 0.047$) of the clinical isolate was significant with CNB oil treatment, but less pronounced than that of RSY150 with similar treatment.

Control cells of both strains (Fig. 3e, g) revealed a smooth and homogeneous surface. There was a significant increase ($p = 0.015$) in cell surface roughness (5.4 ± 1.0 nm) of the clinical isolate following CNB oil exposure compared to control (2.7 ± 1.2 nm), whereas the RSY150 (Fig. 3e–h) showed no significant change ($p = 0.2432$) compared to control cells (treated 4.9 ± 1.08 nm vs. control 3.52 ± 1.77 nm). These results are summarized in Table 4.

For accurate adhesive measurements of the cell surface, 1 μm QI™ AFM images (128 $px^2$ resolution) were captured at the center top of the cells to eliminate artifacts from cell curvature. The adhesive properties of both RSY150 (treated 7.1 ± 0.5 nN vs. control 5.5 ± 0.3 nN) and the clinical isolate (treated 7.6 ± 0.6 nN vs. control 5.7 ± 0.4 nN) were significantly increased (RSY150 $p = 0.0016$, clinical isolate $p = 0.0025$) in response to CNB oil. Figure 3b, f, j and d, h, l shows whole cells, high resolution images, roughness and adhesive forces on the center top of RSY 150 and clinical isolates, respectively, grown for 24 h in the presence of sublethal (62.5 μg/mL) CNB oil. Exposure to CNB oil altered cell shape, increased the average cell volume and roughness (Fig. 3j, f), with a twofold increase in elasticity for RSY150 (treated 7.8 ± 0.9 GPa vs. control 4.7 ± 0.4 GPa) and the clinical isolate (treated 15.0 ± 2.7 GPa vs. control 8.1 ± 0.7 GPa) compared to their respective controls. Table 5 shows a summary of these differences in RSY150 and the clinical isolate.

**Table 4 Comparison of ultrastructural properties of RSY 150 and the clinical isolate from blood**

| | RSY150 | | Clinical isolate | |
|---|---|---|---|---|
| | Control | Treated | Control | Treated |
| Height (nN) | 5.5 ± 0.3 | 7.1 ± 0.5 | 5.7 ± 0.4 | 7.6 ± 0.6 |
| Volume ($\mu m^3$) | 9.4 ± 1.7 | 18.4 ± 4.7 | 16.2 ± 4.7 | 12.7 ± 5.9 |
| Roughness (nm) | 3.5 ± 1.8 | 4.9 ± 1.1 | 2.7 ± 1.2 | 5.4 ± 1.0 |

### CNB oil exposure increases *C. albicans* sensitivity to cell wall disrupting agents

In order to determine whether cell wall integrity was compromised by CNB, RSY150 were exposed to cell wall perturbing agents CR and CFW at 1/2 MIC and 1/4 MIC. Higher sensitivity, reflecting compromised cell wall integrity, was observed in CNB treated cells in comparison to control (Fig. 4a) at 1/2 MIC, progressively decreasing to 1/4 MIC (Fig. 4b). As positive controls, we analyzed mutants lacking genes related to cell wall integrity and to wall organization and biogenesis for their sensitivity to CNB oil treatment. The mutants tested had the following knockouts: *VPS28*, a VPS factor required for vacuolar protein sorting; *CRH11*, encoding a GPI-anchored cell wall transglycosylase; *SSU81*, encoding a function involved in oxidative stress, cell wall biogenesis, and morphogenesis; and *DFG5*, encoding an N-linked mannoprotein of the cell wall and membrane. All these mutants exhibit generalized sensitivity to CNB oil, with lower MIC than the wild type strain (CaSS1), further verifying an impact of CNB on cell wall integrity (Fig. 4c). The associated MIC decreased two-fold, from 125 μg/mL to 61.5 μg/mL, for all mutants.

### CNB oil induces spindle defects in *C. albicans* at MIC and sub-MIC

The RSY150 strain expressing both Tub2-GFP-and Htb-RFP was imaged by LSCM to visualize the mitotic spindle in dividing cells exposed to CNB oil. In untreated cells, the spindle varied in length for the control RSY150 strain, but it consistently spanned the distance between mother and daughter cells (Fig. 5a) and nuclear migration was normal. In contrast, these well-defined mitotic spindles were missing in RSY150 cells exposed to CNB oil at MIC. Furthermore, Tub2-GFP fluorescence in these

**Table 5 Comparison of biophysical properties of RSY150 and the clinical isolate from blood**

| | RSY150 | | Clinical isolate | |
|---|---|---|---|---|
| | Control | Treated | Control | Treated |
| Adhesion (nN) | 5.5 ± 0.3 | 7.1 ± 0.5 | 5.7 ± 0.4 | 7.6 ± 0.6 |
| Elasticity (GPa) | 4.7 ± 0.4 | 7.8 ± 0.9 | 15.0 ± 2.7 | 8.1 ± 0.7 |

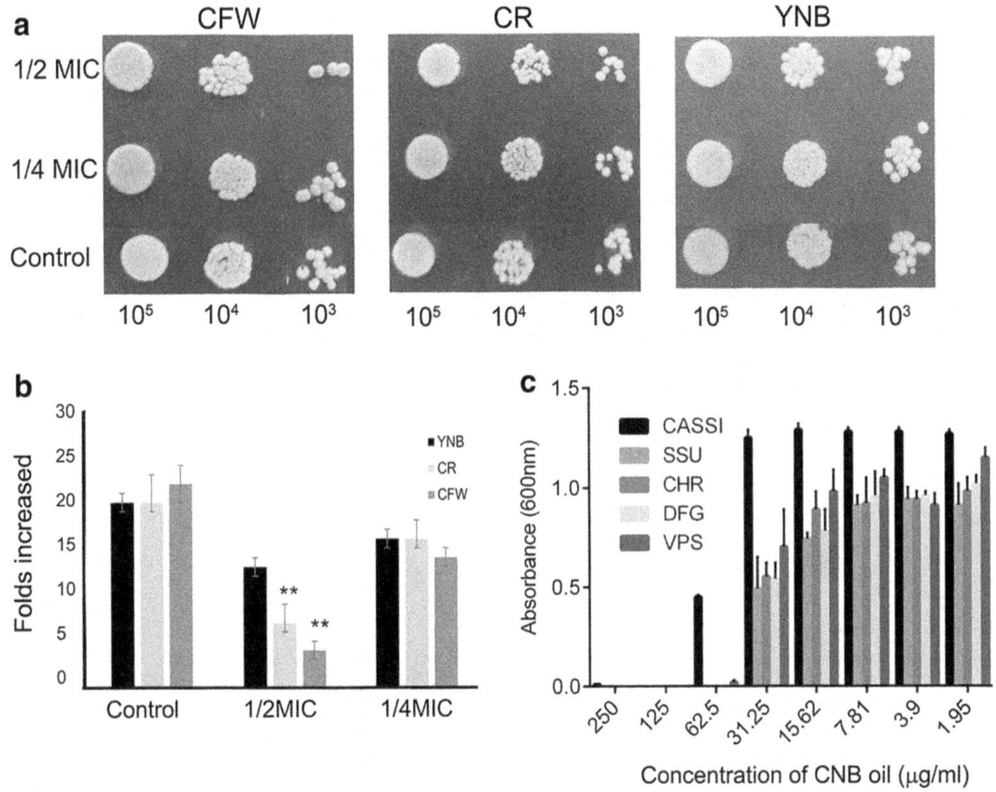


**Fig. 4** CNB treated RSY150 *C. albicans* exhibit a cell wall stress response. Logarithmic phase RSY150 cells were treated with 1/2 and 1/4 MIC CNB oil or without oil for 24 h. **a** Treated and control cells at $10^5$, $10^4$ and $10^3$ CFU were spotted on YNB complete media agar containing CFW and CR and examined after 2 days. Treated cells showed greater sensitivity to cell wall stressors. **b** The number of colonies on the CR and CFW plates show a dose dependent effect that decreases as a function of oil concentration. The 1/2 MIC CNB oil was more effective than 1/4 MIC, with a significant difference ** between 1/2 MIC oil treatment compared to untreated controls ($p < 0.05$). The cell count was normalized with treated cells grown on YNB media agar only. **c** Wild type background strain (CaSS1) and cell wall defective mutants (*SSU81, CRH, DFG* and *VPS*) were exposed to a twofold increasing dilution of CNB oil in a MIC assay, showing overall sensitivity to the CNB oil and indicating a moderate cell wall stress response

cells consistently failed to show the wild type pattern of spindle formation between mother and daughter cells (Fig. 5b). Instead, Tub2-GFP often had reduced fluorescence in cells undergoing nuclear division, with only several fluorescence spots visible, indicating disruption of the mitotic spindle. Cells exposed to sub MIC CNB oil had reduced spindle defects, with most of the cells in recovery phase and observed as fluorescent patches or aggregates of tubulin (Fig. 5c).

RSY150 cells grown to the mid logarithmic phase and exposed to oil for 4 h were imaged by LSCM to view patterns of nuclear division and tubulin distribution in the parent cells and the bud. Budding cells with a single nucleus were considered to be pre-mitosis, representing those in the G1 or S phase. Cells with nuclei traversing the parent and budding cells were considered to be in the anaphase stage of mitosis. Finally, cells having separate nuclei in both parent and budding cells were considered to be at a post-mitotic phase, G2/M. For RSY150 control cells, 52% of the population was in the G1 or S phase; 36% of the cells had undergone mitosis; and the final 10% were in anaphase (Fig. 5b, c). On the other hand, RSY150 treated with MIC of CNB oil displayed noticeable differences in cell cycle distribution, with 39% having not begun mitosis, 48% having completed mitosis, and 14% in anaphase, still undergoing nuclear division (Fig. 5c). Cells treated with 1/2 MIC were at GI phase (38%), whereas the dominant number of cells were at G2/M phase, with no change in the number of cells undergoing anaphase. Cells treated at MIC with CNB oil showed extended cells with pseudohyphal formation, however the effect was less pronounced in cells treated at 1/2 MIC, where large budding cells were observed (Additional file 3: Figure S2a, b). The data clearly indicate that CNB oil induces tubulin depolymerisation leading to cell cycle arrest at anaphase.

Cinnamaldehyde treated *C. albicans* showed similar morphological features to CNB oil treated cells (Additional file 5: Figure S4), with depolymerized tubulin, resulting in an absence of mitotic spindles. Overall the effect was more pronounced in cinnamaldehyde treated

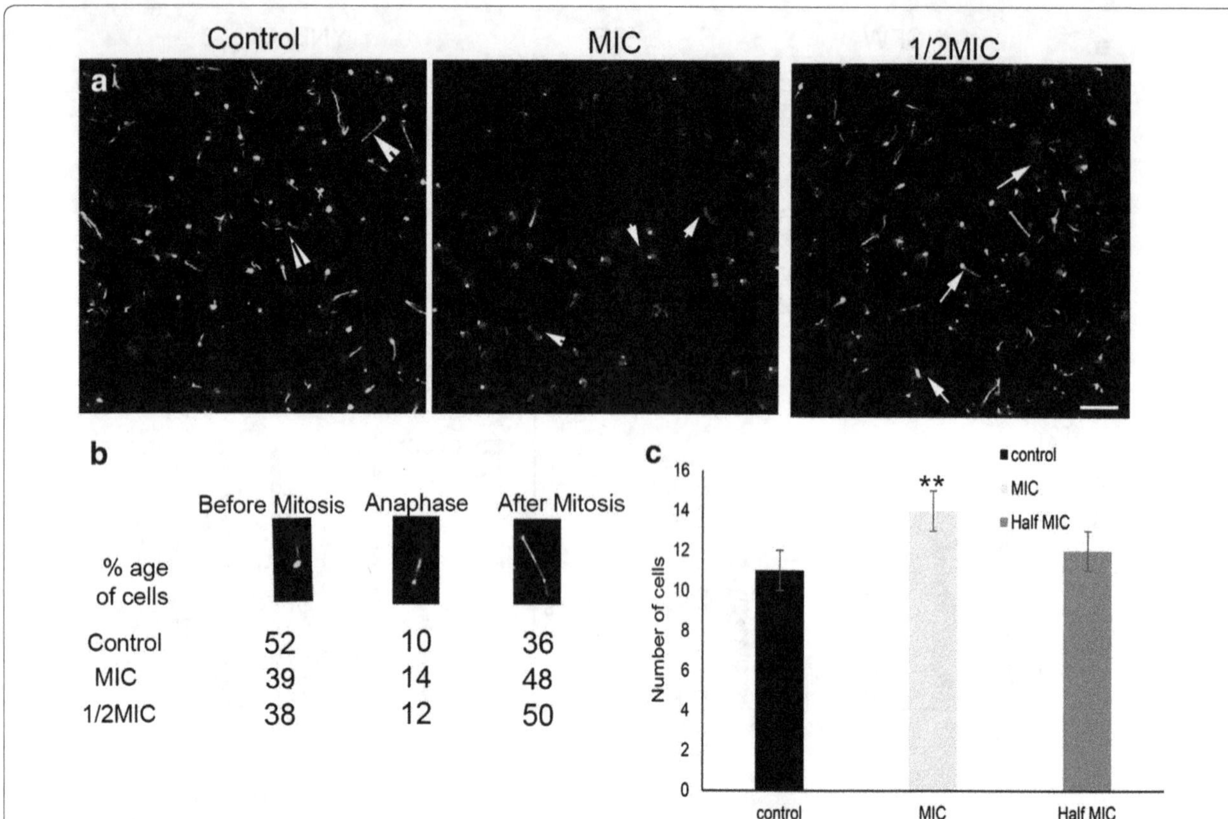


**Fig. 5** CNB oil exposure in RSY150 *C. albicans* caused spindle defects in RSY150 and a delayed cell cycle. Mid logarithmic phase RSY150 cells were treated with MIC and 1/2 MIC oil for 4 h and imaged live by LSCM (Tub2-GFP $\lambda_{ex} = 488$ nm; $\lambda_{em} = 512$ nm and Htb-RFP $\lambda_{ex} = 543$ nm; $\lambda_{em} = 605$ nm). RSY150 were examined for nuclear and spindle localization in budding cells. **a** Spindle morphologies of MIC and 1/2 MIC CNB oil treated cells and untreated control. Control cells showed normal spindle formation during mitosis (arrow heads) whereas the MIC treated cells had a majority of dividing cells lacking a mitotic spindle (small arrow heads). Cells treated with 1/2 MIC of CNB oil had diminished spindle defects. Despite the spindle defect, the cells completed mitosis. Images are representative from three independent experiments. Bar = 5 µm. **b** Cell cycle analysis of cells treated with MIC and 1/2 MIC CNB oil and untreated controls. Of cells treated at MIC CNB oil, 14% were in anaphase in comparison to 10% for control cells, whereas 12% of cells exposed to 1/2 MIC CNB oil were in anaphase. **c** The number of cells counted for treated and control cells in anaphase were plotted. Double asterisks represent statistical significance ($p < 0.05$). For each treatment 250 cells were analyzed

cells as compared to those treated with CNB oil. Cells exposed to cinnamaldehyde at MIC showed a significantly increased percentage of cells in anaphase (15.8%) as compared to control (9.6%), whereas RSY150 cells exposed to linalool were arrested at the G1 phase of the cell cycle, as previously reported [20]. There was complete disruption of the cell wall at MIC after 4 h, with cells having depolymerized tubulin compared to controls, however the cells showed some tubulin fluorescence at 1/2 MIC, with intact cell walls (Additional file 5: Figure S4).

## Discussion

The cell wall of *C. albicans* is the first line of defense against the host immune system and toxic agents, modulating fungal interactions and maintaining cell integrity [52–55]. The *C. albicans* yeast cell wall contains three main components: mannoproteins (~ 39%), β-glucans (~ 59%), and chitin (~ 2%), that contribute to a highly dynamic molecular architecture which is continuously remodelled in response to cell surface interactions [53–55]. The cell wall integrity (CWI) pathway compensates for damage caused by environmental stress, signaling through the MAPK cascade to mediate cell wall biosynthesis and actin organization, regulating cell cycle progression and other necessary events [56–59]. In response to cell wall stressors, for example the antifungal drug caspofungin which is a non-competitive inhibitor of β-1,3-glucan synthase [60–62], activation of the CWI pathway can reinforce the cell wall by increasing chitin levels in *Candida* [63] and *C. glabrata* [64].

### Cell wall and membrane stress

CNB oil treated RSY150 cells in the stationary phase had an irregular chitin distribution found on the lateral wall of budding cells (Fig. 2a). Such effects became pronounced in metabolically active cells grown in the presence of 10% serum (Fig. 2c, f), with intense chitin staining at the hyphal tip. Similar staining patterns have been reported for cells exposed to high sugar or salt concentrations [23], for which hypo-osmotic pressure induced chitin synthesis or polymerization at the growing hyphal tip. Consistent with this idea is the larger cell volume of RSY150 (Fig. 3) and sensitivity of the *SSU18* mutant (Fig. 4c) exposed to CNB oil. Cell wall reinforcement with chitin in response to stress is a well-known phenomenon in yeast [62, 64–67]. It is known that most resistant clinical strains have a higher overall chitin content, reinforcing the cell wall and making them more resistant to cell wall targeted antifungals [68]. As expected, the metabolically active clinical isolate showed more chitin than RSY150 (Additional files 3, 4: Figures S2, S3) for both control and treated cells, confirming cell wall reinforcement in the clinical strain.

Chitin plays an important role in defining the cell wall nanomechanical properties [25, 27, 69], and β-1,3 D-glucan is critical in maintaining cell shape, mechanical rigidity and resistance to osmotic pressure [4, 70]. Exposure to CNB oil produced osmotically fragile, swollen cells with an uneven chitin distribution and increased elasticity, consistent with prior studies [25, 27, 52, 59]. The elasticity values reported in this study are high in comparison, reflecting cross linking from fixation [71]. Inhibition of β-1,3-D-glucan synthase by cinnamaldehyde in *S. cerevisiae* or by caspofungin in *C. albicans* results in reduced cell wall β-1,3-glucan [22] and increased cell wall chitin content, respectively [25]. Furthermore, mutants defective in chitin and β-glucan cross linking (*chr1chr2Δ*) have cell wall elasticity that varies with architecture and composition [60].

In addition to cell wall chitin and glucans, the surface of *C. albicans* is decorated with proteins called adhesins that play a pivotal role in cell communication, adhesion and microbial infection [72, 73]. Many adhesins are mannoproteins that are homogenously dispersed on the cell surface [74, 75]. Cell wall stress, temperature variation, exposure to antifungal agents, host interaction and biofilm formation have been shown to alter the adhesive properties of cells [67, 76]. Adhesive forces in this study are based on the interaction between the cell surface and the hydrophilic, negatively charged silicon nitride AFM tip. Exposure to CNB oil gave rise to a moderate increase in cell adhesion, which may be attributed to cell wall remodelling, as evidenced by concomitant changes in roughness. Changes in adhesion may reflect loss or rearrangement of adhesins, or reorganization of surface sugars [25, 27, 29].

Cell wall stress was characterised not only by changes in chitin distribution and expression, but also polarized growth and moderate sensitivity to cell wall perturbing agents, leading to cell cycle arrest and fungal death [57]. Congo Red and Calcofluor White target chitin and β-1,3-glucan synthases to produce a cytokinetic defect leading to cell cycle arrest [77, 78]. A delayed or arrested cell cycle results in polarized growth with formation of pseudohyphal cells [79–81], as characterized by depolymerized microtubules [35, 82, 83]. Such a morphology was observed in 30% of the cell population exposed to CNB oil, along with defects in microtubule polymerization.

### Spindle defects

Microtubules are central to cell division, forming the mitotic spindle and coordinating nuclear movement [82, 84, 85]. In *Candida*, the pre-mitotic movement of the nuclei through the bud neck and subsequent separation of nuclei are coordinated by microtubules [86]. In normal cells, nuclear division is complete when the buds are still small, such that larger budding cells are never observed without the nucleus [86]. In the control cells, two distinct subcellular microtubule structures were observed, one traversing the mother and budding cells and the other only in the mother cells. Anaphase and G2 cell cycle arrest, as well as a more pronounced filamentous growth, were apparent following CNB exposure (Fig. 5a, c). Cell cycle defects associated with microtubule perturbation were readily visualized in CNB treated RSY150 in which Htb-RFP and Tub2-GFP highlighted the nucleus and microtubules, respectively. Four hours following treatment with CNB at MIC, log phase cells had ablated mitotic spindles, observed as fluorescent patches or aggregates of tubulin (Fig. 5a). Large-budding cells had nuclear staining restricted to the mother cell (Fig. 5a, arrow). These results are consistent with previous observations in which tubulin polymerization inhibitors and *Δkar3/Δkar3* mutants had similar phenotypes [35, 85]. We observe anaphase arrest in cells exposed to CNB at MIC, with 14% of the cells in anaphase as compared to 10% in the control, an effect which is reduced at 1/2 MIC. Cells treated at MIC cinnamaldehyde had 15% of cells in anaphase, slightly less than the *Δkar3/Δkar3* mutant that had 17% of cells in anaphase and exhibited polarized growth [35]. Kar3 may mediate microtubule sliding during nuclear fusion and possibly mitosis, interacting with spindle microtubules to produce an inwardly directed force on the poles that antagonizes the CIP8 and KIP1 outward force [36]. The *Δkar3/Δkar3* mutant (RSY35) is no more resistant to CNB oil than wild type, suggesting that Kar3 is not likely the target of CNB oil.

We observed pseudohyphal growth in CNB oil treated RSY150 at MIC, an effect which became less prominent at 1/2 MIC, a common response of yeast experiencing cell wall stress [87]. For instance, toxins and chemicals that arrest cells in the S phase or mitosis display polarized growth due to activation of the CWI pathway [57, 88, 89]. The CWI pathway activates the MAPK cascade to manage stress, and its components are key elements in controlling the cell cycle. There are three MAPK pathways, namely; the pheromone response pathway, the high osmolality glycerol (HOG) pathway and the *PKC1*-mediated pathway [88, 89]. The increased sensitivity of the *SSU81* knock-out mutant (Fig. 4c) suggests that the HOG pathway is activated in the presence of CNB oil. Under stress there is cross talk between these pathways and cell cycle checkpoints, resulting in a delayed cell cycle [88] and a mixture of yeast and pseudohyphal cells (Additional file 3: Figure S2a, b).

### Impact of CNB oil components

The multisite impact of CNB oil can be attributed to both its major (cinnamaldehyde) and minor components. Cinnamaldehyde has been shown to compromise cell membrane and wall integrity [16], which we observe in this study along with a pronounced spindle defect and anaphase arrest. Minor components, which include limonene (2%), eugenyl acetate (0.6%), linalool (3.9%) and benzyl benzoate (0.6%), also contribute to CNB oil antifungal activity. The synergistic, additive or antagonist activity of these components to modify the overall effect of the oil has yet to be explored. However based on previous reports, these components target the cell membrane and modulate its function, including fluidity and permeability, leading to cell death [16–20]. Limonene (2%) has been reported to affect the cell wall in *S. cerevisiae* [90, 91] and p-cymene (1.4%) inhibits germ tube formation and alters cell membrane integrity [92]. Eugenyl acetate and linalool exhibit anti-virulence activity by inhibiting germ tube formation, and linalool has been reported to arrest cell cycle in the GI phase [20]. Such studies used higher (mg) concentrations of CNB oil components to produce their effects in comparison with their presence in CNB oil at only 0.6-5%. There remains the possibility of synergies between the CNB oil components.

### CNB oil as an antifungal

We propose that the main component of CNB oil, namely cinnamaldehyde (75%), is responsible for the majority of the cell envelope defects observed in this study. The cell envelope mutants and cell stressor assays demonstrate that the cell wall is moderately affected, reinforced by chitin. Further, the sensitivity of all the cell envelope mutants indicates multiple targets for CNB oil, consistent with its chemical heterogeneity. Our data reports for the first time the depolymerisation of β tubulin leading to cell cycle arrest at MIC of CNB oil. The mitotic spindle defect, a novel finding of this study, we attribute to cinnamaldehyde. The effect was strong at MIC and reduced at 1/2 MIC in cells exposed to either CNB oil or cinnamaldehyde. We propose that membrane disruption may lead to aberrant actin filaments, by virtue of actin's association with the membrane, and microtubule depolymerisation that ultimately disrupts the mitotic spindle, causing cell cycle arrest. This notion is supported by a finding in *S. cerevisiae*, for which both cytoplasmic microtubules and actin filaments are needed for spindle orientation [93], but further study is required to determine the role of actin in tubulin depolymerisation.

In summary, we suggest that CNB exerts its antifungal effect by targeting multiple cellular sites, including the cell wall, membrane and cell cycle machinery, ultimately leading to cell death induced by cell cycle arrest, a mechanism which requires further study.

## Conclusions

In conclusion, the CNB oil used in this study consists of a mixture of terpenoids, phenols and aldehydes in varying amounts which target multiple cellular sites. Taken together, our data suggest essential oils may serve as antifungal alternatives or could be used in combination with synthetic antifungal agents to combat antifungal resistance.

## Additional files

**Additional file 1: Table S1.** Chemical composition of CNB oil.

**Additional file 2: Figure S1.** Gas chromatogram of CNB oil. High deviation in RI values for E-cinnamaldehyde and α-caryophyllene result from its high concentration in the CNB oil, resulting in a non-Gaussian behaviour of the peak, and low concentration with late elution, respectively.

**Additional file 3: Figure S2.** (a) Cell cycle stress induced pseudohyphae in CNB oil exposed RSY150. Mid log phase RSY150 after 4 h exposure to CNB oil at MIC and 1/2 MIC, were stained with CFW. Images are epifluorescence (top panel) and bright field (BF; bottom panel). Bar = 5 μm. (b) Quantification of cells in control, MIC and 1/2 MIC from (a). Double asterisks represent $p < 0.05$.

**Additional file 4: Figure S3.**(a) Clinical isolate exposed to CNB oil showed increased chitin content. The clinical isolate from blood at log phase after 4 h exposure to CNB oil at MIC and 1/2 MIC were stained with CFW. Images represent CFW (top panel) and bright field (BF; bottom panel). Bar = 5 μm. (b) Genital clinical isolate with comparable MIC to RSY150 showed a normal chitin distribution.

**Additional file 5: Figure S4.** Spindle morphology of cinnamaldehyde and linalool treated *C. albicans*. Live LSCM of mid log phase cells after 4 h exposure to MIC and 1/2 MIC of cinnamaldehyde and linolool. Cinnamaldehyde treated *C. albicans* at MIC showed a similar spindle morphology of those treated with CNB oil at MIC, whereas linalool treated cells showed a complete absence of tubulin at MIC, with decreased cell size. At 1/2 MIC for both cinnamaldehyde and linalool, tubulin expression appeared as fluorescent spots near the nucleus. Bar = 5 μm.

### Authors' contributions

ZS conducted the AFM experiments, was responsible for MIC determination and associated data analysis. AAG conceived the idea of using essential oils and provided intellectual input into the manuscript. TS designed all experiments and performed LSCM, cell wall stress experiments and wrote an initial draft of the paper. MW provided knock-out strains, JM clinical strains and each provided intellectual input into the data interpretation and manuscript. TESD wrote the initial research proposal, supervised the work, provided intellectual input and edited the manuscript. All authors read and approved the final manuscript.

**Author details**
[1] Department of Chemistry and Biochemistry, University of Regina, 3737 Wascana Parkway, Regina, SK, Canada. [2] Microbiology and Immunology Department, Faculty of Pharmacy, Zagazig University, Zagazig, Egypt. [3] Regina Qu'Appelle Health Region, Regina, SK, Canada. [4] Centre for Structural and Functional Genomics, Concordia University, Montreal, QC, Canada.

**Acknowledgements**
We are grateful to Dr. Richard J. Bennett, Department of Molecular Microbiology and Immunology, Brown University, USA, for providing the RSY150 and RSY35 for this study. We thank Rebecca Jameson for the GC-FID and GC–MS analysis of the CNB oil and Cheghaf Madarati for images of glucanase treated *C. albicans*.

**Competing interests**
The authors declare that they have no competing interests.

**Funding**
This work was supported by National Science and Engineering Research Council (NSERC) grant to TESD (228206-07) and a NSERC Canada Research Chair to MW (950-228957). TESD was supported by the Canada Foundation for Innovation (CFI). TS was partially supported by the Faculty of Science and TESD's CFI IOF. ZS was partially supported by the Faculty of Graduate Studies and Research at the University of Regina.

**References**

1. Clark TA, Slavinski SA, Morgan J, Lott T, Arthington-Skaggs BA, Brandt ME, et al. Epidemiologic and molecular characterization of an outbreak of *Candida parapsilosis* bloodstream infections in a community hospital. J Clin Microbiol. 2004;42(10):4468–72. https://doi.org/10.1128/JCM.42.10.4468-4472.2004.
2. Hajjeh RA, Sofair AN, Harrison LH, Lyon GM, Arthington-Skaggs BA, Mirza SA, et al. Incidence of bloodstream infections due to Candida species and in vitro susceptibilities of isolates collected from 1998 to 2000 in a population-based active surveillance program. J Clin Microbiol. 2004;42(4):1519–27.
3. Ostrosky-Zeichner L, Rex JH, Pappas PG, Hamill RJ, Larsen RA, Horowitz HW, et al. Antifungal susceptibility survey of 2,000 bloodstream Candida isolates in the United States. Antimicrob Agents Chemother. 2003;47(10):3149–54.
4. Odds FC, Brown AJ, Gow NA. Antifungal agents: mechanisms of action. Trends Microbiol. 2003;11(6):272–9.
5. Johnson MD, MacDougall C, Ostrosky-Zeichner L, Perfect JR, Rex JH. Combination antifungal therapy. Antimicrob Agents Chemother. 2004;48(3):693–715.
6. Budzynska A, Sadowska B, Wieckowska-Szakiel M, Rozalska B. Enzymatic profile, adhesive and invasive properties of *Candida albicans* under the influence of selected plant essential oils. Acta Biochim Pol. 2014;61(1):115–21.
7. Goñi P, López P. Antimicrobial activity in the vapour phase of a combination of cinnamon and clove essential oils. Food Chem. 2009;116:982–9.
8. Ahmad A, Khan A, Manzoor N. Reversal of efflux mediated antifungal resistance underlies synergistic activity of two monoterpenes with fluconazole. Eur J Pharm Sci. 2013;48(1–2):80–6. https://doi.org/10.1016/j.ejps.2012.09.016.
9. Bassole IH, Juliani HR. Essential oils in combination and their antimicrobial properties. Molecules. 2012;17(4):3989–4006. https://doi.org/10.3390/molecules17043989.
10. Cowan MM. Plant products as antimicrobial agents. Clin Microbiol Rev. 1999;12(4):564–82.
11. Giordani R, Regli P, Kaloustian J, Mikail C, Abou L, Portugal H. Antifungal effect of various essential oils against *Candida albicans*. Potentiation of antifungal action of amphotericin B by essential oil from *Thymus vulgaris*. Phytother Res. 2004;18(12):990–5. https://doi.org/10.1002/ptr.1594.
12. Negri M, Salci TP, Shinobu-Mesquita CS, Capoci IR, Svidzinski TI, Kioshima ES. Early state research on antifungal natural products. Molecules. 2014;19(3):2925–56. https://doi.org/10.3390/molecules19032925.
13. Li Y, Nie Y, Zhou L, Li S, Tang X, Ding Y, et al. The possible mechanism of antifungal activity of cinnamon oil against *Rhizopus nigricans*. J Chem Pharmaceut Res. 2014;6(5):12–20.
14. Lopez P, Sanchez C, Batlle R, Nerin C. Solid- and vapor-phase antimicrobial activities of six essential oils: susceptibility of selected foodborne bacterial and fungal strains. J Agric Food Chem. 2005;53(17):6939–46. https://doi.org/10.1021/jf050709v.
15. Pires RH, Montanari LB, Martins CH, Zaia JE, Almeida AM, Matsumoto MT, et al. Anticandidal efficacy of cinnamon oil against planktonic and biofilm cultures of *Candida parapsilosis* and *Candida orthopsilosis*. Mycopathologia. 2011;172(6):453–64. https://doi.org/10.1007/s11046-011-9448-0.
16. Khan MS, Ahmad I, Cameotra SS. Phenyl aldehyde and propanoids exert multiple sites of action towards cell membrane and cell wall targeting ergosterol in *Candida albicans*. AMB Express. 2013;3(1):54. https://doi.org/10.1186/2191-0855-3-54.
17. Musthafa KS, Hmoteh J, Thamjarungwong B, Voravuthikunchai SP. Antifungal potential of eugenyl acetate against clinical isolates of Candida species. Microb Pathog. 2016;99:19–29. https://doi.org/10.1016/j.micpath.2016.07.012.
18. Shreaz S, Bhatia R, Khan N, Muralidhar S, Manzoor N, Khan LA. Influences of cinnamic aldehydes on H(+) extrusion activity and ultrastructure of Candida. J Med Microbiol. 2013;62(Pt 2):232–40. https://doi.org/10.1099/jmm.0.036145-0.
19. Taguchi Y, Hasumi Y, Abe S, Nishiyama Y. The effect of cinnamaldehyde on the growth and the morphology of *Candida albicans*. Med Mol Morphol. 2013;46(1):8–13. https://doi.org/10.1007/s00795-012-0001-0.
20. Zore GB, Thakre AD, Jadhav S, Karuppayil SM. Terpenoids inhibit *Candida albicans* growth by affecting membrane integrity and arrest of cell cycle. Phytomedicine. 2011;18(13):1181–90. https://doi.org/10.1016/j.phymed.2011.03.008.
21. Wang GS, Deng JH, Ma YH, Shi M, Li B. Mechanisms, clinically curative effects, and antifungal activities of cinnamon oil and pogostemon oil complex against three species of Candida. J Tradit Chin Med. 2012;32(1):19–24.
22. Bang KH, Lee DW, Park HM, Rhee YH. Inhibition of fungal cell wall synthesizing enzymes by trans-cinnamaldehyde. Biosci Biotechnol Biochem. 2000;64(5):1061–3.
23. Watanabe Y. A one-sided view of kinetochore attachment in meiosis. Cell. 2006;126(6):1030–2. https://doi.org/10.1016/j.cell.2006.09.005.
24. Alsteens D, Beaussart A, El-Kirat-Chatel S, Sullan RM, Dufrene YF. Atomic force microscopy: a new look at pathogens. PLoS Pathog. 2013;9(9):e1003516. https://doi.org/10.1371/journal.ppat.1003516.
25. El-Kirat-Chatel S, Beaussart A, Alsteens D, Jackson DN, Lipke PN, Dufrene YF. Nanoscale analysis of caspofungin-induced cell surface remodelling in *Candida albicans*. Nanoscale. 2013;5(3):1105–15. https://doi.org/10.1039/c2nr33215a.
26. El-Kirat-Chatel S, Beaussart A, Alsteens D, Sarazin A, Jouault T, Dufrene YF. Single-molecule analysis of the major glycopolymers of pathogenic and non-pathogenic yeast cells. Nanoscale. 2013;5(11):4855–63. https://doi.org/10.1039/c3nr00813d.
27. Formosa C, Schiavone M, Martin-Yken H, Francois JM, Duval RE, Dague E. Nanoscale effects of caspofungin against two yeast species, *Saccharomyces cerevisiae* and *Candida albicans*. Antimicrob Agents Chemother. 2013;57(8):3498–506. https://doi.org/10.1128/AAC.00105-13.
28. Formosa C, Schiavone M, Boisrame A, Richard ML, Duval RE, Dague E. Multiparametric imaging of adhesive nanodomains at the surface of *Candida albicans* by atomic force microscopy. Nanomedicine. 2015;11(1):57–65. https://doi.org/10.1016/j.nano.2014.07.008.

29. Francois JM, Formosa C, Schiavone M, Pillet F, Martin-Yken H, Dague E. Use of atomic force microscopy (AFM) to explore cell wall properties and response to stress in the yeast *Saccharomyces cerevisiae*. Curr Genet. 2013;59(4):187–96. https://doi.org/10.1007/s00294-013-0411-0.
30. Canetta E, Walker GM, Adya AK. Nanoscopic morphological changes in yeast cell surfaces caused by oxidative stress: an atomic force microscopic study. J Microbiol Biotechnol. 2009;19(6):547–55.
31. Alsteens D, Aimanianda V, Hegde P, Pire S, Beau R, Bayry J, et al. Unraveling the nanoscale surface properties of chitin synthase mutants of *Aspergillus fumigatus* and their biological implications. Biophys J. 2013;105(2):320–7. https://doi.org/10.1016/j.bpj.2013.05.040.
32. Adams RP. Identification of essential oil components by gas chromatography/mass spectorscopy. Miami: Allured Publishing Corporation; 2007.
33. Golmohammad F, Eikani MH, Maymandi HM. Cinnamon bark volatile oils separation and determination using solid-phase extraction and gas chromatography. Procedia Eng. 2012;42:247–60.
34. Joulain D, König WA. The atlas of spectral data of sesquiterpene hydrocarbons. Hamburg: Verlag, E.B; 1998.
35. Sherwood RK, Bennett RJ. Microtubule motor protein Kar3 is required for normal mitotic division and morphogenesis in *Candida albicans*. Eukaryot Cell. 2008;7(9):1460–74. https://doi.org/10.1128/EC.00138-08.
36. Endow SA, Kang SJ, Satterwhite LL, Rose MD, Skeen VP, Salmon ED. Yeast Kar3 is a minus-end microtubule motor protein that destabilizes microtubules preferentially at the minus ends. EMBO J. 1994;13(11):2708–13.
37. Roemer T, Jiang B, Davison J, Ketela T, Veillette K, Breton A, et al. Large-scale essential gene identification in *Candida albicans* and applications to antifungal drug discovery. Mol Microbiol. 2003;50(1):167–81.
38. CaLS Institute. Performance standards for antimicrobial susceptibility testing: twenty-fourth informational supplement M100-S24. Wayne, PA: CLSI; 2014.
39. Rajput SB, Karuppayil SM. Small molecules inhibit growth, viability and ergosterol biosynthesis in *Candida albicans*. Springerplus. 2013;2(1):26. https://doi.org/10.1186/2193-1801-2-26.
40. Hsu CC, Lai WL, Chuang KC, Lee MH, Tsai YC. The inhibitory activity of linalool against the filamentous growth and biofilm formation in *Candida albicans*. Med Mycol. 2013;51(5):473–82. https://doi.org/10.3109/13693786.2012.743051.
41. Dhamgaye S, Devaux F, Vandeputte P, Khandelwal NK, Sanglard D, Mukhopadhyay G, et al. Molecular mechanisms of action of herbal antifungal alkaloid berberine, in *Candida albicans*. PLoS ONE. 2014;9(8):e104554. https://doi.org/10.1371/journal.pone.0104554.
42. Kumar A, Dhamgaye S, Maurya IK, Singh A, Sharma M, Prasad R. Curcumin targets cell wall integrity via calcineurin-mediated signaling in *Candida albicans*. Antimicrob Agents Chemother. 2014;58(1):167–75. https://doi.org/10.1128/AAC.01385-13.
43. Hammer KA, Carson CF, Riley TV. Antifungal effects of *Melaleuca alternifolia* (tea tree) oil and its components on *Candida albicans*, *Candida glabrata* and *Saccharomyces cerevisiae*. J Antimicrob Chemother. 2004;53(6):1081–5. https://doi.org/10.1093/jac/dkh243.
44. Li L, Aslam M, Rabbi F, Vanderwel MC, Ashton NW, Suh DY. PpORS, an ancient type III polyketide synthase, is required for integrity of leaf cuticle and resistance to dehydration in the moss, *Physcomitrella patens*. Planta. 2017. https://doi.org/10.1007/s00425-017-2806-5.
45. Kwolek-Mirek M, Zadrag-Tecza R. Comparison of methods used for assessing the viability and vitality of yeast cells. FEMS Yeast Res. 2014;14(7):1068–79. https://doi.org/10.1111/1567-1364.12202.
46. Pinto E, Hrimpeng K, Lopes G, Vaz S, Goncalves MJ, Cavaleiro C, et al. Antifungal activity of *Ferulago capillaris* essential oil against Candida, Cryptococcus, Aspergillus and dermatophyte species. Eur J Clin Microbiol Infect Dis. 2013;32(10):1311–20. https://doi.org/10.1007/s10096-013-1881-1.
47. Bhat SV, Booth SC, McGrath SG, Dahms TE. *Rhizobium leguminosarum* bv. viciae 3841 Adapts to 2,4-dichlorophenoxyacetic acid with "auxin-like" morphological changes, cell envelope remodeling and upregulation of central metabolic pathways. PLoS ONE. 2014;10(4):e0123813. https://doi.org/10.1371/journal.pone.0123813.
48. Bhat SV, Booth SC, Vantomme EA, Afroj S, Yost CK, Dahms TE. Oxidative stress and metabolic perturbations in *Escherichia coli* exposed to sublethal levels of 2,4-dichlorophenoxyacetic acid. Chemosphere. 2015;135:453–61. https://doi.org/10.1016/j.chemosphere.2014.12.035.
49. Klis FM, de Koster CG, Brul S. Cell wall-related bionumbers and bioestimates of *Saccharomyces cerevisiae* and *Candida albicans*. Eukaryot Cell. 2014;13(1):2–9. https://doi.org/10.1128/EC.00250-13.
50. Bapat P, Nandy SK, Wangikar P, Venkatesh KV. Quantification of metabolically active biomass using methylene blue dye reduction test (MBRT): measurement of CFU in about 200 s. J Microbiol Methods. 2006;65(1):107–16. https://doi.org/10.1016/j.mimet.2005.06.010.
51. Painting K, Kirsop B. A quick method for estimating the percentage of viable cells in a yeast population, using methylene blue staining. World J Microbiol Biotechnol. 1990;6(3):346–7. https://doi.org/10.1007/bf01201311.
52. Liu Y, Solis NV, Heilmann CJ, Phan QT, Mitchell AP, Klis FM, et al. Role of retrograde trafficking in stress response, host cell interactions, and virulence of *Candida albicans*. Eukaryot Cell. 2014;13(2):279–87. https://doi.org/10.1128/EC.00295-13.
53. Aguilar-Uscanga B, Francois JM. A study of the yeast cell wall composition and structure in response to growth conditions and mode of cultivation. Lett Appl Microbiol. 2003;37(3):268–74.
54. Chaffin WL. *Candida albicans* cell wall proteins. Microbiol Mol Biol Rev. 2008;72(3):495–544. https://doi.org/10.1128/MMBR.00032-07.
55. Free SJ. Fungal cell wall organization and biosynthesis. Adv Genet. 2013;81:33–82. https://doi.org/10.1016/B978-0-12-407677-8.00002-6.
56. Fuchs BB, Mylonakis E. Our paths might cross: the role of the fungal cell wall integrity pathway in stress response and cross talk with other stress response pathways. Eukaryot Cell. 2009;8(11):1616–25. https://doi.org/10.1128/EC.00193-09.
57. Heilmann CJ, Sorgo AG, Mohammadi S, Sosinska GJ, de Koster CG, Brul S, et al. Surface stress induces a conserved cell wall stress response in the pathogenic fungus *Candida albicans*. Eukaryot Cell. 2013;12(2):254–64. https://doi.org/10.1128/EC.00278-12.
58. Walker LA, Maccallum DM, Bertram G, Gow NA, Odds FC, Brown AJ. Genome-wide analysis of *Candida albicans* gene expression patterns during infection of the mammalian kidney. Fungal Genet Biol. 2009;46(2):210–9. https://doi.org/10.1016/j.fgb.2008.10.012.
59. Ene IV, Adya AK, Wehmeier S, Brand AC, MacCallum DM, Gow NA, et al. Host carbon sources modulate cell wall architecture, drug resistance and virulence in a fungal pathogen. Cell Microbiol. 2012;14(9):1319–35. https://doi.org/10.1111/j.1462-5822.2012.01813.x.
60. Imtiaz T, Lee KK, Munro CA, Maccallum DM, Shankland GS, Johnson EM, et al. Echinocandin resistance due to simultaneous FKS mutation and increased cell wall chitin in a *Candida albicans* bloodstream isolate following brief exposure to caspofungin. J Med Microbiol. 2012;61(Pt 9):1330–4. https://doi.org/10.1099/jmm.0.045047-0.
61. Denning DW. Echinocandin antifungal drugs. Lancet. 2003;362(9390):1142–51. https://doi.org/10.1016/S0140-6736(03)14472-8.
62. Walker LA, Munro CA, de Bruijn I, Lenardon MD, McKinnon A, Gow NA. Stimulation of chitin synthesis rescues *Candida albicans* from echinocandins. PLoS Pathog. 2008;4(4):e1000040. https://doi.org/10.1371/journal.ppat.1000040.
63. Lee KK, Maccallum DM, Jacobsen MD, Walker LA, Odds FC, Gow NA, et al. Elevated cell wall chitin in *Candida albicans* confers echinocandin resistance in vivo. Antimicrob Agents Chemother. 2012;56(1):208–17. https://doi.org/10.1128/AAC.00683-11.
64. Cota JM, Grabinski JL, Talbert RL, Burgess DS, Rogers PD, Edlind TD, et al. Increases in SLT2 expression and chitin content are associated with incomplete killing of *Candida glabrata* by caspofungin. Antimicrob Agents Chemother. 2008;52(3):1144–6. https://doi.org/10.1128/AAC.01542-07.
65. Fortwendel JR, Juvvadi PR, Pinchai N, Perfect BZ, Alspaugh JA, Perfect JR, et al. Differential effects of inhibiting chitin and 1,3-{beta}-D-glucan synthesis in ras and calcineurin mutants of *Aspergillus fumigatus*. Antimicrob Agents Chemother. 2009;53(2):476–82. https://doi.org/10.1128/AAC.01154-08.
66. Pfaller M, Riley J. Effects of fluconazole on the sterol and carbohydrate composition of four species of Candida. Eur J Clin Microbiol Infect Dis. 1992;11(2):152–6.
67. Popolo L, Gualtieri T, Ragni E. The yeast cell-wall salvage pathway. Med Mycol. 2001;39(Suppl 1):111–21.

68. Walker LA, Gow NA, Munro CA. Elevated chitin content reduces the susceptibility of Candida species to caspofungin. Antimicrob Agents Chemother. 2013;57(1):146–54. https://doi.org/10.1128/AAC.01486-12.
69. El-Kirat-Chatel S, Mil-Homens D, Beaussart A, Fialho AM, Dufrene YF. Single-molecule atomic force microscopy unravels the binding mechanism of a *Burkholderia cenocepacia* trimeric autotransporter adhesin. Mol Microbiol. 2013;89(4):649–59. https://doi.org/10.1111/mmi.12301.
70. Lesage G, Bussey H. Cell wall assembly in *Saccharomyces cerevisiae*. Microbiol Mol Biol Rev. 2006;70(2):317–43. https://doi.org/10.1128/MMBR.00038-05.
71. Paul BC, El-Ganiny AM, Abbas M, Kaminskyj SG, Dahms TE. Quantifying the importance of galactofuranose in *Aspergillus nidulans* hyphal wall surface organization by atomic force microscopy. Eukaryot Cell. 2011;10(5):646–53. https://doi.org/10.1128/EC.00304-10.
72. Hoyer LL. The ALS gene family of *Candida albicans*. Trends Microbiol. 2001;9:176–80.
73. Lipke PN, Garcia MC, Alsteens D, Ramsook CB, Klotz SA, Dufrene YF. Strengthening relationships: amyloids create adhesion nanodomains in yeasts. Trends Microbiol. 2012;20(2):59–65. https://doi.org/10.1016/j.tim.2011.10.002.
74. Glee PM, Cutler JE, Benson EE, Bargatze RF, Hazen KC. Inhibition of hydrophobic protein-mediated *Candida albicans* attachment to endothelial cells during physiologic shear flow. Infect Immun. 2001;69(5):2815–20. https://doi.org/10.1128/IAI.69.5.2815-2820.2001.
75. Ruiz-Herrera J, Elorza MV, Valentin E, Sentandreu R. Molecular organization of the cell wall of *Candida albicans* and its relation to pathogenicity. FEMS Yeast Res. 2006;6(1):14–29. https://doi.org/10.1111/j.1567-1364.2005.00017.x.
76. Lopez-Ribot JL, Casanova M, Martinez JP, Sentandreu R. Characterization of cell wall proteins of yeast and hydrophobic mycelial cells of *Candida albicans*. Infect Immun. 1991;59(7):2324–32.
77. Roncero C, Duran A. Effect of Calcofluor white and Congo red on fungal cell wall morphogenesis: in vivo activation of chitin polymerization. J Bacteriol. 1985;163(3):1180–5.
78. Kopecka M, Gabriel M. The influence of congo red on the cell wall and (1-3)-beta-D-glucan microfibril biogenesis in *Saccharomyces cerevisiae*. Arch Microbiol. 1992;158(2):115–26.
79. Sudbery PE. Growth of *Candida albicans* hyphae. Nat Rev Micro. 2011;9(10):737–48. http://www.nature.com/nrmicro/journal/v9/n10/suppinfo/nrmicro2636_S1.html.
80. Wightman R, Bates S, Amornrrattanapan P, Sudbery P. In *Candida albicans*, the Nim1 kinases Gin4 and Hsl1 negatively regulate pseudohypha formation and Gin4 also controls septin organization. J Cell Biol. 2004;164(4):581–91. https://doi.org/10.1083/jcb.200307176.
81. Bachewich C, Thomas DY, Whiteway M. Depletion of a polo-like kinase in *Candida albicans* activates cyclase-dependent hyphal-like growth. Mol Biol Cell. 2003;14(5):2163–80. https://doi.org/10.1091/mbc.02-05-0076.
82. Jacobs CW, Adams AE, Szaniszlo PJ, Pringle JR. Functions of microtubules in the *Saccharomyces cerevisiae* cell cycle. J Cell Biol. 1988;107(4):1409–26.
83. Mardon D, Balish E, Phillips AW. Control of dimorphism in a biochemical variant of *Candida albicans*. J Bacteriol. 1969;100:701–7.
84. Huffaker TC, Thomas JH, Botstein D. Diverse effects of beta-tubulin mutations on microtubule formation and function. J Cell Biol. 1988;106(6):1997–2010.
85. Xu D, Jiang B, Ketela T, Lemieux S, Veillette K, Martel N, et al. Genome-wide fitness test and mechanism-of-action studies of inhibitory compounds in *Candida albicans*. PLoS Pathog. 2007;3(6):e92. https://doi.org/10.1371/journal.ppat.0030092.
86. Finley KR, Berman J. Microtubules in *Candida albicans* hyphae drive nuclear dynamics and connect cell cycle progression to morphogenesis. Eukaryot Cell. 2005;4(10):1697–711. https://doi.org/10.1128/ec.4.10.1697-1711.2005.
87. Berman J. Morphogenesis and cell cycle progression in *Candida albicans*. Curr Opin Microbiol. 2006;9(6):595–601. https://doi.org/10.1016/j.mib.2006.10.007.
88. Correia I, Alonso-Monge R, Pla J. MAPK cell-cycle regulation in *Saccharomyces cerevisiae* and *Candida albicans*. Future Microbiol. 2010;5(7):1125–41. https://doi.org/10.2217/fmb.10.72.
89. Correia IS. Editorial. Response to drugs and other chemical stresses in yeasts. OMICS. 2010;14(6):615–7. https://doi.org/10.1089/omi.2010.00ed.
90. Brennan TC, Turner CD, Kromer JO, Nielsen LK. Alleviating monoterpene toxicity using a two-phase extractive fermentation for the bioproduction of jet fuel mixtures in *Saccharomyces cerevisiae*. Biotechnol Bioeng. 2012;109(10):2513–22. https://doi.org/10.1002/bit.24536.
91. Miron D, Battisti F, Silva FK, Lana AD, Pippi B, Casanova B, et al. Antifungal activity and mechanism of action of monoterpenes against dermatophytes and yeasts. Revista Brasileira de Farmacognosia. 2014;24(6):660–7. https://doi.org/10.1016/j.bjp.2014.10.014.
92. Pina-Vaz C, Goncalves Rodrigues A, Pinto E, Costa-de-Oliveira S, Tavares C, Salgueiro L, et al. Antifungal activity of thymus oils and their major compounds. J Eur Acad Dermatol Venereol JEADV. 2004;18(1):73–8.
93. Hwang E, Kusch J, Barral Y, Huffaker TC. Spindle orientation in *Saccharomyces cerevisiae* depends on the transport of microtubule ends along polarized actin cables. J Cell Biol. 2003;161(3):483–8. https://doi.org/10.1083/jcb.200302030.

# 6

# Vegan-mycoprotein concentrate from pea-processing industry byproduct using edible filamentous fungi

Pedro F. Souza Filho[1*], Ramkumar B. Nair[2], Dan Andersson[3], Patrik R. Lennartsson[1] and Mohammad J. Taherzadeh[1]

**Abstract**

**Background:** Currently around one billion people in the world do not have access to a diet which provides enough protein and energy. However, the production of one of the main sources of protein, animal meat, causes severe impacts on the environment. The present study investigates the production of a vegan-mycoprotein concentrate from pea-industry byproduct (PpB), using edible filamentous fungi, with potential application in human nutrition. Edible fungal strains of Ascomycota (*Aspergillus oryzae, Fusarium venenatum, Monascus purpureus, Neurospora intermedia*) and Zygomycota (*Rhizopus oryzae*) phyla were screened and selected for their protein production yield.

**Results:** *A. oryzae* had the best performance among the tested fungi, with a protein yield of 0.26 g per g of pea-processing byproduct from the bench scale airlift bioreactor cultivation. It is estimated that by integrating the novel fungal process at an existing pea-processing industry, about 680 kg of fungal biomass attributing to about 38% of extra protein could be produced for each 1 metric ton of pea-processing byproduct. This study is the first of its kind to demonstrate the potential of the pea-processing byproduct to be used by filamentous fungi to produce vegan-mycoprotein for human food applications.

**Conclusion:** The pea-processing byproduct (PpB) was proved to be an efficient medium for the growth of filamentous fungi to produce a vegan-protein concentrate. Moreover, an industrial scenario for the production of vegan-mycoprotein concentrate for human nutrition is proposed as an integrated process to the existing PPI production facilities.

**Keywords:** Pea-processing byproduct, Edible filamentous fungi, Vegan-mycoprotein concentrate, Meat substitute

## Background

Attributed to the rise in population, urbanization, and income, a steady growth in the consumption of protein from animal sources has been observed in developed countries over the last few decades [1]. Nevertheless, currently around one billion people in the world do not have access to a diet which provides enough protein and energy [2]. Lack of protein can result in severe health problems such as growth failure, muscle weakness and an impaired immune system. Protein-energy malnutrition (PEM) can lead to conditions such as kwashiorkor and marasmus. Additionally, the production of meat has a heavy impact on the environment and makes a large contribution to the eutrophication process [3, 4]. In this context, it is important to find an alternative, cheap, and less resource-consuming protein source to substitute meat or meat products. Fungal organisms such as mushrooms and truffles have traditionally been part of human nutrition largely because of their flavor; however they cannot be considered as an important source of protein in comparison to meat-based sources [1, 5]. Considerable attention has been recently given to the use of filamentous fungi as a commercial human food component; especially due to their high protein content with all the

*Correspondence: pedro.ferreira_de_souza_filho@hb.se
[1] Swedish Centre for Resource Recovery, University of Borås, 50190 Borås, Sweden
Full list of author information is available at the end of the article

essential amino acids to human nutrition, easy digestibility, low-fat content (cholesterol free), and the presence of dietary fibers [6]. The fibre content (6% w/w) is also comparable with other vegetarian protein sources [7].

Several strains of edible filamentous fungi have been recognized as a traditional source of palatable food by many societies around the globe, especially in Asia [8]. *Rhizopus* sp. has been used for centuries in the oriental cuisine in the preparation of fermented food such as tempeh [9]. *Aspergillus oryzae* also has culinary applications for the production of hamanatto, miso and shoyu. *Neurospora intermedia* is used in the preparation of the Indonesian staple food oncom [6]. Similarly, *Monascus purpureus* has been used as coloring and flavoring agent in food and beverages, as in the production of red yeast rice and rice wine [10–12]. Other applications of filamentous fungi include the production of several ingredients for the food and beverage industries, especially enzymes. In recent years, production of vitamins and polyunsaturated fatty acids by these microorganisms has been receiving increased attention [13]. Single-cell protein (SCP) can also be produced by filamentous fungi. An example currently in the market is the filamentous fungus *Fusarium venenatum*, commercialized under the name Quorn™. The fungus is cultivated in a synthetic medium with glucose, ammonium and supplemented with biotin. The costs associated with the substrate and the lack of competition results in a market price for the SCP that is higher than that of meat. Despite the high price, the fungal SCP has found its place in the market as a healthy substitute to meat, with its presence only in developed markets such as Europe and USA [6, 14]. Nevertheless, the SCP from the mycelium of filamentous fungi can be inexpensively produced when using cheap materials as substrates [15]. One such example is the pea processing industry byproduct that is being used in the present study.

Pea (*Pisum sativum*) is the second most important leguminous crop in the world with an annual production above 17 million metric tons, finding its applications mainly as food and feed. Originally from western Asia and northern Africa, its production has spread to over 10 million hectares of farmlands, especially in Russia, China, Canada, Europe, Australia and the United States. Rich in protein, carbohydrate, dietary fiber, vitamins, and minerals, the peas are used to produce food ingredients such as proteins, starches, flours, and fibers [16–18]. Pea proteins have faced a growth in food applications due to their nutritional and functional benefits, including their balanced amino acid profile, positive fat- and water-binding capabilities, emulsification and gelation properties, texture, and nutritional values. Moreover, allergies to pea are less frequent than allergies to other protein-rich grains, like soy [19]. Pea proteins have also been demonstrated as a useful ingredient in the formulation of antihypertensive foods because of their antihypertensive effects [16, 19, 20].

Pea proteins are commercialized in three forms: pea flour, pea protein concentrate, and pea-protein isolate (PPI). The manufacturing of pea flour consists of the dry milling of hulled peas, whereas pea protein concentrate is obtained by dry separation techniques. PPI production generally occurs by isoelectric precipitation at pH around 4.5, followed by a membrane separation technique to increase the protein concentration, such as ultrafiltration and diafiltration. PPI can be used in the preparation of dairy-based beverages, sports and nutritional foods, and other non-dairy sports products, such as vegan style yogurts. Additionally, it can partially replace dairy protein in therapeutic beverages and powders [19, 20]. Despite the high quality of the protein, the pea-processing byproduct (PpB) is considered to have poor functional properties. Therefore, its uses in food applications are limited and it is mainly produced as a byproduct of the protein extraction process [21]. A novel and alternative approach to valorize this pea-processing byproduct (PpB), as discussed in the present study is to convert it into a vegan-mycoprotein concentrate for human food applications, using edible strains of filamentous fungi. Intake of mycoprotein may be beneficial to human health [7, 22–24]. Several studies have investigated the cholesterol-lowering effects of mycoprotein; the results from these studies point to the same direction with reductions in both total and LDL-cholesterol [7, 22]. The greatest benefits are seen in individuals with a higher cholesterol level at baseline and in hypercholesterolaemic subjects. There is a difference between the macronutrients regarding satiety; protein is generally recognised as most satiating, followed by carbohydrates and fat [25]. Compared to other protein sources such as chicken, mycoprotein seems to be more satiating and hence have the possibility to decrease energy intake in subsequent meals [7, 23, 24]. It is possible that fibres in mycoprotein, one-third chitin and two-thirds beta-glucan, might have a specific effect on satiety [7]. Additionally mycoprotein appears to affect the glycaemic response positively [7, 24]. The exact mechanism that explains this is not known, but might be associated with its fibre content [7].

The aim of the present study was hence to convert PpB, a cheap and low-nutritional value byproduct of the PPI production, into a vegan-mycoprotein concentrate for human food applications. Five strains of filamentous fungi, namely *A. oryzae*, *F. venenatum*, *M. purpureus*, *N. intermedia* and *R. oryzae*, were screened for their growth to maximize the protein yield from the PpB. The best cases of the fungal growth were selected and further scaled-up in a bench airlift bioreactor, considering the industrial application potential of the process.

## Methods

### Substrate and enzymes

Pea-processing byproduct (PpB) used for this study was kindly provided by Protein Consulting AB (Sweden). The powder was sieved through a pore size of 0.2–0.25 mm and was characterized using triplicate samples for its carbohydrate, protein, ash, and moisture content. Cellulase cocktail Cellic Ctec2 (Novozymes, Denmark) with 94 FPU/mL activity at 35 °C, amyloglucosidase from *Aspergillus niger* (300 U/mL activity at 35 °C), and α-amylase from *Aspergillus oryzae* (100 U/mg activity at 35 °C) were supplied by Sigma-Aldrich Co. (Germany).

### Microorganisms

Edible food-grade filamentous fungi were used in the present study. The Ascomycota strains were *Neurospora intermedia* CBS 131.92 (Centraalbureau voor Schimmelcultures, Netherlands), *Aspergillus oryzae* var. *oryzae* CBS 819.72, *Monascus purpureus* CBS 109.07, and *Fusarium venenatum* ATCC 20334 (American Type Culture Collection, USA), and the Zygomycota strain was *Rhizopus oryzae* CCUG 61147 (Culture Collection University of Gothenburg, Sweden). All the fungal cultures were maintained on potato dextrose agar (PDA) slants containing (in g/L) potato extract 4; glucose 20; agar 15 and were renewed every 6 months. New PDA plates were prepared via incubation for 3–5 days at 30 °C followed by storage at 4 °C. For preparing spore solution, PDA plates (72 h grown) were flooded with 20 mL sterile distilled water and the spores were released by gently agitating the mycelium with a disposable cell spreader. An inoculum of 3 mL spore suspension (with a spore concentration of $3.9 \times 10^5$–$3.8 \times 10^6$ spores/mL) per liter of the medium was used for the cultivations, unless otherwise specified. For preparing fungal biomass inoculum, the spores were inoculated into 100 mL YPD broth containing (in g/L) glucose 20, peptone 20, and yeast extract 10. The culture was incubated aerobically for 48 h at 35 °C and 125 rpm. The fungal biomass was harvested at the end of the cultivation and used as the inoculum. Dry weight was determined by drying at 105 °C overnight.

### Experimental set-up for fungal cultivation

The cultivations in 250 mL Erlenmeyer flasks were carried out using 100 mL of culture medium consisting only of the PpB substrate dissolved in distilled water. The optimum concentration of the medium was determined by testing the maximum load of PpB substrate which could go through the sterilization operation without causing retrogradation. The term retrogradation refers to the changes in the gelatinized starch during cooling, when the molecules recrystallize. This process leads to the production of a starch with high resistance to the enzymatic attack, thus reducing the effectiveness of the microbial metabolism [26–28]. The Erlenmeyer flasks were kept in a water bath shaker at 35 °C and 150 rpm (with a 9 mm orbital shaking radius). The pH of the sample was adjusted to $5.5 \pm 0.1$ prior to autoclaving by adding HCl 1 M. The enzymes were added according to the substrate at a load of 150 U/g for α-amylase, 163 U/g for amyloglucosidase, and 24 FPU/g for cellulase. At the end of the cultivations, the produced fungal mycelium was collected using a sieve, washed with ultra-pure water, dried at 70 °C and had the weight and protein content analysed. Samples were taken during cultivation to follow the consumption of sugars and the production of metabolites. All the cultivation experiments were conducted in duplicate and the mean values are presented with standard deviations.

### Scaling-up of the fungal cultivation in a bench scale airlift reactor

A 4.5-L airlift bioreactor (Belach Bioteknik, Sweden) with a working volume of 3.5 L was used to scale-up the fungal cultivation process. The entire bioreactor and the draft tube were made of transparent borosilicate glass. An internal loop with cylindrical geometry with 58 mm of diameter, 400 mm of height and 3.2 mm in thickness was used to achieve the airlift-liquid circulation. Aeration at the rate of 0.42 v.v.m. ($volume_{air}/volume_{media}$/min) was maintained throughout the cultivation, using a sintered stainless steel air-sparger with a pore size of 90 μm. Filtration of the inlet air was achieved by passing it through a polytetrafluoroethylene (PTFE) membrane filter (0.1 μm pore size, Whatman, Florham Park, NJ, USA). The cultivation was carried out at the natural pH of PpB media, 6.1 (for 2% substrate) and 6.5 (for 3% substrate) without any adjustments during the cultivation. An enzyme loading of 5 FPU of commercial cellulase enzyme complex (Cellic Ctec 2) per gram substrate was added (filter sterilized) during the start of the cultivation (time 0). The fermentation was carried out at $35 \pm 2$ °C for 48 h, with sample collection at every 12 h.

### Analyses

The fungal spore concentration was measured using a Bürker counting chamber. The total sugar, total solid, suspended solid and volatile solids of the samples were measured according to the National Renewable Energy Laboratory (NREL) methods [29, 30]. The total nitrogen content in the samples was determined by Kjeldahl method applying digestion, distillation, and acid–base titration using the InKjel P digestor and the Behrotest

S1 distiller (Behr Labor-Technik, Germany). Protein was estimated by multiplying the nitrogen content by the nitrogen-to-protein conversion factor of 6.25 [31]. HPLC (Waters 2695, Waters Corporation, Milford, U.S.A.) was used to analyze the components in all liquid fractions. Acetic acid, ethanol, glucose, glycerol, lactic acid, and xylitol were analyzed using an analytical ion exchange column based on hydrogen ions (Aminex HPX-87H, Bio-Rad, USA) operated at 60 °C with 0.6 mL/min of 5 mM $H_2SO_4$ as eluent. Arabinose, galactose, glucose, mannose, and xylose were analyzed using a lead (II)-based column (HPX-87P, Bio-Rad) with two Micro-Guard Deashing (Bio-Rad) precolumns operated at 85 °C with 0.6 mL/min ultrapure water as eluent. A UV absorbance detector (Waters 2487), operating at 210 nm wavelength, was used in series with a refractive index (RI) detector (Waters 2414). Fructose and sucrose in the liquid samples and starch in the PpB media were measured using assay kits of Megazyme (Ireland). The total chemical oxygen demand (COD) was determined using NANOCOLOR® COD 15000 kit, with the photometric determination of the samples using NANOCOLOR® photometers. The viscosities of the samples were determined using a Brookfield digital viscometer-model DV-E (Chemical Instruments AB, Sweden).

#### Statistical analysis

Statistical analysis of the collected data was performed using the software MINITAB® (version 17.1.0). Analyses of variance (ANOVA) of the data used general linear models and a confidence level of 95% was used for all analysis. In the graphs, the average values are presented with an error bar representing one standard deviation. All the results presented in the tables are the average values from duplicate experimental sets and are reported with intervals representing the standard deviation.

### Results and discussion

With an abundance of food and a sedentary lifestyle in many parts of the world there is a need of food that is both nutritious and filling. Consumption of mycoprotein has been shown to improve health in relation to blood cholesterol concentrations, energy intake and glycemic response. Furthermore, it can contribute to satiety and thus decreased energy intake in subsequent meals, something that can increase weight-control [7, 23, 24]. Therefore, mycoprotein can potentially contribute to both prevention and treatment of lifestyle dependent conditions such as obesity and type 2-diabetes; however more research is needed to confirm this. The optimal dose of mycoprotein to boost health also remains unknown. In this study, pea-processing byproduct (PpB) with low nutritional value, obtained as a byproduct of the pea protein isolate process, was used as the substrate for the cultivation of edible ascomycetes and zygomycetes fungi. These fungi are known for their palatability and high protein content, with its potential application as human food components. The results on the characterization of the PpB and the fungal cultivation experiments together with the scale up studies in the airlift reactor are presented and discussed further.

#### Pea-processing byproduct (PpB) substrate characterization

The PpB was presented as a white odorless fine powder with the characterization as presented in Table 1. The PpB substrate was characterized to have a total glucan content of 62.38 ± 0.51% (w/w), more than 90% of this being starch. Protein is the second most common component in the material, amounting 18.19 ± 0.33% in dry weight basis. The C:N ratio of the PpB could be determined from the carbohydrate and protein contents as 10.28 ± 0.20. Retrogradation of the starch was observed to determine the best concentration of the substrate PpB in the medium (distilled water) which would not cause the change in the quality of the liquid, while being autoclaved. The PpB concentrations (dry weight) of 1, 2, 3, 4, and 5% (w/v) were tested. For the substrate loadings of 4 and 5%, gelification of the medium was observed. At 3% substrate loading, the gelification was not as clear as in the previous cases although it was still noticed. For 1 and 2% substrate loadings, there was no observed retrogradation of starch.

Additionally, the viscosity of the PpB suspension was determined. The rheological properties of the cultivation medium have effects on momentum, heat and mass transfer, influencing the fermentation performance. In reactors, the flow properties of the media cause changes in the coalescence of the air bubbles, the bulk mixing, the process control, and the formation of stagnant zones [32]. The 2% PpB suspension was determined to have a viscosity of 1.93 ± 0.15 cP before sterilization. After sterilization in the autoclave, however, the viscosity increased

**Table 1 Characterization of the pea-processing byproduct (PpB)**

| Component | Content (% w/w in dry basis) |
|---|---|
| Protein | 18.19 ± 0.33 |
| Ash | 2.98 ± 0.03 |
| Moisture | 10.54 ± 0.19 |
| Arabinans | 2.61 ± 0.06 |
| Xylans | 0.00 |
| Galactans | 2.30 ± 0.04 |
| Glucans | 62.38 ± 0.51 |
| Of which | |
| Starch | 56.34 ± 2.52 |

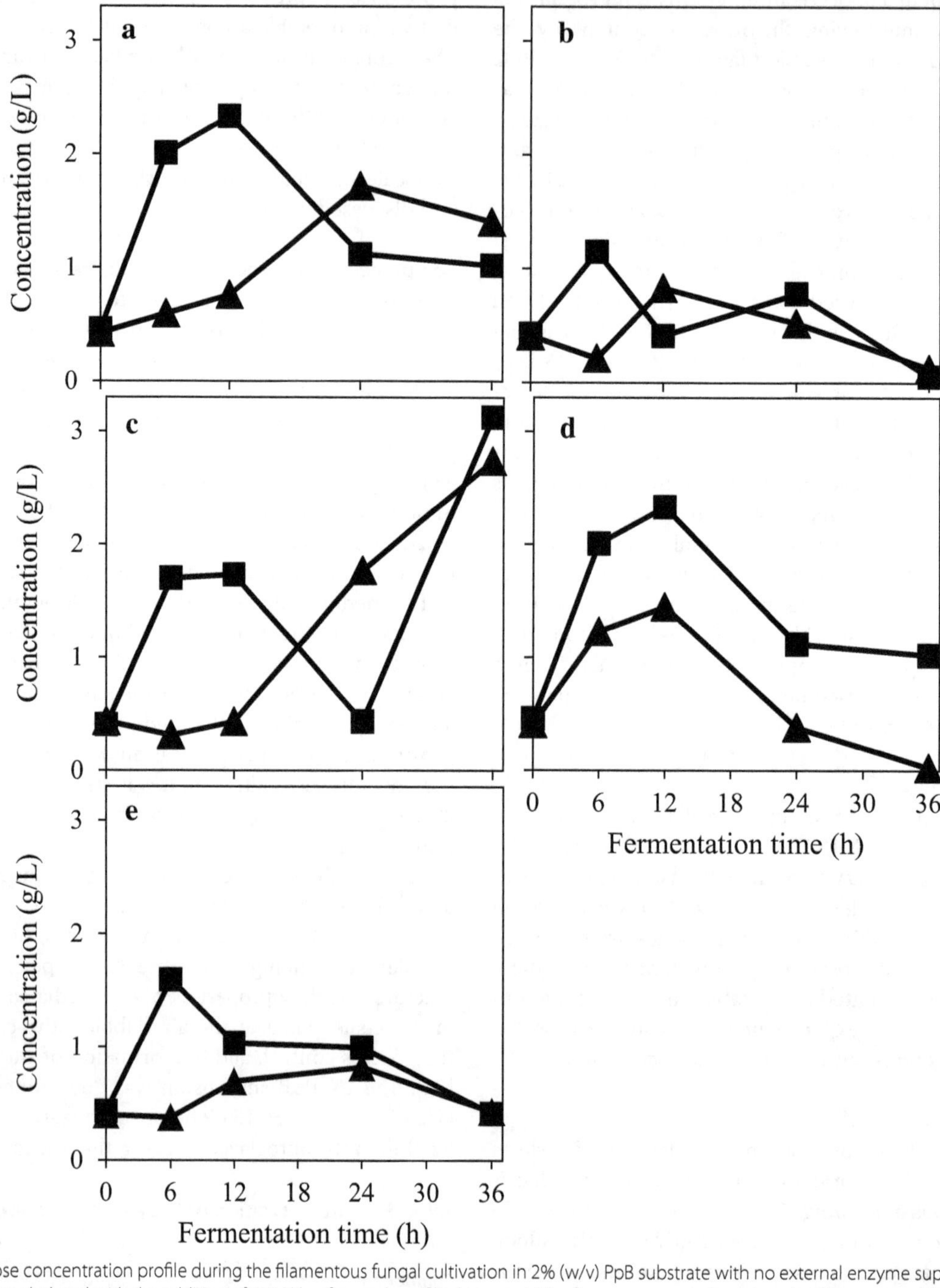


**Fig. 1** Glucose concentration profile during the filamentous fungal cultivation in 2% (w/v) PpB substrate with no external enzyme supplementation (filled triangles) and with the addition of 150 U/g of α-amylase (filled squares). The figure represents fungal strains *M. purpureus* (**a**); *A. oryzae* (**b**); *F. venenatum* (**c**); *N. intermedia* (**d**); and *R. oryzae* (**e**). Coordinates represent the mean values of duplicate tests; with error bars representing standard deviations omitted due to negligible values

to 10.47 ± 0.12 cP, which means the viscosity increased by 441%. The effect of the viscosity of the medium in the performance of bioreactors generally remains unclear. Some studies associate the increase in the viscosity with the reduction of the liquid turbulence, promoting bubble coalescence and the decrease of the gas holdup. On the other hand, other researchers have found an increased gas holdup in viscous media [33].

**PpB as substrate for efficient fungal growth**

The cultivation of four strains of ascomycetes (*N. intermedia*; *A. oryzae*; *M. purpureus*; and *F. venenatum*) and one of the zygomycetes fungi (*R. oryzae*) was examined in a suspension containing 2% (w/v) PpB substrate using 250 mL Erlenmeyer flasks. The tests were carried out with and without α-amylase addition and in duplicate samples. The effect of the cultivation temperature was not studied in this experimental set. Even without enzyme addition, the ascomycete strains showed a good capacity to hydrolyse starch to glucose (Fig. 1a). *M. purpureus* and *R. oryzae* consumption of glucose exceeded the sugar production rate after 24 h. On the other hand, the sugar consumption rates of *A. oryzae* and *N. intermedia* became higher than the sugar production rate after 12 h. Consumption of glucose by *F. venenatum* was always lower than glucose production. On the other hand, when α-amylase was added (Fig. 1b), *M. purpureus* consumed the glucose at a faster rate than its production after the first 12 h of cultivation. *N. intermedia* presented the same behaviour. *R. oryzae* showed an even higher glucose uptake, overtaking the glucose production rate after the first 6 h of cultivation. Glucose profile during *A. oryzae* and *F. venenatum* cultivation oscillated between increasing and decreasing the glucose concentration in the medium during the cultivation, likely because of different glucose production and glucose uptake rates.

*N. intermedia* produced the most ethanol among the microorganisms (4.30 g/L with enzyme addition) while *M. purpureus* cultivation resulted in the lowest final ethanol concentration (0.28 g/L without enzyme addition). *R. oryzae* also converted glucose into lactic acid, yielding 0.66 g/L. α-amylase addition to the media did not result in a significant change in the ethanol and lactic acid production for the ascomycetes strains. However, *R. oryzae* had its ethanol and lactic acid production increased by the addition of the enzyme. Inhibitory effect of these compounds were not considered since the microorganisms have been cultivated in media with concentrations higher than the ones obtained in this study [9, 34]. Biomass production after 36 h is shown in Fig. 2. *M. purpureus* produced significantly less biomass than the other strains ($p$ value $\leq 0.031$), except *F. venenatum* ($p=0.185$). *A. oryzae* produced more biomass than *R. oryzae* ($p=0.074$) and *N. intermedia* ($p=0.038$). The addition of α-amylase to the medium significantly affected only *A. oryzae* ($p=0.003$) and *M. purpureus* ($p=0.038$) growth. When comparing the protein content of the harvested biomass, cultivation of *A. oryzae* resulted in the highest yield of protein per gram of PpB substrate, 0.14 g/g (Table 2).

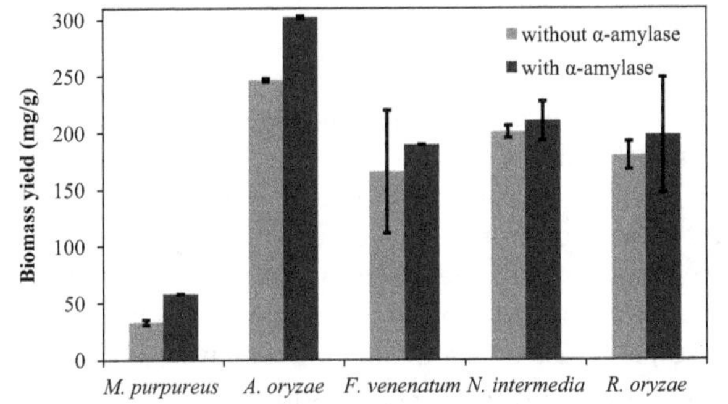


**Fig. 2** Biomass yield; mg dry fungal biomass per gram of PpB substrate after 36 h of cultivation in 2% w/v PpB medium

From the previous results, three strains were being considered the most promising ones. *A. oryzae*, *N. intermedia*, and *R. oryzae* were cultivated in the same medium as before (2% PpB substrate), with the addition of the amyloglucosidase enzyme and the cellulase cocktail to test if a higher biomass yield would be obtained. As observed in Fig. 3, *N. intermedia* was most efficient in consuming the glucose, with its concentration reaching zero in about 18 h. Maximum ethanol concentration reaching 5.98 g/L, was also observed with *N. intermedia*. The fungal protein yield obtained for *A. oryzae* and *R. oryzae* was 0.09 g/g of PpB substrate while for *N. intermedia* it was 0.10 g/g.

**Cultivation of *A. oryzae* in the airlift bioreactor**

*A. oryzae* growth was examined in a bench-scale airlift bioreactor. Since *A. oryzae* is known as an amylase producer [35], for the further experiments the medium was supplemented only with the cellulase cocktail. Four cultivations were run in the bioreactor to test two concentrations of PpB substrate in duplicates: 2 and 3%, resulting in media with pH 6.1 and 6.5 respectively. The results presented in Fig. 4 shows that for both tested

**Table 2 Protein yield from the fungal biomass obtained after 36 h cultivation in 2% pea-processing byproduct (PpB) substrate**

| Enzyme addition | Microorganism | % Protein in dry fungal biomass | Protein yield (g/g PpB substrate) |
|---|---|---|---|
| Without α-amylase | *M. purpureus* | 53.61 | 0.02 |
| | *A. oryzae* | 43.13 | 0.11 |
| | *F. venenatum* | 55.28 | 0.09 |
| | *N. intermedia* | 54.53 | 0.11 |
| | *R. oryzae* | 50.03 | 0.09 |
| With α-amylase | *M. purpureus* | 58.66 | 0.03 |
| | *A. oryzae* | 46.36 | 0.14 |
| | *F. venenatum* | 59.75 | 0.11 |
| | *N. intermedia* | 54.11 | 0.11 |
| | *R. oryzae* | 54.79 | 0.11 |

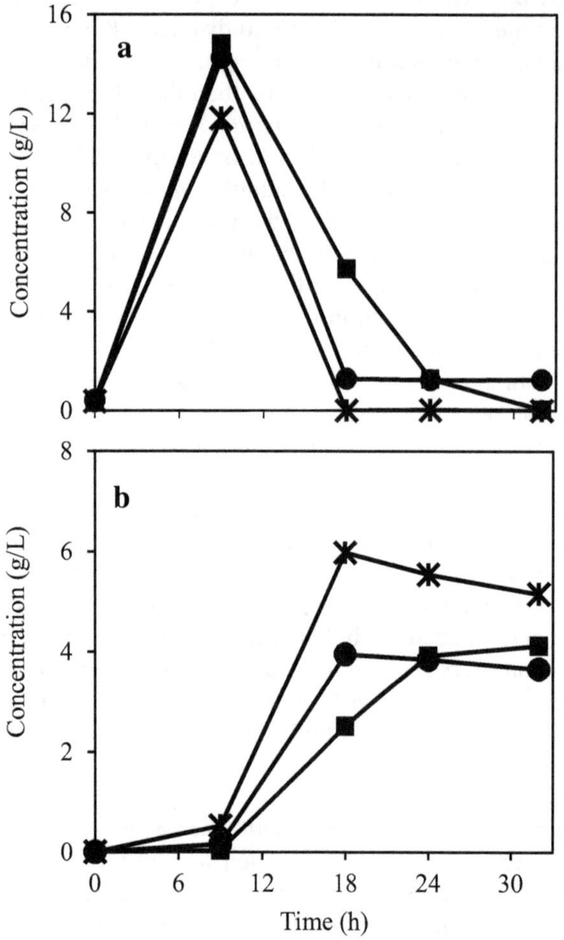


**Fig. 3** Glucose (**a**) and ethanol (**b**) concentration profiles during the filamentous fungal cultivation in 2% (w/v) PpB substrate. Figure represents fungal strains *A. oryzae* (filled squares); *N. intermedia* (asterisk); and *R. oryzae* (filled circles)

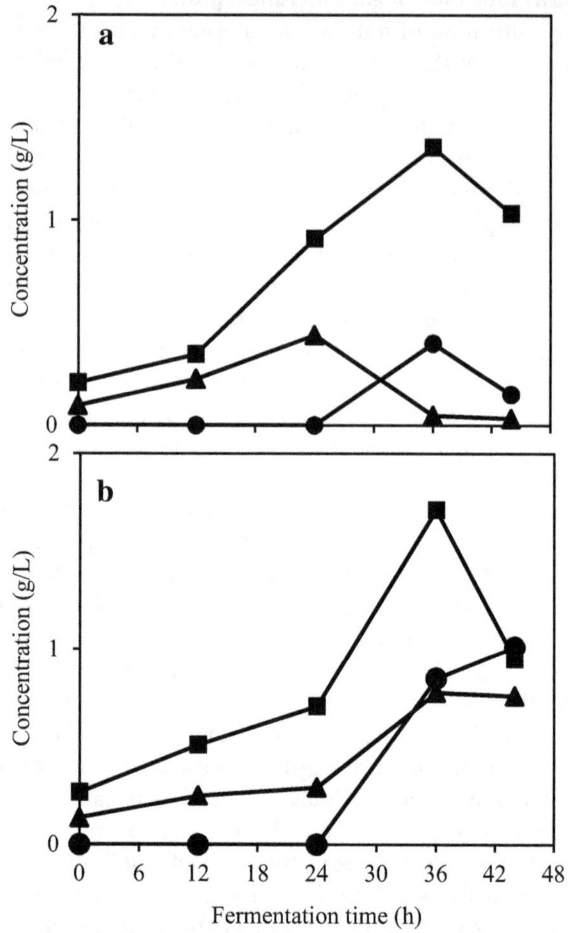


**Fig. 4** Sugar and ethanol concentration profiles during the cultivation of *A. oryzae* in **a** 2% and **b** 3% (w/v) PpB substrate in an airlift bioreactor. Figure represents glucose (filled squares); other sugars (filled triangles); and ethanol (filled circles). Other sugars are presented in xylose equivalent

scenarios glucose concentration increased up to 36 h. Additionally, 44 h were not enough for the total consumption of the sugar. When the pea processing load was 2%, the other sugars were consumed and at 36 h their concentration was below 0.10 g/L. On the other hand, when the initial load of PpB substrate was 3%, the concentration of the sugars (except glucose) reached the maximum value at 36 h. The ethanol profiles were also divergent, with maximum concentration being reached at 36 h (0.40 g/L) for 2% PpB substrate load and at 44 h (1.01 g/L) for 3% PpB substrate load. The obtained protein yields were 0.26 and 0.13 g/g of PpB substrate for 2 and 3% of the substrate, respectively. The yield for 2% PpB medium is almost twice the value obtained in the shake flask experiments. Moreover, working with a low load of the substrate (and low viscosity) resulted in increased protein production. Airlift bioreactors are known for their capacity to operate using higher aeration rates when compared to traditional stirred fermenters [36]. Moreover, efficient oxygen transfer and mixing are obtained when using airlifts [37]. As a result, cultivation in airlift bioreactor yielded low ethanol concentrations and high fungal protein.

#### Process integration at the existing pea-processing industrial facilities

The use of co-products from the industries as important sources of protein for human consumption has been receiving considerable attention. A method of protein production, applicable to a wide range of industrial processes, is to use edible filamentous fungi which could be used as a protein rich meat substitute. Filamentous fungi have been reported as important industrial microorganisms for their capacity to produce numerous metabolites

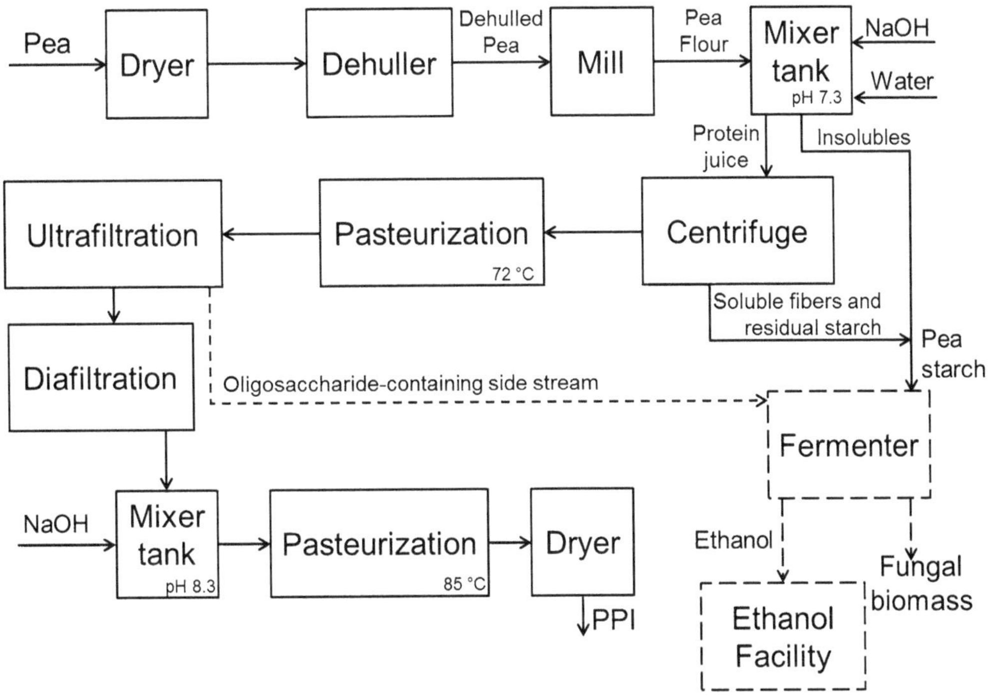


**Fig. 5** Block flow diagram for the production of pea-protein isolate integrated with the production of vegan-mycoprotein concentrate by filamentous fungi. Adapted from [39]

with high market potential [38]. Among these metabolites, the vast assortment of enzymes produced by these microorganisms has special importance. These catalysts are responsible for making filamentous fungi a flexible group able to use a large range of biomaterials as substrate.

The scheme of an industrial process for the production of PPI is presented in Fig. 5 [39] with the proposal of integrating the production of fungal biomass for the valorization of the byproduct. The starch and fibers separated from the pea protein would be fed in a bioreactor. For dilution of the solid material, the low-protein content stream from the ultrafiltration step could be used. This stream is sterile and rich in oligosaccharides, which can also be used by the fungi as substrate. Following the same yields of the bench-scale airlift bioreactor, 1 ton of pea-processing byproduct would consume 50 $m^3$ of water, and would be predicted to produce 680 kg of *A. oryzae* biomass with 260 kg of pure protein under ideal conditions. The biomass, however, would need to be subjected to a heat treatment to reduce its RNA content, resulting in some loss of protein [14]. Moreover, tests for mycotoxin production should be carried out frequently and, in case of detection, application of corrective methods can affect the productivity [14].

The scenario also includes a system in which the broth containing ethanol could be either sent to a distillation unit or to any adjacent ethanol facilities in the vicinity which contain all the necessary installation for ethanol purification [34]. However, the techno-economic feasibility of using PpB as a potential substrate for edible fungal cultivation for feed or food component needs to be verified in detail and hence open for future studies.

## Conclusion

The pea-processing byproduct was proved to be an efficient medium for the growth of filamentous fungi to produce a vegan-protein concentrate. Fungal biomass with about 46 and 54% protein content was obtained from PpB using edible strains of filamentous fungi, *A. oryzae* and *N. intermedia* respectively. Scaling-up of the process to a 4.5 L bench scale airlift bioreactor improved the *A. oryzae* biomass yield, with a total of 0.26 g of fungal protein per g of PpB. Based on the results obtained, an industrial scenario for the production of vegan-mycoprotein concentrate for human nutrition is proposed as an integrated process to the existing PPI production facilities.

**Authors' contributions**

RBN, PRL and MJT conceived the project. RBN performed the experimental and laboratory work. PFSF, RBN, PRL and MJT analyzed and interpreted the data regarding fermentation. DA contributed with the analysis of the data regarding the nutritional aspects. PFSF, RBN and DA wrote the paper. All authors read and approved the final manuscript.

**Author details**
[1] Swedish Centre for Resource Recovery, University of Borås, 50190 Borås, Sweden. [2] Mycorena AB, Stena Center 1A, 411 92 Gothenburg, Sweden. [3] Faculty of Caring Science, Work Life and Social Welfare, University of Borås, 50190 Borås, Sweden.

**Acknowledgements**
Not applicable.

**Competing interests**
The authors declare that they have no competing interests.

**Funding**
This work was financially supported by the Swedish Research Council FORMAS, and the Coordination for the Improvement of Higher Education Personnel, CAPES (Brazil).

**References**

1. Boland MJ, Rae AN, Vereijken JM, Meuwissen MPM, Fischer ARH, van Boekel MAJS, Rutherfurd SM, Gruppen H, Moughan PJ, Hendriks WH. The future supply of animal-derived protein for human consumption. Trends Food Sci Technol. 2013;29(1):62–73.
2. Godfray HCJ, Beddington JR, Crute IR, Haddad L, Lawrence D, Muir JF, Pretty J, Robinson S, Thomas SM, Toulmin C. Food security: the challenge of feeding 9 billion people. Science. 2010;327(5967):812–8.
3. Hartmann C, Siegrist M. Consumer perception and behaviour regarding sustainable protein consumption: a systematic review. Trends Food Sci Technol. 2017;61:11–25.
4. Tukker A, Jansen B. Environmental impacts of products: a detailed review of studies. J Ind Ecol. 2006;10(3):159–82.
5. Sadler MJ. Meat alternatives—market developments and health benefits. Trends Food Sci Technol. 2004;15(5):250–60.
6. Moore D, Chiu SW. Fungal products as food. In: Pointing SB, Hyde KD, editors. Bio-exploitation of filamentous fungi. Hong Kong: Fungal Diversity Press; 2001. p. 223–51.
7. Denny A, Aisbitt B, Lunn J. Mycoprotein and health. Nutr Bull. 2008;33(4):298–310.
8. Smith H, Doyle S, Murphy R. Filamentous fungi as a source of natural antioxidants. Food Chem. 2015;185:389–97.
9. Ferreira JA, Lennartsson PR, Niklasson C, Lundin M, Edebo L, Taherzadeh MJ. Spent sulphite liquor for cultivation of an edible *Rhizopus* sp. BioResources. 2012;7(1):173–88.
10. Chen M-H, Johns MR. Effect of pH and nitrogen source on pigment production by *Monascus purpureus*. Appl Microbiol Biot. 1993;40(1):132–8.
11. Huang Z, Xu Y, Li L, Li Y. Two new *Monascus* metabolites with strong blue fluorescence isolated from red yeast rice. J Agr Food Chem. 2007;56(1):112–8.
12. Tsukahara M, Shinzato N, Tamaki Y, Namihira T, Matsui T. Red yeast rice fermentation by selected *Monascus* sp. with deep-red color, lovastatin production but no citrinin, and effect of temperature-shift cultivation on lovastatin production. Appl Biochem Biotechnol. 2009;158(2):476–82.
13. Archer DB. Filamentous fungi as microbial cell factories for food use. Curr Opin Biotechnol. 2000;11(5):478–83.
14. Wiebe MG. Quorn™ myco-protein—overview of a successful fungal product. Mycologist. 2004;18(1):17–20.
15. Anupama RP. Value-added food: single cell protein. Biotechnol Adv. 2000;18(6):459–79.
16. Hoffmann B, Münch S, Schwägele F, Neusüß C, Jira W. A sensitive HPLC-MS/MS screening method for the simultaneous detection of lupine, pea, and soy proteins in meat products. Food Control. 2017;71:200–9.
17. Stone AK, Avarmenko NA, Warkentin TD, Nickerson MT. Functional properties of protein isolates from different pea cultivars. Food Sci Biotechnol. 2015;24(3):827–33.
18. Tulbek MC, Lam RSH, Wang Y, Asavajaru P, Lam A. Pea: a sustainable vegetable protein crop. In: Nadathur SR, Wanasundara JPD, Scalin L, editors. Sustainable protein sources. San Diego: Academic Press; 2017. p. 145–64.
19. Day L. Proteins from land plants—potential resources for human nutrition and food security. Trends Food Sci Technol. 2013;32(1):25–42.
20. McCarthy NA, Kennedy D, Hogan SA, Kelly PM, Thapa K, Murphy KM, Fenelon MA. Emulsification properties of pea protein isolate using homogenization, microfluidization and ultrasonication. Food Res Int. 2016;89(1):415–21.
21. Ratnayake WS, Hoover R, Warkentin T. Pea starch: composition, structure and properties—a review. Starch-Stärke. 2002;54(6):217–34.
22. Ruxton CHS, McMillan B. The impact of mycoprotein on blood cholesterol levels: a pilot study. Brit Food J. 2010;112(10):1092–101.
23. Williamson DA, Geiselman PJ, Lovejoy J, Greenway F, Volaufova J, Martin CK, Arnett C, Ortego L. Effects of consuming mycoprotein, tofu or chicken upon subsequent eating behaviour, hunger and safety. Appetite. 2006;46(1):41–8.
24. Bottin JH, Swann JR, Cropp E, Chambers E, Ford HE, Ghatei MA, Frost GS. Mycoprotein reduces energy intake and postprandial insulin release without altering glucagon-like peptide-1 and peptide tyrosine-tyrosine concentrations in healthy overweight and obese adults: a randomised-controlled trial. Brit J Nutr. 2016;116(2):360–74.
25. Paddon-Jones D, Westman E, Mattes RD, Wolfe RR, Astrup A, Westerterp-Plantenga M. Protein, weight management, and satiety. Am J Clin Nutr. 2008;87(5):1558S–61S.
26. Chung H-J, Lim HS, Lim S-T. Effect of partial gelatinization and retrogradation on the enzymatic digestion of waxy rice starch. J Cereal Sci. 2006;43(3):353–9.
27. Fredriksson H, Björck I, Andersson R, Liljeberg H, Silverio J, Eliasson AC, Åman P. Studies on α-amylase degradation of retrograded starch gels from waxy maize and high-amylopectin potato. Carbohydr Polym. 2000;43(1):81–7.
28. Frei M, Siddhuraju P, Becker K. Studies on the in vitro starch digestibility and the glycemic index of six different indigenous rice cultivars from the Philippines. Food Chem. 2003;83(3):395–402.
29. Sluiter A, Hames B, Ruiz R, Scarlata C, Sluiter J, Templeton D, Crocker D. Determination of structural carbohydrates and lignin in biomass. Golden, CO: National Renewable Energy Laboratory; 2012.
30. Sluiter A, Ruiz R, Scarlata C, Sluiter J, Templeton D. Determination of sugars, byproducts, and degradation products in liquid fraction process samples. Golden, CO: National Renewable Energy Laboratory; 2008.
31. Ferreira JA, Lennartsson PR, Taherzadeh MJ. Production of ethanol and biomass from thin stillage by *Neurospora intermedia*: a pilot study for process diversification. Eng Life Sci. 2015;15(8):751–9.
32. Kemblowski Z, Kristiansen B. Rheometry of fermentation liquids. Biotechnol Biooeng. 1986;28(10):1474–83.
33. Wu Q, Wang X, Wang T, Han M, Sha Z, Wang J. Effect of liquid viscosity on hydrodynamics and bubble behaviour of an external-loop airlift reactor. Can J Chem Eng. 2013;91(12):1957–63.
34. Mahboubi A, Ferreira JA, Taherzadeh MJ, Lennartsson PR. Value-added products from dairy waste using edible fungi. Waste Manag. 2017;59:518–25.
35. Nair RB, Taherzadeh MJ. Valorization of sugar-to-ethanol process waste vinasse: a novel biorefinery approach using edible ascomycetes filamentous fungi. Bioresour Technol. 2016;221:469–76.
36. Chisti Y, Jauregui-Haza UJ. Oxygen transfer and mixing in mechanically agitated airlift bioreactors. Biochem Eng J. 2002;10(2):143–53.
37. Moo-Young M, Halard B, Allen DG, Burrell R, Kawase Y. Oxygen transfer to mycelial fermentation broths in an airlift fermentor. Biotechnol Bioeng. 1987;30(6):746–53.
38. Ferreira JA, Lennartsson PR, Edebo L, Taherzadeh MJ. Zygomycetes-based biorefinery: present status and future prospects. Bioresource Technol. 2013;135:523–32.
39. Fredrikson M, Biot P, Alminger ML, Carlsson N-G, Sandberg A-S. Production process for high-quality pea-protein isolate with low content of oligosaccharides and phytate. J Agric Food Chem. 2001;49(3):1208–12.

# 7

# How a fungus shapes biotechnology: 100 years of *Aspergillus niger* research

Timothy C. Cairns*, Corrado Nai* and Vera Meyer*

**Abstract**

In 1917, a food chemist named James Currie made a promising discovery: any strain of the filamentous mould *Aspergillus niger* would produce high concentrations of citric acid when grown in sugar medium. This tricarboxylic acid, which we now know is an intermediate of the Krebs cycle, had previously been extracted from citrus fruits for applications in food and beverage production. Two years after Currie's discovery, industrial-level production using *A. niger* began, the biochemical fermentation industry started to flourish, and industrial biotechnology was born. A century later, citric acid production using this mould is a multi-billion dollar industry, with *A. niger* additionally producing a diverse range of proteins, enzymes and secondary metabolites. In this review, we assess main developments in the field of *A. niger* biology over the last 100 years and highlight scientific breakthroughs and discoveries which were influential for both basic and applied fungal research in and outside the *A. niger* community. We give special focus to two developments of the last decade: systems biology and genome editing. We also summarize the current international *A. niger* research community, and end by speculating on the future of fundamental research on this fascinating fungus and its exploitation in industrial biotechnology.

**Keywords:** *Aspergillus niger*, Biotechnology, Industrial biology, Systems biology, Genome editing, Citric acid

## Introduction

For millennia, humanity has practiced rudimental forms of biotechnology: by fermenting starch and sugars present in grains and fruits, ancient civilizations were able to produce bread, beer, wine, and other alcoholic beverages. Prior to the late nineteenth and early twentieth century, these processes were conducted without knowledge of the underlying biological events. Now, brewery and wine making is a well-understood, controlled industrial process. Similarly, in just a century, industrial biotechnology has changed dramatically and flourished, from initial proof-of-principle experimentation in Erlenmeyer flasks, to a multibillion dollar industry producing megatons of useful molecules [1]. Fungal biotechnology is undoubtedly a major contributor and driver of this success. As just one example, the estimated market volume for plant-degrading enzymes from filamentous fungi in 2016 was €4.7 billion, which was expected to reach up to €10 billion within the next decade [2]. In this celebratory historical overview, we outline some of the crucial advances for the filamentous mould *Aspergillus niger* since the very first biotechnological experiments using this fungus 100 years ago.

## 100 years ago: industrial biotech is born

In contrast to what most people might think, citric acid is not—or not anymore—isolated from citrus fruits, but is industrially produced by the filamentous fungus *A. niger*. The process was pioneered by James Currie, a food chemist, who 100 years ago published a study describing the superior properties of *A. niger* for the industrial production of the acid [3]. In particular, Currie showed the necessary growth medium for citric acid biosynthesis, and the ability of the fungus to grow at low pH (2.5–3.5), while still being able to produce high amounts of the metabolite. Moreover, this work demonstrated the direct correlation between amount of substrate in the medium and amount of product, laying the basis for modern-day industrial fermentation of citric acid [3]. In

*Correspondence: t.cairns@tu-berlin.de; corrado.nai@tu-berlin.de; vera.meyer@tu-berlin.de
Department of Applied and Molecular Microbiology, Institute of Biotechnology, Technische Universität Berlin, Gustav-Meyer-Allee 25, 13355 Berlin, Germany

contrast to other species of fungi that had been reported to produce citric acid by 1917, every single strain of *A. niger* that Currie tested could efficiently produce this molecule when grown in sugar solutions. Two years later, the American company Pfizer made a pilot plant for biochemical production of citric acid, and by the mid 1920s, production using *A. niger* fermentation far outweighed extraction from citrus fruits [4].

The citric acid cycle was comprehensively determined over the next several decades, resulting in the award of the Nobel Prize to Hans Krebs and Fritz Lipmann in 1953. The first and final reaction in the cycle involves the formation of citrate from oxaloacetate, acetyl-CoA, and water, by a citrate synthase, ultimately generating chemical energy in the form of adenosine triphosphate (ATP). The objective of industrial microbiologists, including James Currie, was to exploit this cycle, and indeed many other metabolic pathways, to ferment useful molecules.

Although there are variations in fermentation techniques, in general, industrial production of citric acid requires aerobic, submerged growth of *A. niger* in a sugar solution, which is usually derived from inexpensive sources, such as molasses, corn steep liquor, or hydrolysed corn starch, amongst others. After fermentation, *A. niger* is physically removed, usually by filtration, and citric acid is isolated by precipitation of the fermentation mix with calcium hydroxide (lime) to generate calcium citrate salt. Subsequent treatment with sulphuric acid yields the citric acid product.

The widespread applications of citric acid are shown by current figures regarding this metabolite: in 2007, worldwide production was estimated at 1.6 million tons, with an estimated value of $2.6 billion in 2014 and predicted to rise to $3.6 billion by 2020 [1, 5]. As a weak acid, it can be used as an antioxidant, preservative, acidulant, pH-regulator, or flavour in food and beverages, as well as comparable applications in the pharmaceutical and cosmetics industries. Citric acid is currently predominantly produced in China, which accounts for approximately 60% of global production [1]. However, *A. niger* industrial applications are not just limited to the production of citric acid; as a prolific secretor, numerous industrially relevant enzymes and other molecules are produced by this fungus. Below, we summarize some of the key developments in the field over the last century.

## A historical snapshot of *A. niger* research

The fundamental and applied scientific discoveries using *A. niger* over the last 100 years are extremely diverse. As Currie wrote in 1917: 'Few concise statements can be made concerning the metabolism of an organism capable of producing such a variety of chemical transformations as *Aspergillus niger*' [3]. Nevertheless, some trends regarding the history of the *A. niger* research field can, in general terms, be deciphered. We conducted a survey of the *A. niger* literature since Currie's seminal study by interrogating the PubMed database [6] for any publication containing '*Aspergillus niger*' in the title. The resulting articles (>3000, see Additional file 1: Table S1) were divided into five 20-year periods based on their publication date, and the 20 most common words from the available titles during each time period were visualized as word clouds (Fig. 1). Although querying the PubMed database for '*Aspergillus niger*' in abstract and keywords resulted in more returned manuscripts (>8700 hits), we decided to limit our word-cloud analysis specifically to titles. We applied this restriction as searching among manuscript abstracts returned a majority of hits where researchers used *A. niger* in simple growth assays to validate efficacy of putative antifungals. While of interest, these research efforts (which were extremely common from 1977 onwards) are not, specifically, interested in *A. niger* biology per se.

Our analysis of 100 years of *A. niger* publications indicated, unsurprisingly, that 'citric acid' and 'fermentation' were amongst the most common returned words in titles from every period (Fig. 1). Clearly, Currie's discovery of conditions for maximizing citric acid production [3] was indeed a biotechnological revolution, and one that would constitute a major focus of research over the next century. Indeed, *A. niger* research has rapidly grown over the past 40 years (Fig. 2). Although by no means exhaustive, distinct historical trends become apparent from our analysis (Fig. 1).

### The foundations of *A. niger* research

From the period 1937–1956, *A. niger* researchers were predominantly concerned with the impact of micro/macro nutrients in cultivation media and growth parameters for optimized citric acid fermentation (Fig. 1, [7, 8]). Most studies specifically utilized submerged culture, which obviously reflects the need to control growth, morphology, and, ultimately, citric acid production during fermentation (Additional file 1: Table S1 and [9]). After the award of Hermann Muller's Nobel Prize in 1946 for work on mutations via X-ray irradiation, the late 1940s and early 1950s saw the first mutagenesis studies in biological sciences deployed by the *A. niger* community in order to generate strains with improved citric acid yields [10]. The number of UV, X-ray, and chemical mutagenesis efforts increased from 1957 to 1976, with most focusing on enhancing citric acid production (e.g. [11] and Fig. 1).

Notably, the period from 1957 to 1976 saw a paradigm shift amongst the *A. niger* biotechnological research community, with the widespread realization that this

**Fig. 1** A historical snapshot of *A. niger* research. The PubMed database was interrogated for any publication that contained *'Aspergillus niger'* in the title. Publication titles were assigned to periods of 20 years. As 1917–1936 only returned 7 manuscripts, this period was omitted. Word clouds were generated online (https://worditout.com/word-cloud/create), and the top 20 words, not including *'Aspergillus niger'* or non-technical terms (e.g. prepositions etc.), are depicted. Size of each word is proportional to relative frequency amongst all the titles in that period. Retrieved articles for each period: 112 (1937–1956), 481 (1957–1976), 642 (1977–1996), 1789 (1997–2017)

fungus was not exclusively applicable for citric acid fermentation, but was also a prolific producer of useful enzymes (Fig. 1). A rapid growth of studies, facilitated by technological advances in chromatography, conducted purification and enzymatic analysis of diverse range of *A. niger* proteins, including various oxidases, dehydrogenases, hydrolases, cellulases, and pectinases, amongst others (Fig. 1, [12–16]). These discoveries were not only useful for industrial production of enzymes, but also significantly contributed to fundamental understanding of enzyme function. One notable pair of isoenzymes first purified in the late 1960s and early 1970s was *A. niger* glucoamylases [17, 18]. These discoveries would eventually lead to the widespread application of *A. niger* glucoamylases in the fermentation, food, and beverage industries, where these enzymes catalyse the saccharification of partially processed starch to glucose. Indeed, glucoamylase research features heavily in publications from 1977 to present day (Fig. 1). Interestingly, a recent business report showed that the glycoamylase market played a significant role for the success of leading multinational biotech companies, including AB Enzymes, Amano Enzyme, DSM, Genencor, Novozymes, and Verenium [19].

Outside the field of industrial microbiology, there was early interest in the 1970s in the clinical spectrum of disease caused by the *Aspergillus* genus, with authors first associating *A. niger* spore inhalation with the onset of asthma [20]. The serious threat of fungal disease both with regards to human health and crop destruction is now much better understood [21]. Indeed, fungal infections affect and estimated 1.2 billion people globally, resulting in approximately 1.5–2 million deaths per year. As a consequence of limited antifungal therapeutic options and timely diagnostics, mortality rates can be extremely high, reaching up to 90% in case of immunocompromised patient and/or drug-resistant strains causing invasive aspergillosis or other systemic mycoses [2]. Indeed, the fungicide market to control fungal growth in

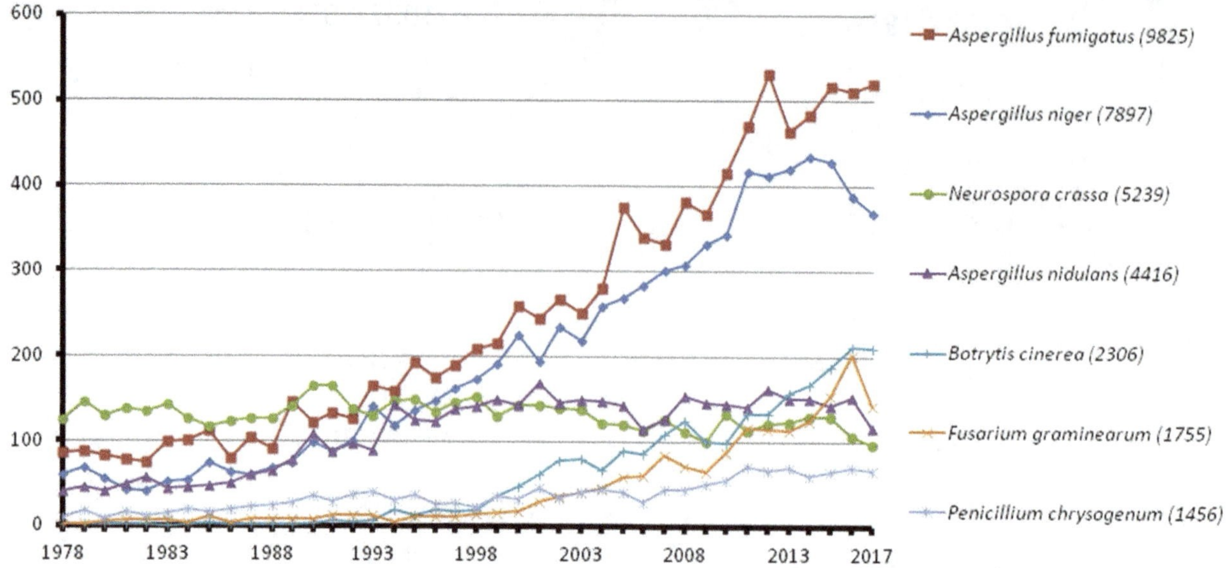


**Fig. 2** Total number or paper published on PubMed for selected filamentous fungi over the last 40 years. The literature database was queried for title, abstract and keywords with the full species name (in brackets: total results for 1978–2017). As comparison: *Saccharomyces cerevisiae* gave more than 110,000 hits with around 2000–4000 hits/year over the last 25 years (Additional file 2: Table s2)

agriculture and medicine was estimated at €10 billion in 2014 [2].

### The dawn of molecular biology and the first *A. niger* transformations

A revolution in biotechnology occurred between 1977 and 1996 with the advent of molecular biology. These technological developments and rapid discoveries in *A. niger* molecular biology and genetics are reflected by a heavy community-wide focus on genes, cloning, and sequence analyses (Fig. 1). Arguably the most fundamental technique for molecular analyses of any organism is the ability to transform exogenous DNA into target cells and integrate them into recipient genomes. In a seminal study, Peter Punt and Cees van den Hondel utilised a hygromycin resistance gene from *Escherichia coli*, encoding a phosphotransferase, as a dominant selectable marker in *A. niger* and *A. nidulans* [22]. In addition to becoming one of the most commonly used dominant selectable markers in fungal transformation, this work pioneered plasmid mediated cassette integration in filamentous fungal genomes using the vector pAN7-1. Moreover, the authors further validated the use of the *A. nidulans* glyceraldehyde-3-phosphate dehydrogenase (*gpd*) promoter, and the *trpC* terminator. In an alternative approach, other studies optimized a homologous transformation system using the orotidine 5′-phosphate decarboxylase gene *pyrG* [23, 24]. This recyclable auxotrophic marker would ultimately facilitate several hundred studies of *A. niger* gene function, and is still used today, most obviously as a transformation system, but also as a convenient locus for cassette integration during heterologous and homologous gene expression studies [25].

Introduction of such molecular studies began to enable industrial microbiologists to obviate a diverse range of technical challenges when working with filamentous fungi. As just one example, heterologous expression of a porcine pancreatic phospholipase gene in *A. niger* by the lab of David Archer initially did not produce detectable protein, which was revealed to be due to degradation of this enzyme by intracellular and extracellular *A. niger* proteases [26]. The authors therefore used PCR and restriction endonuclease cloning approaches, which were cutting-edge at the time (and indeed still practiced in most molecular laboratories today), to express a prophospholipase-glucoamylase fusion protein in a protease-deficient *A. niger* strain. This recombinant approach enabled secreted concentrations of fusion protein at 10 μg/mL.

The utilisation for *A. niger* for production of a diverse range of enzymes also continued between 1977 and 1996, as exemplified by numerous efforts during this period for the fermentation and purification of proteases (Fig. 1). Fungal proteases are active at a broad range of abiotic conditions (e.g. pH, temperature) and consequently are now applied in food, laundry detergent, and pharmaceutical processes, amongst others (reviewed in [27]).

## The last 20 years: rapid developments in the *A. niger* research field

The development of the *A. niger* molecular toolkit accelerated from 1997 to 2017. Notable milestones include the generation of nonhomologous end joining (NHEJ) mutants in a collaborative effort between our lab and the lab of Arthur Ram [28], which enable increased targeting efficiency of exogenous DNA cassettes with recipient genomes during fungal transformation. Filamentous fungal NHEJ mutants were first described in the model organism *Neurospora crassa* in 2004 [29] and the drastic increase in cassette targeting rates (up to 100% of transformed fungal cells) led to the rapid application of this tool by researchers in the *Aspergillus* genus, including *A. niger* [28, 30].

The application of *A. niger* as a cell factory for useful enzymes continued to rapidly expand between 1997 and 2017 (Fig. 1). Phytases, for example, were first marketed in 1991, and are used to improve the nutritional content of animal feed by generating inorganic phosphorus from phytic acid, which is the major form of organic phosphorus in plant seed [31]. The biotechnological production market of phytases is estimated to be over €150 million per year, with *A. niger* one of the most commonly used microorganisms [32].

In addition to homologous or heterologous production of single industrially relevant proteins, *A. niger* and other Aspergilli have been increasingly harnessed for synthesis of diverse enzyme mixtures over the last decade. One critical application of such enzyme cocktails is the degradation of plant polysaccharides, whereby cellulose, hemicellulose, and pectin can be broken down into oligosaccharides and monosaccharides and the number of *A. niger* genes predicted to encode proteins capable of plant biomass degradation is over 170 [33]. Moreover, the transcription factors that regulate these genes are being rapidly elucidated, for example the amylolytic regulator AmyR (the first regulator identified in *A. niger*) [34], pectinolytic regulator RhaR [35], and hemi-cellulolytic regulator XlnR [36, 37], amongst several others (reviewed in [33]). Integrating knowledge of transcription factor networks with a comprehensive understanding of upstream molecular sensors and signalling cascades may enable the engineering of *A. niger* isolates with increased plant biomass degradation capabilities. The impact of such microbial cell factories will enable the renewable generation of starting material for production of biofuels and other industrial processes from plant material. Consequently, future *A. niger* strains may enable transition from our current fossil-based economy to a bio-based economy, whereby future fuel is generated from renewable resources.

In general, between 1996 and present day, much progress has been made for enzyme expression using *A. niger*, with higher titres of secreted proteins increasingly possible (e.g. 30 g/L for glucoamylase is common) (reviewed in [38, 39]). These advances have been achieved by several now routine approaches, including codon optimisation of non-fungal genes, and, in some instances, use of a fusion carrier protein [38]. However, a significant bottleneck that inhibits maximum yield of secreted proteins is an incomplete understanding of filamentous fungal secretion. In the model postulated by Taheri-Talesh using *A. nidulans* [40], secretion and polar growth are physically coupled at the hyphal tip (first showed in *A. niger* by Han Wösten et al. in 1991 [41]). Currently, however, the underlying mechanisms of fungal protein secretion are not understood as an integrated system, with numerous outstanding questions hampering rational strain engineering. For example, by investigating secretome in concentric zones of *A. niger* colonies it has been observed that secretion and growth can be uncoupled [42]. How can this phenomenon be further exploited to further increase secretion yields without affecting hyphal tip growth? What molecular signals and regulators control and limit protein secretion? We speculate that delivering a systems-level understanding of the molecular and cellular mechanisms that underpin fungal secretion will one of the major research goals over the next 20 years.

Several studies over the past decade have interrogated the effect of filamentous microscopic and macroscopic morphologies on secretion during industrial applications. For example, *A. niger* hyperbranching phenotypes have been generated to study the underlying morphogenetic gene network controlling polar growth, and, elsewhere, secretion or the macromorphology of *A. niger* has been modified by the addition of micro-particles to submerged media [43, 44]. These genetic or microbiological approaches offer increasingly accurate control of hyphal branch length, enabling optimization of fermentation culture viscosity, and also minimising *A. niger* sensitivity to shear stress. Intriguingly, a promising avenue of research for maximising industrial protein titres comes from analyses of secretion in *A. oryzae*, which has conclusively demonstrated that secretion also occurs at septal junctions [45, 46]. This might conceivably be harnessed by industrial microbiologists as a secondary secretion route to maximize secretion of useful enzymes in Aspergillus species, including *A. niger*. Another level of complexity behind *A. niger* secretion was uncovered by pioneering studies of the labs of Han Wösten, Arthur Ram and Cees van den Hondel: the discovery of heterogeneity of fungal secretion on the cell [47], hyphal [48] and colony level [49]. This work laid the conceptual framework for follow-up studies on population heterogeneity in model and industrial Aspergilli [50–52].

Outside the field of industrial microbiology, one alarming observation is that the last century of *A. niger* research spans the discovery of penicillin, for which Alexander Fleming, Ernst Chain, and Howard Florey were awarded the Nobel Prize in 1945, and the subsequent global emergence of drug resistant pathogenic microbes. Currently, as the incidence of drug resistance increases, the number of chemical compounds approved for use in agriculture or the clinic is decreasing [53]. Given that microbial secondary metabolites are a rich source of novel bioactive molecules [54], the focus of our lab and others over the last 5 years has been to establish *A. niger* as an industrial platform strain for drug discovery and natural product production. This objective is based on the assumption that the high intracellular glycolytic flux towards citric acid (and amino acids derived thereof) can be exploited and redirected into non-ribosomal peptide synthesis. Recent work has demonstrated we could indeed genetically engineer *A. niger* to heterologously overexpress a non-ribosomal peptide synthetase (NRPS) from *Fusarium* spp. in several g/L amounts [55]. This proof of principle experiment, which utilized the highly optimised and titratable synthetic Tet-on gene switch [25] to produce the antimicrobial cyclohexadepsipeptide enniatin, will hopefully pave the way for future production of multiple secondary metabolites. Indeed, more recently, we could generate *A. niger* isolates expressing truncated enniatin NRPS enzymes or with key domains positionally exchanged, to generate new-to-nature molecules at high titers (e.g. 1.3 g/L) [56]. Excitingly, some of these new molecules demonstrated enhanced antiparasitic activity when compared to existing drugs [56]. Given the high diversity of fungal secondary metabolite genes, and their frequent transcriptional silence under laboratory conditions, expression using *A. niger* as a heterologous host, and the molecular approaches validated by these two studies, hold great promise for discovery of new chemical leads for compound development in agriculture and the pharmaceutical industry. This is further supported by recent work that has applied viral DNA sequences (encoding, for example, the 2A peptide), in order to enable polycistronic gene expression in *A. niger* [57, 58]. These studies provide proof of principle that complex secondary metabolites, which require multiple enzymes for their biosynthesis, can be produced by polycistronic gene switches in *A. niger*.

To summarize, one century after James Currie's ground-breaking work on biotechnological citric acid production, his assessment still generally holds true: it is difficult to make comprehensive statements about the production capabilities of *A. niger* owing to its metabolic versatility, and the many as-yet undisclosed metabolic pathways. However, the dawn of molecular biology, and recent breakthroughs in synthetic biology for *A. niger*, have ultimately engineered a multipurpose cell factory out of a citric acid producer. *A. niger* is the most versatile filamentous fungal platform strain which can now be exploited to produce acids, proteins, enzymes, and medicinal drugs (Table 1). Two developments that occurred over the last decade promise to open entirely new avenues of scientific study using *A. niger*: genome sequencing and the introduction of genome editing. The following sections look at these two developments in more detail.

## 2007: the genome sequence of *Aspergillus niger* is released

The first filamentous fungal genome to be published was that of the model ascomycete *Neurospora crassa* in 2003, 160 years after the discovery of this species in a Paris bakery [59]. Several hallmarks of fungal genomes were reported in this ground-breaking draft, in particular (1) contiguous gene clusters for secondary metabolite biosynthesis; (2) defence from parasitic mobile genetic elements via repeat induced point mutation; (3) variations in telomeric gene content when compared to telomere-distal chromosome regions; (4) and the presence of two putative RNA silencing pathways, amongst others [59]. It was in this context that the first *A. niger* genome was released in 2007 [60], by which time three Aspergillus genomes were also publicly available: *A. nidulans* [61], *A. oryzae* [62], and *A. fumigatus* [63]. These genomes were representative of the broad utilities and challenges posed by the *Aspergillus* genus: model organism, food producer, and human-infecting fungus, respectively. The publication of the *A. niger* genome was thus the first, and, arguably, the most important, industrial *Aspergillus* genome to be sequenced. For this draft assembly, Herman Pel and colleagues used the enzyme producing isolate CBS 513.88, which is a derivative of *A. niger* NRRL 3122, a strain generated by classical mutagenesis for glucoamylase A production [60]. This was therefore the first global analyses of the *A. niger* genome repertoire that had been harnessed in industrial applications for many decades, and remains one of most comprehensively annotated genome resource for the *A. niger* community.

Numerous explanations of the suitability of *A. niger* for industrial applications were identified from the estimated 34 Mb genome, with its estimated 14,165 coding genes. For example, the authors predicted various gene duplication events that may have led to expansion of genes necessary for the production of the citrate precursor oxaloacetate [60], an observation that explains the remarkable capability for citric acid production by

**Table 1 Selection of (multi)national companies exploiting *A. niger* for the production of important industrial compounds. Modified after Fiedler et al. 2013 [79]**

| Company | Headquarter | Products |
|---|---|---|
| Adcuram | Germany | Citric acid |
| AB Enzymes | Germany | Glucoamylase |
| ADM | USA | Citric acid |
| Agennix | Germany | Lactoferrin |
| Amano Enzyme Co. Ltd. | Japan | β-Galactosidase, Glucoamylase, Glucose oxidase, Hemicellulase, Proteases |
| Anhui BBCA Biochemical | China | Citric acid |
| BASF | Germany | Hemicellulase, Phytase |
| Biocon | India | Cellulase, Hemicellulase, Pectinase |
| Cangzhou Kangzhuang Chemical | China | Glucoamylase |
| Cargill | USA | Citric acid |
| Christian Hansen | Denmark | Chymosin |
| COFCO | China | Citric acid |
| DSM | The Netherlands | Arabinase, Asparaginase, Catalase, Cellulase, β-Galactosidase, Glucoamylase, Glucose oxidase, Hemicellulase, Lactoferrin, Lipase, Pectinase, Phytase, Proteases, Xylanase |
| Dupont IB | The Netherlands | Catalase, β-Galactosidase, Glucoamylase, Glucose oxidase, Hemicellulase, Lipase, Pectinase |
| Dyadic | USA | Cellulase, Glucoamylase, Glucose oxidase |
| Gadot Biochemical Industries | Israel | Citric acid |
| Genencor INT | USA | Cellulase, Hemicellulase, β-Galactosidase |
| Haihang Industry | China | Cellulase |
| Iwata Chemical Co. Ltd | Japan | Citric acid |
| Jungbunzlauer | Switzerland | Citric acid |
| Mitsubishi Foods Co. Ltd. | Japan | Proteases |
| Megazyme | USA | Catalase, Inulinase, Glucosidase |
| Novozymes | Denmark | Asparaginase, Catalase, β-Galactosidase, Glucoamylase, Hemicellulase, Lipase, Pectinase, Phytase, Proteases |
| Shandong Longda Bio-Products | China | Glucoamylase, Pectinase |
| Shin Nihon Chemical Co. Ltd. | Japan | Arabinase, Catalase, Cellulase, β-Galactosidase, Hemicellulase, Proteases |
| Tate & Lyle | UK | Citric acid |
| Verenium | USA | Glucoamylase, Proteases |

*A. niger*. Indeed, this hypothesis has been supported by a very recent comprehensive comparative genomic analysis of black aspergilli in a community effort led by Ronald de Vries [64], which was published exactly 10 years after the public release of *A. niger* genome sequence. With regards to the nutritional versatility of *A. niger*, genes encoding putative nutrient transporters were enriched in *A. niger* when compared to *A. nidulans* or *A. fumigatus*. These genes were presumed to enable uptake or sensing of diverse carbon and nitrogen sources [60].

In the context of novel bioactive molecule discovery, numerous putative secondary metabolite loci were identified based on the presence of either a polyketide synthase (PKS) or NRPS encoding gene [60]. Intriguingly, the vast majority of these clusters lacked either an experimentally verified or predicted biosynthetic product, thus indicating the potential for novel pharmaceutical discovery using *A. niger* and other Aspergilli [65, 66]. Indeed, these observations have been corroborated by recent comparative genomic analyses of the *Aspergillus* genus [64], which indicate that *A. niger* CBS 513.88 has 57 predicted secondary metabolite clusters, whilst another study predicts 81 secondary metabolite clusters [67], the highest of all Aspergillus genomes analyzed so far. The latter study follows an extensive manual annotation approach and likely more precisely predicts the actual secondary metabolite repertoire of *A. niger*. Taken together, these exemplar discoveries highlight how the *A. niger* draft genome provided the first global explanations for the many industrially relevant phenotypes of this organism, and facilitated a new era of forward genetics in this species. Moreover, in the immediate aftermath of this revolutionary resource for the *A. niger* community, comparative genomic analyses amongst the aspergilli would also redefine species concepts [68], interrogate sexual compatibility [60], and the nature of fungal virulence [69], amongst other critical developments (reviewed in [70]).

DNA sequencing technology and analyses are now sufficiently accurate and high throughput to be routinely applied to answer a diverse range of fundamental research questions in *A. niger*. One recent and notable example is the so called bulk segregant analyses developed by Arthur Ram's lab, which was used to identify a single nucleotide polymorphism (SNP) responsible for a nonacidifying phenotype of a UV-mutated isolate [71]. In the bulk segregant approach, the *A. niger* parasexual cycle is used to cross a wild-type strain with a mutant of interest, and haploid segregants with the phenotype of interest are identified. Subsequently, DNA from these segregants and parental isolates are sequenced to identify SNPs. Amongst these isolates, the SNP that is conserved in all segregants, yet absent in the wild-type isolate, is responsible for the mutant phenotype. Fascinatingly, they demonstrated that the nonacidifying mutant phenotype was due to a lack of citric acid secretion, and the SNP was located in the gene encoding the putative methyltransferase LaeA [71]. This protein is a component of the velvet complex, which regulates light responses, development, and secondary metabolism in Aspergilli [72]. Consequently, these systems genetics approaches have shed light on the link between LaeA, citric acid, and secondary metabolism in *A. niger*.

#### Analyses of *A. niger* genome sequences identify several challenges that are yet to be comprehensively resolved

Several pitfalls to industrial applications of *A. niger* were also highlighted from the publication of the draft genome [60]. Unsurprisingly, numerous predicted protease encoding genes were identified, many of which contained a secretion peptide, which undoubtedly pose a significant challenge to industrial protein production. Additionally, gene clusters predicted to biosynthesize the mycotoxins fumonisin and ochratoxin A were present in CBS 513.88. Subsequent metabolomic analyses led by Jens Frisvad and his colleagues suggest up to 10% of industrially used *A. niger* isolates are able to produce these potential carcinogens [73]. In addition to potential issues with toxicity, production of unwanted secondary metabolites might confound production efforts of heterologous metabolites or new-to-nature compounds, as these molecules will be produced under similar conditions, and will likely be co-extracted during proof-of-concept of scale-up stages. However, with genome editing technology (see below), it should be possible to tackle this problem by removing mycotoxin clusters [2].

A more general problem that became apparent from the *A. niger* genome sequence, however, was that functional predictions were only possible for approximately half of the putative 14,165 putative open reading frames [60]. Subsequent release of additional *A. niger* genomes [74, 75] and the continued improvement and refinement of online genome databases and analyses portals [76–80] have not drastically increased rates of gene functional annotation. Indeed, genome mining of the acidogenic isolate ATCC 1015 [74] using the publicly available analyses portal FungiDB [79] indicates that 4491 (approximately 41%) of predicted genes encode products that are annotated as a 'hypothetical protein', which also lack any functional prediction based on Gene Ontology (GO) or Interpro Domains.

This genomic 'black box' presents several challenges for systems level understanding of *A. niger* and rational strain engineering for industrial applications. Firstly, while several thousand 'hypothetical' genes are transcriptionally active during a diverse range of experimental conditions, many of which model industrial processes [79], the incentive to study these genes using time and labour-intensive loss-of-function approaches is very low. This is further complicated by functional redundancy, where deletion of a single gene has no measurable impact. This problem has been partially obviated by continued molecular tool development in *A. niger*, as highly optimized, titratable, and inducible/repressible promoters have been developed [81]. These molecular tools enable expression of a gene above the native levels, leading to measurable phenotypic effects in so called 'gain-of-function' approaches, with the added advantage that it is possible to functionally characterize essential genes. However, these strategies are unlikely to have the necessary throughput for functional characterization of several thousand genes.

Secondly, assigning functional prediction to the genomic 'black box' using gene orthology is also problematic, as model (or reference) organisms can be misleading. Indeed, inferring function from unicellular yeasts, such as *Saccharomyces cerevisiae*, or other Aspergilli, such as *A. nidulans*, is at best advisory, and at worst misleading, with genes and encoded products playing different roles between organisms [2, 82]. Recent applications of genome editing in filamentous fungi arguably hold the greatest promise for rapid gene functional characterization in *A. niger* [83, 84], with the potential to lead to comprehensive, systems level understanding of this organism, or to engineer new synthetic or semi-synthetic derivatives for highly optimized industrial applications.

### A new era: genome editing with CRISPR/Cas

Actually described 30 years ago in 1987 [85], CRISPR (clustered regularly interspaced palindromic repeats) elements have been universally exploited along with associated endonucleases (for example proteins of the Cas family) for about 5 years. Early DNA sequencing of bacterial genome revealed short repetitive sequences with unknown functions in *E. coli* [85], which were then discovered across many bacterial species and two decades later revealed as an adaptive defence mechanism against bacteriophages [86]. While studying ways to reduce susceptibility of starter cultures for yoghurt production against phages, industrial biotechnologists observed that cultures previously exposed to a virus where resistant upon a second encounter with the same [86]—the reason being the specific recognition, double-strand cut and inactivation of the invading DNA by the CRISPR/Cas9 system. In a seminal 2012 paper, it was shown that the system is programmable to cut any DNA sequence with high specificity [87], which generated a watershed momentum, given the ability of the endonuclease to retain its activity in many different organisms (fungi, insects, mice, humans, plants etc.). The use of the genome editing technology in filamentous fungi has been recently reviewed [88]. While delivery of DNA encoding the components of the system (endonuclease, guide RNA), or in vitro generated components themselves, into the fungal cell remains a challenge due to the fungal cell wall, and still requires common protocols as protoplasting, different strategies have been developed to increase efficiency in species such as *A. niger* and other *Aspergillus* spp., *Alternaria alternata, Coprinopsis cinerea, Ustilago maydis, Trichoderma reesei, Neurospora crassa, Penicillium chrysogenum, Myceliophthora thermophila, Beauveria brassiana* (reviewed in [88]). These strategies include the use of efficient promoters like *trpC, gpdA* or RNA polymerase III promoters like *U6* for in vivo expression of endonuclease and/or gRNA, codon optimization of Cas9, transient expression of endonuclease, integration of endonuclease into the host genome, delivery of in vitro synthesized gRNA or purified endonuclease, and others. Also, considerations concerning the use of appropriate markers (following transformation of CRISPR/Cas components or genome editing) and lethal or unwanted (e.g. off-target) effects of the endonuclease have been addressed in filamentous fungi. For the latter, specificity of genome editing can be increased by (1) favouring host's homology-directed repair (HDR) over non-homologous end-joining (NHEJ) pathway following DNA cut, or (2) generating DNA double-strand breaks with long overhangs by using either a modified Cas9 (nickase Cas9—a.k.a. nCas9—able to cut only one DNA strand and thus following duplexing with two distinct gRNA generating long "sticky ends") [89].

With the proper strategy, CRISPR/Cas technology looks like a spider able to catch a couple of flies; indeed, a broad spectrum of opportunities arises for strain engineering. One recent example illustrates the reach of the technology when applied to *A. niger* and other filamentous fungi. Kuivanen et al. [90] used CRISPR/Cas technology to delete multiple genes in *A. niger* for the biotechnological production of the platform chemical galactaric acid. Derived from D-galacturonic acid, the main component of the natural polymer pectin, galactaric acid is used as precursor for Nylon and in skin-care cosmetics [90]. Although *A. niger* can hydrolyze pectin, D-galacturonic acid is also a precursor for the fungal galacturonic acid pathway, and galactaric acid can be catabolized by an unknown pathway. The authors deleted seven genes involved in catabolism of D-galacturonic acid and galactaric acid in *A. niger* using a strategy involving in vitro synthesized gRNA and plasmid-encoded Cas9. With such an engineered *A. niger* strain, the authors showed the digestion of pectin-rich biomass into galactaric acid in a single process [90].

## The *Aspergillus niger* community as of today

In an effort to map the landscape of international research groups currently working on *A. niger*, we retrieved all PubMed articles with '*Aspergillus niger*' in title, abstract, and keywords published during the last 5 years (2013–2017), resulting in 2068 hits (Additional file 3: Table S3). Members of the community were defined by having authored at least 5 articles over the last 5 years. This list of researchers was then manually curated to highlight group leaders/PIs, and collaboration among research groups (as determined by at least one co-authorship). We focused on this relative short time span to ensure mapping of researchers actively working on *A. niger*, and included abstract and keywords in the search filter to expand the results to the community of those scientists not only studying *A. niger* biology, but also investigating the fungus in other relevant areas (e.g. bioremediation, geomicrobiology, pathogenicity, toxin production and food safety, agricultural microbiology). Based on previously reported information [91–93] we also compiled a list of (multinational) companies using *A. niger* as a workhorse for the production of citric acid and enzymes (Table 1). Our mapping of both basic and applied research on *A. niger* (Fig. 3, Table 1) shows that the community is diverse and geographically dispersed, yet (at least for the basic research community) well-connected.

## Future challenges for the *A. niger* community

This historical overview has covered some of the scientific trends and key discoveries that have occurred in the field of *A. niger* biotechnology. Clearly, there are a diverse range of other industrially relevant fungi and bacteria that have also undergone revolutionary advances since their first use by early industrial microbiologists in the late nineteenth and early twentieth centuries. What does the future of industrial biotechnology hold for *A. niger*, and other microbial cell factories? In an

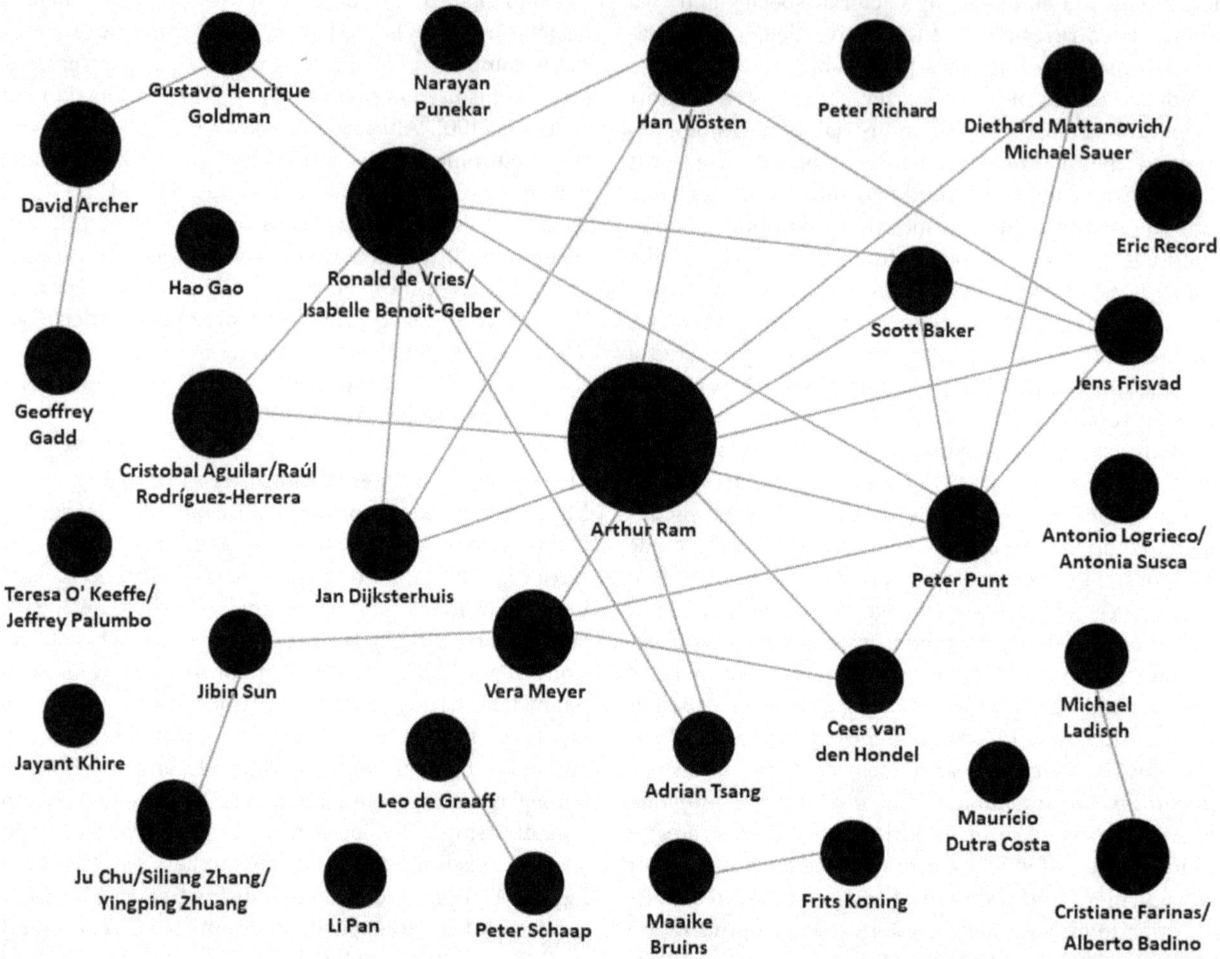


**Fig. 3** The *A. niger* research community is diverse and well-connected. Network of the community of researchers investigating *A. niger* biology or using the fungus as model organism based on our literature survey. Articles with '*Aspergillus niger*' in title, abstract or keywords published in 2013–2017 were retrieved from PubMed. Scientists with at least 5 last (co-)authorships were selected (Additional file 3: Table 3). Size of circles roughly indicates number of published articles. Connections indicate collaborations as indicated by at least one co-authorship. Scientists are located mainly in Europe, followed by China, South America, North America, and India. The position of the circles is arbitrary. Please note that some connections/collaborations might be missing due to the stringency of our PubMed search

attempt to answer this question, we generated predictions for common research themes and topics over the next 20 years of *A. niger* research (Fig. 4). This speculation indicated a hypothesized future focus on synthetic biology (including generation of mycotoxin-free isolates), network analysis (including genomics, gene expression and metabolomics), increased applications of co-cultivation technology and CRISPR-Cas9 genome editing, and a continued focus on secondary metabolism, fermentation, citric acid production, enzymes, and glucoamylase research (Fig. 4).

In general, three key components are becoming increasingly available in the biotechnologist's toolbox; (1) publicly available and well-annotated genome data for several *A. niger* strains, and open-source bioinformatic pipelines that enable sophisticated comparative genomic and other analyses by non-coders; (2) guideline cultivation protocols with engineered macroscopic morphologies of *A. niger*, which facilitate improved product titres; (3) a versatile suite of molecular techniques for high-throughput gene functional analyses, including genome editing. What more has to come to fully understand and optimally exploit *A. niger*? In our opinion, the following outstanding questions need to be addressed by the community in the near future:

1. How can community efforts maintain and increase the quality and usability (from data deposition to analysis) of fungal datasets in light of the increasing amount of omics and literature data generated? How

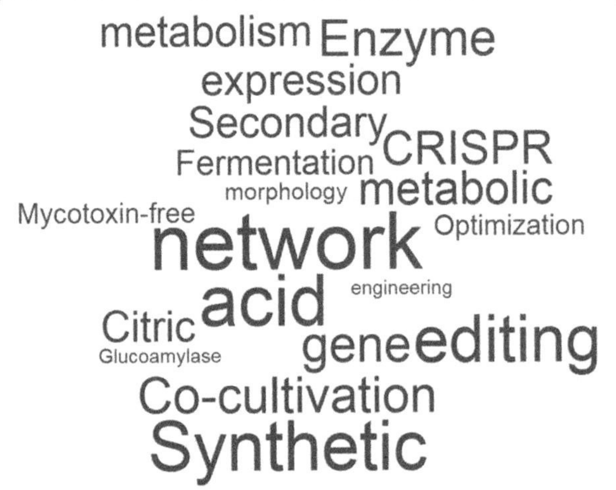


**Fig. 4** Predicting the next 20 years of *A. niger* research. Each author of this review independently predicted and independently weighted approximately 20 words that they hypothesize will commonly feature in future manuscript titles that also contain the words '*Aspergillus niger*'. The word cloud was generated online as described in Fig. 1

can we verify the accuracy of predictive algorithms to assign function to hypothetical genes?

2. How can we solve the "hypothetical protein problem", and how can we assign function to those putative proteins? What is the best and easiest way to integrate genome, transcriptome, proteome, and metabolome data for powerful comparative omics approaches?
3. How can we generate accurate genome-wide metabolic networks that also are integrated with other omics data?
4. How can miniaturised cultivation in microtiter plate (or smaller) size be adapted for *A. niger* and other filamentous fungi for high-throughput screenings?
5. How can cell heterogeneity be investigated, so that variations in metabolism and gene-expression are not averaged over a whole colony or mycelium, and which new tools will foster these single-cell approaches?
6. How can stable and reproducible growth of *A. niger* in mixed cultures be achieved, both in academic research (e.g. to investigate activation of secondary metabolism) and industrial processes (e.g. for efficient enzyme production)? Which co-cultivation approaches and tools for the current paradigmatic shift in microbial cultivation [94] can be specifically adapted/developed for *A. niger* and other filamentous fungi?
7. Would concerted efforts to construct a genome-wide deletion and/or overexpression library of *A. niger*, similar to that existing for *S. cerevisiae* [95], *N. crassa* [96] and *Schizosaccharomyces pombe* [97], be sufficiently helpful for the community to warrant the substantial investment of research funds and resources?
8. How can a minimal *A. niger* genome be defined, and with which approach should it be generated? Which secondary metabolite clusters should be included or omitted?
9. Which synthetic biology tools to regulate different metabolic pathways in parallel can be developed or implemented, e.g. for the construction of genetic circuits to optimize metabolic fluxes for efficient production of enzymes, organic acids, or secondary metabolites?
10. Can *A. niger*, or other filamentous fungi, be exploited for "space biotechnological" purposes, as essential companions of astronauts for the autonomous production of food, enzymes, antibiotics, or for use in terraforming efforts?

## Conclusion

Given the tremendous advances in the knowledge of *A. niger* biology over the last century, and the concomitant development of bioinformatics, cultivation, and molecular tools which are now at disposal of the community, this industrial fungus has the potential to remain one of the most versatile fungal platform microorganism. *A. niger* offers a chassis for products which cannot be produced in easier to handle bacterial systems, and it is able to produce not only proteins and enzymes at high concentrations, but also pharmaceuticals which are beneficial for human and animal health. Indeed, we predict that *A. niger* will be one of the key organisms involved in the next industrial revolution: the change from a fossil-based economy to a bio-economy. At this pace, we are excited to witness what the future will bring.

**Authors' contributions**

TC, CN and VM co-wrote the manuscript. The authors apologize to the authors of many excellent discoveries in the field of *A. niger* research that we did not discuss in the context of this review. All authors read and approved the final manuscript.

**Competing interests**

The authors declare that they have no competing interests.

**Funding**

TC and CN gratefully acknowledge Postdoctoral funding from Technische Universität Berlin.

**References**

1. Show PL, Oladele KO, Siew QY, Aziz Zakry FA, Lan JC-W, Ling TC. Overview of citric acid production from *Aspergillus niger*. Front Life Sci. 2015;8:271–83.
2. Meyer V, Andersen MR, Brakhage AA, Braus GH, Caddick MX, Cairns TC,

et al. Current challenges of research on filamentous fungi in relation to human welfare and a sustainable bio-economy: a white paper. Fungal Biol Biotechnol. 2016;3:6.
3. Currie JN. The citric acid fermentation of *Aspergillus niger*. J Biol Chem. 1917;31:15–37.
4. Penicillin Production through Deep-tank Fermentation. https://www.acs.org/content/acs/en/education/whatischemistry/landmarks/penicillin.html. Accessed 9 Apr 2018.
5. Citric Acid Market worth USD 3.6 Billion USD by 2020. https://www.marketsandmarkets.com/PressReleases/citric-acid.asp. Accessed 9 Apr 2018.
6. Sayers EW, Barrett T, Benson DA, Bryant SH, Canese K, Chetvernin V, et al. Database resources of the National Center for Biotechnology Information. Nucleic Acids Res. 2009;37(Suppl):1.
7. Tomlinson N, Campbell JJ, Trussell PC. The influence of zinc, iron, copper, and manganese on the production of citric acid by *Aspergillus niger*. II. Evidence for the essential nature of copper and manganese. J Bacteriol. 1951;61:17–25.
8. Kitos PA, Campbell JJR, Tomlinson N. Influence of temperature on the trace element requirements for citric acid production by *Aspergillus niger*. Appl Microbiol. 1953;1:156–9.
9. Shu P, Johnson MJ. Effect of the composition of the sporulation medium on citric acid production by *Aspergillus niger* in submerged culture. J Bacteriol. 1947;54:161–7.
10. Gardner JF, James LV, Rubbo SD. Production of citric acid by mutants of *Aspergillus niger*. Microbiology. 1956;14:228–37.
11. Trumphy BH, Millis NF. Nutritional requirements of an *Aspergillus niger* mutant for citric acid production. Microbiology. 1963;30:381–93.
12. Clarke AE, Stone BA. Properties of a β-(1 → 4)-glucan hydrolase from *Aspergillus niger*. Biochem J. 1965;96:802–7.
13. Cain RB. The identity of shikimate dehydrogenase and quinate dehydrogenase in *Aspergillus niger*. Biochem J. 1972;127:15P.
14. Tsuge H, Natsuaki O, Ohashi K. Purification, properties, and molecular features of glucose oxidase from *Aspergillus niger*. J Biochem. 1975;78:835–43.
15. Toraya T, Fujimura M, Ikeda S-I, Fukui S, Yamada H, Kumagai H. Affinity chromatography of amine oxidase from *Aspergillus niger*. Biochim Biophys Acta Protein Struct. 1976;420:316–22.
16. Mill PJ. The pectic enzymes of *Aspergillus niger*. A mercury-activated exopolygalacturonase. Biochem J. 1966;99:557–61.
17. Lineback DR, Russell IJ, Rasmussen C. Two forms of the glucoamylase of *Aspergillus niger*. Arch Biochem Biophys. 1969;134:539–53.
18. Pazur JH, Knull HR, Cepure A. Glycoenzymes: structure and properties of the two forms of glucoamylase from *Aspergillus niger*. Carbohydr Res. 1971;20:83–96.
19. Glucoamylase Industry: 2017 Global Market Demand, Growth, Trends and 2022 Forecast Report. https://marketersmedia.com/glucoamylase-industry-2017-global-market-demand-growth-trends-and-2022-forecast-report/259805. Accessed 9 Apr 2018.
20. Johnson TF, Reisman RE, Arbesman CE. Late onset asthma due to inhalation of *Aspergillus niger*. Clin Exp Allergy. 1975;5:397–401.
21. Cairns TC, Studholme DJ, Talbot NJ, Haynes K. New and improved techniques for the study of pathogenic fungi. Trends Microbiol. 2015;24:35–50.
22. Punt PJ, Oliver RP, Dingemanse MA, Pouweisa PH, van den Handel CAMJJ. Transformation of *Aspergillus* based on the hygromycin B resistance marker *from Escherichia coli*. Gene. 1987;56:117–24.
23. Goosen T, Bloemheuvel G, Gysler C, de Bie DA, van den Broek HWJ, Swart K. Transformation of *Aspergillus niger* using the homologous orotidine-5′-phosphate-decarboxylase gene. Curr Genet. 1987;11:499–503.
24. van Hartingsveldt W, Mattern IE, van Zeijl CMJ, Pouwels PH, van den Hondel CAMJJ. Development of a homologous transformation system for *Aspergillus niger* based on the pyrG gene. Mol Gen Genet. 1987;206:71–5.
25. Meyer V, Wanka F, van Gent J, Arentshorst M, van den Hondel CAMJJ, Ram AFJ. Fungal gene expression on demand: An inducible, tunable, and metabolism-independent expression system for *Aspergillus niger*. Appl Environ Microbiol. 2011;77:2975–83.
26. Roberts IN, Jeenes DJ, MacKenzie DA, Wilkinson AP, Sumner IG, Archer DB. Heterologous gene expression in *Aspergillus niger*: a glucoamylase-porcine pancreatic prophospholipase A2 fusion protein is secreted and processed to yield mature enzyme. Gene. 1992;122:155–61.
27. de Souza PM, de Assis Bittencourt ML, Caprara CC, de Freitas M, de Almeida RPC, Silveira D, et al. A biotechnology perspective of fungal proteases. Braz J Microbiol. 2015;46:337–46.
28. Meyer V, Arentshorst M, El-Ghezal A, Drews AC, Kooistra R, van den Hondel CAMJJ, et al. Highly efficient gene targeting in the *Aspergillus niger* kusA mutant. J Biotechnol. 2007;128:770–5.
29. Ninomiya Y, Suzuki K, Ishii C, Inoue H. Highly efficient gene replacements in *Neurospora strains* deficient for nonhomologous end-joining. Proc Natl Acad Sci. 2004;101:12248–53.
30. da Silva Ferreira ME, Kress MRVZ, Savoldi M, Goldman MHS, Härtl A, Heinekamp T, et al. The akuBKU80 mutant deficient for nonhomologous end joining is a powerful tool for analyzing pathogenicity in *Aspergillus fumigatus*. Eukaryot Cell. 2006;5:207–11.
31. Greiner R, Konietzny U. Phytase for food application. Food Technol Biotechnol. 2006;44:125–40.
32. Haefner S, Knietsch A, Scholten E, Braun J, Lohscheidt M, Zelder O. Biotechnological production and applications of phytases. Appl Microbiol Biotechnol. 2005;68:588–97.
33. Culleton H, Mckie V, De Vries RP. Physiological and molecular aspects of degradation of plant polysaccharides by fungi: What have we learned from *Aspergillus*? Biotechnol J. 2013;8:884–94.
34. Petersen KL, Lehmbeck J, Christensen T. A new transcriptional activator for amylase genes in *Aspergillus*. Mol Gen Genet. 1999;262:668–76.
35. Gruben BS, Zhou M, Wiebenga A, Ballering J, Overkamp KM, Punt PJ, et al. *Aspergillus niger* RhaR, a regulator involved in L-rhamnose release and catabolism. Appl Microbiol Biotechnol. 2014;98:5531–40.
36. van Peij NN, Visser J, de Graaff LH. Isolation and analysis of *xlnR*, encoding a transcriptional activator co-ordinating xylanolytic expression in *Aspergillus niger*. Mol Microbiol. 1998;27:131–42.
37. van Peij NN, Gielkens MM, de Vries RP, Visser J, de Graaff LH. The transcriptional activator XlnR regulates both xylanolytic and endoglucanase gene expression in *Aspergillus niger*. Appl Environ Microbiol. 1998;64:3615–9.
38. Lubertozzi D, Keasling JD. Developing *Aspergillus* as a host for heterologous expression. Biotechnol Adv. 2009;27:53–75.
39. Punt PJ, van Biezen N, Conesa A, Albers A, Mangnus J, van den Hondel CAMJJ. Filamentous fungi as cell factories for heterologous protein production. Trends Biotechnol. 2002;20:200–6.
40. Taheri-Talesh N, Horio T, Araujo-baza L, Dou X, Espeso EA, Pen MA, et al. The tip growth apparatus of *Aspergillus nidulans*. Mol Biol Cell. 2008;19:1439–49.
41. Wosten HAB, Moukha SM, Sietsma JH, Wessels JGH. Localization of growth and secretion of proteins in *Aspergillus niger*. J Gen Microbiol. 1991;137:2017–23.
42. Krijgsheld P, Altelaar AFM, Post H, Ringrose JH, Müller WH, Heck AJR, et al. Spatially resolving the secretome within the mycelium of the cell factory *Aspergillus niger*. J Proteome Res. 2012;11:2807–18.
43. Kwon MJ, Nitsche BM, Arentshorst M, Jorgensen TR, Ram AF, Meyer V. The transcriptomic signature of RacA activation and inactivation provides new insights into the morphogenetic network of *Aspergillus niger*. PLoS ONE. 2013;8:e68946.
44. Wucherpfennig T, Lakowitz A, Driouch H, Krull R, Wittmann C. Customization of *Aspergillus niger* morphology through addition of talc micro particles. J Vis Exp. 2012;61:4023.
45. Read ND. Exocytosis and growth do not occur only at hyphal tips. Mol Microbiol. 2011;81:4–7.
46. Hayakawa Y, Ishikawa E, Shoji J, Nakano H, Kitamoto K. Septum-directed secretion in the filamentous fungus *Aspergillus oryzae*. Mol Microbiol. 2011;81:40–55.

47. de Bekker C, Bruning O, Jonker MJ, Breit TM, Wösten HA. Single cell transcriptomics of neighboring hyphae of *Aspergillus niger*. Genome Biol. 2011;12:R71.
48. Vinck A, Terlou M, Pestman WR, Martens EP, Ram AF, van den Hondel CAMJJ, et al. Hyphal differentiation in the exploring mycelium of *Aspergillus niger*. Mol Microbiol. 2005;58:693–9.
49. Levin AM, de Vries RP, Conesa A, De Bekker C, Talon M, Menke HH, et al. Spatial differentiation in the vegetative mycelium of *Aspergillus niger*. Eukaryot Cell. 2007;6:2311–22.
50. Bleichrodt RJ, Vinck A, Read ND, Wösten HAB. Selective transport between heterogeneous hyphal compartments via the plasma membrane lining septal walls of *Aspergillus niger*. Fungal Genet Biol. 2015;82:193–200.
51. Masuo S, Komatsuzaki A, Takeshita N, Itoh E, Takaaki O, Zhou S, et al. Spatial heterogeneity of glycogen and its metabolizing enzymes in *Aspergillus nidulans* hyphal tip cells. Fungal Genet Biol. 2018;110:48–55.
52. Bleichrodt RJ, van Veluw GJ, Recter B, Maruyama JI, Kitamoto K, Wösten HAB. Hyphal heterogeneity in *Aspergillus oryzae* is the result of dynamic closure of septa by Woronin bodies. Mol Microbiol. 2012;86:1334–44.
53. Schäberle TF, Hack IM. Overcoming the current deadlock in antibiotic research. Trends Microbiol. 2014;22:165–7.
54. Harvey AL, Edrada-Ebel R, Quinn RJ. The re-emergence of natural products for drug discovery in the genomics era. Nat Rev Drug Discov. 2015;14:111–29.
55. Richter L, Wanka F, Boecker S, Storm D, Kurt T, Vural Ö, et al. Engineering of *Aspergillus niger* for the production of secondary metabolites. Fungal Biol Biotechnol. 2014;1:4.
56. Steiniger C, Hoffmann S, Mainz A, Kaiser M, Voigt K, Meyer V, et al. Harnessing fungal nonribosomal cyclodepsipeptide synthetases for mechanistic insights and tailored engineering. Chem Sci. 2017;8(11):7834–43.
57. Schuetze T, Meyer V. Polycistronic gene expression in *Aspergillus niger*. Microb Cell Fact. 2017;16:162.
58. Geib E, Brock M. ATNT: an enhanced system for expression of polycistronic secondary metabolite gene clusters in *Aspergillus niger*. Fungal Biol Biotechnol. 2017;4:13.
59. Galagan JE, Calvo SE, Borkovich KA, Selker EU, Read NO, Jaffe D, et al. The genome sequence of the filamentous fungus *Neurospora crassa*. Nature. 2003;422:859–68.
60. Pel HJ, de Winde JH, Archer DB, Dyer PS, Hofmann G, Schaap PJ, et al. Genome sequencing and analysis of the versatile cell factory *Aspergillus niger* CBS 513.88. Nat Biotechnol. 2007;25:221–31.
61. Galagan JE, Calvo SE, Cuomo C, Ma LJ, Wortman JR, Batzoglou S, et al. Sequencing of *Aspergillus nidulans* and comparative analysis with *A. fumigatus* and *A. oryzae*. Nature. 2005;438:1105–15.
62. Machida M, Asai K, Sano M, Tanaka T, Kumagai T, Terai G, et al. Genome sequencing and analysis of *Aspergillus oryzae*. Nature. 2005;438:1157–61.
63. Nierman WC, Pain A, Anderson MJ, Wortman JR, Kim HS, Arroyo J, et al. Genomic sequence of the pathogenic and allergenic filamentous fungus *Aspergillus fumigatus*. Nature. 2005;438:1151–6.
64. de Vries RP, Riley R, Wiebenga A, Aguilar-Osorio G, Amillis S, Uchima CA, et al. Comparative genomics reveals high biological diversity and specific adaptations in the industrially and medically important fungal genus *Aspergillus*. Genome Biol. 2017;18:28.
65. Bignell E, Cairns TC, Throckmorton K, Nierman WC, Keller NP. Secondary metabolite arsenal of an opportunistic pathogenic fungus. Philos Trans R Soc B Biol Sci. 2016;371:pii: 20160023.
66. Kjærbølling I, Vesth TC, Frisvad JC, Nybo JL, Theobald S, Kuo A, et al. Linking secondary metabolites to gene clusters through genome sequencing of six diverse *Aspergillus* species. Proc Natl Acad Sci. 2018;115:E753–61.
67. Inglis DO, Binkley J, Skrzypek MS, Arnaud MB, Cerqueira GC, Shah P, et al. Comprehensive annotation of secondary metabolite biosynthetic genes and gene clusters of *Aspergillus nidulans*, *A. fumigatus*, *A. niger* and *A. oryzae*. BMC Microbiol. 2013;13:91.
68. Rokas A, Payne G, Fedorova ND, Baker SE, Machida M, Yu J, et al. What can comparative genomics tell us about species concepts in the genus *Aspergillus*? Stud Mycol. 2007;59:11–7.
69. Soanes DM, Alam I, Cornell M, Wong HM, Hedeler C, Paton NW, et al. Comparative genome analysis of filamentous fungi reveals gene family expansions associated with fungal pathogenesis. PLoS ONE. 2008;3:e2300.
70. Scazzocchio C. Fungal biology in the post-genomic era. Fungal Biol Biotechnol. 2014;1:7.
71. Niu J, Arentshorst M, Nair PDS, Dai Z, Baker SE, Frisvad JC, et al. Identification of a classical mutant in the industrial host *Aspergillus niger* by systems genetics: LaeA is required for citric acid production and regulates the formation of some secondary metabolites. G3 Genes Genomes. Genetics. 2016;6:193–204.
72. Bayram O, Krappmann S, Ni M, Bok JW, Helmstaedt K, Valerius O, et al. VelB/VeA/LaeA complex coordinates light signal with fungal development and secondary metabolism. Science. 2008;320:1504–6.
73. Frisvad JC, Larsen TO, Thrane U, Meijer M, Varga J, Samson RA, et al. Fumonisin and ochratoxin production in industrial *Aspergillus niger* strains. PLoS ONE. 2011;6:e23496.
74. Andersen MR, Salazar MP, Schaap PJ, Van De Vondervoort PJI, Culley D, Thykaer J, et al. Comparative genomics of citric-acid-producing *Aspergillus niger* ATCC 1015 versus enzyme-producing CBS 513.88. Genome Res. 2011;21:885–97.
75. Gong W, Cheng Z, Zhang H, Liu L, Gao P, Wang L. Draft genome sequence of *Aspergillus niger* strain An76. Genome Announc. 2016;4:e01700–15.
76. Gilsenan JM, Cooley J, Bowyer P. CADRE: The Central Aspergillus Data REpository 2012. Nucleic Acids Res. 2012;40:D660–6.
77. Wortman JR, Gilsenan JM, Joardar V, Deegan J, Clutterbuck J, Andersen MR, et al. The 2008 update of the *Aspergillus nidulans* genome annotation: a community effort. Fungal Genet Biol. 2009;46(Suppl):1.
78. Cerqueira GC, Arnaud MB, Inglis DO, Skrzypek MS, Binkley G, Simison M, et al. The Aspergillus Genome Database: multispecies curation and incorporation of RNA-Seq data to improve structural gene annotations. Nucleic Acids Res. 2014;42:D705–10.
79. Stajich JE, Harris T, Brunk BP, Brestelli J, Fischer S, Harb OS, et al. FungiDB: an integrated functional genomics database for fungi. Nucleic Acids Res. 2012;40:D675–81.
80. Grigoriev IV, Nikitin R, Haridas S, Kuo A, Ohm R, Otillar R, et al. MycoCosm portal: gearing up for 1000 fungal genomes. Nucleic Acids Res. 2014;42:D699–704.
81. Wanka F, Cairns T, Boecker S, Berens C, Happel A, Zheng X, et al. Tet-on, or Tet-off, that is the question: advanced conditional gene expression in *Aspergillus*. Fungal Genet Biol. 2016;89:72–83.
82. Kwon MJ, Arentshorst M, Roos ED, van den Hondel CAMJJ, Meyer V, Ram AFJ. Functional characterization of Rho GTPases in Aspergillus niger uncovers conserved and diverged roles of Rho proteins within filamentous fungi. Mol Microbiol. 2011;79:1151–67.
83. Sarkari P, Marx H, Blumhoff ML, Mattanovich D, Sauer M, Steiger MG. An efficient tool for metabolic pathway construction and gene integration for *Aspergillus niger*. Bioresour Technol. 2017;245:1327–33.
84. Nødvig CS, Nielsen JB, Kogle ME, Mortensen UH. A CRISPR-Cas9 system for genetic engineering of filamentous fungi. PLoS ONE. 2015;10:e0133085.
85. Ishino Y, Shinagawa H, Makino K, Amemura M, Nakata A. Nucleotide sequence of the *iap* gene, responsible for alkaline phosphatase isozyme conversion in *Escherichia coli*, and identification of the gene product. J Bacteriol. 1987;169:5429–33.
86. Barrangou R, Fremaux C, Deveau H, Richards M, Boyaval P, Moineau S, et al. CRISPR provides acquired resistance against viruses in prokaryotes. Science. 2007;315:1709–12.
87. Jinek M, Chylinski K, Fonfara I, Hauer M, Doudna JA, Charpentier E. A programmable dual-RNA-guided DNA endonuclease in adaptive bacterial immunity. Science. 2012;337:816–21.
88. Deng H, Gao R, Liao X, Cai Y. CRISPR system in filamentous fungi: current achievements and future directions. Gene. 2017;627:212–21.
89. Zetsche B, Gootenberg JS, Abudayyeh OO, Slaymaker IM, Makarova KS, Essletzbichler P, et al. Cpf1 is a single RNA-guided endonuclease of a class 2 CRISPR-Cas system. Cell. 2018;163:759–71.

90. Kuivanen J, Wang Y-MJ, Richard P. Engineering *Aspergillus niger* for galactaric acid production: elimination of galactaric acid catabolism by using RNA sequencing and CRISPR/Cas9. Microb Cell Fact. 2016;15:210.
91. Ward OP. Production of recombinant proteins by filamentous fungi. Biotechnol Adv. 2011;30:1119–39.
92. Meyer V. Genetic engineering of filamentous fungi–progress, obstacles and future trends. Biotechnol Adv. 2008;26:177–85.
93. Fiedler MRM, Nitsche BM, Franziska W, Meyer V. *Aspergillus*: a cell factory with unlimited prospects. In: Gupta VK, Schmoll M, Maki M, editors. Applications of microbial engineering. Boca Raton: CRC Press; 2013.
94. Nai C, Meyer V. From axenic to mixed cultures: Technological advances accelerating a paradigm shift in microbiology. Trends Microbiol. 2017:pii: S0966-842X(17)30253-6 (article in press).
95. Winzeler EA, Shoemaker DD, Astromoff A, Liang H, Anderson K, Andre B, et al. Functional characterization of the *S. cerevisiae* genome by gene deletion and parallel analysis. Science. 1999;285:901–6.
96. Dunlap JC, Borkovich KA, Henn MR, Turner GE, Sachs MS, Glass NL, et al. Enabling a community to dissect an organism: overview of the *Neurospora* functional genomics project. Adv Genet. 2007;57:49–96.
97. Kim D-U, Hayles J, Kim D, Wood V, Park H-O, Won M, et al. Analysis of a genome-wide set of gene deletions in the fission yeast *Schizosaccharomyces pombe*. Nat Biotechnol. 2010;28:617–23.

8

# Genome-wide analysis of cytochrome P450s of *Trichoderma* spp.: annotation and evolutionary relationships

Sonia Chadha[1], Sayaji T. Mehetre[1], Ravindra Bansal[1], Alan Kuo[2], Andrea Aerts[2], Igor V. Grigoriev[2], Irina S. Druzhinina[3] and Prasun K. Mukherjee[1*]

**Abstract**

**Background:** Cytochrome P450s form an important group of enzymes involved in xenobiotics degradation and metabolism, both primary and secondary. These enzymes are also useful in industry as biotechnological tools for bioconversion and a few are reported to be involved in pathogenicity. *Trichoderma* spp. are widely used in industry and agriculture and are known for their biosynthetic potential of a large number of secondary metabolites. For realising the full biosynthetic potential of an organism, it is important to do a genome-wide annotation and cataloguing of these enzymes.

**Results:** Here, we have studied the genomes of seven species (*T. asperellum*, *T. atroviride*, *T. citrinoviride*, *T. longibrachiatum*, *T. reesei* , *T. harzianum* and *T. virens*) and identified a total of 477 cytochrome P450s. We present here the classification, evolution and structure as well as predicted function of these proteins. This study would pave the way for functional characterization of these groups of enzymes and will also help in realization of their full economic potential.

**Conclusion:** Our CYPome annotation and evolutionary studies of the seven *Trichoderma* species now provides opportunities for exploration of research-driven strategies to select *Trichoderma* species for various applications especially in relation to secondary metabolism and degradation of environmental pollutants.

## Background

*Trichoderma* (Hypocreales, Ascomycota, Dikarya) species are among the most common fungi frequently isolated as mycotrophs from various fungi and as saprotrophs from free soil, soil litter, dead wood and rhizosphere, and includes more than 256 accepted species [1, 2]. These fungi are economically important due to their ability to produce enzymes of industrial importance, ability to kill/inhibit many plant pathogenic fungi, to boost plant immunity and promote plant growth, in addition to their ability to produce a plethora of secondary metabolites [3, 4]. A few species/strains are known to be opportunistic human pathogens [5]. *Trichoderma* spp. are thus ideal candidates for genome-wide studies to further augment their biotechnological applications. The first species to be sequenced is *Trichoderma reesei*, industrial source of cellulases and hemicellulases [6]. This was soon followed by whole genome sequencing of two strongly mycoparasitic species, viz. *T. atroviride* and *T. virens* [7]. A comparative analysis of the mycoparasitic species i.e., *T. atroviride* and *T. virens* with that of weaker mycoparasitic species *T. reesei* yielded novel information on the genome-scale differences between these species. In general, the mycoparasitic species are enriched in genes involved mycoparasitism and secondary metabolism [1, 7, 8]. Four more species, i.e., *T. asperellum* and *T. harzianum* (biocontrol species) and *T. longibrachiatum* and *T. citrinoviride* (opportunistic human pathogens) were subsequently sequenced by US Department of Energy Joint Genome Initiative (Mycocosm [9]; http://jgi.doe.gov/

*Correspondence: prasunmukherjee1@gmail.com
[1] Nuclear Agriculture and Biotechnology Division, Bhabha Atomic Research Centre, Trombay, Mumbai 400085, India
Full list of author information is available at the end of the article

fungi). However, detailed analyses of these four genomes are awaited.

Cytochrome P450 genes (CYPs) are found in the genomes of prokaryotes and lower and higher eukaryotes. CYPs constitute a large superfamily of heme-thiolate proteins involved in the metabolism of a wide variety of both exogenous and endogenous compounds [10]. CYPs are heme b containing monooxygenases which were recognized and defined as a distinct class of hemoproteins [11]. Cyp proteins catalyze the regio-, chemo- and stereospecific oxidation of a vast number of substrates under mild reaction conditions, thus accomplishing chemical transformations. These functions make them important players in xenobiotic degradation and in primary and secondary metabolism. A few such enzymes are also reported to be involved in pathogenicity of plant pathogenic fungi [12–16]. Their diverse functional properties reflect their biological roles and make them important candidates for extensive investigation to explore diverse aspects of P450 functions and regulation as well as for biotechnological applications [17, 18].

Cytochrome P450s are categorized into two main classes, B (initially assigned as Bacterial) and E (initially assigned as Eukaryotic). Bacterial P450s with three component systems [an FAD-containing flavoprotein (NADPH or NADH-dependent reductase), an iron sulphur protein, and the P450 hemeprotein] and the fungal P450 nor (nitric oxide reductase). Clan CYP 55 belong to the 'B'-class [19]. All the other known P450s from distinct systems, including eukaryotic and bacterial P450s, belong to the 'E'-class. The eukaryotic microsomal P450 system contains two components, the NADPH:P450 oxidoreductase (POR), a flavoprotein containing both FAD and FMN, and the P450 monooxygenase containing the heme domain. The prokaryotic (bacterial) soluble P450 monooxygenase P450BM3 (Cyp102) exists as a single protein with both heme and flavin functional domains.

The complete CYP complement of one organism, called CYPome, is a collection of CYP genes in the genome of that species [20]. The current state of knowledge on P450 evolution in eukaryotes points to CYP51 as the ancestral P450, which is believed to have led to the evolution of all the present day P450 families [21]. The expansion and diversification of CYPomes may also provide information on fungal evolutionary adaptation to ecological niches. A key development affecting applied P450 research is the need to define and annotate ever-expanding genomic information. Various web-based resources have been developed to probe and assign various orphan CYPs in numerous genomes, owing to the identification of conserved motifs responsible for oxygen and heme-binding. These databases reveal that enormous number of sequence-diverse P450s is yet to be discovered and explored for functions and diverse activities in all kingdoms. One of the most commonly used resources includes the Nelson database (http://drnelson.uthsc.edu/cytochromeP450.html) [21]. The grouping scheme for CYPs is based on amino acid sequence similarity [22]. The original nomenclature for CYPs is based upon amino acid identity where Cyp proteins with at least 40% identity are placed in the same family [22, 23]. However, due to various evolutionary mechanisms, a straight forward nomenclature might be difficult, therefore, family definition is recommended by integrating phylogeny and protein evolution [24]. To each family, Cyp number is designated according to their taxonomic groups. Fungal Cyp families are numbered as Cyp51-Cyp69, Cyp501-Cyp699 and Cyp5001-Cyp6999. With rapid increase in discoveries of new Cyp proteins through genome sequencing, Nelson database lacks efficiency to annotate all Cyp proteins. For higher-level grouping of families identified via the sequence similarity-based scheme, CYP clan system was first developed and then applied to classify metazoan CYPs [25]. The CYP clan approach places all Cyp families with a monophyletic origin into a single clan and has been successfully applied to classify Cyp families in fungi [26]. For example, if new Cyps had equal identity to two or more Cyp families, they can be tentatively assigned to a clan in which these families belong. A site dedicated to filamentous fungi has been developed that includes comprehensive information on P450 clans and families (http://p450.riceblast.snu.ac.kr) [27]. In filamentous fungi, CYPs are involved in various physiological processes including fitness, resistance to xenobiotics and biosynthesis of a vast array of secondary metabolites with applications in biomedical, agricultural and industrial fields [28–31].

Keeping in view the wide spectrum of biotechnological applications of *Trichoderma* species, and the important role that CYPs play in the biology of fungi, we decided to annotate and make an inventory of the CYPome in the seven species of *Trichoderma* that have been sequenced by JGI. Annotation of these genes would help in commercial exploitation of these proteins. Earlier, the CYPome of several fungal species have been analysed in detail, e.g., *Aspergillus nidulans* [29], *Phanerochete chrysosporium* [32], *Mycosphaerella graminicola* [33] and *Grosmannia clavigera* [34]. However, this subject has not been covered in earlier analyses of *Trichoderma* genomes, except for the inclusion of *T. reesei* in a broad analysis of fungal CYPomes [35]. Moreover, a detailed phylogenetic analysis of *Trichoderma* CYPome could advance our understanding of the evolutionary processes of cytochrome P450 proteins in fungi.

## Results

### CYP proteins in *Trichoderma*

*Trichoderma* CYPome embodies a group of cytochrome P450 diverse proteins which are predicted to participate in a spectrum of functions involved in primary, secondary and xenobiotic metabolism. A total of 595 cytochrome P450 proteins have been identified in seven *Trichoderma* species. These entries were further analysed for the presence of full cytochrome P450 domain which led to the selection of a total 477 Cyp proteins (Table 1) for the detailed study. Entries with incomplete sequences and domains are listed in Additional file 1: Table S1.

Analysis showed that *T. harzianum* genome harbours the highest number of Cyps (101), followed by *T. virens* (90), *T. asperellum* (62), *T. atroviride* (57), *T. citrinoviride* (57), *T. reesei* (57) *and T. longibrachiatum* (53). The number of Cyp proteins for families Cyp5080, Cyp52, Cyp534, Cyp535, Cyp541 and Cyp618 were found conserved among seven *Trichoderma* species. Cytochrome P450 families Cyp504, Cyp505, Cyp5080, Cyp51, Cyp52, Cyp528, Cyp534, Cyp535, Cyp539, Cyp541, Cyp548, Cyp570, Cyp58, Cyp584, Cyp61, Cyp618, Cyp620, Cyp65 and Cyp671 were found ubiquitously present in *Trichoderma* suggesting a conserved role of these proteins. In *Trichoderma*, Cyp families Cyp5039, Cyp5044, Cyp5046, Cyp5049, Cyp5055, Cyp5057, Cyp5060, Cyp5128, Cyp5129, Cyp5134, Cyp5168, Cyp5181, Cyp5246, Cyp5262, Cyp5268, Cyp5292, Cyp5296, Cyp5320, Cyp5334, Cyp5390 and Cyp5391 didn't have any matches in Fungal Cytochrome P450 Database (FCPD). The cytochrome P450 families unique to *Trichoderma* were identified as Cyp5039, Cyp5049, Cyp5055, Cyp5057, Cyp5128, Cyp5129, Cyp5134, Cyp5268, Cyp5292, Cyp5296, Cyp5390 and Cyp5391. These families were predicted to be involved in both xenobiotic and secondary metabolism (Table 2).

### Abundance and diversity of cytochrome P450 family/clan

Identified cytochrome P450 proteins were annotated and classified into 85 families (Fig. 1) and 37 clans (Fig. 2). *Trichoderma* species showed diversity in the number of annotated Cyp families (Table 1, Figs. 1, 2, 3). The numbers of annotated Cyp families among *Trichoderma* species ranged from 36 (*T. atroviride*) to 67 (*T. harzianum*). Annotated CYP clans were also found to be diverse in *Trichoderma* (Fig. 3). The highest numbers of CYP clans were identified in *T. harzianum* (31) and *T. virens* (31). *T. asperellum* and *T. atroviride* contained 25 and 22 clan types respectively. Clans CYP52 and CYP65 were found to be most abundant with 55 and 56 protein entries, respectively (Fig. 3). The number of proteins in the most abundant clans CYP52 and CYP65 ranged from 6 to 12 among *Trichoderma* species. Clan CYP673 was identified only in *T. virens* and *T. harzianum,* containing 1 and 2 members respectively, and was found to be absent in other five species. Similarly, clan CYP56 proteins were found to be unique to *T. asperellum*, *T. harzianum* and *T. virens* with single entries in each species. Clan540 proteins were found absent in *T. citrinoviride*, *T. longibrachiatum* and *T. reesei*. Clans CYP5042, 642, 659 and 677 were identified only in *T. virens* and were absent in all other species.

### Phyletic distribution of CYP families and clans in *Trichoderma*

The genome-wide comparisons and annotations of P450s have allowed us to further develop the relationships among Cyp families in different *Trichoderma* species. To demonstrate the divergence of the primary sequences and evolutionary relationships of cytochrome P450 families in *Trichoderma*, a detailed phylogenetic analysis was carried out using 477 aligned Cyp protein sequences. The phylogenetic tree depicting evolutionary relationships among *Trichoderma* cytochrome P450 proteins are illustrated in Fig. 4. Further, the distribution of different CYP

**Table 1 Taxonomic distribution of putative CYPs in seven *Trichoderma* species**

| Species | Genome size (Mb) | No. of predicted genes | Total Cyp proteins | Proteins with complete sequences | Clan type | Family type | Families with no FCPD matches |
|---|---|---|---|---|---|---|---|
| *T. asperellum* | 37.46 | 12,586 | 73 | 62 | 25 | 40 | 7 |
| *T. atroviride* | 36.10 | 11,863 | 69 | 57 | 22 | 36 | 5 |
| *T. citrinoviride* | 33.48 | 9397 | 75 | 57 | 23 | 41 | 6 |
| *T. longibrachiatum* | 32.24 | 10,792 | 68 | 53 | 21 | 38 | 4 |
| *T. reesei* | 34.10 | 9129 | 70 | 57 | 23 | 42 | 4 |
| *T. harzianum* | 40.98 | 14,095 | 118 | 101 | 31 | 67 | 12 |
| *T. virens* | 39.00 | 12,427 | 122 | 90 | 31 | 59 | 12 |

**Table 2 Phylogenetic clustering of *Trichoderma* CYP families and clans**

| Phylogenetic group ID | Total entries | CYP family | CYP Clan | Putative functions |
|---|---|---|---|---|
| 1 | 33 | Cyp5044[a], Cyp5078, Cyp5080, Cyp5104, Cyp528, Cyp531, Cyp532, Cyp5320[a], Cyp631 | CYP528, CYP531, CYP532 | Xenobiotic metabolism |
| 2 | 19 | Cyp535, Cyp570 | CYP507 | Xenobiotic metabolism |
| 3 | 3 | Cyp673 | CYP673 | |
| 4 | 11 | Cyp5055[a], Cyp5057[a], Cyp5262[a], Cyp537, Cyp62, Cyp684 | CYP537, CYP62 | Xenobiotic metabolism Secondary metabolism |
| 5 | 35 | Cyp5039[a], Cyp5094, Cyp5128[a], Cyp5129[a], Cyp5292[a], Cyp551, Cyp552, Cyp58, Cyp677, Cyp680, Cyp682 | CYP58, CYP677 | Secondary metabolism Xenobiotic metabolism |
| 6 | 10 | Cyp5246[a], Cyp53 | CYP53 | Xenobiotic metabolism |
| 7 | 3 | Cyp630 | CYP630 | Primary metabolism |
| 8 | 23 | Cyp574, Cyp5076, Cyp5168[a], Cyp671 | CYP574 | Secondary metabolism |
| 9 | 17 | Cyp548 | CYP548 | Xenobiotic metabolism |
| 10 | 56 | Cyp5117, Cyp561, Cyp563, Cyp65 | CYP65 | Secondary metabolism |
| 11 | 3 | Cyp627 | CYP627 | |
| 12 | 62 | Cyp5049[a], Cyp52, Cyp5296[a], Cyp538, Cyp539, Cyp584, Cyp587, Cyp655 | CYP52, CYP59 | Xenobiotic metabolism |
| 13 | 19 | Cyp5181[a], Cyp5334[a], Cyp534, Cyp613, Cyp685 | CYP534, CYP613 | Xenobiotic metabolism |
| 14 | 39 | Cyp5134[a], Cyp526, Cyp5390[a], Cyp617, Cyp618 | CYP526, CYP547 | Secondary metabolism |
| 15 | 36 | Cyp505, Cyp5099, Cyp540, Cyp541 | CYP505, CYP540, CYP56 | Primary metabolism |
| 16 | 9 | Cyp504 | CYP504 | Xenobiotic metabolism |
| 17 | 51 | Cyp5046[a], Cyp5068, Cyp5268[a], Cyp530, Cyp5391[a], Cyp620, Cyp621 | CYP530, CYP533 | Xenobiotic metabolism |
| 18 | 1 | Cyp5042 | CYP5042 | |
| 19 | 18 | Cyp503, Cyp5090, Cyp559, Cyp611, Cyp635, Cyp636, Cyp641, Cyp642 | CYP54, CYP550, CYP559, CYP642, CYP657, CYP659 | Secondary metabolism |
| 20 | 29 | Cyp5060[a], Cyp51, Cyp55, Cyp61 | CYP51, CYP55, CYP61 | Primary metabolism |

[a] Corresponding clans for these families are absent in FCPD

clans and families in 20 phylogenetic groups with their putative functions are summarized in Table 2.

Evolutionary studies differentiated 477 cytochrome P450 proteins from 7 *Trichoderma* species into 20 phylogenetic groups (Fig. 4). Group 1 consisted of a total of 33 Cyp proteins from clans CYP528, CYP531 and CYP532. In *Trichoderma,* clan CYP531 consists of five Cyp families including Cyp5078, Cyp5080, Cyp5104, Cyp531 and Cyp631. Group 2 consisted of total 19 protein members belonging to clan CYP507. Members of clan CYP507 have been predicted to be involved in xenobiotic metabolism in Pezizomycotina [36]. In FCPD, clan CYP507 consists of four Cyp families including Cyp 507, Cyp525, Cyp535 and Cyp570. Of these four families, only Cyp535 and Cyp570 families are present in *Trichoderma* species. Group 2 containing clan CYP507 proteins was further differentiated into two sub-groups containing families Cyp535 (7 proteins) and Cyp570 (12 proteins) respectively. In *Trichoderma,* all 19 proteins belonging to clan 507 are grouped together in group 2 suggesting conserved putative role of Cyp535 and Cyp570 in xenobiotics metabolism. Clan CYP673 in group 3 consists of only three members-two from *T. harzianum* and one from *T. virens.*

Group 4 consists of 11 proteins from 2 clans (CYP537 and CYP62). In FCPD, clan CYP537 consists of two families: Cyp537 and Cyp577. In *Trichoderma,* Cyp577 family is absent and Cyp537 proteins are present only in *T. asperellum, T. atroviride, T. citrinoviride* and *T. harzianum.* In group 4, all identified members of clan CYP62 grouped together. Clan CYP62 in FCPD consists of three CYP families including CYP62, CYP626 and CYP684. In *Trichoderma,* one Cyp62 (*T. harzianum*) and three Cyp684 proteins (one each) were identified in *T.* atroviride, *T. harzianum* and *T. virens.* Group 4 also contained Cyp50555, Cyp5057 and Cyp5262 proteins. The corresponding clans for these three families are absent in FCPD. Protein Cyp5262 was grouped together with members of clan CYP537, whereas Cyp50555 and Cyp5057 proteins formed a separate subgroup in Group

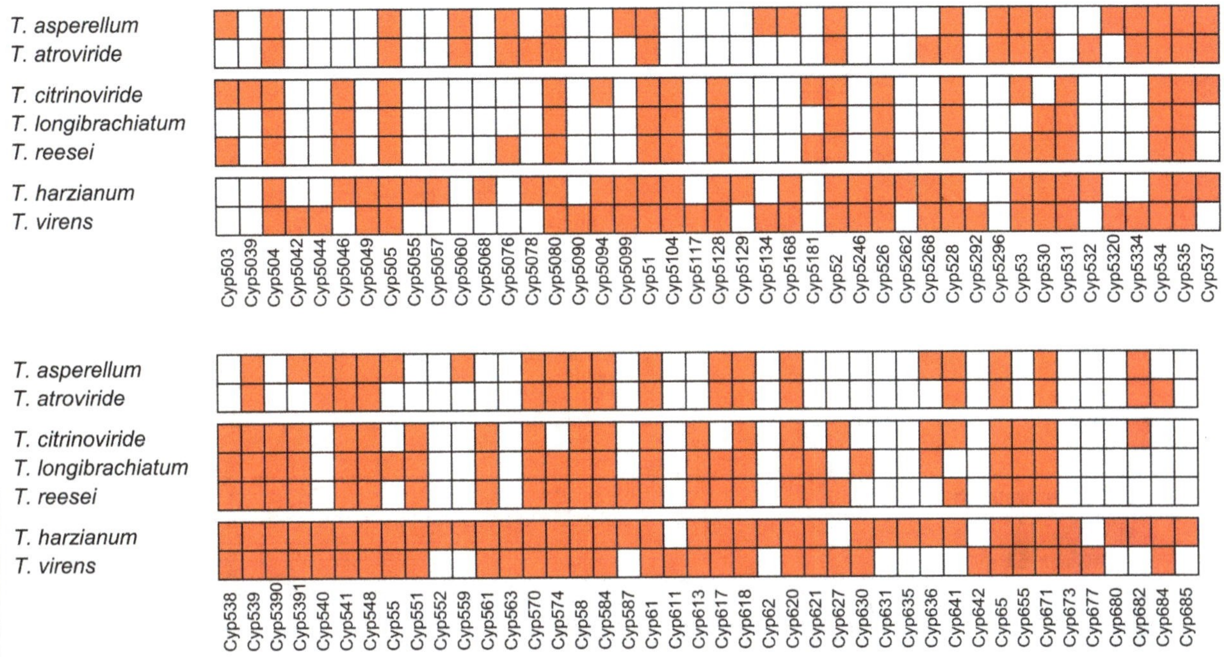


**Fig. 1** Cytochrome P450 families identified in *Trichoderma*

4. Group 5 contained total 35 proteins belonging to clans CYP58 and CYP677, which includes diverse Cyp families Cyp5039, Cyp5094, Cyp5128, Cyp5129, Cyp5292, Cyp551, Cyp552, Cyp58, Cyp677, Cyp680 and Cyp682. Clans for Cyp families 5039, 5128, 5129 and 5292 are not available in FCPD. *Trichoderma* has only one Cyp677 protein i.e., in *T. virens* which was grouped closely with Cyp5292 in phyletic Group 5. A total of 26 proteins belonging to clan CYP58 are identified in *Trichoderma.* Cyp58 family had a single member in all *Trichoderma* species analysed except in *T. virens* (2 proteins). All 7 members of CYP53 clan were grouped together in Group 6. Cyp53 family was found in all *Trichoderma* species except *T. longibrachiatum.* These proteins are involved in xenobiotic metabolism. The group 6 also contained three Cyp5246 proteins, clan for this family is absent in FCPD. Family Cyp5246 is present only in *T. harzianum* and *T. virens.* Members of Cyp53 and Cyp5246 families were differentiated in two clear sub-groups. Group 7 consists of only three proteins belonging to clan CYP630; one each from *T. harzianum, T. longibrachiatum* and *T. virens.* The group 8 consists of 23 proteins from clan CYP574 including families Cyp5076, Cyp574 and Cyp671. Four members of Cyp5168 family were also clustered in group 8.

Group 9 consists of all 17 proteins of clan CYP548. In *Trichoderma,* Cyp548 family is ubiquitously present in all seven species, where *T. asperellum* and *T. atroviride* contained four and three proteins respectively followed by two each in *T. citrinoviride, T. harzianum, T. longibrachiatum, T. reesei* and *T. virens.* These proteins are known to be involved in xenobiotic metabolism. The second largest phylogenetic Group 10 has 56 Cyps from the clan CYP65 which are involved in secondary metabolism. It comprised of families Cyp5117, Cyp561, Cyp563 and Cyp65. Group 11 consists of three Cyp627 proteins.

In *Trichoderma,* group 12 is the largest with 62 Cyp proteins. These Cyps from clans CYP52 and CYP59 were differentiated separately into two sub-groups. Clans CYP52 and CYP59 involve members of Cyp52, Cyp538, Cyp539, Cyp584, Cyp587 and Cyp655 families. Two entries of Cyp587 family belonging to clan CYP59 were grouped together with two proteins each from Cyp5049 and Cyp5296 families. The corresponding clan for Cyp5049 and Cyp5296 families were found to be absent in FCPD. In group 12, Cyp proteins of clan CYP52 were grouped together in the separate sub-group. Group 13 contained 19 Cyps belonging to clans CYP534 and CYP613. Two Cyp proteins belonging to family Cyp5181 were also present in group 13. Protein members of groups 12 and 13 were predicted to be involved in xenobiotic metabolism (Table 2).

Group 14 consists of proteins belonging to clans CYP526 and CYP547 which were differentiated separately into two sub-groups. In *Trichoderma,* 2 Cyp families of clan CYP547 were identified that includes Cyp617 (7) and Cyp618 (7). Cyp5134 proteins were grouped together in sub-group containing clan CYP526 proteins. Group 15 consists of 36 proteins involved in primary

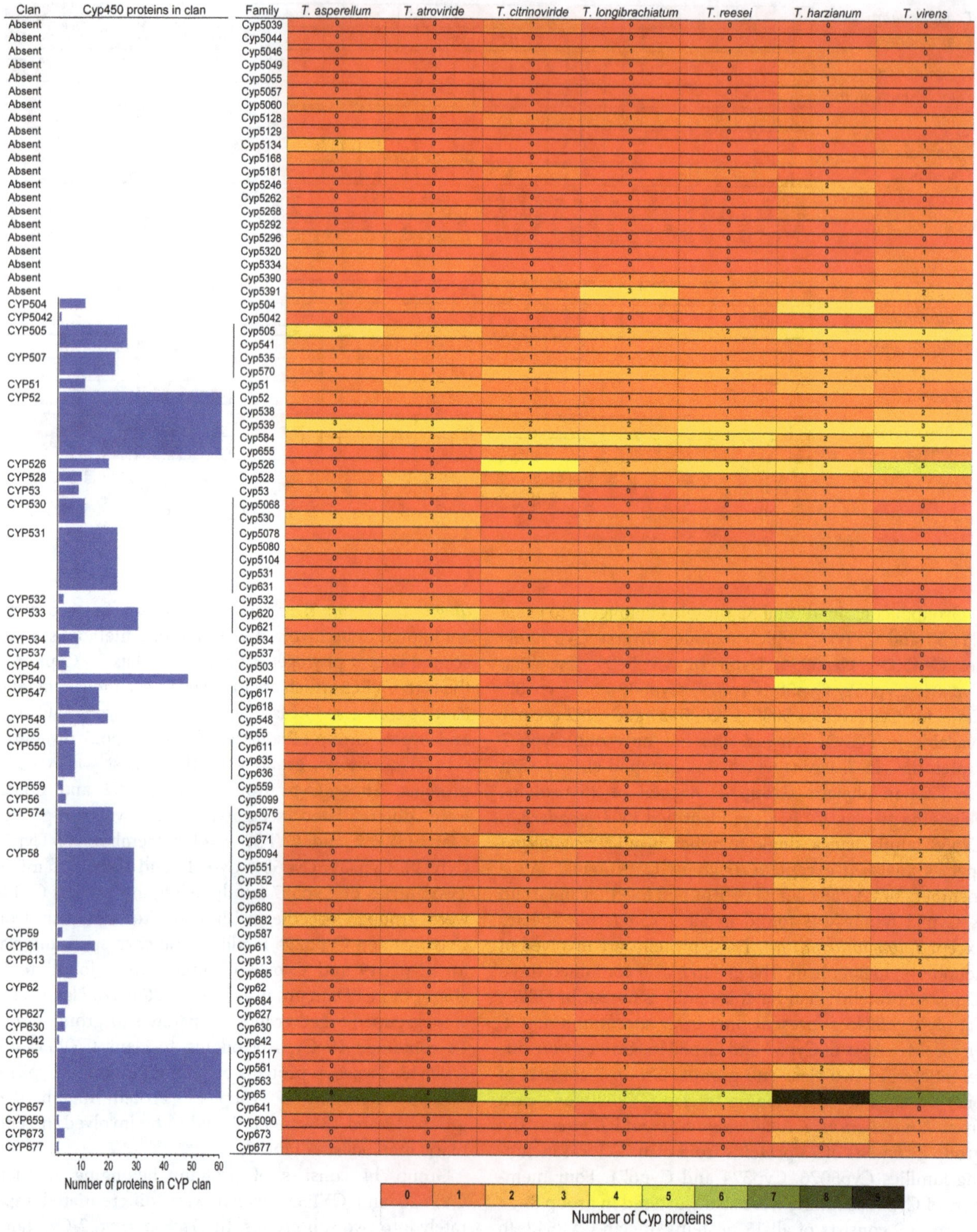

**Fig. 2** Abundance of cytochrome P450 families and clans in *Trichoderma*. The heatmap displays the abundance of Cyp protein families among *Trichoderma* species. Blue bar represent the number of Cyp proteins in a clan

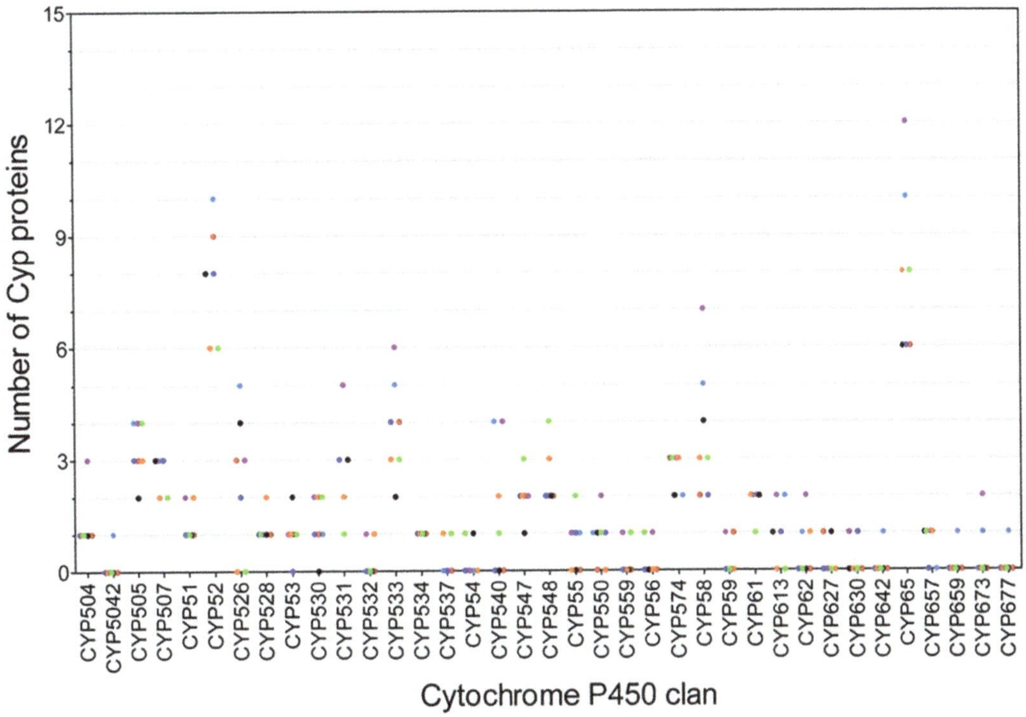


**Fig. 3** Diversity of cytochrome P450 clans among *Trichoderma* species. Scatter-plot presents number of proteins in CYP clans in *T. asperellum* (orange), *T. atroviride* (green), *T. citrinoviride* (black), *T. harzianum* (purple), *T. longibrachiatum* (dark blue), *T. reesei* (red) and *T. virens* (light blue)

metabolism that includes members of clans CYP505 (15), CYP540 (11), CYP541 (7) and Cyp5099 (3). All three Cyp5099 proteins belonging to clan CYP56 family were included in this group. These proteins were identified only in *T. asperellum*, *T. harzianum* and *T. virens*. Cyp5099 proteins were found closely related to Cyp540 proteins and together formed a separate sub-group. Another sub-group contained all proteins belonging to clan CYP505 which includes Cyp505 and Cyp541 families. All nine protein members of clan CYP504 were clustered together in group 16. These proteins are known to be involved in xenobiotic metabolism. *Trichoderma* species contain single copy of Cyp504 protein except *T. harzianum* which contains three copies of Cyp504 protein involved in phenylacetate catabolism [37].

Group 17 is the third largest Cyp group consisting of 51 Cyps from clans CYP530 and CYP533. In this group, CYP533 is the most dominant clan followed by CYP530. Clans CYP530 and CYP533 include Cyp families Cyp530 (8 proteins) and Cyp5068 (1 protein), and Cyp620 (23 proteins) and Cyp621 (4 proteins) respectively. This group also contained Cyp5046 (4), Cyp5391 (8) and Cyp5268 (3) proteins. The corresponding clans for these families are absent in FCPD. Group 18 contains one Cyp5042 protein of *T. virens*. Group19 includes 18 proteins belonging to clans CYP54, CYP550, CYP559, CYP642, CYP657 and CYP659. These clans are involved in secondary metabolism. A total of 29 proteins from 7 *Trichoderma* species corresponding to three clans including CYP51, CYP55 and CYP61 were clustered together in group 20. These are known to be involved in primary metabolism. In this group, CYP51 and CYP61 families dominate with 9 and 15 members respectively. Further, all proteins belonging to Cyp51 were grouped together in group 20. This suggests that Cyp51 protein which is involved in primary metabolism (sterol biosynthesis) is diversified only to a lesser extent in *Trichoderma*. In comparison to some of the ascomycetous fungi, which carry multiple CYP51 proteins, *T. atroviride* and *T. harzianum* contained two copies each, whereas *T. asperellum*, *T. citrinoviride*, *T. longibrachiatum*, *T. reesei* and *T. virens* contained only single copy of Cyp51 protein.

### Characteristic motifs of the *Trichoderma* CYP families

Several signature motifs are conserved in fungal Cyp proteins as per pervious findings [26, 35, 37, 38]. In *Trichoderma*, we identified the characteristic signature motifs of CYP super family AGXDTT, EXXR, PERW and FXXGXRXCXG for each phylogenetic group (Fig. 5). These motifs are functionally essential for the Cyp proteins. Conserved motif FXXGXRXCXG (also known as CXG) is designated as a heme-binding domain [26, 29, 39]

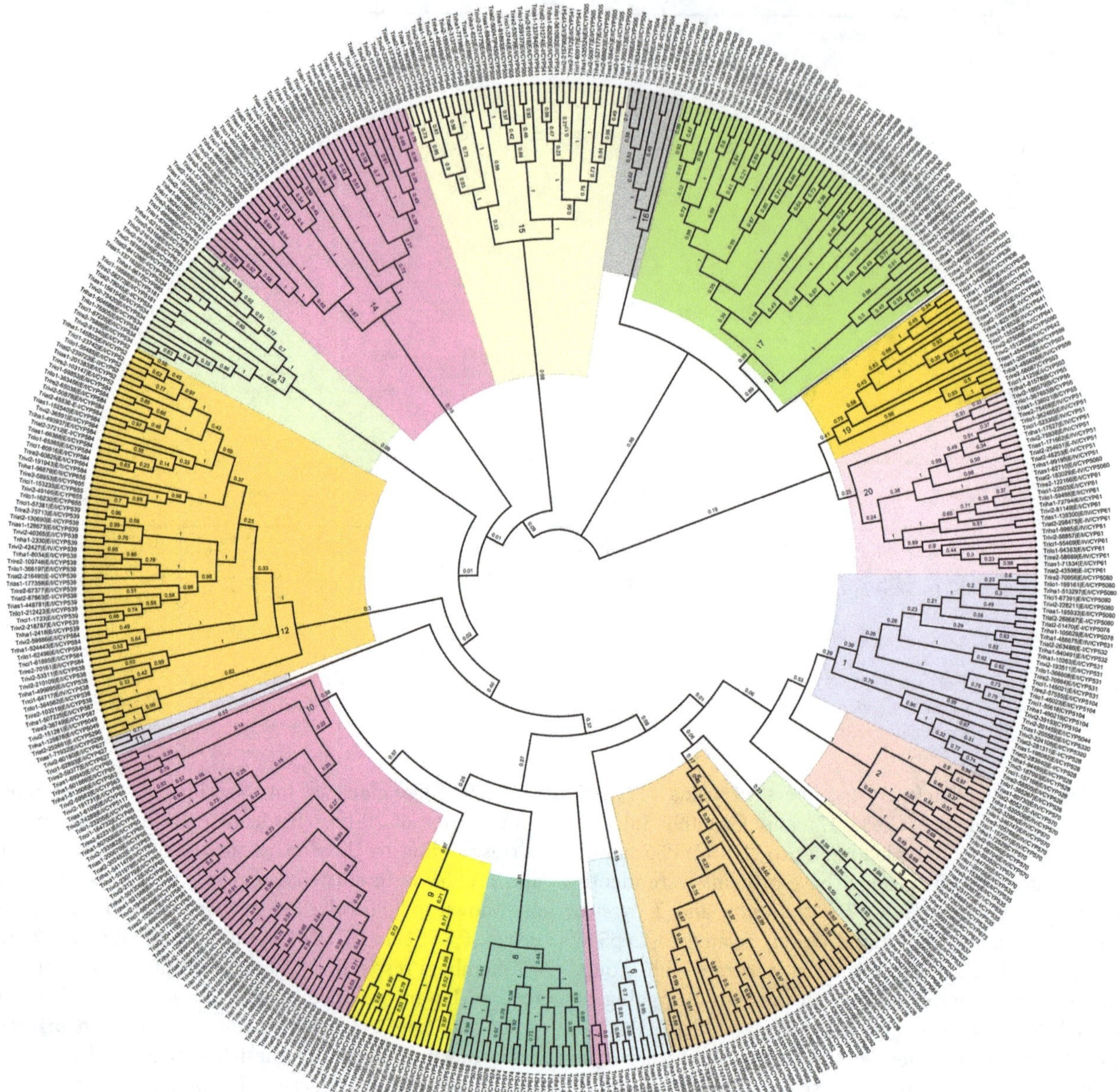

**Fig. 4** Evolutionary relationships of cytochrome P450 proteins among *Trichoderma* species. Phylogenetic tree was constructed inferred using the minimum evolution method[72] using MEGA5 software. Phylogenetic groups (1–20) and bootstrap frequencies are shown in the tree. Tree includes Cyp proteins from all seven *Trichoderma species* including *T. asperellum*, *T. atroviride*, *T. citrinoviride*, *T. harzianum*, *T. longibrachiatum*, *T. reesei* and *T. virens*. Each phylogenetic group is indicated by a specific color

and includes a conserved cysteine residue that binds to the Fe of the heme. In *Trichoderma*, the cysteine residue of the P450 signature CXG motif is invariantly conserved in all P450s, whereas two glycine and one phenylalanine residues were also found to be conserved among majority of phylogenetic groups, which are in accordance with previous reports [37, 40]. In phylogenetic groups 13, 16 and 19, Cyp proteins contain glutamate/aspartate, tyrosine and glycine respectively instead of a phenylalanine residue. Another variant of FXXGXRXCXG motif was found in groups 1, 6 and 20 where first amino acid residue of the motif was either phenylalanine or tryptophan.

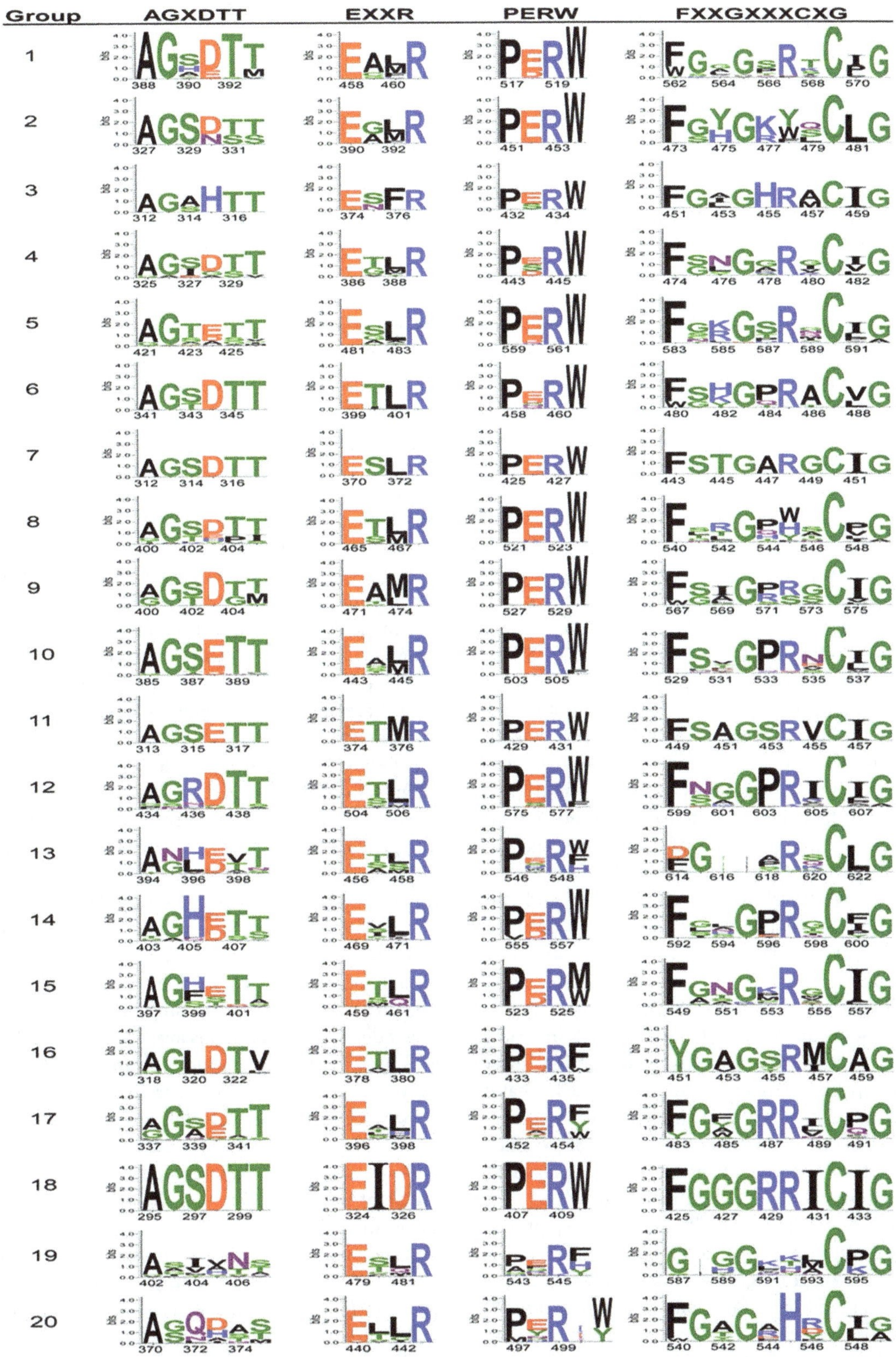


**Fig. 5** Conserved signature motifs of *Trichoderma* Cyp proteins for 20 phylogenetic groups

Further, in groups 5, 8, 12 and 20, FXXGXRXCXG and FXXGXRXCXA variants were identified. Conserved motif EXXR is present in helix K, on the proximal side of heme and probably is involved in the stabilization of the core structure of Cyp proteins [26, 35, 39]. In *Trichoderma* motif EXXR, glutamic acid and arginine residues were found to be highly conserved, whereas two middle 'XX' residues were found to be highly variable. These results are in concurrence with previously reported literature for fungal cytochrome P450 proteins [35, 37, 40, 41]. Another conserved motif of cytochrome P450 protein family is PERW (known as PER) which forms E-R-R triad in Cyp proteins [26]. In *Trichoderma*, we found PERW as the predominant signature, in accordance with previous reports in fungi [27, 35]. Motif PERW was found to be relatively conserved in *Trichoderma* with few exceptions that mainly includes phylogenetic groups 19 and 20. Group 19 consists of Cyp proteins from clans CYP54, CYP550, CYP559, CYP642, CYP657 and CYP659, which have been predicted to play role in secondary metabolism. High diversity of PER motif of this group could be attributed to the evolving functions of Cyp P450 protein members. Phylogenetic group 20 consisting of clans CYP51, CYP61 and CYP55 includes Cyp proteins belonging to both class E (CYP51 and CYP61) and B (CYP55). In this group, variant of PERW motif was identified where clan CYP55 proteins (class B) contained amino acid residues K/E/Q between PER and W/Y. The absence of an amino acid residue between arginine and tryptophan residues in "PERW" motif in all class E Cyp proteins indicate the early functional divergence of PERW motif in class B and E cytochrome P450 proteins. These results provide an insight on the structure–function relationships in such a diverse and complex Cyp protein families. Further, we also identified conserved motif, AGXDTT in *Trichoderma* cytochrome P450 proteins. Motif AGXDTT contributes to oxygen binding and activation [35]. The oxygen-binding domain (AGXDTT) was found to be highly variable in *Trichoderma* cytochrome P450 proteins. The terminal threonine residue in AGXDTT motif involved in the formation of the enzyme's critical oxygen-binding pocket was found to be replaced predominately by valine in phylogenetic group 16. Other amino acid residues that replaced terminal threonine in different groups included serine or methionine. For motifs AGXDTT and CXG, Cyp proteins in phyletic groups 13, 19 and 20 (Table 2) were relatively less conserved, suggesting divergence of these Cyp protein sequences and their functions in *Trichoderma*. We found that the conserved signature motifs and their variants identified in *Trichoderma* showed few exceptions to previous reports. These results suggest Cyp signature motifs have evolved in *Trichoderma* to accommodate enormously wide range of substrate specificities and their substrate-binding regions.

**Cytochrome P450s associated with secondary metabolism related gene clusters**

A survey of the genomes of seven *Trichoderma* spp. revealed that of the 477 cytochrome P450 genes present in the seven genomes, as many as 100 genes are associated with putative secondary metabolism related gene clusters namely NRPS, PKS, NRPS-PKS, NRPS-like, and terpene cyclase clusters (Additional file 1: Table S2).

## Discussion

*Trichoderma* species are the champions of opportunistic success [1]. They can be found virtually in all ecological niches, both terrestrial and aquatic. These fungi are capable of parasitizing a wide range of fungal and oomycetes species. Many species are known to colonize the rhizosphere and roots, both externally and internally [3]. Some are reported to be endophytes [42] while a few are aggressive parasites on cultivated mushrooms [43]. A few species are known to be opportunistic human pathogens while some strains are nematode-parasite, demonstrating their ability to parasitize members of animal kingdom [1]. Several *Trichoderma* strains are plant growth enhancers and some can colonize composts [44]. A few strains are known to be xenobiotics degraders. Most species are prolific producers of a wide range of secondary metabolites, with a total of more than 1000 compounds chemically characterized [45]. Cytochrome P450s are important for cells to perform a wide variety functions like primary and secondary metabolism, xenobiotic degradation and cellular defence (e.g., in interaction with other fungi). Recently, a *T. virens* P450 (TvCyt2; Protein Id. 190045) has been shown to be involved in biocontrol and plant growth promotion [46]. Basidiomycetes are capable of metabolizing a wide range of endogenous and exogenous compounds by using cytochrome P450s [47]. Great deal of information is available on the role of P450s in degradation of lignins and polyaromatic hydrocarbons by white rot fungus *Phanerochaete chrysosporium* and brown-rot fungus *Postia placenta*, as well as medicinal mushrooms like *Coriolus versicolor* and *Lentinula edodes* [48–51]. Role of P450s in colonization of living wood by the plant pathogen *Heterobasidion irregulare* is also well established [52].

In the present study, *Trichoderma* CYPome from seven *Trichoderma* species viz. *T. asperellum*, *T. atroviride*, *T. citrinoviride*, *T. harzianum*, *T. longibrachiatum*, *T. reesei* and *T. virens* is annotated. Overall, our analysis identified a total of 477 CYPs in these genomes. To provide support for the annotation process, the identified CYPs were also examined for conserved CYP domain. Our analysis of the CYPome has identified 12 families unique to *Trichoderma*. All the *Trichoderma* species examined are a rich source of Cyp proteins (55 in *T. longibrachiatum* to 100 in *T. harzianum*).

In *Trichoderma*, clan CYP52 consisted of families Cyp52, Cyp538, Cyp539, Cyp584 and Cyp655. Cyp52 family is found only amongst *Candida*-related species of fungi and these proteins catalyze the conversion of fatty acids and alkanes to alpha, omega-dicarboxylic acids [53]. The number of Cyp61 proteins was conserved in all *Trichoderma* species and these proteins were also grouped together in Group 20. Cyp61 proteins are involved in primary metabolism. In *Saccharomyces cerevisiae*, CYP61 codes for sterol 22 desaturase [54], which is involved in later stages of the ergosterol pathway in metabolizing Ergosta-5,7,24(28)-trienol to Ergosta-5,7,22,24(28)-tetraenol by introducing a C-22(23) double bond in the sterol side chain. Since Cyp61 is involved in the later stages of ergosterol pathway, it is considered to have evolved as a result of duplication and diversification of the CYP51 gene. In ascomycetes and basidiomycetes, clan CYP51 is involved in sterol biosynthesis and is recognized as the housekeeping CYP, and has been a popular antifungal target for the control of fungal diseases in humans and crop plants [29–31, 55]. In comparison to some of the ascomycetous fungi, which carry multiple CYP51 genes, *T. atroviride and T. harzianum* contained two copies whereas *T. asperellum, T. citrinoviride, T. longibrachiatum, T. reesei* and *T. virens* contained only single copy of Cyp51 protein. In addition, all members of both clans CYP51 and CYP61 which are involved in primary metabolism (sterol biosynthesis) are grouped together in group 20, suggesting that both Cyp51 and Cyp61 proteins are diversified only to a lesser extent in *Trichoderma.*

Motif analysis led to the identification of four signature motifs in phylogenetic groups, which correspond to the conserved tertiary structure and enzyme functions in spite of the wide sequence diversity and functions of Cyp proteins. Modifications found in the heme-binding domain FXXGXRXCXG are more frequently found in CYPs with catalytic activity, often not requiring oxygen [29]. These results indicate Cyp members of groups 5, 8, 12, 13, 16, 19 and 20 may have novel catalytic activities in *Trichoderma*. Some P450s showed variations of the signature motifs mainly in AGXDTT, EXXR and FXXGXRXCXG motifs. These results are in accordance with previous reports [38, 41] where it was proposed that these P450s variations may be due to misaligned sequences or that the P450s are missing the invariant residues at these motifs. In our study, phylogenetic group 10 containing protein families of clan CYP65 showed highly conserved motifs, suggesting functional conservation of CYP65 clan in analysed *Trichoderma* species. All members of clan CYP65 are involved in secondary metabolism. CYP65 is reported to catalyze the epoxidation reaction during the synthesis of trichothecenes [56, 57] and radicicol [37]. Identification of conserved and variable CYP motif signatures among and within phylogenetic groups in the present study may provide us information on CYP evolution, structure, and function in *Trichoderma* and have application in classification of proteins in gene expression analysis [58].

Cyp56 clan, found to be unique to *T. asperellum, T. harzianum* and *T. virens* (mycoparasites) has been characterized earlier in yeast [59, 60]. Members of Cyp56 clan are involved in meiotic spore wall biogenesis, particularly in dityrosine biosynthesis [59–61]. Members of the clan CYP507, CYP530, CYP531, CYP532 and CYP548 are known to be involved in xenobiotics metabolism [36]. Abundance of these proteins in *Trichoderma* may be related to the ability of these fungi to metabolize a wide range of xenobiotics, including many fungicides. Similarly, ability of *Trichoderma* spp. to produce a plethora of secondary metabolites could be linked to the abundance of P450s belonging to the clan CYP574, CYP58 and CYP65 proteins that have been implicated in trichothecene biosynthesis [62]. In *T. harzianum*, three copies of Cyp504 protein are present as compared to single copy in other *Trichoderma* species. Expansion of Cyp504 proteins in *T. harzianum* suggest important role of Cyp504 protein in xenobiotic metabolism. Further, the family members of Cyp504 were also reported to be up-regulated during cuticle infection by insect pathogenic fungi *Metarhizium anisopliae* and *M. acridum* [63]. Cyp505 family was found to be expanded in *T. asperellum, T. harzianum* and *T. virens* where these species contained three Cyp505 proteins each. Cyp505 proteins are membrane-associated fatty acid hydroxylase [64]. Cyp528 family has only one protein entry in all *Trichoderma* species analysed except *T. atroviride* where family Cyp528 consisted of two proteins. Similarly, Cyp58 family has a single protein entry in all *Trichoderma* species analysed except in *T. virens* where family Cyp58 consisted of two proteins. Previous studies also showed expansion of clan Cyp58 proteins in fungi [36]. In *Trichoderma*, the increase in CYPome size of *T. harzianum* and *T. virens* may be due to the expansion of certain *CYP* gene families or the presence of novel genes that are essential for their lifestyle. Previous reports have associated expansions of the fungal CYP families with the evolution of various fungal traits including pathogenicity [65]. Our phylogenetic analysis showed uneven distribution of CYP group sizes in *Trichoderma* species, which are in concordance with extreme expansions and contractions of certain CYP families in the course of evolution. Expansion of CYP clans in different *Trichoderma* species could aid them in more competent survival in their respective habitats.

*Trichoderma* spp. are prolific producers of secondary metabolites, many with antimicrobial, anticancer and plant growth-promoting properties [45] Cyps are known to play central role in biosynthesis if many, if not most of the secondary metabolites of plant and microbial origin. Till date, however, only a handful of *Trichoderma* Cyps have been investigated for their role in biosynthesis of secondary metabolites [46, 66, 67]. Our present findings suggest that more than 20% of the catalogued Cyps from *Trichoderma* are part of putative secondary metabolism-related gene clusters. There is a need for systematic studies on the functions of these Cyps which would lead to the discovery of novel pathways, metabolites and intermediates with greater biotechnological significance.

## Conclusion

*Trichoderma* CYPome described in our study is by combining information generated from existing databases, predicting conserved domains and identifying structural motifs in each hypothetical protein. By following internationally recognized nomenclature system, we have identified novel CYP clans and families unique to *Trichoderma*. Phylogenetic analysis elucidated distribution of Cyp families and clans in different evolutionary groups and their probable functions in metabolism or biosynthesis based on the comparisons with CYPomes of other organisms. The number of these proteins correlates with the genome size and many are species-specific. Unfortunately, the functions of none of these proteins are known. One reason being a lack of systematic studies and annotation of these proteins. Our CYPome annotation and evolutionary studies of seven *Trichoderma* species now provides opportunities for exploration of research-driven strategies to select *Trichoderma* species for various applications especially in relation to secondary metabolism and degradation of environmental pollutants. Several of these proteins could also have biotechnological applications like biotransformation and synthesis of pharmaceutically important drugs.

## Methods

### Sequence data

Sequences of Cytochrome P450s were retrieved from the Joint Genome Institute (JGI) fungal genome database MycoCosm (http://genome.jgi-psf.org/programs/fungi/index.jsf) for all the species of genus *Trichoderma*. The species included were *T. asperellum* (CBS 433.97) v1.0, *T. atroviride* (IMI 206040) v2.0, *T. citrinoviride* v4.0, *T. harzianum* (CBS 226.95) v1.0, *T. longibrachiatum* (ATCC 18648) v3.0, *T. reesei* (QM 6a) v2.0 and *T. virens* (Gv29-8) v2.0.

### Annotation of CYPs

The annotation pipeline of the CYPome in the *Trichoderma* species was done in a two-step procedure of identification and annotation. The identification step of CYP family was performed by using Conserved Domain Database (CDD); the cut-off of positive hits was set at *E*-value of $10^{-2}$. Entries with incomplete sequences and domain were manually removed from the data. Cyp proteins with complete conserved cytochrome P450 domains were further subjected to the annotation procedure using the Nelson's P450 database against all named fungal cytochrome P450s (http://blast.uthsc.edu) with the *E*-value of $10^{-4}$ [68]. For annotation, sequence similarity cut-off of 40% was used. For few entries, we have followed criteria of the phenomenon called family creep that allows sequences less than 40% to be included in a family. For such entries, we have used sequence similarity cut-off of 30% and above. These predicted CYPs were then assigned to the corresponding family and clan types based on their highest homology according to the International P450 Nomenclature Committee Databases used by Nelson (http://drnelson.uthsc.edu/CytochromeP450.html) [21] and the fungal cytochrome P450 database (http://p450.riceblast.snu.ac.kr) [27] respectively.

### Structural feature analysis of CYP protein sequences

Presence of cytochrome P450 conserved domain was confirmed using conserved domain database [69]. To reveal phylogenetic group-specific conservation pattern of cytochrome P450 proteins, structural features were explored. To identify cyp conserved signature motifs, multiple protein sequence alignments for each phylogenetic group were built by MAFFT program [70] using E-INS-i iterative refinement method. Alignments were further refined and viewed using AliView [71]. Consensus logos of the alignments were automatically generated by WebLogo 3 program [72] and used for visualization of the conservation of signature motifs for each phylogenetic group. The generated logos were used for the analysis.

### Phylogenetic reconstruction of CYPs

After removal of redundant and incomplete sequences, the protein sequences were aligned using MUSCLE [73]. The evolutionary history was inferred using the minimum evolution method [74]. The bootstrap consensus tree inferred from 1000 replicates was taken to represent the evolutionary history of the taxa analysed [75]. The evolutionary distances were computed using the Poisson correction method [76] and are in the units of the number of amino acid substitutions per site. The rate variation among sites was modelled with a gamma distribution (shape parameter = 1). The ME tree was searched using the close-neighbor-interchange (CNI) algorithm [77] at a search level of 1. The neighbor-joining algorithm [78] was used to generate the initial tree. Evolutionary analyses were conducted in MEGA5 [79]. Phylogenetic trees were visualized with FigTree v1.1.2 [80].

### Identification of cytochrome P450s associated with secondary metabolism related gene clusters

A genome-wide survey was done to identify cytochrome P450s associated (presence in the vicinity) with secondary metabolism-related gene clusters, viz., NRPS, PKS, PKS/NRPS, NRPS-like and terpene cyclase clusters either manually (*T. reesei, T. virens and T. atroviride* [81, 82] or using automated pipeline on the respective genome pages (for *T. citrinoviride, T. longibrachiatum, T. asperellum* and *T. harzianum*).

#### Authors' contributions

PKM, SC and ID conceptualized and framed the work. SC designed the study and performed CYP manual annotation, phylogenetics, function and motif predictions, analysed data, prepared figures and supplementary data. STM performed data mining and contributed to CYP function prediction. RB identified the association of the genes with secondary metabolism-related gene clusters. AK, AA and IVG at JGI did the whole genome sequencing and automated annotation. SC, PKM and ID wrote the manuscript. PKM coordinated this study. All authors read and approved the final manuscript.

#### Author details

[1] Nuclear Agriculture and Biotechnology Division, Bhabha Atomic Research Centre, Trombay, Mumbai 400085, India. [2] U.S. Department of Energy Joint Genome Institute, Walnut Creek, CA 94598, USA. [3] Research Area Biochemical Technology, Institute of Chemical and Biological Engineering, TU Wien, 1060 Vienna, Austria.

#### Acknowledgements

The authors thank Head, Nuclear Agriculture and Biotechnology Division, Bhabha Atomic Research Centre, Mumbai for encouragement and support. The work conducted by the U.S. Department of Energy Joint Genome Institute, a DOE Office of Science User Facility, was supported by the Office of Science of the U.S. Department of Energy under Contract No. DE-AC02-05CH11231.

#### Competing interests

The authors declare that they have no competing interests.

#### References

1. Druzhinina I, Seidl-Seiboth V, Herrera-Estrella A, Horwitz BA, Kenerley CM, Monte E, Mukherjee PK, Zeilinger S, Grigoriev IV, Kubicek CP. *Trichoderma*: the genomics of opportunistic success. Nat Rev Microbiol. 2011;9:749–59.
2. Bissett J, Gams W, Jaklitsch W, Samuels GJ. Accepted *Trichoderma* names in the year 2015. IMA Fungus. 2015;6:263–95.
3. Mukherjee PK, Horwitz BA, Singh US, Mukherjee M, Schmoll M. *Trichoderma* in agriculture, industry and medicine: an overview. In: Mukherjee PK, Horwitz BA, Singh US, Mukherjee M, Schmoll M, editors. *Trichoderma*: biology and applications. Oxfordshire: CABI; 2013. p. 152–72.
4. Mukherjee PK, Horwitz BA, Herrera-Estrella A, Schmoll M, Kenerley CM. *Trichoderma* research in the genome era. Annu Rev Phytopathol. 2013;51:105–29.
5. Hatvani L, Manczinger L, Vágvölgyi C, Kredics L. *Trichoderma* as a human pathogen. In: Mukherjee PK, Horwitz BA, Singh US, Mukherjee M, Schmoll M, editors. *Trichoderma*: Biology and Applications. Oxfordshire: CABI; 2013. p. 152–72.
6. Martinez D, Berka RM, Henrissat B, Saloheimo M, Arvas M, Baker SE, Chapman J, Chertkov O, Coutinho PM, Cullen D, Danchin EG, Grigoriev IV, et al. Genome sequencing and analysis of the biomass-degrading fungus *Trichoderma reesei* (syn. *Hypocrea jecorina*). Nat Biotechnol. 2008;26:553–60.
7. Kubicek CP, Herrera-Estrella A, Seidl-Seiboth V, Martinez DA, Druzhinina IS, Thon M, Zeilinger S, Casas-Flores S, Horwitz BA, Mukherjee PK, et al. Comparative genome sequence analysis underscores mycoparasitism as the ancestral life style of *Trichoderma*. Genome Biol. 2011;12:R40.
8. Schmoll M, Dattenböck C, Carreras-Villaseñor N, Mendoza-Mendoza A, Tisch D, Alemán MI, Baker SE, Brown C, Cervantes-Badillo MG, Cetz-Chel J, Cristobal-Mondragon GR, et al. The genomes of three uneven siblings: footprints of the lifestyles of three *Trichoderma* species. Microbiol Mol Biol Rev. 2016;80:205–327.
9. Grigoriev IV, Nikitin R, Haridas S, Kuo A, Ohm R, Otillar R, Riley R, Salamov A, Zhao X, Korzeniewski F, Smirnova T. MycoCosm portal: gearing up for 1000 fungal genomes. Nucleic Acids Res. 2014;42(D1):D699–704.
10. Degtyarenko KN. Structural domains of P450-containing monooxygenase systems. Protein Eng. 1995;8:737–47.
11. Klingenberg M. Pigments of rat liver microsomes. Arch Biochem Biophys. 1958;75:376–86.
12. Han Y, Liu X, Benny U, Kistler HC, VanEtten HD. Genes determining pathogenicity to pea are clustered on a supernumerary chromosome in the fungal plant pathogen *Nectria haematococca*. Plant J. 2001;25:305–14.
13. Siewers V, Viaud M, Jimenez-Teja D, Collado IG, Gronover CS, Pradier JM, Tudzynsk B, Tudzynski P. Functional analysis of the cytochrome P450 monooxygenase gene bcbot1 of *Botrytis cinerea* indicates that botrydial is a strain-specific virulence factor. MPMI. 2005;18:602–12.
14. Fan J, Urban M, Parker JE, Brewer HC, Kelly SL, Hammond-Kosack KE, Fraaije BA, Liu X, Cools HJ. Characterization of the sterol 14α-demethylases of *Fusarium graminearum* identifies a novel genus-specific CYP51 function. New Phytol. 2013;198:821–35.
15. Takaoka S, Kurata M, Harimoto Y, Hatta R, Yamamoto M, Akimitsu K, Tsuge T. Complex regulation of secondary metabolism controlling pathogenicity in the phytopathogenic fungus *Alternaria alternata*. New Phytol. 2014;202:1297–309.
16. Zhang DD, Wang XY, Chen JY, Kong ZQ, Gui YJ, Li NY, Bao YM, Dai XF. Identification and characterization of a pathogenicity-related gene VdCYP1 from *Verticillium dahliae*. Sci Rep. 2016;6:27979.
17. Bernhardt R. Cytochromes P450 as versatile biocatalysts. J Biotechnol. 2006;124:128–45.
18. Urlacher VB, Eiben S. Cytochrome P450 monooxygenases: perspectives for synthetic application. Trends Biotechnol. 2006;24:324–30.
19. Kizawa H, Tomura D, Oda M, Fukamizu A, Hoshino T, Gotoh O, Yasui T, Shoun H. Nucleotide sequence of the unique nitrate/nitrite inducible cytochrome P450 cDNA from *Fusarium oxysporum*. J Biol Chem. 1991;266:10632–7.
20. Lamb DC, Skaug T, Song HL, Jackson CJ, Podust LM, Waterman MR, Kell DB, Kelly DE, Kelly SL. The cytochrome P450 complement (CYPome) of *Streptomyces coelicolor* A3 (2). J Biol Chem. 2002;277:24000–5.
21. Nelson DR. Cytochrome P450 and the individuality of species. Arch Biochem Biophys. 1999;369:1–10.
22. Nebert D, Adesnik M, Coon MJ, Estabrook RW, Gonzalez FJ, Guengerich FP, Gunsalus IC, Johnson EF, Kemper B, Levin W, et al. The P450 gene superfamily: recommended nomenclature. DNA. 1987;6:1–11.
23. Nelson DR, Koymans L, Kamataki T, Stegeman JJ, Feyereisen R, Waxman DJ, Waterman MR, Gotoh O, Coon MJ, Estabrook RW, Gunsalus IC, Nebert DW. P450 superfamily: update on new sequences, gene mapping, accession numbers and nomenclature. Pharmacogenet. 1996;6:1–42.
24. Nelson D, Werck-Reichhart DA. P450-centric view of plant evolution. Plant J. 2011;66(1):194–211.
25. Nelson DR. Metazoan cytochrome P450 evolution. Comp Biochem Physiol C Pharmacol Toxicol Endocrinol. 1998;121(1-3):15–22.
26. Deng J, Carbone I, Dean RA. The evolutionary history of cytochrome P450 genes in four filamentous Ascomycetes. BMC Evol Biol. 2007;7(1):10–30.
27. Park J, Choi J, Ahn K, Park B, Park J, Kang S, Lee Y-H. Fungal cytochrome P450 database. BMC Bioinform. 2008;9:402.

28. Hoffmeister D, Keller NP. Natural products of filamentous fungi, enzymes, genes and their regulation. Nat Prod Rep. 2007;24:393–416.
29. Kelly DE, Kraševec N, Mullins J, Nelson DR. The CYPome (cytochrome P450 complement) of *Aspergillus nidulans*. Fungal Genet Biol. 2009;46:S53–61.
30. Kelly SL, Kelly DE. Microbial cytochromes P450: biodiversity and biotechnology. Where do cytochrome P450 come from, what do they do and what can they do for us? Philos Trans R Soc Lond B Biol Sci. 2013;368(1612):20120476.
31. Ichinose H. Metabolic diversity and cytochromes P450 of fungi. In: Yamazaki H, editor. Fifty years of cytochrome P450 research. Tokyo: Springer; 2014. p. 187–205.
32. Syed K, Yadav JS. P450 monooxygenases (P450ome) of the model white rot fungus *Phanerochaete chrysosporium*. Crit Rev Microbiol. 2012;38(4):339–63.
33. Newsome AW, Nelson D, Corran A, Kelly SL, Kelly DE. The cytochrome P450 complement (CYPome) of *Mycosphaerella graminicola*. Biotechnol Appl Biochem. 2013;60:52–64.
34. Lah L, Haridas S, Bohlmann J, Breuil C. The cytochromes P450 of *Grosmannia clavigera*: genome organization, phylogeny, and expression in response to pine host chemicals. Fungal Genet Biol. 2013;50:72–81.
35. Chen W, Lee MK, Jefcoate C, Kim SC, Chen F, Yu JH. Fungal cytochrome p450 monooxygenases: their distribution, structure, functions, family expansion, and evolutionary origin. Genome Biol Evol. 2014;6(7):1620–34.
36. Moktali V, Park J, Fedorova-Abrams ND, Park B, Choi J, Lee YH, Kang S. Systematic and searchable classification of cytochrome P450 proteins encoded by fungal and oomycete genomes. BMC Genom. 2012;13(1):525.
37. Sirim D, Widmann M, Wagner F, Pleiss J. Prediction and analysis of the modular structure of cytochrome P450 monooxygenases. BMC Struct Biol. 2010;10:34.
38. Sezutsu H, Le Goff G, Feyereisen R. Origins of P450 diversity. Philos Trans R Soc B. 2013;368:20120428.
39. Werck-Reichhart D, Feyereisen R. Cytochromes P450: a success story. Genome Biol. 2000;1:6.
40. Gotoh O. Substrate recognition sites in cytochrome P450 family 2 (CYP2) proteins inferred from comparative analyses of amino acid and coding nucleotide sequences. J Biol Chem. 1992;267:83–90.
41. Syed K, Mashele SS. Comparative analysis of P450 signature motifs EXXR and CXG in the large and diverse kingdom of fungi: identification of evolutionarily conserved amino acid patterns characteristic of P450 Family. PLoS ONE. 2014;9(4):e95616.
42. Bailey BA, Melnick RL. The endophytic *Trichoderma*. In: Mukherjee PK, Horwitz BA, Singh US, Mukherjee M, Schmoll M, editors. *Trichoderma*: biology and applications. Oxfordshire: CABI; 2013. p. 152–72.
43. Komon-Zelazowska M, Bissett J, Zafari D, Hatvani L, Manczinger L, Woo S, Lorito M, Kredics L, Kubicek CP, Druzhinina IS. Genetically closely related but phenotypically divergent *Trichoderma* species cause green mold disease in oyster mushroom farms worldwide. Appl Environ Microbiol. 2007;73(22):7415–26.
44. Zaidi NW, Singh US. *Trichoderma* in plant health management. In: Mukherjee PK, Horwitz BA, Singh US, Mukherjee M, Schmoll M, editors. *Trichoderma*: biology and applications. Oxfordshire: CABI; 2013. p. 152–72.
45. Zeilinger S, Gruber S, Bansal R, Mukherjee PK. Secondary metabolism in *Trichoderma*—chemistry meets genomics. Fungal Biol Rev. 2016;30(2):74–90.
46. Ramírez-Valdespino CA, Porras-Troncoso MD, Corrales-Escobosa AR, Wrobel K, Martínez-Hernández P, Olmedo-Monfil V. Functional characterization of TvCyt2, a member of the p450 monooxygenases from *Trichoderma virens* relevant during the association with plants and mycoparasitism. Mol Plant Microbe Interact. 2018;31:289–98.
47. Ichinose H. Cytochrome P450 of wood-rotting basidiomycetes and biotechnological applications. Biotechnol Appl Biochem. 2013;60:71–81.
48. Bhattacharya SS, Syed K, Shann J, Yadav JS. A novel P450-initiated biphasic process for sustainable biodegradation of benzo[*a*]pyrene in soil under nutrient-sufficient conditions by the white rot fungus *Phanerochaete chrysosporium*. J Hazard Mater. 2013;261:675–83.
49. Wang J, Yamamoto R, Yamamoto Y, Tokumoto T, Dong J, Thomas P, Hirai H, Kawagishi H. Hydroxylation of bisphenol A by hyper lignin-degrading fungus *Phanerochaete sordida* YK-624 under non-ligninolytic condition. Chemosphere. 2013;93:1419–23.
50. Akiyama R, Kajiwara S, Shishido K. Catalytic reaction of basidiomycete *Lentinula edodes* cytochrome P450, Le. CYP1 enzyme produced in yeast. Biosci Biotechnol Biochem. 2004;68:79–84.
51. Ichinose H, Wariishi H, Tanaka H. Identification and heterologous expression of the cytochrome P450 oxidoreductase from the white-rot basidiomycete *Coriolus versicolor*. Appl Microbiol Biotechnol. 2002;59:658–64.
52. Mgbeahuruike AC, Kovalchuk A, Ubhayasekera W, Nelson DR, Yadav JS. CYPome of the conifer pathogen *Heterobasidion irregulare*: inventory, phylogeny, and transcriptional analysis of the response to biocontrol. Fungal Biol. 2017;121:158–71.
53. Eschenfeldt WH, Zhang Y. H, Samaha H, Stols L, Eirich LD, Wilson CR. Donnelly MI. Transformation of fatty acids catalyzed by cytochrome P450 monooxygenase enzymes of *Candida tropicalis*. Appl Environ Microbiol. 2003;69:5992–9.
54. Skaggs BA, Alexander JF, Pierson CA, Schweitzer KS, Chun KT, Koegel C, Barbuch R, Bard M. Cloning and characterization of the *Saccharomyces cerevisiae* C-22 sterol desaturase gene, encoding a second cytochrome P-450 involved in ergosterol biosynthesis. Gene. 1996;169:105–9.
55. Becher R, Wirsel SG. Fungal cytochrome P450 sterol 14 alpha-demethylase (CYP51) and azole resistance in plant and human pathogens. Appl Microbiol Biotechnol. 2012;95:825–40.
56. Keller G, Turner NP, Bennett JW. Fungal secondary metabolism-from biochemistry to genomics. Nat Rev Microbiol. 2005;3:937–47.
57. Ward TJ, Bielawski JP, Kistler HC, Sullivan E, O'Donnell K. Ancestral polymorphism and adaptive evolution in the trichothecene mycotoxin gene cluster of phytopathogenic *Fusarium*. Proc Natl Acad Sci USA. 2002;99:9278–83.
58. Jensen S, Shen L, Liu J. Combining phylogenetic motif discovery and motif clustering to predict co-regulated genes. Bioinformatics. 2005;21:3832–9.
59. Melo NR, Moran GP, Warrilow AGS, Dudley E, Smith SN, Sullivan DJ, Lamb DC, Kelly DE, Coleman DC, Kelly SL. CYP56 (*Dit2p*) in *Candida albicans*: characterization and investigation of its role in growth and antifungal drug susceptibility. Antimicrob Agents Chemother. 2008;52:3718–24.
60. Crešnar B, Petrič S. Cytochrome P450 enzymes in the fungal kingdom. Biochim Biophys Acta. 2011;1814:29–35.
61. Briza P, Eckerstorfer M, Breitenbach M. The sporulation-specific enzymes encoded by the DIT1 and DIT2 genes catalyze a two-step reaction leading to a soluble LL-dityrosine-containing precursor of the yeast spore wall. Proc Natl Acad Sci USA. 1994;91(10):4524–8.
62. Brown DW, McCormick SP, Alexander NJ, Proctor RH, Desjardins AE. Inactivation of a cytochrome PJ450 is a determinant of trichothecene diversity in *Fusarium* species. Fungal Genet Biol. 2002;36:224–33.
63. Gao Q, Jin K, Ying SH, Zhang Y, Xiao G, Shang Y, Duan Z, Hu X, Xie XQ, Zhou G, Peng G, Luo Z, Huang W, Wang B, Fang W, Wang S, Zhong Y, Ma LJ, St Leger RJ, Zhao GP, Pei Y, Feng MG, Xia Y, Wang C. Genome sequencing and comparative transcriptomics of the model entomopathogenic fungi *Metarhizium anisopliae* and *M. acridum*. PLoS Genet. 2011;7(1):e1001264.
64. Kitazume T, Takaya N, Nakayama N, Shoun H. *Fusarium oxysporum* fatty-acid subterminal hydroxylase (CYP505) is a membrane-bound eukaryotic counterpart of *Bacillus megaterium* cytochrome P450BM3. J Biol Chem. 2000;275:39734–40.
65. Soanes DM, Alam I, Cornell M, Wong HM, Hedeler C, Paton NW, Rattray M, Hubbard SJ, Oliver SG, Talbot NJ. Comparative genome analysis of filamentous fungi reveals gene family expansions associated with fungal pathogenesis. PLoS ONE. 2008;3(6):e2300.
66. Malmierca MG, Cardoza RE, Alexander NJ, McCormick SP, Hermosa R, Monte E, Gutiérrez S. Involvement of *Trichoderma* trichothecenes in the biocontrol activity and induction of plant defense-related genes. Appl Environ Microbiol. 2012;78:4856–68.
67. Bansal R, Sherkhane PD, Oulkar D, Khan Z, Banerjee K, Mukherjee PK. The viridin biosynthesis gene cluster of *Trichoderma virens* and its conservancy in the bat white-nose fungus *Pseudogymnoascus destructans*. Chem Select. 2018;3:1289–93.
68. Nelson DR. The cytochrome P450 homepage. Hum Genom. 2009;4:59–65.
69. Marchler-Bauer A, Lu S, Anderson JB, Chitsaz F, Derbyshire MK, DeWeese-Scott C, Fong JH, Geer LY, Geer RC, Gonzales NR, et al. CDD: a conserved domain database for the functional annotation of proteins. Nucleic Acids

Res. 2011;39:225–9.
70. Katoh K, Rozewicki J, Yamada KD. MAFFT online service: multiple sequence alignment, interactive sequence choice and visualization. Brief Bioinform. 2017. https://doi.org/10.1093/bib/bbx108.
71. Larsson A. AliView: a fast and lightweight alignment viewer and editor for large datasets. Bioinformatics. 2014;30:3276–8.
72. Crooks GE, Hon G, Chandoniam JM, Brennerm SE. WebLogo: a sequence logo generator. Genome Res. 2004;2004(14):1188–90.
73. Edgar RC. MUSCLE: a multiple sequence alignment method with reduced time and space complexity. BMC Bioinform. 2004;5:113.
74. Rzhetsky A, Nei MA. Simple method for estimating and testing minimum evolution trees. Mol Biol Evol. 1992;9:945–67.
75. Felsenstein J. Confidence limits on phylogenies: an approach using the bootstrap. Evolution. 1985;39:783–91.
76. Zuckerkandl E, Pauling L. Evolutionary divergence and convergence in proteins. In: Bryson V, Vogel HJ, editors. Evolving genes and proteins. New York: Academic Press; 1965. p. 97–166.
77. Nei M, Kumar S. Molecular evolution and phylogenetics. New York: Oxford University Press; 2000.
78. Saitou N, Nei M. The neighbor-joining method: a new method for reconstructing phylogenetic trees. Mol Biol Evol. 1987;4:406–25.
79. Tamura K, Peterson D, Peterson N, Stecher G, Nei M, Kumar S. MEGA5: molecular evolutionary genetics analysis using maximum likelihood, evolutionary distance, and maximum parsimony methods. Mol Biol Evol. 2011;28:2731–9.
80. Rambaut A. FigTree v1.4.3: Tree figure drawing tool. 2016. http://tree.bio.ed.ac.uk/software/figtree/. Accessed 5 Feb 2018.
81. Bansal R, Mukherjee PK. Identification of novel gene clusters for secondary metabolism in *Trichoderma* genomes. Microbiology. 2016;85:185–90.
82. Bansal R, Mukherjee PK. The terpenoid biosynthesis toolkit of *Trichoderma*. Nat Prod Commun. 2016;11:431–4.

# Whole-genome sequencing reveals highly specific gene targeting by in vitro assembled Cas9-ribonucleoprotein complexes in *Aspergillus fumigatus*

Qusai Al Abdallah, Ana Camila Oliveira Souza, Adela Martin-Vicente, Wenbo Ge and Jarrod R. Fortwendel*

**Abstract**

**Background:** CRISPR/Cas9-based genome editing is quickly becoming a powerful tool within the field of fungal genetics. Adaptation of CRISPR/Cas9 systems are allowing for rapid and highly efficient gene targeting within fungi. We recently reported the adaptation of a simple CRISPR/Cas9 system for gene deletion that is effective across multiple genetic backgrounds of *Aspergillus fumigatus*. This system employs in vitro assembly of Cas9 ribonucleoproteins (RNPs) coupled with micro-homology repair templates for gene deletion. Although highly efficient at gene targeting in wild type genetic backgrounds of *A. fumigatus*, the potential for our system to produce unwanted off-target mutations has not been addressed.

**Results:** Next-generation Illumina sequencing was used to identify genome mutations among transformants isolated from standard (no Cas9) and Cas9-mediated integration of a hygromycin deletion cassette. Two different concentrations of Cas9 were utilized to examine the association of Cas9 concentration with total numbers and types of genomic mutations. For each of the three test groups (zero, low, and high Cas9), three transformants were sequenced and compared to the parent strain. Bioinformatics analyses revealed the average number of total mutations to be similar among all three test groups. *A. fumigatus* transformation using standard, non-Cas9-mediated methods resulted in an average of $373 \pm 28$ mutations. In comparison, transformation with in vitro assembled Cas9-RNPs using either high (1 μg/μl) or low (0.5 μg/μl) levels of Cas9 resulted in an average of $326 \pm 19$ and $395 \pm 69$ mutations, respectively. In all cases, the vast majority of mutations identified were intergenic. No correlation between the amount of Cas9 utilized for transformation and the overall number of mutations was found. Finally, the specific type of mutation introduced during the transformation process was not Cas9-dependent, as both single-nucleotide polymorphisms and insertion/deletion events were not significantly different between the experimental groups.

**Conclusions:** CRISPR/Cas9-based genome editing in *A. fumigatus* using in vitro assembled RNPs coupled with micro-homology templates is a reliable method of gene targeting. This system is highly efficient and is not associated with increased off-target mutations caused by introduction of the Cas9 nuclease.

**Keywords:** *Aspergillus fumigatus*, CRISPR/Cas9, Genome editing, Off-target mutation

## Background

Gene deletion in *A. fumigatus* wild type strains is plagued by low homologous recombination rates and typically requires gene-targeting cassettes that contain ≥ 1000 base pairs of homology to be cloned upstream and downstream of a selection marker. The problem of low homologous recombination rates can be circumvented by using *A. fumigatus* strains mutated to have defective non-homologous end joining (NHEJ) DNA repair pathways [1, 2]. Although these strains have increased gene

*Correspondence: jfortwen@uthsc.edu
Department of Clinical Pharmacy and Translational Science, University of Tennessee Health Science Center, Memphis, TN, USA

targeting efficiencies, deletion cassettes still require 500–1000 bp regions of flanking homology and the defective NHEJ pathway(s) should be restored to ensure subtle genetic interactions between the targeted and NHEJ loci do not complicate phenotype interpretation. Finally, standard gene deletion methods in either wild type or NHEJ-defective backgrounds also rely on either multi-step cloning or overlap extension PCR techniques to build gene-targeting cassettes.

To improve gene targeting and genome editing in *A. fumigatus*, CRISPR/Cas9 gene editing technology has recently been implemented. In CRISPR-mediated genome editing, an RNA-directed Cas9 DNA nuclease is employed to recognize and cleave specific DNA sequences after forming a ribonucleoprotein (RNP) complex with a guide RNA (gRNA) [3]. This gRNA is a duplex that is composed of a CRISPR RNA (crRNA) and a transactivating CRISPR RNA (tracrRNA) [3]. The crRNA contains a 20-base region designated as the "protospacer", which guides specific DNA cleavage by binding to the complementary protospacer in the target genome [4]. However, Cas9-mediated DNA cleavage occurs only if the protospacer is followed by an "NGG" protospacer adjacent motif (PAM) in the target genome [5]. Several CRISPR/Cas9 systems have been developed in *Aspergillus* species [6–11]. In most of these organisms, the Cas9 enzyme and gRNA are introduced via expression constructs that are either contained within autonomously replicating plasmid or are integrated into the genome. Those that employ plasmids, control Cas9 activity through the presence or absence of selective agents in the medium for plasmid maintenance whereas those designed for integration typically rely on regulatable promoters. CRISPR/Cas9 systems are becoming ever more sophisticated in *Aspergillus*, as evidenced by recent work showing that highly-efficient marker-free gene editing can be accomplished in *A. nidulans*, *A. niger*, and *A. oryzae* strains defective in NHEJ [11]. In addition, highly efficient multi-site targeting is now possible in each of these *Aspergillus* species [11]. In *A. fumigatus*, the original use of CRISPR/Cas9 involved strains that constitutively expressed Cas9 from an integrated construct [7]. A later iteration in *A. fumigatus* employed autonomously replicating plasmids for Cas9 and gRNA expression and also utilized selectable marker cassettes ("repair templates" in CRISPR/Cas9 terminology) with microhomology regions that are only 35–50 bp in length, showing that efficient gene targeting can be accomplished in *A. fumigatus* with only small regions of DNA homology [12]. Although CRISPR/Cas9 gene targeting appears highly efficient in *A. fumigatus*, the systems in place thus far rely on genetically altering strains to express the required Cas9 nuclease and/or gRNA components or on building DNA-based constructs for expression of these components [7, 12].

We have reported the adaptation of a CRISPR/Cas9 gene editing system that utilizes in vitro assembled Cas9-RNPs coupled with microhomology repair templates [13]. Rather than genetically altering strains to express Cas9 or gRNAs, in vitro assembly relies on generating the Cas9 RNPs in a test tube before introducing them, along with a repair template (if required), into cells prepared for transformation. In this system, the gRNA was formed in vitro by incubating a mixture of equal molar amounts of crRNA and tracrRNA until a complex is formed. The crRNA and tracrRNA are purchased separately and then assembled into a gRNA complex so that the crRNA can be re-designed for each new protospacer the user desires to target. Next, purified Cas9 enzyme was mixed with the crRNA-tracrRNA complexes and incubated to allow for the formation of Cas9 RNPs. Cas9 concentrations of 1 and 0.5 μg/μl were utilized based on optimization experiments in our laboratory. The Cas9 RNP complexes were then used for standard transformation of *A. fumigatus* protoplasts along with 2 μg of hygromycin resistance cassette that is flanked by 35 base pair homology regions as a repair template. The major advantages of this in vitro assembly system are the simplicity (i.e., does not require strain construction) and the potentially portability from strain to strain. Our system generated nearly 100% gene targeting in the ΔakuB$^{\Delta Ku80}$ mutant, increased gene deletion frequencies in the wild type strain Af293 from the typical ~ 5% up to ~ 74%, and produced gene targeting efficiencies of ~ 90% in a clinical isolate [13]. Although gene targeting was greatly improved by our method, and is generally highly efficient in CRISPR/Cas9 methods developed for other the *Aspergillus* species, the potential for off-target mutations has largely not been addressed. Multiple recent studies have highlighted the potential for CRISPR/Cas9-based gene editing systems to induce unwanted off-target mutations [14–16]. This is typically believed to occur by promiscuous induction of double strand DNA breaks at sites other than the intended protospacer. To ensure the reliability of our in vitro assembly system, we sought to examine the impact of exogenous Cas9 addition during the transformation process on the induction of genomic mutations in *A. fumigatus* using transformants from the studies we previously performed in the Af293 genetic background.

## Results

To examine the potential for off-target mutations induced by the transient presence of gRNAs and the Cas9 nuclease in our system, we completed next-generation whole-genome sequencing on a subset of transformants isolated from our previous study and performed a comparative analysis of the relative numbers and characteristics of genomic mutations. In our previous study, reporting the optimization of in vitro assembled Cas9 RNPs for gene targeting in *A. fumigatus*, we generated multiple transformants from three basic experimental designs. In the first, we utilized the wild type reference isolate, Af293, and performed a transformation using standard protoplasting protocols to ectopically integrate a hygromycin selection cassette, *hygR* [13]. In the second and third experiments, we used Af293 to perform targeted CRISPR/Cas9-mediated integrations of the same *hygR* cassette to delete the coding region of a polyketide synthase, *pksP* [13]. The *pksP* locus was chosen for protocol optimization as it allowed for colony color-based identification of homologous integrations. The major difference between the second and third experiments of our previous study was the use of either high (1 μg/μl) or low (0.5 μg/μl) concentrations of the Cas9 nuclease, respectively. Therefore, the experimental groups for our whole genome analyses in the current study included three isolates from each of these three transformation conditions: no Cas9 (standard transformation protocol), 0.5 μg/μl Cas9 and 1 μg/μl Cas9 (Table 1). As a reference, we also sequenced the parent strain, Af293. *A. fumigatus* Af293 is the genome reference isolate with a well-annotated genome. However, this parent isolate was re-sequenced to account for genomic mutations that may have arisen during the repetitive sub-culturing of this strain in our laboratory. Average genome coverage ranged from 38× to 62× for all isolates (Table 2).

**Table 1 Strains used in this study**

| Strain name | Background |
|---|---|
| Af293 | Wild type |
| NC1 | *hygR* (no Cas9) |
| NC2 | *hygR* (no Cas9) |
| NC3 | *hygR* (no Cas9) |
| HC4 | Δ*pksP-hygR* (1 μg/μl Cas9) |
| HC5 | Δ*pksP-hygR* (1 μg/μl Cas9) |
| HC6 | Δ*pksP-hygR* (1 μg/μl Cas9) |
| LC7 | Δ*pksP-hygR* (0.5 μg/μl Cas9) |
| LC8 | Δ*pksP-hygR* (0.5 μg/μl Cas9) |
| LC9 | Δ*pksP-hygR* (0.5 μg/μl Cas9) |

For strains names, NC = no Cas9, HC = high Cas9, and LC = low Cas9. Indicated at the right are the concentrations of Cas9 used to produce each mutant strain. *hygR* = hygromycin resistance cassette; *pksP* = polyketide synthase; Δ*pksP-hygR* = *pksP* locus replaced by *hygR*

Comparative bioinformatics analyses revealed that multiple genomic mutations were present in the transformants of all strains, regardless of the presence or absence of Cas9. In the experimental group lacking Cas9, a group average of 373 ± 28 mutations were identified (Table 2). Considered alone, this finding demonstrates the potential mutagenic nature of *A. fumigatus* protoplast transformation. This standard transformation protocol may induce intense, albeit temporary, cellular stress as it requires the enzymatic digestion of the cell wall to release membrane-bound protoplasts followed by recovery on an osmotically stabilized agar medium. Transformants from both the high (1 μg/μl) and low (0.5 μg/μl) Cas9 concentration experiments displayed total numbers of genomic mutations similar to the no Cas9 control, with an average of

**Table 2 Cas9-mediated gene deletion is not associated with increased genomic mutations *in A. fumigatus***

| | NC1 | NC2 | NC3 | HC4 | HC5 | HC6 | LC7 | LC8 | LC9 |
|---|---|---|---|---|---|---|---|---|---|
| Average coverage | 53× | 62× | 60× | 58× | 54× | 61× | 56× | 49× | 38× |
| Total mutations | 396 | 342 | 385 | 345 | 326 | 307 | 345 | 366 | 474 |
| Average total mutations | 373 ± 28 SD | | | 326 ± 19 SD ($p > 0.05$) | | | 395 ± 69 SD ($p > 0.05$) | | |
| *Analysis based on type of mutation identified* | | | | | | | | | |
| SNPs | 371 | 318 | 363 | 321 | 314 | 292 | 326 | 347 | 439 |
| Indels | 25 | 24 | 22 | 24 | 12 | 15 | 19 | 19 | 35 |
| *Analysis based on location of mutation* | | | | | | | | | |
| Intergenic | 380 (96%) | 331 (97%) | 379 (98%) | 331 (96%) | 320 (98%) | 301 (98%) | 339 (98%) | 359 (98%) | 446 (94%) |
| Average intergenic | 363 ± 28 SD | | | 317 ± 15 SD ($p = 0.04$) | | | 381 ± 57 SD ($p > 0.05$) | | |
| Coding region | 16 (4%) | 11 (3%) | 6 (2%) | 14 (4%) | 6 (2%) | 6 (2%) | 6 (2%) | 7 (2%) | 28 (6%) |
| Average coding region | 11 ± 5 SD | | | 9 ± 5 SD ($p > 0.05$) | | | 14 ± 12 SD ($p > 0.05$) | | |

Displayed are the total and average number of mutations among the three experimental groups: no ("NC"—0 μg/μl), low ("LC"—0.5 μg/μl) and high ("HC"—1 μg/μl) levels of Cas9. For the intergenic and coding region mutation rows, the numbers in parentheses represent the percent (%) of total mutations. For the average mutations per group, the mean ± standard deviation (SD) is provided. The Student's *t* test assuming unequal variance was used for statistical comparisons of the Cas9 (HC or LC) and the no-Cas9 (NC) groups and *p* values are presented

326 ± 19 and 395 ± 69, respectively (Table 2). Among all transformants, only a small subset of the identified mutations were located within coding regions of the genome, as the vast majority (> 96% for all isolates) were intergenic (Table 2). A single transformant (LC9) from the low Cas9 concentration experiment was found to contain relatively higher levels of total mutations, with a total of 474, and a slightly increased distribution of these mutations into coding regions (~ 5.9% of total compared to an average of ~ 2.6% for all other transformants). However, as this transformant is an isolate from the low Cas9 concentration experiment, this apparent increase in total number of mutations and distribution towards coding regions is not likely associated with the activity of Cas9. Further supporting this assertion, we found no statistically significant difference in the total number of mutations identified among the experimental groups (Table 2). Therefore, our analyses revealed no Cas9 concentration-dependent increase in induction of genomic mutations. The mutations identified in our study are likely induced by the transformation process, including cell wall digestion.

To further examine if the presence of Cas9 during transformation may influence the type of mutation introduced, we analyzed additional characteristics of the identified genomic variations. Those mutations identified within coding regions displayed comparable distributions among the categories defined in Table 3. In general, most mutations were located within the 3′ UTRs of genes regardless of the presence or absence of Cas9. The only discrepancy noted was that all transformants within the high and low concentration Cas9 groups displayed a low number of non-synonymous mutations identified within coding regions, whereas the standard non-Cas9 transformants were free of non-synonymous mutations (Table 3). However, similar to our findings with total numbers of mutations, the low concentration Cas9 transformants contained more non-synonymous mutations (~ 0.8% of total mutations) than those from the high concentration Cas9 experiment (~ 0.5% of total mutations). Thus, we interpret these mutations not as a consequence of Cas9 presence but as variability among isolates resulting from the standard protoplast transformation process.

Cas9-induced double strand breaks (DSBs) can be repaired by two major pathways in the cell: the NHEJ DNA repair pathway or homology directed repair [17]. The NHEJ repair pathway is error prone and induces small insertion and deletion (indel) events into the genome [17]. Therefore, a negative consequence of Cas9-mediated gene editing could be the promiscuous induction of DSBs resulting in increased indels throughout the affected transformant(s). To see if this specific type of off-target mutation may exist in our collection, we also analyzed transformants from each experimental group for the disproportional generation of indels versus single nucleotide polymorphisms (SNPs). To do this, the total genomic mutations (intergenic and coding region) were classified into either SNPs or indels and the resulting numbers were averaged for each experimental group. Our data indicated that the generation of SNPs was favored over indels (1–80 nucleotides in length) within each experimental group and no significant differences in the relative amounts of either mutation were noted (Fig. 1). When expressed as a percent of total mutations, the non-Cas9 mediated transformation generated strains containing 6.4% indels, whereas the high and low concentration Cas9 transformants had 5.2 and 6.2% indels, respectively. Therefore, the concentration of Cas9 is not associated with a disproportional increase in the number of indels between experimental groups. Although our protospacers were designed to have minimal off-site complementarity, some level of similarity between our designed sequences and distant areas of the genome is unavoidable. To ensure that off-target complementarity of our gRNA complexes was not promiscuously driving

**Table 3 Cas9-mediated gene deletion does not cause alterations in the types of coding region mutations in *A. fumigatus***

| | 0 µg/µl Cas9 | | | 1 µg/µl Cas9 | | | 0.5 µg/µl Cas9 | | |
|---|---|---|---|---|---|---|---|---|---|
| | NC1 | NC2 | NC3 | HC4 | HC5 | HC6 | LC7 | LC8 | LC9 |
| 3′ UTR | 9 | 9 | 2 | 9 | 2 | – | 1 | 1 | 10 |
| 5′ UTR | 3 | – | 1 | 2 | – | 1 | 1 | – | 1 |
| Frameshift | 1 | – | – | 1 | – | – | – | – | – |
| Intron | 2 | 2 | 2 | 1 | 1 | 1 | 2 | 3 | 2 |
| Non-synonymous | – | – | – | 1 | 1 | 3 | 2 | 1 | 6 |
| Start lost | – | – | – | – | 1 | – | – | – | – |
| Synonymous | – | – | – | – | 1 | 1 | – | 1 | 9 |
| Splice region | 1 | – | 1 | – | – | – | – | 1 | – |

Shown are the numbers and types of identified mutations located within coding regions of transformants within the three experimental groups

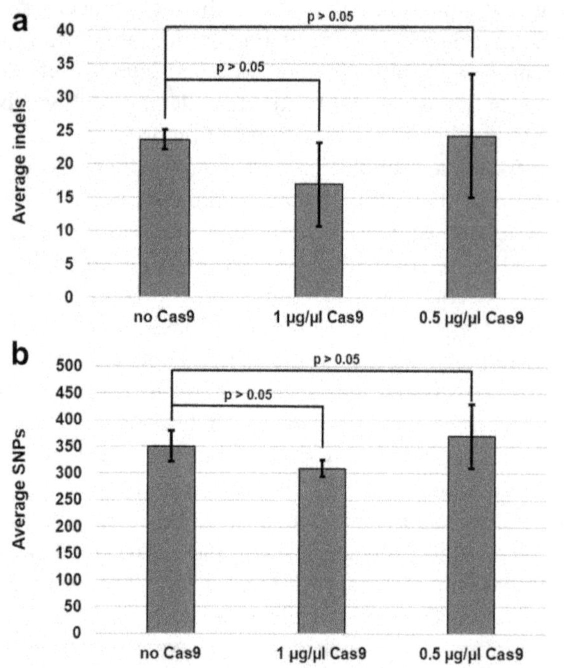


**Fig. 1** Cas9-mediated gene deletion does not cause a disproportional increase in SNPs or indels in *A. fumigatus*. Segregation of genomic mutations into SNPs (**b**) and indels (**a**) revealed that the concentration of Cas9 was not positively associated with an increase in a specific subset of mutation. Shown are the average number of mutation events within each experimental group. Student's *t* test assuming unequal variance was utilized for statistical comparison between the "no Cas9" and either the low (0.5 µg/µl) or high (1 µg/µl) Cas9 group

Cas9 to generate unwanted DSBs and subsequent off-target mutations, we finally interrogated each transformant genome for variations surrounding ten potential off-target sites. These off-target sites were defined as areas where the protospacer had twelve or more base pairs of complementarity. For both the 5′ and 3′ protospacer, our analysis found zero genomic mutations in these areas (data not shown). Together, these data indicate that CRISPR/Cas9 editing by our methods is highly specific in *A. fumigatus*.

## Discussion

The implementation of novel gene editing technologies in *A. fumigatus*, like CRISPR/Cas9, is a critical step toward making significant advances in studies related to virulence and antifungal drug resistance in this important human pathogen. We have shown that our system, relying on in vitro assembled Cas9-RNP complexes, produces highly efficient gene targeting in multiple genetic backgrounds of *A. fumigatus* [13]. Because it does not rely on strains that have been genetically engineered to increase homologous recombination rates or to express CRISPR/Cas9 components, this system can be utilized to study gene and pathway function across many isolates. Multiple studies have indicated that CRISPR/Cas9-based gene editing is potentially associated with off-target mutations whereas many studies have also described this system as highly specific [18]. Off-target mutations could be due to promiscuous activity of the Cas9 nuclease, yet more often seem to be caused by non-specific binding of gRNA [18]. Using the system we have adopted for *A. fumigatus*, both of these potential issues are properly addressed [13]. The Cas9 nuclease is introduced as a purified protein and, therefore, is only transiently present. Compared to systems that constitutively or conditionally express Cas9, this should reduce the potential for unwanted mutagenesis of the genome generated by promiscuous Cas9 activity. In fact, in a few yeast and parasite studies, expression of CRISPR/Cas9 components was associated with toxicity [19–21]. Our system avoids this problem entirely. Additionally, using the rules for PAM selection, protospacer design, and RNP assembly that we have previously reported [13], we were able to generate Cas9-transformants that did not show increased genomic variability. Because the genomic variation identified in our study is not correlated with CRISPR/Cas9 methods, our data also suggest that either the single sub-culture of a strain of *A. fumigatus* within the laboratory or the stress induced by our standard protoplast transformation procedure induces genomic variation.

Although whole genome sequencing of transformants has proven useful for the analysis of off-target effects in many systems, this technique does have the limitation of being unable to identify Cas9-induced DSBs that are perfectly repaired. With this caveat in mind, whole genome sequencing has been successfully applied to characterize the potential for off-target effects of Cas9-based transformations in the plant pathogen *Ustilago maydis* [22]. This study relied on expressing Cas9 from a strong constitutive promoter on a self-replicating plasmid. After transformation, strains could then be cured of this plasmid to avoid continuous passage and growth in the presence of Cas9. Whole genome sequencing of transformants acquired in this study revealed that none of the identified genome mutations were likely to be due to Cas9 activity mediated by gRNA binding [22]. Our results support the same conclusion when purified Cas9 is added exogenously to *A. fumigatus*. It is of note, however, that our study only investigated two concentrations of Cas9, did not examine the effects of varying amounts of crRNA or tracrRNA or the ratios of Cas9 to tracrRNA and crRNA, and only investigated off-target effects upon targeting of only one gene. It is possible that by targeting a different protospacer or by significantly altering the Cas9 RNP composition, a different outcome might have been observed. However, even though off-target effects might increase under these other conditions, further efforts that try to minimize them can be pursed. For example, bioinformatic tools are now available to interrogate genomes for potential off-target sites, including Cas-OFFinder and Cas-Designer [23, 24]. Databases like these can aid in protospacer selection and gRNA design to minimize the potential for off-target mutations when targeting a new genomic locus. Also, multiple studies into ways to minimize off-target mutations in CRISPR/Ca9 systems have been published and new techniques are constantly being pursued. One technique might be to limit the time of Cas9 activity in the cell via use of photoactivatable split-Cas9 [25]. Our system accomplishes limited Cas9 activity through introduction of the Cas9 enzyme. However, if required, the specificity of our system could also be further bolstered by employing high-fidelity or rationally engineered Cas9 enzymes with increased specificity or through the use of truncated versions of single-gRNAs [26–28].

## Conclusions

The data provided here demonstrate that CRISPR/Cas9-mediated gene targeting, using our in vitro assembled Cas9-RNP system, does not cause an increase in genomic variation over standard transformation protocols. We also identified no disproportional Cas9-dependent increase in SNPs or indels among treated strains. Therefore, Cas9-mediated gene deletion using in vitro assembled Cas9-RNPs coupled with microhomology repair templates is a reliable method for generating targeted mutations in *A. fumigatus*.

## Methods

### Strains and culture conditions

Strains used for this study are listed in Table 1. Strain Af293 is the *A. fumigatus* reference genome isolate [29] and all transformant strains employed for whole genome sequencing were generated, as part of a previous study, in this genetic background [13]. All strains were maintained on glucose minimal media (GMM) agar [30], supplemented with hygromycin (150 µg/ml) for selection. Conidia were harvested in water from three-day old plates and enumerated by hemocytometer.

### Genomic DNA extraction

Genomic DNA was extracted following a slight modification of previous published protocols, using the Qiagen DNeasy Plant Mini Kit [31]. Briefly, strains were inoculated in GMM broth at a conidial density of $10^6$ conidia/ml and incubated for 20 h at 37 °C with shaking at 250 rpm. Mycelia were harvested by vacuum filtration and 300 mg of a semi-dry mycelial mat was crushed under liquid nitrogen. The resulting mycelial powder was resuspended in 800 µl of buffer AP1 and 8 µl of RNase A and vigorously vortexed. Following 3 h of incubation at 65 °C, the fungal lysate was centrifuged for 5 min at 20,000×*g* and the supernatant was transferred to a new 1.5 ml tube. Next, 260 µl of buffer P3 were added to the supernatant and the mixture was incubated for 5 min on ice, then centrifuged for 5 min at 20,000×*g*. The supernatant was then transferred to QIAshredder spin column (700 µl at a time) and centrifuged for 2 min at 20,000×*g*. The flow-through was collected into a 2 ml tube without disturbing the pellet, and 1.5 volumes of buffer AW1 was added and mixed by pipetting. The mixture was transferred (700 µl at a time) into a DNeasy Mini spin column and the genomic DNA was allowed to bind to the column membrane by centrifuging for 1 min at ≥ 6000×*g*. The genomic DNA was washed first with 700 µl of AW2 buffer and centrifuged for 1 min at ≥ 6000×*g*, followed by a second washing step using 300 µl of Buffer AW2 and centrifugation at 20,000×*g* for 5 min. The second washing step was essential to remove any residual ethanol on the membrane before elution step. The spin column was transferred to a new 1.5 ml tube. For the elution of genomic DNA, 100 µl Buffer 5 mM Tris–HCl (pH 8.5) was added to the center of the column and the column was incubated at room temperature for 5 min, followed by a centrifugation step for 1 min at ≥ 6000×*g*. Final DNA concentrations were quantified using Nanodrop and Qubit Fluorometer, following the manufacturer's protocol.

### Library preparation and bioinformatics analyses

Library preparations and genome sequencing reactions were performed at the University of Alabama at Birmingham Heflin Center for Genomic Science. The Qiagen QIAseq FX DNA prep kit was used for library preparations, following the manufacturer's instructions. Paired end 300 base pair sequencing reads were generated on the Illumina MiSeq following standard protocols. Bioinformatics services were provided by code4DNA (www.code4DNA.com). Reads from each sample were aligned using bwa mem (v0.7.15) to the *A. fumigatus* reference genome build A_fumigatus_Af293_version_s03-m05-r05 downloaded from AspGD.org [32]. Samtools (v1.3) fixmate and rmdup were used to remove PCR duplicates [33]. Sequence mutations were called using FreeBayes (v1.1.0) with the haploid population-based model [34]. Low quality (QUAL < 30) and low depth (DP < 10) mutations were filtered out using VCFtools (v0.1.15) which was also used to remove variant calls where no sequence reads were available for at least one sample. The population.vcf file was split into individual samples using VCFtools vcf-subset. Mutations were annotated using snpEff (v4.3r) and VCFtools vcf-isec was used to select mutations found in each affected samples but not in the Ref. [35].

### Abbreviations

CRISPR: clustered regularly interspaced short palindromic repeats; crRNA: CRISPR RNA; DSB: double strand break; GMM: glucose minimal media; gRNA: guide RNA; Indel: insertion/deletion; RNP: ribonucleoprotein; SNP: single-nucleotide polymorphism; tracrRNA: trans-activating crRNA.

### Authors' contributions

QAA performed the transformation of strains to obtain isolates for genome sequencing and isolated genomic DNA for sequencing studies. WG cultured and prepared the selected strains for analysis. JRF contracted for the genomic sequencing and bioinformatics analyses. QAA, WG, AMV, ACOS and JRF contributed to data interpretation and manuscript preparation. All authors read and approved the final manuscript.

### Acknowledgements

The authors would like to thank code4DNA (www.code4DNA.com) for contributing written summaries of how bioinformatics analyses were performed.

### Competing interests

The authors declare that they have no competing interests.

### Funding

This work was supported by NIH Grant R01AI106925 to J.R.F.

### References

1. Krappmann S, Sasse C, Braus GH. Gene targeting in *Aspergillus fumigatus* by homologous recombination is facilitated in a nonhomologous end-joining-deficient genetic background. Eukaryot Cell. 2006;5:212–5.
2. da Silva Ferreira ME, Kress MRVZ, Savoldi M, Goldman MHS, Härtl A, Heinekamp T, Brakhage AA, Goldman GH. The akuBKU80 mutant deficient for nonhomologous end joining is a powerful tool for analyzing pathogenicity in *Aspergillus fumigatus*. Eukaryot Cell. 2006;5:207–11.
3. Sander JD, Joung JK. CRISPR-Cas systems for editing, regulating and targeting genomes. Nat Biotechnol. 2014;32:347–55.
4. Shah SA, Erdmann S, Mojica FJM, Garrett RA. Protospacer recognition motifs. RNA Biol. 2013;10:891–9.
5. Mojica FJM, Díez-Villaseñor C, García-Martínez J, Almendros C. Short motif sequences determine the targets of the prokaryotic CRISPR defence system. Microbiology. 2009;155:733–40.
6. Nødvig CS, Nielsen JB, Kogle ME, Mortensen UH. A CRISPR–Cas9 system for genetic engineering of filamentous fungi. PLoS ONE. 2015;10:e0133085. https://doi.org/10.1371/journal.pone.0133085.
7. Fuller KK, Chen S, Loros JJ, Dunlap JC. Development of the CRISPR/Cas9 system for targeted gene disruption in *Aspergillus fumigatus*. Eukaryot Cell. 2015;14:1073–80.
8. Katayama T, Tanaka Y, Okabe T, Nakamura H, Fujii W, Kitamoto K, Maruyama J. Development of a genome editing technique using the CRISPR/Cas9 system in the industrial filamentous fungus *Aspergillus oryzae*. Biotechnol Lett. 2016;38:637–42.
9. Zhang C, Meng X, Wei X, Lu L. Highly efficient CRISPR mutagenesis by microhomology-mediated end joining in *Aspergillus fumigatus*. Fung Genet Biol. 2016;86:47–57.
10. Weyda I, Yang L, Vang J, Ahring BK, Lübeck M, Lübeck PS. A comparison of *Agrobacterium*-mediated transformation and protoplast-mediated transformation with CRISPR–Cas9 and bipartite gene targeting substrates, as effective gene targeting tools for *Aspergillus carbonarius*. J Microbiol Methods. 2017;135:26–34.
11. Nødvig CS, Hoof JB, Kogle ME, Jarczynska ZD, Lehmbeck J, Klitgaard DK, Mortensen UH. Efficient oligo nucleotide mediated CRISPR–Cas9 gene editing in *Aspergilli*. Fung Genet Biol. 2018. https://doi.org/10.1016/j.fgb.2018.01.004.
12. Zhang C, Meng X, Wei X, Lu L. Highly efficient CRISPR mutagenesis by microhomology-mediated end joining in *Aspergillus fumigatus*. Fung Genet Biol. 2016;86(Supplement C)):47–57.
13. Al Abdallah Q, Ge W, Fortwendel JR. A simple and universal system for gene manipulation in *Aspergillus fumigatus in vitro*-assembled Cas9-guide RNA ribonucleoproteins coupled with microhomology repair templates. mSphere. 2017. https://doi.org/10.1128/msphere.00446-17.
14. Zhang X-H, Tee LY, Wang X-G, Huang Q-S, Yang S-H. Off-target effects in CRISPR/Cas9-mediated genome engineering. Mol Ther Nucleic Acids. 2015;4:e264. https://doi.org/10.1038/mtna.2015.37.
15. Schaefer KA, Wu W-H, Colgan DF, Tsang SH, Bassuk AG, Mahajan VB. Unexpected mutations after CRISPR–Cas9 editing *in vivo*. Nat Methods. 2017;14:547–8.
16. Zhang Q, Xing H-L, Wang Z-P, Zhang H-Y, Yang F, Wang X-C, Chen Q-J. Potential high-frequency off-target mutagenesis induced by CRISPR/Cas9 in *Arabidopsis* and its prevention. Plant Mol Biol. 2018;96:445–56.
17. Lieber MR. The mechanism of double-strand DNA break repair by the nonhomologous DNA end joining pathway. Ann Rev Biochem. 2010;79:181–211.
18. O'Geen H, Yu AS, Segal DJ. How specific is CRISPR/Cas9 really? Cur Opin Chem Biol. 2015;29:72–8.
19. Ryan OW, Skerker JM, Maurer MJ, Li X, Tsai JC, Poddar S, Lee ME, DeLoache W, Dueber JE, Arkin AP, Cate JHD. Selection of chromosomal DNA libraries using a multiplex CRISPR system. eLife. 2014;3:e03703. https://doi.org/10.7554/elife.03703.
20. Generoso WC, Gottardi M, Oreb M, Boles E. Simplified CRISPR–Cas genome editing for *Saccharomyces cerevisiae*. J Microbiol Methods. 2016;127:203–5.
21. Jiang W, Brueggeman AJ, Horken KM, Plucinak TM, Weeks DP. Successful transient expression of Cas9 and single guide RNA genes in *Chlamydomonas reinhardtii*. Eukaryot Cell. 2014;13:1465–9.
22. Schuster M, Schweizer G, Reissmann S, Kahmann R. Genome editing in *Ustilago maydis* using the CRISPR–Cas system. Fungal Genet Biol. 2016;89:3–9.
23. Park J, Bae S, Kim J-S. Cas-Designer: a web-based tool for choice of CRISPR–Cas9 target sites. Bioinformatics. 2015;31:4014–6.
24. Bae S, Park J, Kim J-S. Cas-OFFinder: a fast and versatile algorithm that searches for potential off-target sites of Cas9 RNA-guided endonucleases. Bioinformatics. 2014;30:1473–5.
25. Nihongaki Y, Kawano F, Nakajima T, Sato M. Photoactivatable CRISPR–Cas9 for optogenetic genome editing. Nat Biotechnol. 2015;33:755–60.
26. Kleinstiver BP, Pattanayak V, Prew MS, Tsai SQ, Nguyen NT, Zheng Z, Joung JK. High-fidelity CRISPR–Cas9 nucleases with no detectable genome-wide off-target effects. Nature. 2016;529:490–5.
27. Slaymaker IM, Gao L, Zetsche B, Scott DA, Yan WX, Zhang F. Rationally engineered Cas9 nucleases with improved specificity. Science. 2016;351:84–8.
28. Fu Y, Sander JD, Reyon D, Cascio VM, Joung JK. Improving CRISPR–Cas nuclease specificity using truncated guide RNAs. Nat Biotechnol. 2014;32:279–84.
29. Nierman WC, Pain A, Anderson MJ, Wortman JR, Kim HS, Arroyo J, Berriman M, Abe K, Archer DB, Bermejo C, Bennett J, Bowyer P, Chen D, Collins M, Coulsen R, Davies R, Dyer PS, Farman M, Fedorova N, Fedorova N, Feldblyum TV, Fischer R, Fosker N, Fraser A, García JL, García MJ, Goble A, Goldman GH, Gomi K, Griffith-Jones S, Gwilliam R, Haas B, Haas H, Harris D, Horiuchi H, Huang J, Humphray S, Jiménez J, Keller N, Khouri H, Kitamoto K, Kobayashi T, Konzack S, Kulkarni R, Kumagai T, Lafton A, Latgé J-P, Li W, Lord A, Lu C, Majoros WH, May GS, Miller BL, Mohamoud Y, Molina M, Monod M, Mouyna I, Mulligan S, Murphy L, O'Neil S, Paulsen I, Peñalva MA, Pertea M, Price C, Pritchard BL, Quail MA, Rabbinowitsch E, Rawlins N, Rajandream M-A, Reichard U, Renauld H, Robson GD, de Córdoba SR, Rodríguez-Peña JM, Ronning CM, Rutter S, Salzberg SL, Sanchez M, Sánchez-Ferrero JC, Saunders D, Seeger K, Squares R, Squares S, Takeuchi M, Tekaia F, Turner G, de Aldana CRV, Weidman J, White O, Woodward J, Yu J-H, Fraser C, Galagan JE, Asai K, Machida M, Hall N, Barrell B, Denning DW. Genomic sequence of the pathogenic and allergenic filamentous fungus *Aspergillus fumigatus*. Nature. 2005;438:1151–6.
30. Shimizu K, Keller NP. Genetic involvement of a cAMP-dependent protein kinase in a G protein signaling pathway regulating morphological and chemical transitions in *Aspergillus nidulans*. Genetics. 2001;157:591–600.

31. Hagiwara D, Takahashi H, Watanabe A, Takahashi-Nakaguchi A, Kawamoto S, Kamei K, Gonoi T. Whole-genome comparison of *Aspergillus fumigatus* strains serially isolated from patients with Aspergillosis. J Clin Microbiol. 2014;52:4202–9.
32. Li H, Durbin R. Fast and accurate short read alignment with Burrows–Wheeler transform. Bioinformatics. 2009;25:1754–60.
33. Li H, Handsaker B, Wysoker A, Fennell T, Ruan J, Homer N, Marth G, Abecasis G, Durbin R. The sequence alignment/map format and SAMtools. Bioinformatics. 2009;25:2078–9.
34. Garison E, Marth G. Haplotype-based variant detection from short-read sequencing. 2012. arXiv preprint: arXiv:1207.3907[q-bio.GN].
35. Cingolani P, Platts A, Wang LL, Coon M, Nguyen T, Wang L, Land SJ, Lu X, Ruden DM. A program for annotating and predicting the effects of single nucleotide polymorphisms, SnpEff. Fly. 2012;6:80–92.

# 10

# Truncation of the transcriptional repressor protein Cre1 in *Trichoderma reesei* Rut-C30 turns it into an activator

Alice Rassinger[1], Agnieszka Gacek-Matthews[2,3], Joseph Strauss[2], Robert L. Mach[1] and Astrid R. Mach-Aigner[1*]

**Abstract**

**Background:** The filamentous fungus *Trichoderma reesei* (*T. reesei*) is a natural producer of cellulolytic and xylanolytic enzymes and is therefore industrially used. Many industries require high amounts of enzymes, in particular cellulases. Strain improvement strategies by random mutagenesis yielded the industrial ancestor strain Rut-C30. A key property of Rut-C30 is the partial release from carbon catabolite repression caused by a truncation of the repressor Cre1 (Cre1-96). In the *T. reesei* wild-type strain a full *cre1* deletion leads to pleiotropic effects and strong growth impairment, while the truncated *cre1-96* enhances cellulolytic activity without the effect of growth deficiencies. However, it is still unclear which function Cre1-96 has in Rut-C30.

**Results:** In this study, we deleted and constitutively expressed *cre1-96* in Rut-C30. We found that the presence of Cre1-96 in Rut-C30 is crucial for its cellulolytic and xylanolytic performance under inducing conditions. In the case of the constitutively expressed Cre1-96, the cellulase activity could further be improved approximately twofold. The deletion of *cre1-96* led to growth deficiencies and morphological abnormalities. An in silico domain prediction revealed that Cre1-96 has all necessary properties that a classic transactivator needs. Consequently, we investigated the cellular localization of Cre1-96 by fluorescence microscopy using an eYFP-tag. Cre1-96 is localized in the fungal nuclei under both, inducing and repressing conditions. Furthermore, chromatin immunoprecipitation revealed an enrichment of Cre1-96 in the upstream regulatory region of the main transactivator of cellulases and xylanases, Xyr1. Interestingly, transcript levels of *cre1-96* show the same patterns as the ones of *xyr1* under inducing conditions.

**Conclusions:** The findings suggest that the truncation turns Cre1 into an activating regulator, which primarily exerts its role by approaching the upstream regulatory region of *xyr1*. The conversion of repressor proteins to potential activators in other biotechnologically used filamentous fungi can be applied to increase their enzyme production capacities.

**Keywords:** Carbon catabolite repression, *Trichoderma reesei*, Cre1, Gene regulation, Transcription factor, Cellulases, Xylanases, Chromatin

## Background

Cellulose and hemicellulose are the most abundant biopolymers in plants. After the industrial processing of trees, crops and other plants, which are grown for food and other purposes, a lot of cellulosic and hemicellulosic waste accumulates [1]. The quality and composition of this waste can be quite versatile, depending on the branch of industry they originate from. However, they all share a significant, unused carbohydrate content that can be utilized for the production of valuable products [1]. The main challenge for an economic utilization of these waste products is the efficient conversion of cellulose-rich biomass to products such as (ligno)cellulosic ethanol [2]. One main limitation is the extraction of monomeric and dimeric sugars such as cellobiose, D-glucose and D-xylose from cellulose and hemicellulose [3]. The rigidity of the

*Correspondence: astrid.mach-aigner@tuwien.ac.at
[1] Institute of Chemical, Environmental and Bioscience Engineering, TU Wien, Gumpendorfer Str. 1a, 1060 Vienna, Austria
Full list of author information is available at the end of the article

structure of cellulose and hemicellulose requires first mechanical and chemical treatment, which demand high temperatures, harsh chemicals and create an ecologically difficult disposable waste stream. Secondly, hydrolysis of cellulose and hemicellulose is done enzymatically. For the hydrolysis in industrial scale, the main bottleneck is an affordable price on bulk amounts of cellulose and hemicellulose degrading enzymes [3]. The filamentous fungus *Trichoderma reesei* (*T. reesei)* is one of the top producers of such enzymes (e.g. cellobiohydrolases (EC 3.2.1.91), endoglucanases (EC 3.2.1.4), endo-β-1,4-xylanases (EC 3.2.1.8), β-xylosidases (EC 3.2.1.37) (reviewed in [4])) in industry. Those enzymes are moderately expressed in the presence of cellulose and the hemicellulose xylan and stronger by the respective degradation products. Surprisingly, lactose also triggers the expression of these enzymes even though it is not present in the natural environment of the fungus. Although the exact induction mechanism is not fully understood, the uptake of lactose by a permease is necessary for the activation of cellulase gene expression [5].

Anyhow, the enzyme formation is limited by carbon catabolite repression (CCR) in the presence of high concentrations of easily metabolizable monomeric carbohydrates, such as D-glucose or D-xylose [6]. The uptake of D-glucose enables the fungus to rapidly gain energy; hence, the degradation of complex biopolymers by the cellulolytic and xylanolytic enzymes is shut down. The CCR mechanism is well conserved amongst various organisms ranging from bacteria to humans. Based on the sequence homologies to CreA from *Aspergillus* species, the Carbon catabolite repressor protein Cre1 (encoded by *cre1*) was described as the regulator of CCR in *T. reesei* during the 1990ies [7]. Cre1 is a $C_2H_2$ zinc finger protein and binds to a 5′-SYGGRG-3′ motif within upstream regulatory regions (URR) of cellulase and xylanase encoding genes (e.g. *cbh1* [8], *xyn1* [9]). Its regulon also comprises sugar transporters, developmental processes, and parts of the chromatin remodelling machinery such as nucleosome positioning [10, 11]. Most notably, Cre1 acts negatively on the transcription of the main and essential transactivator of cellulolytic and xylanolytic enzyme expression, Xyr1 [12]. Thus, Xyr1 is also a subject to CCR mediated by Cre1 [13]. With regards to the industry-scale production of hydrolytic enzymes, top producing *T. reesei* strains became a necessity. Random mutagenesis yielded the mutant strain Rut-C30, which achieves enzyme yields of 20 g/L [14]. The nowadays used industrial *T. reesei* strains (yielding up to 100 g/L [15]) are based on Rut-C30 and thus share a similar genetic background. Predominately, this includes a truncation of Cre1, which led to partial de-repression from CCR on D-glucose [16]. Nevertheless, with regards to the wild-type system, we refer in this manuscript to D-glucose as a repressing condition. In 2014, Mello-de-Sousa and colleagues used the *T. reesei* wild-type strain to demonstrate that this truncated Cre1 (Cre1-96) positively influences cellulase expression, while the full deletion of *cre1* leads to strong pleiotropic effects and growth impairment [17]. The enhancement of cellulase expression by Cre1-96 was attributed to a chromatin opening in the URR of cellulase-encoding genes and also of the *xyr1* gene. However, the impact of Cre1-96 was never studied directly in Rut-C30. The exact regulatory mechanism of Cre1-96 and its role as a putative new transcription factors in industrial strains still remain to be elucidated. In this study, we investigated the effects of a *cre1-96* deletion in Rut-C30 on its growth behaviour, the enzymatic activities and the transcriptional profiles of cellulase- and xylanase-encoding genes (*cbh1*, *xyn1*) and of *xyr1*. To determine the subcellular localization of the putative transcription factor, the nuclear import was examined under cellulase inducing and repressing conditions. Moreover, we performed chromatin immunoprecipitation and nuclease digestion to learn which genes are targeted by Cre1-96 and what is its impact on the DNA accessibility within the URR of its target genes. Finally, we constitutively expressed *cre1-96* in Rut-C30 and examined the impact on the cellulolytic activities.

## Results

### Deletion and constitutive expression of *cre1-96* in *T. reesei* Rut-C30

To identify the function of Cre1-96 in Rut-C30, the encoding gene was deleted from the genome. Therefore, a deletion cassette was integrated by homologous recombination at the *cre1-96* locus, resulting in a gene replacement of *cre1-96* in Rut-C30. Two *cre1-96* deletion strains were identified by diagnostic PCR (Additional file 1: Figure S1). Both deletion strains were used throughout this study and are in the following termed Rut-C30Δ*cre1-96* (1) and (2) in the figures. In the parent strain Rut-C30, the structural gene of *cre1-96* was put under the control of the *tef1* promoter. The homologous integration of this expression cassette at the *cre1-96* locus was again verified by diagnostic PCR and the resulting strain is in the following termed Rut-C30OE*cre1-96* (Additional file 2: Figure S2).

### Cre1-96 is required for cellulolytic and xylanolytic performance of Rut-C30

To investigate a possible impact of Cre1-96 on cellulase and xylanase gene expression, the *cre1-96* deletion strain and its parent strain Rut-C30 were grown on plates containing lactose or carboxymethylcellulose (CMC) to resemble cellulase-inducing conditions (Fig. 1). Further,

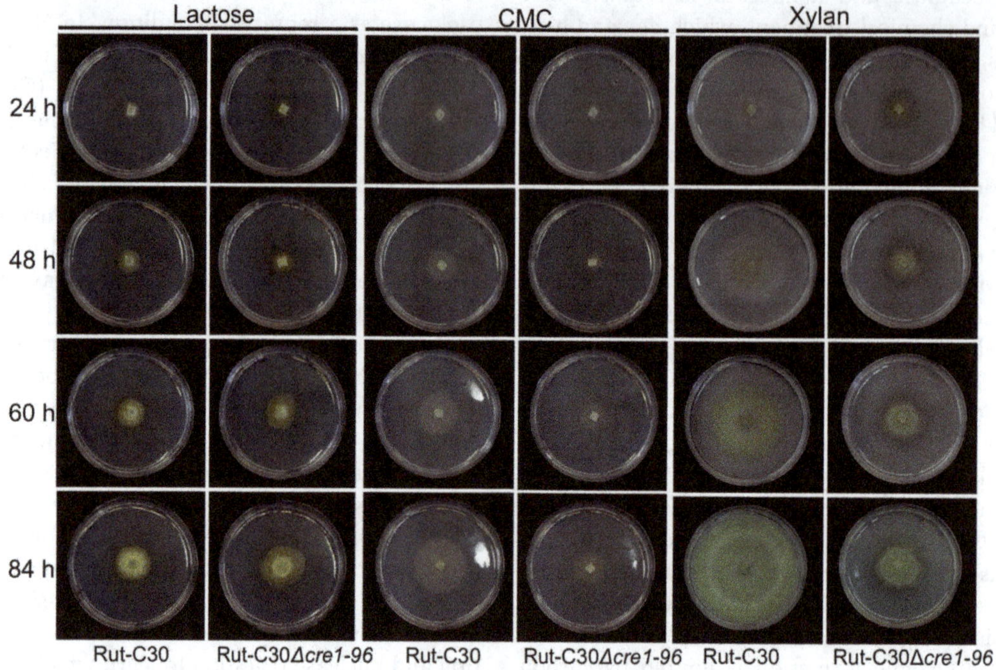


**Fig. 1** Growth behaviour of Rut-C30Δ*cre1-96* under cellulase inducing conditions. The *T. reesei* strains Rut-C30 and Rut-C30Δ*cre1-96* were pre-grown on MEX plates and were then transferred in biological duplicates to MA medium plates supplemented with 1% (w/v) lactose, CMC or xylan. Plates were incubated at 30 °C and pictures were taken after 24, 48, 60 and 84 h

they were grown on xylan for induction of xylanase expression (Fig. 1), on a non-inducing condition (glycerol) and on a repressing condition (D-glucose) (Additional file 3: Figure S3). Photos of the plates were taken after 24, 48, 60 and 84 h of growth. On lactose, no clear differences in growth were obtained between the two tested strains at any time point (Fig. 1). However, the radial colony formation seemed to be abnormal after 60 and 84 h when *cre1-96* was absent (Fig. 1). On CMC and xylan, growth deficiencies were observed in the *cre1-96* deletion strain at all times points in comparison to the parent strain. The colony was clearly reduced in size, while no influence on sporulation was visible (Fig. 1). On glycerol and D-glucose no obvious growth reduction was visible at any time point. However, the spore pigmentation changed in colour intensity (from yellow to light yellow or white) on glycerol and in shade (from green or yellow to brownish) on D-glucose comparing the *cre1-96* deletion strain to Rut-C30 after 60 and 84 h (Additional file 3: Figure S3).

To learn whether the slower growth of the strains carrying the *cre1-96* deletion results from less cellulase and xylanase activity, we tested supernatants from cultivations under inducing conditions (lactose) but also under repressing conditions by enzymatic assays. Supplementary, the abundance of *cre1-96* transcript was determined under inducing conditions. In contrast to the growth experiments on plates, the biomass formation in the liquid cultures was now also reduced on lactose in the Δ*cre1-96* strains (Fig. 2a). For this reason, the obtained cellulolytic and xylanolytic activities (Fig. 2b, c) were normalized to the biomass. Normalized to the biomass, the *cre1-96* deletion caused a complete loss of cellulolytic and of xylanolytic activity at earlier time points (36 and 48 h) and a strong reduction is observed after 60 h (Fig. 2b, c). Expression of *cre1-96* itself was equally high at all time points under inducing conditions and necessary for the enzyme production (Fig. 2e). Obviously, the presence of *cre1-96* is needed for a good performance in cellulase and xylanase production. Importantly, cellulolytic activities were also lost when D-glucose is used as the carbon source, which is not the case in the parent strain Rut-C30 (Fig. 2d). This reflects that in Rut-C30 the production of cellulases and xylanases is positively influenced by the presence of Cre1-96 regardless if inducing or repressing conditions are prevailing.

#### Cre1-96 influences the transcript formation of *cbh1*, *xyn1* and *xyr1*

The findings on the reduced enzyme activities prompted us to examine whether Cre1-96 regulates Cre1-target genes on the transcriptional level under inducing conditions. Therefore, we measured the transcript levels of *cbh1*, *xyn1* and *xyr1* on lactose in Rut-C30 and both

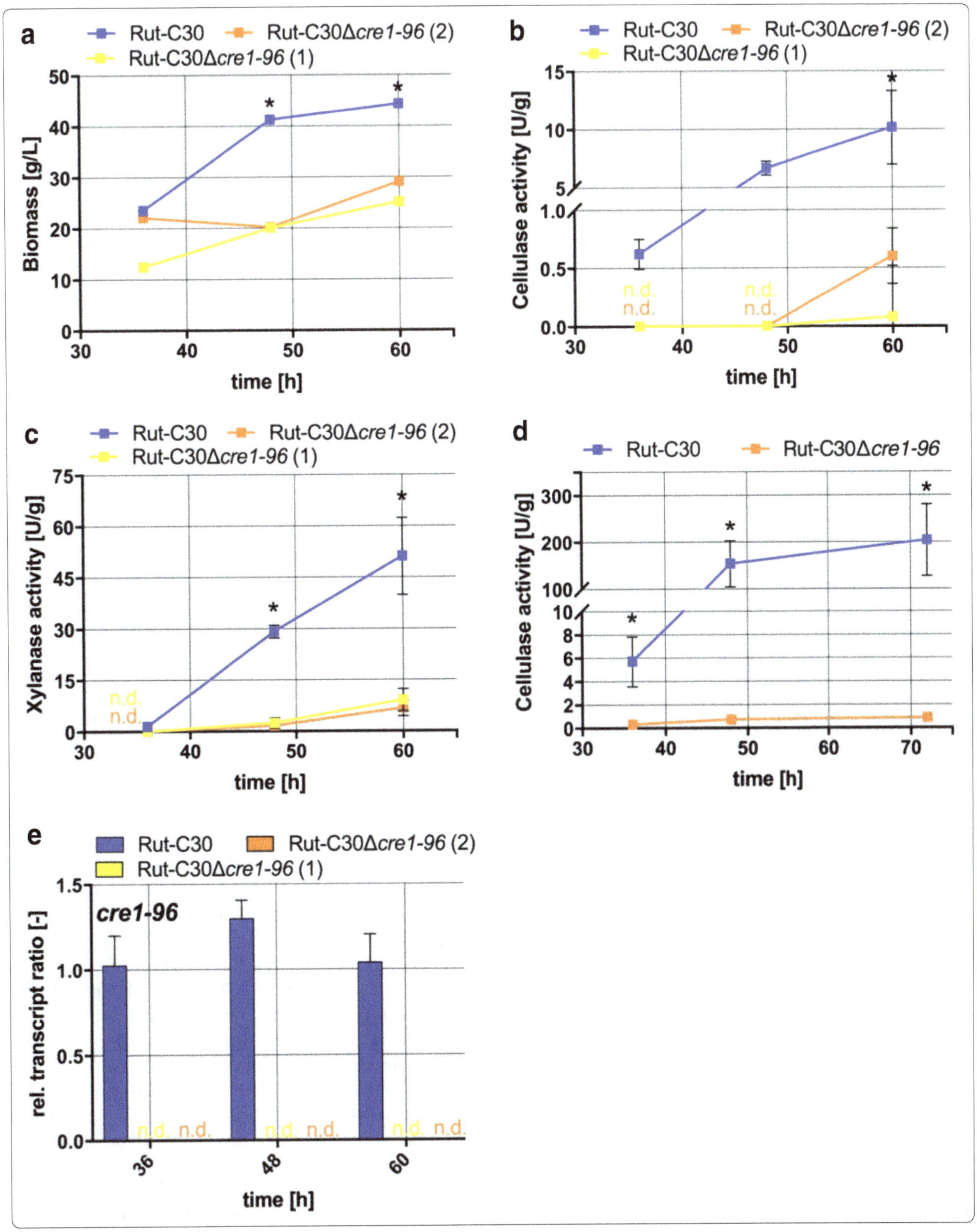
a
Rut-C30
Rut-C30Δcre1-96 (2)
Rut-C30Δcre1-96 (1)
Biomass [g/L]
time [h]
b
Cellulase activity [U/g]
n.d.
time [h]
c
Xylanase activity [U/g]
n.d.
time [h]
d
Rut-C30
Rut-C30Δcre1-96
Cellulase activity [U/g]
time [h]
e
rel. transcript ratio [-]
cre1-96
n.d.
time [h]

(See figure on previous page.)

**Fig. 2** Cellulolytic and xylanolytic activities in absence and presence of Cre1-96. *T. reesei* strains Rut-C30 (blue squares) and both Rut-C30Δ*cre1-96* strains (yellow and orange squares) were cultivated in liquid medium supplemented with 1% (w/v) lactose or D-glucose for 36, 48 and 60 h. The endo-cellulolytic on lactose (**b**) and on glucose (**d**) as well as the xylanolytic activities on lactose (**c**) in the culture supernatants were measured in biological and technical duplicates and normalized to the biomass measured as wet weight (**a**). The enzymatic activities are given as means and the error bars indicate the standard deviations. The values were statistically analysed by an unpaired two-tailed *t* test in a confidence interval of 95%, and asterisks indicate significant differences. **e** Relative *cre1-96* transcript ratios were analysed for both deletion strains and Rut-C30 grown on lactose. Transcript analysis was performed in biological and technical duplicates by qPCR, data were normalized to the housekeeping genes *sar1* and *act*, and referred to the transcript level of Rut-C30 at 36 h. The relative transcript ratios are given as means and the error bars indicate the standard deviations. Error bars are not shown for standard deviations ≤ 3.5%. All values were statistically analysed in a confidence interval of 95%; 'n.d.' means not detected

*cre1-96* deletion strains. In the case of *cbh1*, the transcript levels were significantly reduced in the deletion strains compared to the parent strain at all time points (Fig. 3a). In the case of *xyn1*, the transcript levels were also significantly reduced in the deletion strains compared to its parent strain (Fig. 3b). Generally spoken, the *cbh1* and *xyn1* transcriptional profiles matched the measured enzymatic activities in Rut-C30 and both deletion strains. Interestingly, the transcript levels of *xyr1* were also reduced in the deletion strains compared to its parent strains after 36 and 48 h (Fig. 3c), but not any more at the later time point (60 h). To summarize, Cre1-96 has an impact on the formation of *cbh1* and *xyn1* transcript levels, and also on those of the main activator Xyr1 under inducing conditions.

### Cre1-96 only indirectly regulates genes involved in the lactose metabolism

The observed differences in the biomass formation in liquid culture on lactose (compare Fig. 2a) gave rise to the possibility that the lactose metabolism could be altered in the *cre1-96* deletion strains. Several genes are necessary for the conversion of lactose to D-galactose and D-glucose. The lactose hydrolysis depends on the extracellular β-galactosidase Bga1 and on the D-xylose reductase Xyl1. A deletion of *xyl1* results in reduced growth on lactose, which is explained by low transcript levels of *bga1* [18]. Here, we investigated the genes coding for the D-xylose reductase (*xyl1*), the β-galactosidase (*bga1*), the galactokinase (*gal1*) and a lactose specific permease (*Tre3405*) [5]. A previous study demonstrated that Xyr1 is involved in the regulation of some lactose metabolism genes by activating *xyl1* and *bga1*, but not *gal1* transcription [19]. As it seems that Cre1-96 has an influence on the *xyr1* transcript formation (compare Fig. 3c), it is very likely that also genes involved in the lactose metabolism are affected by the *cre1-96* deletion. Significantly reduced transcripts of the *xyl1*, *bga1* and *Tre3405* genes were detected in the *cre1-96* deletion strain (Fig. 4a–c), whereas *gal1* transcripts accumulated to similar levels (Fig. 4d). Altogether, this suggests that Cre1-96 acts directly on *xyr1* transcript formation and thereby indirectly influences the transcript levels of *xyl1*, *bga1* and *Tre3405*.

### Cre1-96 fulfils the requirements for a transcription factor

Enzymatic measurements and transcript analysis suggested that Cre1-96 exerts a positive effect on cellulase and xylanase gene expression, on the transcript formation of cellulase and xylanase-encoding genes, and most importantly on *xyr1*. To be considered as an activator, some properties need to be fulfilled. First of all, Cre1-96 needs to bind the DNA of its target genes, which is supported by previously reported in vivo footprinting experiments and in vitro protein-DNA binding studies for Cre1-96 and Cre1 [7, 17, 20]. A second essential prerequisite is its localization in the nucleus, at least transiently. In silico domain analysis revealed that Cre1-96 has a putative bipartite nuclear localization signal (NLS) (TVIK – linker – RPYK) located at amino acids (aa) positions 33–63 (Fig. 5a). This bipartite NLS was found with a score of 5 in the case of both proteins, Cre1 and Cre1-96. Scores ranging from 3 to 5 suggest that the protein can be localized both in the nucleus and the cytoplasm. Besides this, alignment of Cre1-96 and Cre1 to homologues from other filamentous fungi revealed further conserved domains or amino acids (Additional file 4: Figure S4). In Cre1-96 a part of a zinc finger domain was identified at 59–79 aa, and a putative transactivation domain (25–37 aa) as well (Fig. 5b). The conserved sequence parts, which are missing in Cre1-96 compared to the full-length Cre1, are a part of the full zinc finger binding domain (87–109 aa), stretches of acidic amino acids (121–129 aa, 243–246 aa, 359–374 aa), two other conserved domains (256–289 aa and 317–325 aa), the nuclear export signal (NES, 304–312 aa), a C-terminal repression domain (317–343 aa) and the phosphorylation site at Ser241 [20] (Fig. 5a). To summarize the in silico analysis, the truncated protein Cre1-96 has lost many potentially important domains but still contains all domains that are essential for a transcription factor, i.e. a DNA-binding domain, NLS

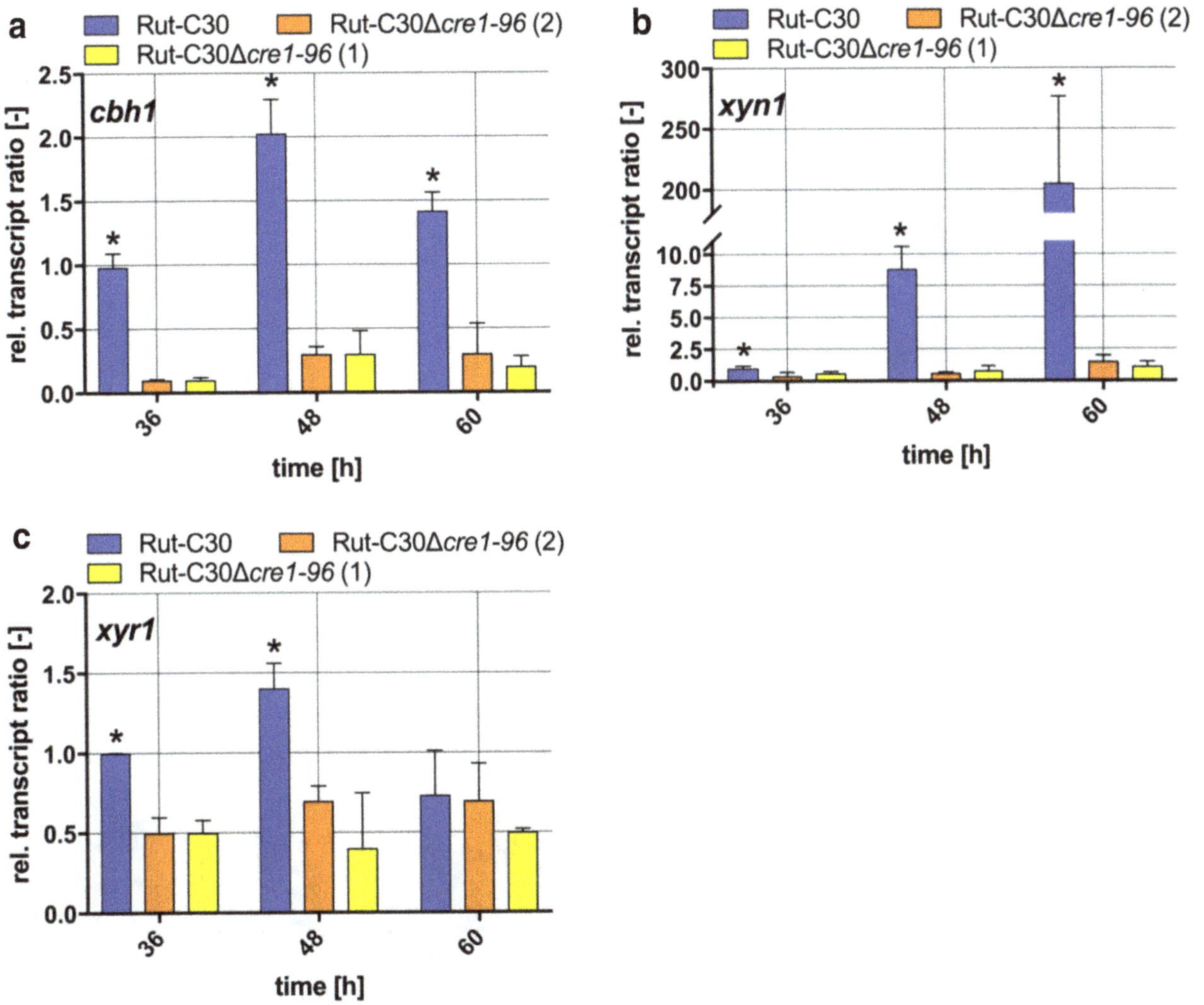


**Fig. 3** Transcript levels of *cbh1*, *xyn1* and *xyr1* in absence and presence of Cre1-96. *T. reesei* strains Rut-C30 (blue bars) and both Rut-C30Δ*cre1-96* strains (yellow and orange bars) were cultivated in liquid medium supplemented with 1% (w/v) lactose for 36, 48 and 60 h. Transcript analyses of *cbh1* (**a**), *xyn1* (**b**) and *xyr1* (**c**) were performed in biological and technical duplicates by qPCR, data were normalized to the housekeeping genes *sar1* and *act*, and referred to the respective transcript levels of Rut-C30 at 36 h. The relative transcript ratios are given as means and the error bars indicate the standard deviations. All values were statistically analysed in a confidence interval of 95% and asterisks indicate significant differences

and one N-terminal acidic region potentially functioning as activator domain.

To monitor the localization of Cre1-96 in the fungal hyphae, a strain expressing an eYFP-tagged Cre1-96 was generated and cultivated in liquid medium containing D-glucose or lactose. It should be noted that we used QM6a for analysis of the nuclear transport of Cre1-96 to exclude any cross-genetic effects resulting from the other mutations present in Rut-C30. Confocal fluorescence microscopy was performed using a droplet of the liquid culture and the visualization of the fungal nuclei was achieved with Hoechst staining. The localization of Cre1-96 was determined by the detection of fluorescence emission of eYFP. Merging of the eYFP signal and the nuclei fluorescence emissions revealed the presence of Cre1-96 in the nuclei of *T. reesei*. Nuclear localization of Cre1-96 was observed under both, repressing (Fig. 6a) and inducing conditions (Fig. 6b), similar to the full length Cre1 [21].

### Cre1-96 targets Cre1-binding sites within the URR of *xyr1*

To learn where the transcription factor Cre1-96 is targeted to, we performed chromatin immunoprecipitation (ChIP) followed by qPCR analyses. For this purpose, the strain expressing eYFP-tagged Cre1-96 was used. As an initial control we tested cellulase activities and biomass formation in tagged and untagged strains to exclude any impact of the eYFP-tag.

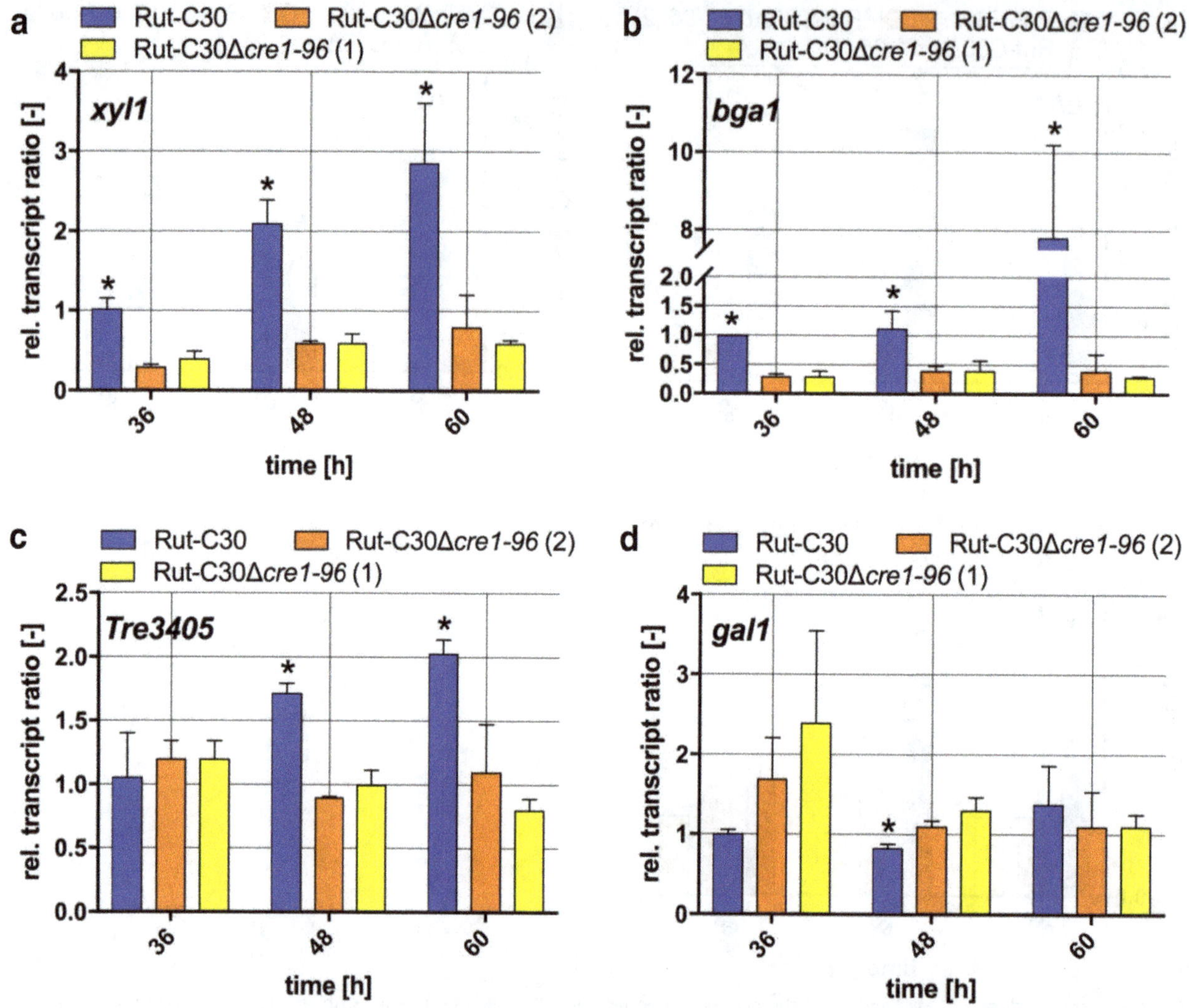


**Fig. 4** Transcript levels of *xyl1*, *bga1* and *Tre3405* in absence and presence of Cre1-96. *T. reesei* strains Rut-C30 (blue bars) and both Rut-C30Δ*cre1-96* strains (yellow and orange bars) were cultivated in liquid medium supplemented with 1% (w/v) lactose for 36, 48 and 60 h. Transcript analyses of *xyl1* (**a**), *bga1* (**b**) and *Tre3405* (**c**) were performed in biological and technical duplicates by qPCR, data were normalized to the housekeeping genes *sar1* and *act*, and referred to the respective transcript levels of Rut-C30 at 36 h. The relative transcript ratios are given as means and the error bars indicate the standard deviations. All values were statistically analysed in a confidence interval of 95% and asterisks indicate significant differences

The results of these preliminary experiments show no impact of the tag on Cre1-96 function (Additional file 5: Figure S5). As the nuclear localization of Cre1-96 was observed after 16 h cultivation in liquid malt extract (MEX) medium supplemented with D-glucose, this condition was chosen for the ChIP experiment. An enrichment of Cre1-96 was identified with anti-GFP antibodies (please note that they are able to bind eYFP) and qPCR. Since we had already indications that *xyr1* is a target of Cre1-96, specific primers were chosen for the analysis of Cre1-96 associated DNA within the URR of *xyr1*. The relative amount of Cre1-96 targeted DNA is almost threefold enriched in this target region compared to the non-target housekeeping gene *sar1* indicating that indeed the truncated Cre1-96 protein might directly activate *xyr1* transcription (Fig. 7).

### Chromatin accessibility is only moderately affected by a *cre1-96* deletion

Previous reports demonstrated a role of Cre1-96 in promoting chromatin accessibility [17]. Hence, we analysed the chromatin accessibility in the URR of Cre1-96 target genes (i.e. *xyr1*, *xyn1* and *cbh1)* in the *cre1-96* deletion strain and its parent strain under inducing conditions. Both strains were cultivated in liquid medium on lactose. The fungal mycelium was harvested after 36, 48

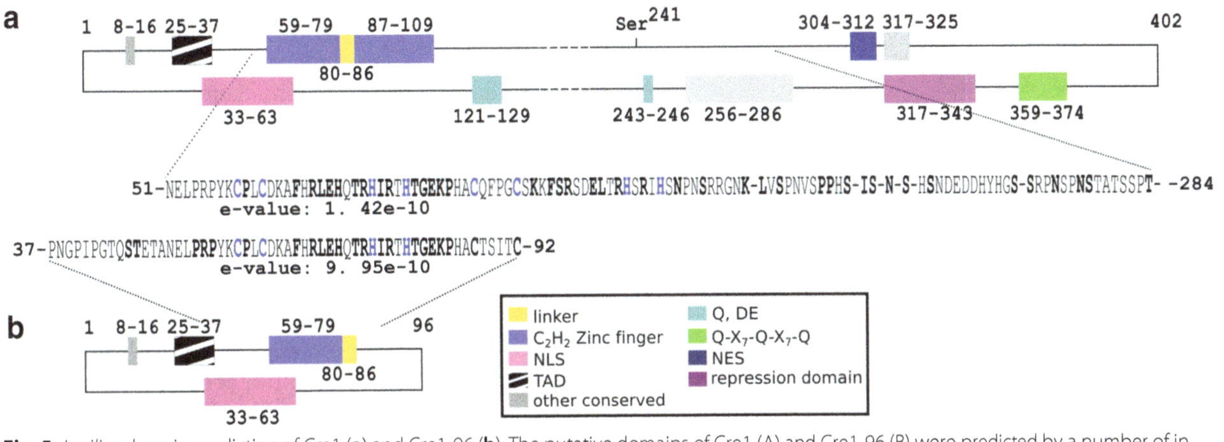

Fig. 5 *In silico* domain prediction of Cre1 (**a**) and Cre1-96 (**b**). The putative domains of Cre1 (A) and Cre1-96 (B) were predicted by a number of in silico prediction tools and alignment algorithms as described in the Methods section. Numbers indicate the amino acid (aa) positions and coloured boxes indicate identified domains: blue, $C_2H_2$ zinc finger; yellow, linker; pink, nuclear localization signal (NLS); black-white striped, transactivation domain (TAD); grey, other conserved region; turquoise, Q (glutamine), DE (aspartic (D) and glutamic acid (E)); green, Q-$X_7$ Q-$X_7$-Q; dark blue, nuclear export signal (NES) and violet, repression domain. Phosphorylation site Ser241 according to [20] is marked. The predicted zinc finger domain superfamily are presented as amino acid sequence and the zinc finger binding sites (cysteine (C) or histidine (H)) are coloured in blue. The e-value is given below the amino acid sequence

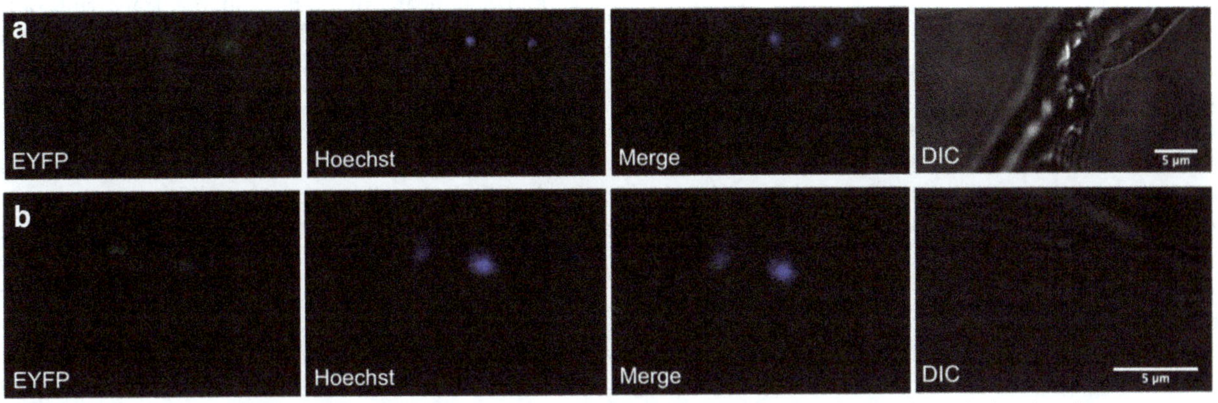

**Fig. 6** Confocal microscopy analysis of the localization of Cre1-96. The *T. reesei* strain QM6a*cre1-96::eyfp* was cultivated in liquid medium supplemented with 1% (w/v) D-glucose (**a**) or 1% (w/v) lactose (**b**). For the visualization of fungal nuclei a Hoechst staining was performed. The following pictures were imaged to localize Cre1-96::eYFP in the fungal hyphae: detection of eYFP (EYFP), detection of the nuclei within the fungal cells (Hoechst), overlay of eYFP and Hoechst emissions (Merge), and brightfield image (DIC). Scales are given in the DIC pictures

and 60 h, followed by chromatin accessibility real-time PCR (CHART-PCR). In the case of *xyr1*, significant differences in the chromatin accessibility were just found after 60 h (Fig. 8a). However, there is no relation between the chromatin status and the transcript level suggesting that chromatin accessibility as measured by our assays is not changing with transcriptional activity. In the case of *cbh1*, significant opening of chromatin in Rut-C30 went along with higher transcript level compared to the deletion strain (Fig. 8b). However, this could be only observed for one time point (i.e. 48 h). Finally, the chromatin accessibility in the *xyn1* URR did differ between Rut-C30 and the *cre1-96* deleted strain at two time points of investigation (Fig. 8c). However, again a transcription-related change in accessibility could not be observed.

### Constitutively expressed *cre1-96* enhances cellulase activity

Based on above findings, we have solid indication that Cre1-96 is a necessary activating regulator for cellulase gene expression in Rut-C30. For benefits towards biotechnological applications, we constructed a *T. reesei* strain having a constitutively expressed *cre1-96* under the control of the *tef1* promoter (in the following termed

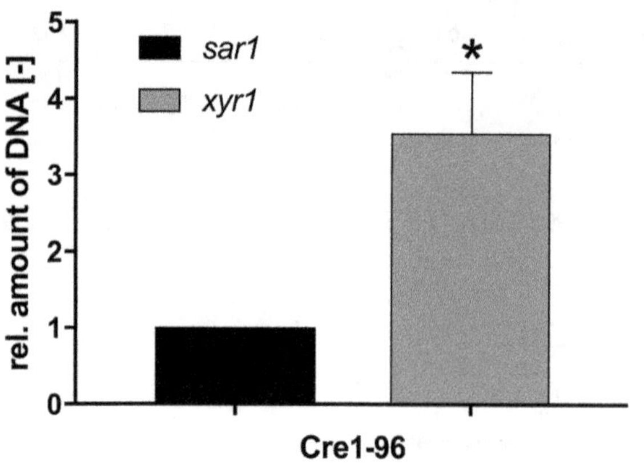


**Fig. 7** Chromatin immunoprecipitation of Cre1-96::eYFP. The *T. reesei* strain QM6a*cre1-96::eyfp* was cultivated for 16 h in 20 mL MEX medium supplemented with 1% (w/v) D-glucose. After crosslinking, the chromatin was enzymatically fragmented by MNase treatment and Cre1-96::eYFP targeted DNA was enriched using anti-GFP antibodies. The relative amount of DNA was measured by qPCR. The immunoprecipitated DNA was normalized to the input control. The resulting ratio for the *xyr1* gene (grey bar) was further normalized to the ratio of housekeeping gene *sar1*, which is set to 1 (black bar). The values were statistically analysed in a confidence interval of 95% and the asterisk indicates a significant difference

Rut-C30OE*cre1-96)*, which was cultivated in parallel with the deletion strain and the parent strain in liquid medium on lactose. Cellulase activities were subsequently measured in the culture supernatants. Rut-C30OE*cre1-96* had a constant increase in the cellulase activity over time, and it was significantly higher than in the other two strains from 48 h of incubation time on (Fig. 9a). After 60 h of incubation, Rut-C30OE*cre1-96* outcompeted Rut-C30 in cellulolytic performance almost twofold. With regard to the growth on cellulase inducing substrates, we observed a similar (24 and 48 h) or a slightly faster growth (60 and 84 h) of Rut-C30OE*cre1-96* compared to Rut-C30 on CMC plates (Fig. 9b). On lactose, no visible differences in colony size were observed amongst all three strains (Fig. 9b). Similar growth was observed under non-inducing (glycerol) and repressing conditions (D-glucose) (Additional file 6: Figure S6). Importantly, Rut-C30OE*cre1-96* did not show the non-radial growth that was observed for Rut-C30Δ*cre1-96* (Additional files 7 and 8: Tables S1 and S2).

## Discussion

The carbon catabolite repressor Cre1 represses transcription of its targets genes by binding their URR. Surprisingly, if truncated, Cre1-96 is still able to bind DNA but converts to a putative activator. Like Cre1/Cre1-96, the transcription factor PacC, which is involved in pH-regulation in *A. nidulans* [22], is a zinc finger protein. PacC is processed by the Pal signalling pathway under alkaline pH and subsequently, moves into the nucleus. In its truncated form it acts as repressor of acidic-expressed genes. Even if its molecular action, namely competition for DNA binding, is different from Cre1-96, in both cases a truncated zinc finger protein acts as transcription factor. The detailed findings of the Cre1 repressor truncation in *T. reesei* are discussed below.

When *cre1* is exchanged for *cre1-96* in the wild-type strain QM6a, higher *cbh1, cbh2* and *xyr1* transcript levels were obtained compared to a full deletion of *cre1* [17]. However, Rut-C30 that natively carries Cre1-96 had even higher transcript levels of those genes than the QM6a-CREI$_{96}$ strain. Therefore, we deleted *cre1-96* in Rut-C30 to study the effects on transcript and corresponding enzyme levels.

We observed growth deficiencies, i.e. a slower growth, reduced biomass formation and growth abnormalities, in the Rut-C30 strain lacking *cre1-96*. Under cellulase inducing conditions, we found reduced growth in the *cre1-96* deletion strain compared to its parental strain on CMC plates and lactose liquid cultures (compare Figs. 1, 2a) and significant differences in the transcript ratios of genes involved in the lactose metabolism (compare Fig. 4). This indicates a change either in uptake of degradation products into the cell by transporters or in the enzymatic activity required for the conversion of CMC or lactose into an inducing substance (e.g. transglycosylation by BGLI). We did not observe differences in growth on lactose between the parent and the *cre1-96* deletion strain in the case of cultivation on plates while we did observe differences in the case of cultivation in liquid medium (i.e. determination of the mycelial biomass weight). Interestingly, Cánovas and colleagues found that the biomass accumulation from plates does usually not correlate with the radial growth diameter [23] so these abnormalities cannot be explained by the current model. Particularly, a non-radial growth of fungal hyphae was observed on lactose in the *cre1-96* deletion strain (compare Fig. 1). At this point it has to be mentioned that, a *cre1* deletion in the *T. reesei* strain QM6a leads to strongly impaired growth and morphological changes [8]. Portnoy and colleagues identified several genes involved in hyphal development (e.g. RAS1, PhiA, MedA), which are regulated by Cre1 on D-glucose [10]. Also in other filamentous fungi, like *Neurospora crassa* (*N. crassa*), CRE-1 seems to influence the hyphal growth and polarity because the enzyme activity of an involved cAMP protein kinase A is dependent on CRE-1 [24]. Altogether, this implies that Cre1-96 might exert additional functions (similar to Cre1), besides its role in cellulase and hemicellulase gene expression. Another aspect worth considering was reported by dos Santos Castro and colleagues. RNA

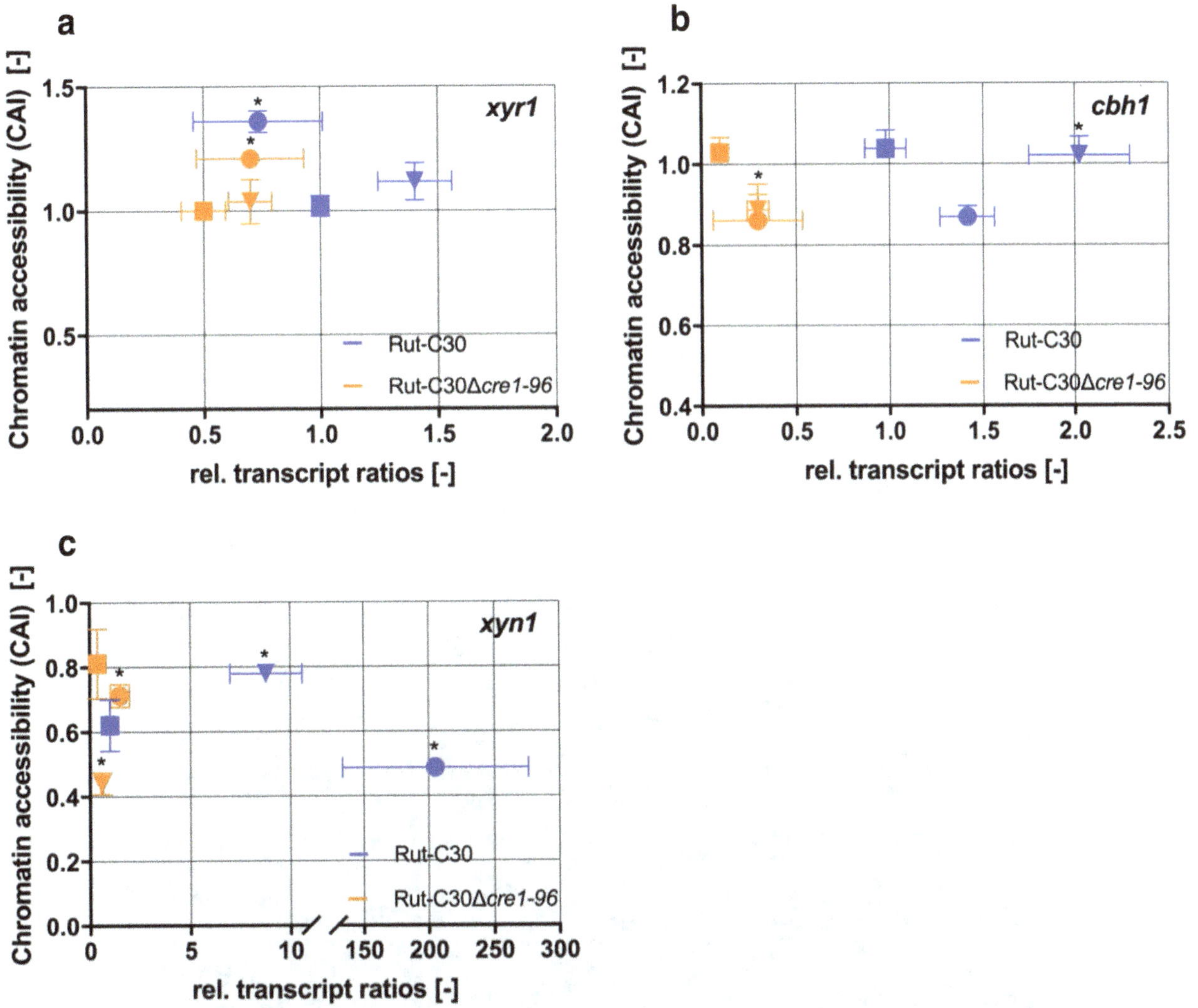


**Fig. 8** Chromatin accessibility under cellulase inducing conditions. *T. reesei* strains Rut-C30 (blue symbols) and Rut-C30Δ*cre1-96* (orange symbols) were cultivated in liquid medium supplemented with 1% (w/v) lactose for 36 (squares), 48 (triangles) and 60 h (dots). CHART-PCR and transcript analysis of *cbh1* (**b**), *xyn1* (**c**) and *xyr1* (**a**) were performed in biological and technical replicates. Both sets of data were normalized to the housekeeping genes *sar1* and *act*. The relative transcript ratios are given as means and plotted on the x-axis and the chromatin accessibility indices (CAI) are plotted on the y-axis. Error bars indicate the standard deviations. All CAI values were statistically analysed in a confidence interval of 95% and asterisks indicate significant differences amongst the chromatin-related data

sequencing analysis of *T. reesei* QM9414 under repressing (D-glucose) and inducing (cellulose, α-sophorose) conditions [25] indicated that several MFS permeases are differentially expressed on D-glucose. Notably, amongst the strongest down-regulated genes in the absence of Cre1 are proteins involved in cellular transport, such as MFS permeases [10]. This indicates that Cre1 and most probably Cre1-96 might also play a role in the sugar uptake in the cell.

With regard to the cellulase and xylanase activity, we observed either a loss or strong reduction in enzymatic activities in the *cre1-96* deletion strain compared to its parental strain (compare Fig. 2b, c). In Rut-C30, the transcript profile of *xyr1* relates to the profile of *cbh1* (compare Fig. 3), which is in full agreement with previously published results [26]. Most interestingly, the transcription profile of *cre1-96* relates to the profile of *xyr1* (compare Figs. 2d, 3c). Thus, Cre1-96 might have an effect on the regulation of *xyr1* transcription.

The fluorescence microscopy revealed that Cre1-96 is under repressing and inducing conditions present in the nucleus. A carbon source-dependent shuttling of Cre1 between cytosol and nucleus was proposed by Lichius and colleagues [21]. In silico analysis suggests that Cre1

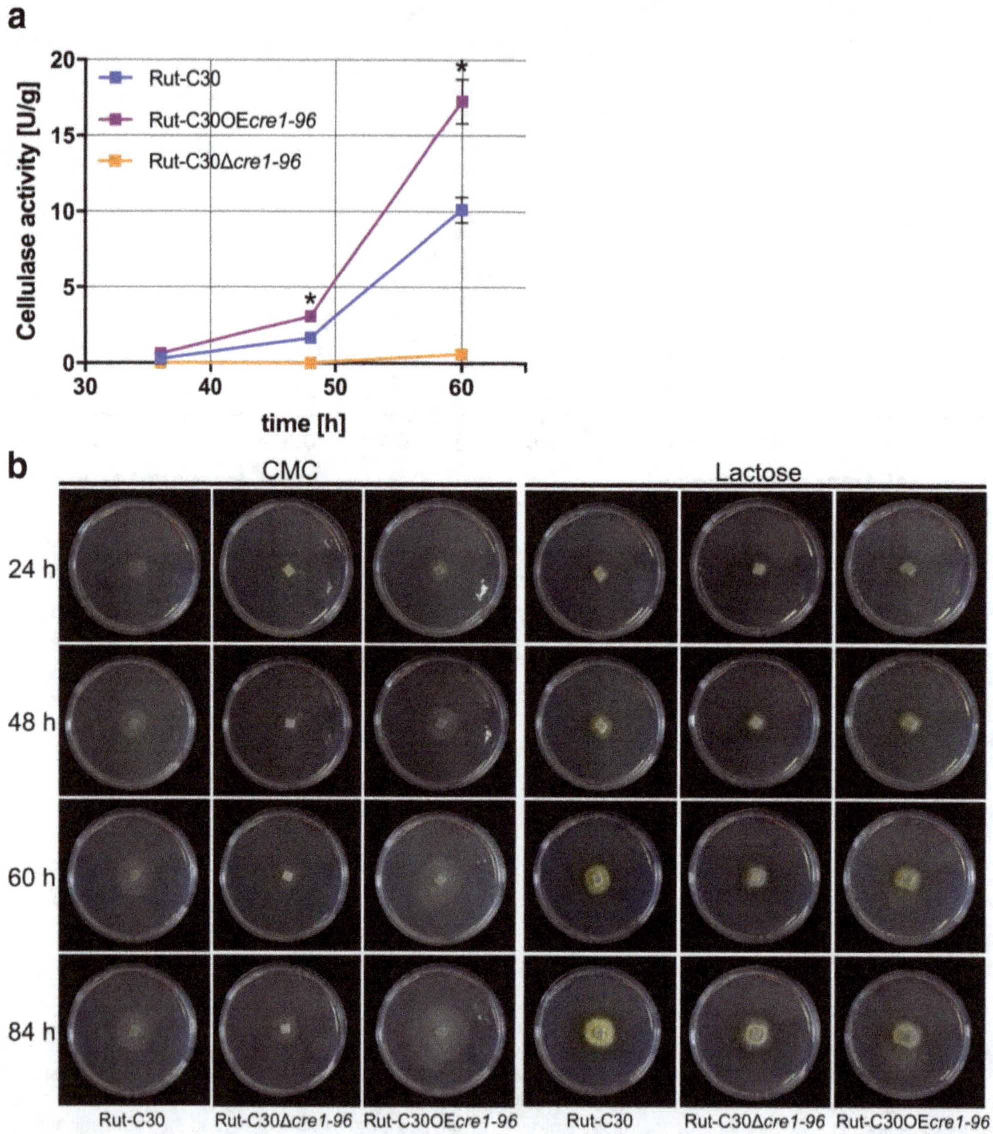


**Fig. 9** Cellulase activity in presence of a constitutively expressed Cre1-96. **a** *T. reesei* strains Rut-C30 (blue squares), Rut-C30OE*cre1-96* (purple squares) and Rut-C30Δ*cre1-96* (orange squares) were cultivated in liquid medium supplemented with 1% (w/v) lactose for 36, 48 and 60 h. The endo-cellulolytic activities in the culture supernatants were measured in biological and technical duplicates and normalized to the biomass. The enzymatic activities are given as means and the error bars indicate the standard deviations. Error bars are not shown for standard deviations ≤ 3.5%. The values were statistically analysed by an ordinary one-way ANOVA and a Tukey's posthoc test in a confidence interval of 95%, and asterisks indicate significant differences. **b** For the growth assays, the *T. reesei* strains Rut-C30, Rut-C30OE*cre1-96* and Rut-C30Δ*cre1-96* were pre-grown on MEX plates and were then transferred to MA medium plates supplemented with 1% (w/v) CMC or 1% (w/v) lactose. Plates were incubated at 30 °C and pictures were taken after 24, 48, 60 and 84 h

has a nuclear export signal (NES) at amino acids positions 304–312 (LPSLRNLSL, predicted by using [27]). Cre1-96 lacks this putative NES due to its truncation and therefore, could remain inside the nucleus regardless the used carbon source. Besides this, Cre1-96 possesses a putative N-terminal transactivation domain as Cre1 does, but Cre1-96 importantly lacks the C-terminus of Cre1 (compare Fig. 5) that highly likely mediates repression as it was described for CreA in *Aspergillus nidulans* [28]. Taken together the extended residence in the nucleus, the presence of a putative transactivating domain and the lack of the repression domain, would explain the positive impact of Cre1-96 on the cellulase activity in Rut-C30 compared to the *cre1-96* deletion strain under cellulase inducing conditions.

Notably, Cre1-96 also lacks the previously identified phosphorylation site Ser241 [20]. In the case of Cre1, Ser241 needs to be phosphorylated for an efficient DNA binding under repressing conditions. However, Cziferszky and colleagues reported that GST fusion proteins of Cre1 fragments without Ser241 bind in vitro regardless the tested condition. Besides this, previously published in vivo footprinting results also supported the capability of Cre1-96 to bind DNA [17].

With regard to the targeting of Cre1-96, we found that Cre1-96 is enriched on its DNA binding sites in the *xyr1* URR. However, the chromatin accessibility in the *xyr1* URR is not significantly different. Neither for *cbh1* or *xyn1* any cohesive trend could be observed. Noteworthy, an earlier reported nucleosomal mapping of *cbh1* and *cbh2* promoter regions showed no positioned nucleosomes under repressing and inducing conditions in Rut-C30 [29, 30]. This lack of positioned nucleosomes is the likely explanation of the similarity in DNA accessibility observed in our experiments between conditions of expressed and non-expressed genes.

Anyhow, we propose here that a truncation of Cre1 positively influences the transactivator Xyr1 and thus enhances cellulolytic performance and phenotypically converts the carbon catabolite repressor into an activator.

## Conclusions

Due to a truncation the Cre1 repressor can turn into an activator as seen in Cre1-96, which now functions to activate cellulase and xylanase expression. Cre1-96 meets all requirements for a transcription factor. It localizes to the nucleus and directly binds to the URR of target genes, in particular of the main transactivator of the mentioned enzymes, Xyr1, and most probably exerts thereby its activation role. Our findings encourage testing this strategy to increase enzymatic performance in other filamentous fungi, which contain functional Cre1 homologues.

## Methods

### Fungal strains

The *T. reesei* strains QM6a$\Delta$*tmus53* [31], the QM6a$\Delta$*tmus53* bearing an eYFP-tagged Cre1-96 (referred to in the text as QM6a*cre1-96::eyfp*, this study), Rut-C30$\Delta$*tmus53* (referred to in the text as Rut-C30, VTT Finland), two *cre1-96* deletion strains Rut-C30$\Delta$*tmus53*$\Delta$*cre1-96* (referred to in the text as Rut-C30$\Delta$*cre1-96* (1) and (2), this study) and the *cre1-96* constitutively expressing strain Rut-C30$\Delta$*tmus53*OE*cre1-96* (referred to in the text as Rut-C30OE*cre1-96*, this study) were maintained on malt extract (MEX) agar plates containing 0.1% (w/v) peptone from casein at 30 °C. Uridine was added to a final concentration of 5 mM for all Rut-C30 strains. For strain selection hygromycin B was added to a final concentration of 113 U/mL for the QM6a-related strains and 56.5 U/mL for the Rut-C30-related strains. Homokaryon selection was carried out on MEX/peptone/hygromycin B plates, with uridine if applicable, and 0.1% (w/v) Igepal C-60.

### Growth conditions

If not indicated otherwise in a Methods section, for cultivation experiments $10^6$ conidia spores per mL were incubated in 100 mL Erlenmeyer flasks on a rotary shaker (180 rpm) at 30 °C for 60 h in 30 mL of MA medium supplemented with 0.1% (w/v) peptone and 1% (w/v) lactose or 1% (w/v) D-glucose as sole carbon source. If not stated otherwise, all strains were cultivated in triplicates and were harvested after 36, 48 and 60 h of cultivation. Fungal mycelia were separated from the supernatant by filtering with Miracloth (EMD Millipore, part of Merck KGaA, Darmstadt, Germany). Mycelia grown on lactose were weighted immediately before shock freezing and wet weight was used as the biomass reference for the enzymatic assays. Frozen mycelia were used for genomic DNA extraction, RNA extraction and for chromatin digestion. Culture supernatants were used for the measurement of enzymatic activities in technical duplicates.

For the comparison of the growth behaviour on plates, the strains Rut-C30, Rut-C30$\Delta$*cre1-96* (1) and (2) and Rut-C30OE*cre1-96* were pre-grown on solid MEX media with 0.1% (w/v) peptone and were transferred in biological duplicates onto MA agar plates supplemented with 1% (w/v) lactose, CMC, xylan, glycerol or D-glucose for 84 h. Due to the same growth behaviour of the duplicates, only one replicate is displayed in the figures. In the case of the deletion strains, Rut-C30$\Delta$*cre1-96* (2) is shown.

### Plasmid construction

*Escherichia coli* strain Top10 (Invitrogen, part of Life Technologies, Paisley, UK) was used for all cloning purposes throughout this study and grown on LB medium at 37 °C. Generation of competent *E. coli* cells and subsequent transformation was performed according to standard protocols using $CaCl_2$. If applicable, ampicillin and hygromycin B were added to final concentrations of 100 mg/mL and 113 U/mL, respectively.

PCRs for all cloning purposes were performed with Pwo DNA Polymerase (peqlab VWR, Radnor, Pennsylvania, USA) or Phusion High-Fidelity DNA Polymerase (Thermo Scientific, Waltham, Massachusetts, USA) according to the manufacturer's instructions. All used primers were purchased from Sigma Aldrich and are listed in Table 1.

For the construction of the *cre1-96* deletion cassette the 5′-flank of *cre1-96* was amplified by PCR using chromosomal DNA of *T. reesei* QM6a$\Delta$*tmus53* (identical

**Table 1 Primers used for strain construction in this study**

| Primer name | Sequence 5′–3′ | References |
|---|---|---|
| 5′cre1_NotI fwd | GCGGCCGCTGGAGGTGACGAGAAGAAAAATTCAGG | This study |
| 5′cre1_XmaI rev | CCCGGGAGTCAAAAAGCAAGTACGCGACGTTG | This study |
| hph_XmaI_BamHI | CCCGGGTTGGATCCAGGGAGACGAGGTTGTGATGAATAC | This study |
| hph_SpeI rev | ACTAGTAAGTAGCACCGCTGTCGTCTG | This study |
| 5Pcre1_NotI fwd | GCGGCCGCAGCCAAGACTCAGCATAAAGAGGTTG | This study |
| 5Pcre1_XmaI rev | CCCGGGAGGTACCAAACAAAGCGAGCAAGTAC | This study |
| ptef_BspEI fwd | TCCGGATGTGTGACAGCTCGCGCAG | This study |
| ptef_NdeI rev | CATATGTGACGGTTTGTGTGATGTAGCGTG | This study |
| cre1-96_NdeI fwd | CATATGATGCAACGAGCACAGTCTGCC | This study |
| Cre1-96_BamHI rev | GGATCCTTAGAAAAAAAAGCAGGTAATGGAGGTGC | This study |
| cre1-96_BspEI fwd | TCCGGAATGCAACGAGCACAGTCTGCC | This study |
| cre1-96-TAA_NdeI rev | CATATGGAAAAAAAAGCAGGTAATGGAGGTGC | This study |
| linker_NheI rev | GCTAGCGCGGGGGGCGCAC | This study |
| linker_NdeI fwd | CATATGCACAACATGGTCAAGCAGAAGC | This study |
| YFP_NheI fwd | GCTAGCATGGTCAGCAAGGGCGAGG | This study |
| YFP_BamHI rev | GGATCCCTTGTACAGCTCGTCCATGCCG | This study |

sequence to Rut-C30) as template with the primers 5′creI_NotI fwd and 5′creI_XmaI rev. The PCR product was subcloned into pJET1.2 (Thermo Scientific) by blunt end ligation using T4 DNA ligase (Thermo Scientific) yielding pJET1.2-5′-cre1. The 3′-flank of *cre1* and a hygromycin B resistance cassette were amplified by PCR using chromosomal DNA of *T. reesei* QM6a-Cre1$_{96}$ [17] as template with the primers hph_XmaI_BamHI and hph_SpeI rev and was inserted into pJET1.2 by blunt end ligation yielding pJET1.2-hph. The hygromycin B resistance cassette bears the constitutive promoter of the *pki* gene, the hygromycin B structural gene and the terminator of *cbh2* [32]. The subcloned 5′-flank of *cre1* was recovered by NotI/XmaI digestion and was inserted into the NotI/XmaI-digested vector pJET1.2-hph. The resulting plasmid was termed pJET1.2-5hph3cre1. Subsequently, the vector pJET1.2-5hph3cre1 was cut by NotI and SpeI and the cassette was inserted into a NotI/SpeI-digested derivative pMS plasmid yielding pMS*-5hph3cre1. The orientation of the hygromycin B resistance gene and the *pki* promoter were in the opposite orientation as the 5′-flank and the 3′-flank of *cre1*. This was determined by plasmid sequencing (Microsynth, Balgach, Switzerland).

For the constitutive expression of *cre1-96* the promoter of the *tef1* gene was used. For this purpose, the promoter region (1500 bp upstream of ATG) of *tef1* (p*tef1*) was amplified by PCR using chromosomal DNA of *T. reesei* QM6aΔ*tmus53* as template with the primers ptef_BspEI fwd and ptef_NdeI rev. The structural gene *cre1-96* was amplified using the primers cre1-96_NdeI fwd and cre1-96_BamHI rev. Both PCR fragments were subcloned into pJET1.2 (Thermo Scientific), yielding pJET1.2-Ptef and pJET1-2-cre1-96. Both plasmids were digested by BspEI/NdeI, the p*tef1* fragment was isolated and ligated into the BspEI/NdeI-digested pJET1.2-cre1-96 to yield pJET1.2-Ptefcre1-96. The 5′-flank of *cre1* started at -1500 bp until 2400 bp to avoid a residual background of the native *cre1* promoter. This 5′-flanking region was amplified using the primers 5Pcre1_NotI fwd and 5Pcre_XmaI rev and was subcloned into pJET1.2 by blunt end ligation using T4 DNA ligase (Thermo Scientific) yielding pJET1.2-5Pcre1. The 3′-flank of *cre1* and a hygromycin B resistance cassette was constructed as described for the *cre1-96* deletion cassette. The subcloned 5′-flank of *cre1* was recovered by NotI/XmaI digestion of pJET1.2-5Pcre1 and was inserted into the NotI/XmaI-digested vector pJET1.2-hph. The resulting plasmid was termed pJET1.2-5′cre1-hph. The plasmid pJET1.2-Ptefcre1-96 was BspEI/BamHI-digested, the fragment Ptefcre1-96 isolated and ligated by cohesive ends with a XmaI/BamHI-digested pJET1.2-5′cre1-hph to yield pJET1.2-3Ptefcre1-96. Subsequently, this plasmid was cut by NotI and SpeI and the cassette was inserted into a NotI/SpeI-digested derivative pMS plasmid yielding the final plasmid pMS*-Ptef-cre1-96. The correct orientation and sequence of the plasmid were confirmed by sequencing (Microsynth).

For construction of pMS*-*cre1-96::eyfp* the coding sequence of *cre1-96*, a linker and *eyfp* were amplified by PCR using chromosomal DNA from *T. reesei* QM6a-Cre1$_{96}$ and the plasmid pCD-EYFP [33] as templates and the following primers: cre1-96_BspEI fwd and cre1-96-TAA_NdeI rev to amplify *cre1-96* from QM6a-Cre1$_{96}$; linker_NdeI fwd and linker_NheI rev to amplify the linker sequence from the pCD-EYFP plasmid; YFP_NheI fwd and YFP_BamHI to amplify the coding sequence of *eyfp* from the pCD-EYFP plasmid. Importantly, the fluorescent tag was fused to the C-terminus of the Cre1-96 because this was reported to be necessary for proper recruitment and import in the case of the full length Cre1 [21]. The PCR products were subcloned into pJET1.2, yielding pJET1.2-cre1-96(-TAA), pJET1.2-linker and pJET1.2-YFP. The first two plasmids were digested with BspEI/NdeI, the cre1-96 fragment was isolated and ligated into the BspEI/NdeI-digested recipient vector pJET1.2-linker to generate the plasmid pJET1.2-cre1-96-linker. The insert cre1-96-linker was recovered by BspEI/NheI digestion and cloned into the BspEI/NheI-digested pJET1.2-YFP to yield pJET1.2-cre1-96::eyfp. A BspEI/BamHI double digest of pJET1.2-cre1-96::eyfp recovered the cre1-96::eyfp insert, which was cloned into the XmaI/BamH-digested vector pJET1.2-5hph3cre1 yielding pJET1.2-5cre1-96::eyfp. Finally, the 5cre1-96::eyfp was recovered by NotI/SpeI digestion of pJET1.2-5cre1-96::eyfp and was cloned into a NotI/SpeI-digested derivative pMS plasmid yielding the final plasmid.

#### Fungal protoplast transformation

Protoplast transformation of *T. reesei* was performed as described by Gruber et al. [34]. For the gene replacement of *cre1*, the plasmid pMS*-*cre1-96::eyfp* was linearized by NotI digestion and transformed into *T. reesei* QM6aΔ*tmus53*. For the deletion of *cre1-96*, the plasmid pMS*-5hph3cre1 was linearized by NotI digestion and transformed into *T. reesei* Rut-C30Δ*tmus53*. For the constitutive expression of *cre1-96* under the control of the promoter of *tef1*, the plasmid pMS*-*ptef::cre1-96* was transformed into *T. reesei* Rut-C30Δ*tmus53*. Each transformation reaction was added to 40 mL melted, 50 °C warm MEX agar containing 1.2 M D-sorbitol. This mixture was poured into 4 sterile petri dishes, which were incubated at 30 °C for at least 2 h for protoplast regeneration. Appropriate amount of hygromycin B was added to 40 mL melted, 50 °C warm MEX agar containing 1.2 M D-sorbitol and was poured as a 10 mL-overlay on all 4 plates. Transformation plates were further incubated at 30 °C for 2–4 days until colonies were visible. The resulting candidates were subjected to 3 rounds of homokaryon selection by streaking.

#### Isolation of genomic DNA

Genomic DNA was isolated from approximately 50 mg mycelium in 1 mL CTAB buffer (1.4 M NaCl, 100 mM Tris–HCl pH 8.0, 10 mM EDTA, 2% (w/v) CTAB) by homogenization using a FastPrep(R)-24 cell disrupter (MP Biomedicals, Santa Ana, California, USA) followed by a phenol/chloroform extraction. RNA was degraded using RNaseA (Thermo Scientific). DNA was precipitated with isopropanol, washed with 70% (w/v) ethanol, and dissolved in distilled $H_2O$.

#### Diagnostic PCR analysis

100 ng of chromosomal DNA was used as template in a 25-μL-PCR using GoTaq® G2 polymerase (Promega, Madison, Wisconsin, USA) according to manufacturer's instructions. Primer sequences are provided in Additional file 7: Table S1 and Additional file 8: Table S2. For subsequent agarose gel electrophoresis of DNA fragments a GeneRuler 1 kb DNA Ladder (Thermo Scientific) was applied for estimation of fragment size. DNA sequencing was performed at Microsynth.

#### RNA extraction and reverse transcription

Fungal mycelia were homogenized in 1 mL of peqGOLDTriFast DNA/RNA/protein purification system reagent (peqlab VWR, Radnor, Pennsylvania, USA) using a FastPrep(R)-24 cell disrupter (MP Biomedicals). RNA was isolated according to the manufacturer's instructions, and the concentration was measured using the NanoDrop 1000 (Thermo Scientific). Reverse transcription of the isolated mRNA was carried out using the RevertAidTM H Minus First Strand cDNA Synthesis Kit (Thermo Scientific) according to the manufacturer's instructions.

#### Transcript analysis

Quantitative PCR (qPCR) was performed in a Rotor-Gene Q system (Qiagen, Hilden, Germany). Reactions were performed in technical duplicates or triplicates. The amplification mixture (final volume 15 μL) contained 7.5 μL 2 × iQ SYBR Green Mix (Bio-Rad, Hercules, California, USA), 100 nM forward and reverse primer, and 2.5 μL cDNA (diluted 1:20). Primer sequences and cycling conditions are provided in Table 2. Data normalization using *sar1* and *act* as reference genes and calculations were performed as previously published [35].

#### In silico prediction of protein domains

The protein sequences of Cre1 (JGI *Trichoderma reesei* QM6a v2.0 Database, Protein ID 120117) and Cre1-96 (JGI *Trichoderma reesei* Rut C-30 v1.0 Database, Protein ID 23706) were obtained from the respective genome databases [36, 37]. The identification of the DNA binding domain (i.e. $C_2H_2$ zinc finger and linker sequence) was achieved by using the NCBI conserved domain search [38]. Multiple sequence alignment of Cre1-96, Cre1 and its homologues of

**Table 2 Primer used for qPCR**

| Primer name | Sequence 5′–3′ | References |
|---|---|---|
| actfw | TGAGAGCGGTGGTATCCACG | [35] |
| actrev | GGTACCACCAGACATGACAATGTTG | [35] |
| sar1fw | TGGATCGTCAACTGGTTCTACGA | [35] |
| sar1rev | GCATGTGTAGCAACGTGGTCTTT | [35] |
| xyr1f | CCCATTCGGCGGAGGATCAG | [35] |
| xyr1r | CGAATTCTATACAATGGGCACATGGG | [35] |
| taqxyn1 f | CAGCTATTCGCCTTCCAACAC | [13] |
| taqxyn1 r | CCAAAGTTGATGGGAGCAGAA | [13] |
| cbh1f | GATGATGACTACGCCAACATGCTG | [12] |
| cbh1r | ACGGCACCGGGTGTGG | [12] |
| cre1_a_f | ACCTCCTGAATCCAACGTCGG | [17] |
| cre1_a_r | TGGGTGCGAATGTGCCTGG | [17] |
| bga1f | CGTTTGATCCTTTCGGCGGCT | [19] |
| bga1r | CCAAAGGTCATGTATATGTTGAAGATGGTC | [19] |
| gal1f | GGAGGCATGGACCAGGC | [19] |
| gal1r | GACATGCTTGTTGGAGGTGACG | [19] |
| xorf | CTGTGACTATGGCAACGAAAAGGAG | [19] |
| xorr | CACAGCTTGGACACGTGAAGAG | [19] |
| st RT 1 | CCGTCTACCGTCTGTTGTGC | [5] |
| st RT 2 | GAAGTAGGAAAGAACCGCATTG | [5] |

*Aspergillus nidulans* (*A. nidulans*, NCBI Accession ID: XP_663799.1), *Aspergillus niger* (*A. niger*, NCBI Accession ID: XP_001399519.1), *Neurospora crassa* (*N. crassa*, NCBI Accession ID: XP_961994.1), *Trichoderma atroviride* (*T. atroviride*, NCBI Accession ID: XP_013941427.1), *Trichoderma virens* (*T. virens*, NCBI Accession ID: XP_013956509.1) and *Saccharomyces cerevisiae* (*S. cerevisiae*, NCBI Accession ID: NP_011480.1) was conducted using Clustal Omega [39] and identified conserved amino acids and protein domains. Prediction of the NLS was achieved by applying the NLS Mapper [40] on Cre1 and Cre1-96. For the in silico identification of the transactivation domain, the Nine Amino Acids Transactivation Domain (9aaTAD) Prediction Tool [41] was used [42]. As search specification, the less stringent pattern was chosen as the most adequate pattern for both proteins.

### Confocal microscopy

The localization of the eYFP-labelled Cre1-96 was determined by confocal microscopy and image processing using Fiji [43]. Samples were prepared from liquid cultures. Therefore, $10^6$ spores per mL of QM6a*cre1-96::eyfp* were used to inoculate 20 mL of MA medium supplemented with 1% (w/v) D-glucose and 1% (w/v) lactose and incubated at 30 °C and 180 rpm for 16 h. A 10-μL sample was taken and embedded between two glass coverslips (24 × 60, 24 × 24). For the nuclear staining, 4 μL of a 1:10-diluted (distilled water) Hoechst 34580 stain (Thermo Scientific, 5 mg/mL in DMSO) was added before putting the glass coverslip on top of the sample and incubated for 10 min in darkness. Live-cell imaging was performed using a Nikon C1 confocal laser scanning unit sitting on top of a Nikon Eclipse TE2000-E inverted microscope base (Nikon Inc., Melville, New York, USA). An argon ion laser emitting a wavelength of 488 nm excited fluorescent proteins and Hoechst stained nuclei. The emission wavelength was detected with a photomultiplier in a range of 500–530 nm. Laser intensity and illumination time were kept the same for all samples. Pictures were taken as a single picture configuration at a resolution of 1024 × 1024 pixels.

### Enzyme assays

Endo-xylanolytic and endo-cellulolytic activities of cultivation supernatants were measured with Xylazyme AX tablet assay and Azo-CMC-Cellulose assay (both Megazyme International Ireland, Wicklow, Ireland) according to the manufacturer's instructions. For the comparison of cellulolytic activities of Rut-C30OE*cre1-96* and the *cre1-96* deletion, only Rut-C30Δ*cre1-96* (2) was used due to similar results of previous experiments of this study (e.g. transcript analysis).

### Chromatin immunoprecipitation (ChIP) and quantitative PCR analysis

*T. reesei* strain QM6a*cre1-96::eyfp* was grown for 16 h in MEX supplemented with 1% (w/v) D-glucose at 30 °C at 180 rpm. Crosslinking was performed with 1% (w/v) formaldehyde for 15 min at room temperature and gentle shaking every 2–3 min. Quenching was performed by the addition of 125 mM glycine at room temperature for 5 min and gently shaking. Mycelia were filtered by Miracloth, washed with distilled water, dry-pressed between sheets of Whatman paper and frozen in liquid nitrogen. The chromatin shearing and the ChIP protocol were performed according to [44] with the following adaptions. An amount of 100–200 mg of fungal mycelia was grinded in liquid nitrogen and suspended in MNase digestion buffer (50 mM Hepes–KOH pH 7.5, 50 mM NaCl, 1 mM $CaCl_2$, 5 mM $MgCl_2$, 1 mM PMSF, 1× fungal protease inhibitors (Sigma, St. Louis, Missouri, USA)). Chromatin shearing was enzymatically performed by using 0.4 U MNaseI (Sigma,) on 200 μL mycelia aliquots at 37 °C for 13 min. The reaction was stopped by adding 100 μL Lysis Buffer v2 (50 mM Hepes–KOH pH 7.5, 255 mM NaCl, 12 mM EDTA, 2% (w/v) Triton-X100, 0.2% (w/v) Na-deoxcholate, 1 mM PMSF, 1× fungal protease inhibitors (Sigma)). For the precipitation of the protein-antibody complex, an Anti-GFP antibody (ChIP grade; Abcam, Cambridge, UK) and Dynabeads® Protein A magnetic

**Table 3 Primer used for ChIP-qPCR**

| Primer name | Sequence 5′–3′ | Reference |
|---|---|---|
| sar1 3UTR f | TGACGGGGAGAACATGTGCTC | This study |
| sar1 3UTR r | ATGCGACTCCCACAAGTGGTG | This study |
| ChIP_xyr1 upstream f | TACACAAGAGCAATGGCCCTAGC | This study |
| ChIP_xyr1 upstream r | TGGATGGATGGAGAACGGGATG | This study |

beads (Thermo Scientific) were used. The obtained conjugate was washed 3 times with a low salt buffer (0.1% (w/v) SDS, 1% (w/v) Triton X-100, 2 mM EDTA pH 8.0, 20 mM Tris–HCl pH 8.0, 150 mM NaCl), once with a final wash buffer (0.1% (w/v) SDS, 1% (w/v) Triton X-100, 2 mM EDTA pH 8.0, 20 mM Tris–HCl pH 8.0, 500 mM NaCl) and once with TE buffer. Then, samples were eluted in TES buffer (10 mM Tris–HCl pH 8.0, 1 mM EDTA, 1% (w/v) SDS). Protein-bound DNA was treated with Proteinase K (Thermo Scientific) and DNA samples were purified using the MiniElute PCR Purification Kit (Qiagen) according to the manufacturer's protocol. The precipitated DNA was quantified by qPCR performed in iCycler Thermal Cycler (Bio-Rad) and the use of a standard curve. A reaction volume of 25 µL including the following compounds: 2× iQ SYBR® Green Supermix (Bio-Rad), 10 µM primers and 5 µL of immunoprecipitated and input DNA (1:5 diluted in EB) or genomic DNA for the standard curve. The annealing temperature was 60 °C and the primer sequences are provided in Table 3. The qPCR cycling protocol and the adequate amounts of reagents were chosen as recommend in the manufacturer's instructions. All experiments were performed in biological and technical duplicates.

### Chromatin accessibility real-time PCR (CHART-PCR) assays

DNaseI digestions of chromatin and subsequent qPCR analyses were carried out as described before [17]. To be noted, only one of both deletion strains (Rut-C30Δ*cre1-96* (2)) was used for this analysis due to similar results from previous experiments of this study (e.g. transcript analysis). The qPCR analyses of the DNaseI-treated samples were performed to measure the relative abundance of DNA of the target regions. PCRs were performed in triplicates in a Rotor-Gene Q system (Qiagen) using the reaction mixture (final volume 20 µL) and the cycling conditions as described before [17]. Primer sequences are provided in Table 4. The amount of intact input DNA of each sample was calculated by comparing the threshold values of the PCR amplification plots with a standard curve generated for each primer set using serial dilutions of genomic, undigested DNA. The chromatin accessibility index (CAI) was defined as: $CAI = (Dc1 + Dc2)/2Ds$, where Ds is the amount of intact

**Table 4 Primer used for CHART-PCR**

| Primer name | Sequence 5′–3′ | References |
|---|---|---|
| epiactinTr_f | CTTCCCTCCTTTCCTCCCCCTCCAC | [17] |
| epiactinTr_r | GCGACAGGTGCACGTACCCTCCATT | [17] |
| episar1Tr_f | GTCAGGAAATGCCGCACAAGCAAGA | [17] |
| episar1Tr_r | TGTGTTTTACCGCCTTGGCCTTTGG | [17] |
| epixyr1_1Tr_f | CCTTTGGCCATCTACACAAGAGCAA | [45] |
| epixyr1_1Tr_r | CGCAATTTTTATTGCTGTTCGCTTC | [45] |
| epicbh1_1Tr_f | AAGGGAAACCACCGATAGCAGTGTC | [46] |
| epicbh1_1Tr_r | TTTCACTTCACCGGAACAAACAAGC | [46] |
| epixn1_1Tr_f | GCACTCCAAGGCCTTCTCCTGTACT | [46] |
| epixyn1_1Tr_r | TAGATTGAACGCCACCCGCAATATC | [46] |

DNA detected for each target region, and Dc1 and Dc2 are the amounts of intact DNA detected for the promoter regions of *sar1* and *act*, respectively, which were used as reference genes for normalization.

## Additional files

**Additional file 1: Figure S1.** Deletion of *cre1-96* in *T. reesei* Rut-C30. (A) Rut-C30 was transformed with the plasmid pMS*-5hph3cre1 that bears the deletion cassette consisting of the hygromycin resistance gene under the *pki* promoter and the terminator of *cbh2* (dark grey arrow, *hph*) to replace the native *cre1-96* gene (light grey arrow, *cre1-96*). (B) Agarose gel electrophoresis of diagnostic PCR was performed. Primer pairs added to the respective PCR are indicated on top of the gel, the strain of which the genomic DNA was used as template is indicated below each lane. Candidate strains (Δ*cre1-96* (1) and (2)) yielded expected fragments with the primer pair 1F and 1R or 3F and 3R, and no fragment in case of primer pair 2F and 2R. Rut-C30 was applied as negative control in the case of the PCR using primer pair 1F and 1R and as positive control in the PCR using primer pair 2F and 2R. Water added to the respective PCR in a no template control PCR (NTC). A DNA ladder (L) was included for estimation of fragment size.

**Additional file 2: Figure S2.** Constitutive expression of *cre1-96* in *T. reesei* Rut-C30. (A) Rut-C30 was transformed with the plasmid pMS*-*ptef::cre1-96* that bears the *tef1* promoter (white bar, p*tef1*), the *cre1-96* gene (light grey arrow, *cre1-96*), and the marker cassette (dark grey bar, *hph*). The latter consists of the hygromycin resistance gene under the *pki* promoter and the terminator of *cbh2*. (B) Agarose gel electrophoresis of diagnostic PCR was performed. Primer pairs added to the respective PCR are indicated on top of the gel, the strain of which the genomic DNA was used as template is indicated below each lane. A candidate strain (OE*cre1-96*) yielded expected fragments with all three primer pairs. Rut-C30 was applied as negative control in case of the PCR using primer pair 1F and 1R as well as 2F and 2R and as a positive control in the PCR using primer pair 3F and 3R. A DNA ladder (L) was included for estimation of fragment size.

**Additional file 3: Figure S3.** Growth behaviour of Rut-C30Δ*cre1-96* on glycerol and D-glucose. The *T. reesei* strains Rut-C30 and Rut-C30Δ*cre1-96* (2) were pre-grown on MEX plates and were then transferred to MA medium plates supplemented with 1 % (w/v) glycerol or D-glucose. Plates were incubated at 30 °C and pictures were taken after 24, 48, 60 and 84 hours.

**Additional file 4: Figure S4.** Multiple sequence alignment of Cre1 homologues. Multiple sequence alignment of *T. reesei* Cre1, Cre1-96 and Cre1 homologues of *A. nidulans*, *A. niger*, *N. crassa*, *T. atroviride*, *T. virens* and *S. cerevisiae* was conducted using Clustal Omega (http://www.ebi.ac.uk/

Tools/msa/clustalo/). Protein sequences were retrieved from respective genome databases. The alignment revealed conserved amino acids and protein domains based on sequence similarities.

**Additional file 5: Figure S5.** Cellulase activity and biomass formation of QM6a*cre1-96* and QM6a*cre1-96::eyfp* on D-glucose. The *T. reesei* strains QM6a*cre1-96* (blue bar, Cre1-96) and QM6a*cre1-96::eyfp* (purple bar, Cre1-96::eYFP) were cultivated in triplicates for 45 hours in MA medium supplemented with 1 % (w/v) D-glucose. Cellulase activities (A) of the culture supernatants were measured in technical duplicates and the biomass (B) was collected by filtration with miracloth and is depicted as dry weight on the y-axis.

**Additional file 6: Figure S6.** Growth behaviour of Rut-C30OE*cre1-96* on glycerol and D-glucose. The *T. reesei* strains Rut-C30, Rut-C30Δ*cre1-96* (2) and Rut-C30OE*cre1-96* were pre-grown on MEX plates and were then transferred to MA medium plates supplemented with 1 % (w/v) glycerol and D-glucose. Plates were incubated at 30 °C and pictures were taken after 24, 48, 60 and 84 hours.

**Additional file 7: Table S1.** Primers used for the diagnostic PCR of Rut-C30Δ*cre1-96* (1) and (2).

**Additional file 8: Table S2.** Primers used for the diagnostic PCR of Rut-C30OE*cre1-96*.

### Abbreviations

aa: amino acid; CAI: chromatin accessibility index; CCR: carbon catabolite repression; CHART-PCR: chromatin accessibility real-time PCR; ChIP: chromatin immunoprecipitation; CMC: carboxymethylcellulose; Cre1: carbon catabolite repressor 1; CreA: carbon catabolite repressor A; EB: elution buffer; eYFP: enhanced yellow fluorescent protein; LB: lysogeny broth; MA: Mandels-Andreotti; MEX: malt extract; NES: nuclear export signal; NLS: nuclear localization signal; qPCR: quantitative PCR; TAD: transactivation domain; URR: upstream regulatory regions; Xyr1: xylanase regulator 1.

### Authors' contributions

AR constructed the plasmids and the fungal strains, performed the growth experiments, the chromatin analyses (ChIP and CHART-PCR), the confocal microscopy, the in silico domain analysis, the transcript analyses, the enzymatic assays, and drafted this manuscript. AGM contributed to the chromatin analyses (ChIP). JS participated in the conception of the study, analysis of data and revision of the manuscript. RLM participated in the conception of the study. ARMA participated in the conception of the study, supervised the experiments, and revised the manuscript. All authors read and approved the final manuscript.

### Author details

[1] Institute of Chemical, Environmental and Bioscience Engineering, TU Wien, Gumpendorfer Str. 1a, 1060 Vienna, Austria. [2] Fungal Genetics and Genomics Lab, Department of Applied Genetics and Cell Biology, BOKU-University of Natural Resources and Life Sciences, Konrad Lorenz Str. 24, 3430 Tulln/Donau, Austria. [3] Institute of Microbiology, University of Veterinary Medicine Vienna, Veterinärplatz 1, 1210 Vienna, Austria.

### Acknowledgements

We gratefully acknowledge Markku Saloheimo from VTT Finland for providing the Rut-C30Δ*tmus53* strain. We kindly thank Lena Studt and Harald Berger for the support of the ChIP experiment. We appreciate the help from Daniel Kiesenhofer, Lisa Kappel and Alexander Lichius with the confocal microscope.

### Competing interests

The authors declare that they have no competing interests.

### Funding

This work was funded by two grants from the Austrian Science Fund (FWF): V232 and P24851 given to ARMA. Work at the Fungal Genetics and Genomics Lab was funded by Grant LS12-009 of the NFB Life Science Fund to JS.

### References

1. Singh A, Mishra P. Overview of problems and potential. In: Microbial pentose utilization—current applications in biotechnology. vol. 33; 1995. p. 1–3.
2. Kumar R, Singh S, Singh OV. Bioconversion of lignocellulosic biomass: biochemical and molecular perspectives. J Ind Microbiol Biotechnol. 2008;35(5):377–91.
3. Singh A, Mishra P. Extraction of pentosans from lignocellulosic materials. In: Microbial pentose utilization—current applications in biotechnology. vol. 33; 1995. p. 71–98.
4. Aro NPT, Penttilä M. Transcriptional regulation of plant cell wall degradation by filamentous fungi. FEMS Microbiol Rev. 2005;29:719–39.
5. Ivanova C, Bååth JA, Seiboth B, Kubicek CP. Systems analysis of lactose metabolism in *Trichoderma reesei* identifies a lactose permease that is essential for cellulase induction. PLoS ONE. 2013;8(5):e62631.
6. Mach-Aigner AR, Pucher ME, Mach RL. D-Xylose as a repressor or inducer of xylanase expression in *Hypocrea jecorina* (*Trichoderma reesei*). Appl Environ Microbiol. 2010;76(6):1770–6.
7. Strauss J, Mach RL, Zeilinger S, Hartler G, Stoffler G, Wolschek M, et al. Cre1, the carbon catabolite repressor protein from *Trichoderma reesei*. FEBS Lett. 1995;376(1–2):103–7.
8. Nakari-Setälä T, Paloheimo M, Kallio J, Vehmaanperä J, Penttilä M, Saloheimo M. Genetic modification of carbon catabolite repression in *Trichoderma reesei* for improved protein production. Appl Environ Microbiol. 2009;75(14):4853–60.
9. Mach RL, Strauss J, Zeilinger S, Schindler M, Kubicek CP. Carbon catabolite repression of xylanase I (*xyn1*) gene expression in *Trichoderma reesei*. Mol Microbiol. 1996;21(6):1273–81.
10. Portnoy T, Margeot A, Linke R, Atanasova L, Fekete E, Sandor E, et al. The CRE1 carbon catabolite repressor of the fungus *Trichoderma reesei*: a master regulator of carbon assimilation. BMC Genomics. 2011;12:269.
11. Ries L, Belshaw NJ, Ilmén M, Penttilä ME, Alapuranen M, Archer DB. The role of CRE1 in nucleosome positioning within the *cbh1* promoter and coding regions of *Trichoderma reesei*. Appl Microbiol Biotechnol. 2014;98(2):749–62.
12. Stricker AR, Grosstessner-Hain K, Würleitner E, Mach RL. Xyr1 (xylanase regulator 1) regulates both the hydrolytic enzyme system and D-xylose metabolism in *Hypocrea jecorina*. Eukaryot Cell. 2006;5(12):2128–37.
13. Mach-Aigner AR, Pucher ME, Steiger MG, Bauer GE, Preis SJ, Mach RL. Transcriptional regulation of *xyr1*, encoding the main regulator of the xylanolytic and cellulolytic enzyme system in *Hypocrea jecorina*. Appl Environ Microbiol. 2008;74(21):6554–62.
14. Montenecourt BS, Eveleigh DE. Production and characteriation of high yielding cellulase mutants of *Trichoderma reesei*. TAPPI J. 1979;28:101–8.
15. Ward M. Improving secreted enzyme production by *Trichoderma reesei*. In: 9th International workshop on *Trichoderma* and *Gliocladium*: 2006; Vienna.
16. Ilmén M, Thrane C, Penttilä M. The glucose repressor gene *cre1* of *Trichoderma*: isolation and expression of a full-length and a truncated mutant form. Mol Gen Genetics. 1996;251(4):451–60.
17. Mello-de-Sousa TM, Gorsche R, Rassinger A, Pocas-Fonseca MJ, Mach RL, Mach-Aigner AR. A truncated form of the Carbon catabolite repressor 1 increases cellulase production in *Trichoderma reesei*. Biotechnol Biofuels. 2014;7(1):129.
18. Seiboth B, Gamauf C, Pail M, Hartl L, Kubicek CP. The D-xylose reductase of *Hypocrea jecorina* is the major aldose reductase in pentose and D-galactose catabolism and necessary for beta-galactosidase and cellulase induction by lactose. Mol Microbiol. 2007;66(4):890–900.
19. Stricker AR, Steiger MG, Mach RL. Xyr1 receives the lactose induction signal and regulates lactose metabolism in *Hypocrea jecorina*. FEBS Lett. 2007;581(21):3915–20.
20. Cziferszky A, Mach RL, Kubicek CP. Phosphorylation positively regulates DNA binding of the carbon catabolite repressor Cre1 of *Hypocrea jecorina* (*Trichoderma reesei*). J Biol Chem. 2002;277(17):14688–94.

21. Lichius A, Seidl-Seiboth V, Seiboth B, Kubicek CP. Nucleo-cytoplasmic shuttling dynamics of the transcriptional regulators XYR1 and CRE1 under conditions of cellulase and xylanase gene expression in *Trichoderma reesei*. Mol Microbiol. 2014;94(5):1162–78.
22. Tilburn J, Sarkar S, Widdick DA, Espeso EA, Orejas M, Mungroo J, et al. The Aspergillus PacC zinc finger transcription factor mediates regulation of both acid- and alkaline-expressed genes by ambient pH. EMBO J. 1995;14(4):779–90.
23. Cánovas D, Studt L, Marcos AT, Strauss J. High-throughput format for the phenotyping of fungi on solid substrates. Sci Rep. 2017;7:4289.
24. Sun J, Glass LN. Identification of the CRE-1 cellulotytic regulon in *Neurospora crassa*. PLoS ONE. 2011;6(9):e25654.
25. dos Santos Castro L, Pedersoli WR, Antonieto AC, Steindorff AS, Silva-Rocha R, Martinez-Rossi NM, et al. Comparative metabolism of cellulose, sophorose and glucose in *Trichoderma reesei* using high-throughput genomic and proteomic analyses. Biotechnol Biofuels. 2014;7(1):41.
26. Derntl C, Gudynaite-Savitch L, Calixte S, White T, Mach RL, Mach-Aigner AR. Mutation of the Xylanase regulator 1 causes a glucose blind hydrolase expressing phenotype in industrially used *Trichoderma* strains. Biotechnol Biofuels. 2013;6(1):62.
27. NetNES 1.1 Server. http://www.cbs.dtu.dk/services/NetNES/. Accessed 31 May 2017.
28. Ries L, Beattie SR, Espeso EA, Cramer RA, Goldman GH. Diverse Regulation of CreA Carbon Catabolite Repressor in *Aspergillus nidulans*. Genetics. 2016;203(1):335–52.
29. Ries L, Belshaw NJ, Ilmén M, Penttilä ME, Alapuranen M, Archer DB. The role of CRE1 in nucleosome positioning within the *cbh1* promoter and coding regions of *Trichoderma reesei*. Appl Microbiol Biotechnol. 2014;98(2):749–62.
30. Zeilinger S, Schmoll M, Pail M, Mach RL, Kubicek CP. Nucleosome transactions on the *Hypocrea jecorina* (*Trichoderma reesei*) cellulase promoter *cbh2* associated with cellulase induction. Mol Genet Genomics. 2003;270(1):46–55.
31. Steiger MG, Vitikainen M, Uskonen P, Brunner K, Adam G, Pakula T, et al. Transformation system for *Hypocrea jecorina* (*Trichoderma reesei*) that favors homologous integration and employs reusable bidirectionally selectable markers. Appl Environ Microbiol. 2011;77(1):114–21.
32. Mach RL, Schindler M, Kubicek CP. Transformation of *Trichoderma reesei* based on hygromycin B resistance using homologous expression signals. Curr Genet. 1994;25(6):567–70.
33. Derntl C, Kiesenhofer DP, Mach RL, Mach-Aigner AR. Novel strategies for genomic manipulation of *Trichoderma reesei* with the purpose of strain engineering. Appl Environ Microbiol. 2015;81(18):6314–23.
34. Gruber F, Visser J, Kubicek CP, de Graaff LH. Cloning of the *Trichoderma reesei pyrG* gene and its use as a homologous marker for a high-frequency transformation system. Curr Genet. 1990;18(5):447–51.
35. Steiger MG, Mach RL, Mach-Aigner AR. An accurate normalization strategy for RT-qPCR in *Hypocrea jecorina* (*Trichoderma reesei*). J Biotechnol. 2010;145(1):30–7.
36. JGI *Trichoderma reesei* v2.0 Genome Database. http://genome.jgi.doe.gov/Trire2/Trire2.home.html. Accessed 28 April 2014.
37. JGI *Trichoderma reesei* Rut C-30 v1.0 Genome Database. http://genome.jgi.doe.gov/TrireRUTC30_1/TrireRUTC30_1.home.html. Accessed 28 April 2014.
38. NCBI Conserved Domain Search. http://www.ncbi.nlm.nih.gov/Structure/cdd/wrpsb.cgi. Accessed 01 June 2014.
39. Clustal Omega. http://www.ebi.ac.uk/Tools/msa/clustalo/. Accessed 13 June 2017.
40. cNLS Mapper. http://nls-mapper.iab.keio.ac.jp/cgi-bin/NLS_Mapper_form.cgi. Accessed 08 September 2014.
41. Nine Amino Acids Transactivation Domain (9aaTAD) Prediction Tool. http://www.med.muni.cz/9aaTAD/. Accessed 25 May 2017.
42. Piskacek S, Gregor M, Nemethova M, Grabner M, Kovarik P, Piskacek M. Nine-amino-acid transactivation domain: establishment and prediction utilities. Genomics. 2007;89(6):756–68.
43. Schindelin J, Arganda-Carreras I, Frise E, Kaynig V, Longair M, Pietzsch T, et al. Fiji: an open-source platform for biological-image analysis. Nat Methods. 2012;9:676–82.
44. Lando D, Endesfelder U, Berger H, Subramanian L, Dunne PD, McColl J, et al. Quantitative single-molecule microscopy reveals that CENP-ACnp1 depostiton occurs during G2 in fission yeast. Open Biol. 2012;2(7):120078.
45. Mello-de-Sousa TM, Rassinger A, Derntl C, Poças-Fonseca MJ, Mach-Aigner AR, Mach RL. The relation between chromatin status, Xyr1 and cellulase expression in *Trichoderma reesei*. Curr Genomics. 2016;17:1–8.
46. Mello-de-Sousa TM, Rassinger A, Pucher ME, dos Santos Castro L, Persinoti GF, Silva-Rocha R, et al. The impact of chromatin remodelling on cellulase expression in *Trichoderma reesei*. BMC Genomics. 2015;16:588.

# 11

# Identification of the decumbenone biosynthetic gene cluster in *Penicillium decumbens* and the importance for production of calbistrin

Sietske Grijseels[1†], Carsten Pohl[2†], Jens Christian Nielsen[3], Zahida Wasil[1], Yvonne Nygård[2], Jens Nielsen[3,4], Jens C. Frisvad[1], Kristian Fog Nielsen[1], Mhairi Workman[1], Thomas Ostenfeld Larsen[1], Arnold J. M. Driessen[2] and Rasmus John Normand Frandsen[1*]

**Abstract**

**Background:** Filamentous fungi are important producers of secondary metabolites, low molecular weight molecules that often have bioactive properties. Calbistrin A is a secondary metabolite with an interesting structure that was recently found to have bioactivity against leukemia cells. It consists of two polyketides linked by an ester bond: a bicyclic decalin containing polyketide with structural similarities to lovastatin, and a linear 12 carbon dioic acid structure. Calbistrin A is known to be produced by several uniseriate black Aspergilli, *Aspergillus versicolor*-related species, and Penicillia. *Penicillium decumbens* produces calbistrin A and B as well as several putative intermediates of the calbistrin pathway, such as decumbenone A-B and versiol.

**Results:** A comparative genomics study focused on the polyketide synthase (PKS) sets found in three full genome sequence calbistrin producing fungal species, *P. decumbens, A. aculeatus* and *A. versicolor*, resulted in the identification of a novel, putative 13-membered calbistrin producing gene cluster (*calA* to *calM*). Implementation of the CRISPR/Cas9 technology in *P. decumbens* allowed the targeted deletion of genes encoding a polyketide synthase (*calA*), a major facilitator pump (*calB*) and a binuclear zinc cluster transcription factor (*calC*). Detailed metabolic profiling, using UHPLC-MS, of the Δ*calA* (PKS) and Δ*calC* (TF) strains confirmed the suspected involvement in calbistrin productions as neither strains produced calbistrin nor any of the putative intermediates in the pathway. Similarly analysis of the excreted metabolites in the Δ*calB* (MFC-pump) strain showed that the encoded pump was required for efficient export of calbistrin A and B.

**Conclusion:** Here we report the discovery of a gene cluster (*calA-M*) involved in the biosynthesis of the polyketide calbistrin in *P. decumbens*. Targeted gene deletions proved the involvement of CalA (polyketide synthase) in the biosynthesis of calbistrin, CalB (major facilitator pump) for the export of calbistrin A and B and CalC for the transcriptional regulation of the *cal*-cluster. This study lays the foundation for further characterization of the calbistrin biosynthetic pathway in multiple species and the development of an efficient calbistrin producing cell factory.

**Keywords:** *Penicillium decumbens*, Calbistrin, Secondary metabolite, Decalin, Polyketide, Biosynthesis

*Correspondence: rasf@bio.dtu.dk
†Sietske Grijseels and Carsten Pohl have contributed equally to this work
[1] Department of Biotechnology and Biomedicine, Technical University of Denmark, 2800 Kgs. Lyngby, Denmark
Full list of author information is available at the end of the article

## Background

Filamentous fungi are generally prolific producers of secondary metabolites, which possess a wide range of different biological activities. It is a widely accepted view that secondary metabolites serve an important role for the producing fungi to survive in their respective ecological niches, yet many of these small-molecules are also of great importance to humans. Prominent examples of medical use of secondary metabolites include the antibacterial penicillin, the cholesterol-lowering agent lovastatin/compactin and the antifungal griseofulvin. Today fungal secondary metabolites continue to serve as an important source of small-molecules for the discovery of novel drugs.

The amounts of secondary metabolites that are naturally produced by fungi are often far below the amounts necessary for profitable industrial-scale production of the given compound. Traditionally, native fungal production strains have been optimized via strategies relying on random mutagenesis coupled with screening for strains with improved production levels and fermentations properties. The most well-known example being the optimization of penicillin production, where strain improvement programs have succeeded in increasing titers and productivity by at least three orders of magnitude [1]. Recent advances in our understanding of the metabolic pathways for the production of secondary metabolites, full genome sequences, and improvements in genetic engineering tools now allow rational strain improvement by metabolic engineering for enhancing the natural product yield [2–4]. However, in order to employ such techniques, the biosynthetic genes and/or regulatory elements for production of a given compound first have to be identified and characterized. Over the past decades, the genetic basis for production of numerous fungal secondary metabolites has been elucidated, by linking production to gene clusters or genes encoding key-enzyme responsible for biosynthesis of the carbon backbone of the respective secondary metabolites. Still, for the vast majority of the secondary metabolites known today, the biosynthetic pathway and genetic basis remains unknown.

The secondary metabolites calbistrin A has been reported to possess a number of interesting bioactivities such as antifungal active against *Candida albicans* [5], 3-hydroxy-3-methyl-glutaryl-coenzyme A reductase inhibition in mammalian cells [6] and cytotoxic toward both healthy and leukemic human cells [7]. Calbistrin A and the related B and C are produced by several uniseriate black Aspergilli, *Aspergillus versicolor*-related species and Penicillia species [8, 9]. Among the Penicillia, the recently genome sequenced *Penicillium decumbens* [10] is interesting because it produces calbistrin A and C and also accumulates several metabolites that are structurally related to calbistrins, namely decumbenone A, B and C [11] (Fig. 1a). All calbistrins are predicted to consist of two individual polyketide chains linked by an ester bond: a decalin containing heptaketide (C14 chain) and a linear dioic acid (also termed dicarboxylic acid) structure formed from a hexaketide (C12 chain) [8]. The calbistrins show structural similarities to the natural cholesterol lowering statins, such as lovastatin produced by *Monascus ruber* [12] and *A. terreus* [13] and compactin produced by *P. solitum* [14–16]. Compactin and lovastatin are both known to consist of two separately synthesized polyketides, a decalin structure formed from a nonaketide (C18 chain) and an ester bound linear diketide (C4 chain) attached to the decalin structure at the same position as seen in calbistrins (Fig. 1). Biosynthesis of the two natural statins is well documents in literature, and formation of the decalin structure has been shown to proceed via an enzymatic intramolecular [4+2] Diels–Alder cycloaddition, catalyzed by the polyketide synthase (PKS) responsible formation of the nonaketide backbone of these molecules [17].

Motivated by the reported activities of calbistrin A, the interesting structural similarities and differences between the calbistrins and naturally occuring statins we set out to elucidate the genetic and enzymatic basis for biosynthesis of calbistrins. We chose to perform a comparative genomic analysis of known calbistrin producers, which resulted in the identification of a putative biosynthetic gene cluster (*cal*) for production of calbistrin. To prove the suggested involvement of the identified genes we next developed a transformation protocol and a CRISPR/Cas9 based system for targeted genetic modification of *P. decumbens*. This system allowed us to efficiently delete three genes in the putative *cal*-cluster and analyze the metabolic effects. Deletion of a putative PKS (*calA*) and a transcription factor (*calC*) resulted in the complete abolishment of calbistrin biosynthesis, while deletion of a putative efflux pump (*calB*) significantly reduced extracellular levels of calbistrin A and C. The presented results lay the foundation for the future optimization and development of an efficient cell factory for the production of calbistrins.

## Results

### Chemical analysis reveals the presence of calbistrins and related compounds in extracts of *P. decumbens*

Ultra high performance liquid chromatography-high resolution-mass spectrometry (UHPLC-HRMS) analysis of ethyl acetate extracts of the *P. decumbens* wild-type (WT) cultured on Czapek yeast autolysate medium (CM) showed that calbistrin A and calbistrin C were produced under these culture conditions (Fig. 1b). Previous studies of calbistrins have shown that the $[M+H]^+$ ions are

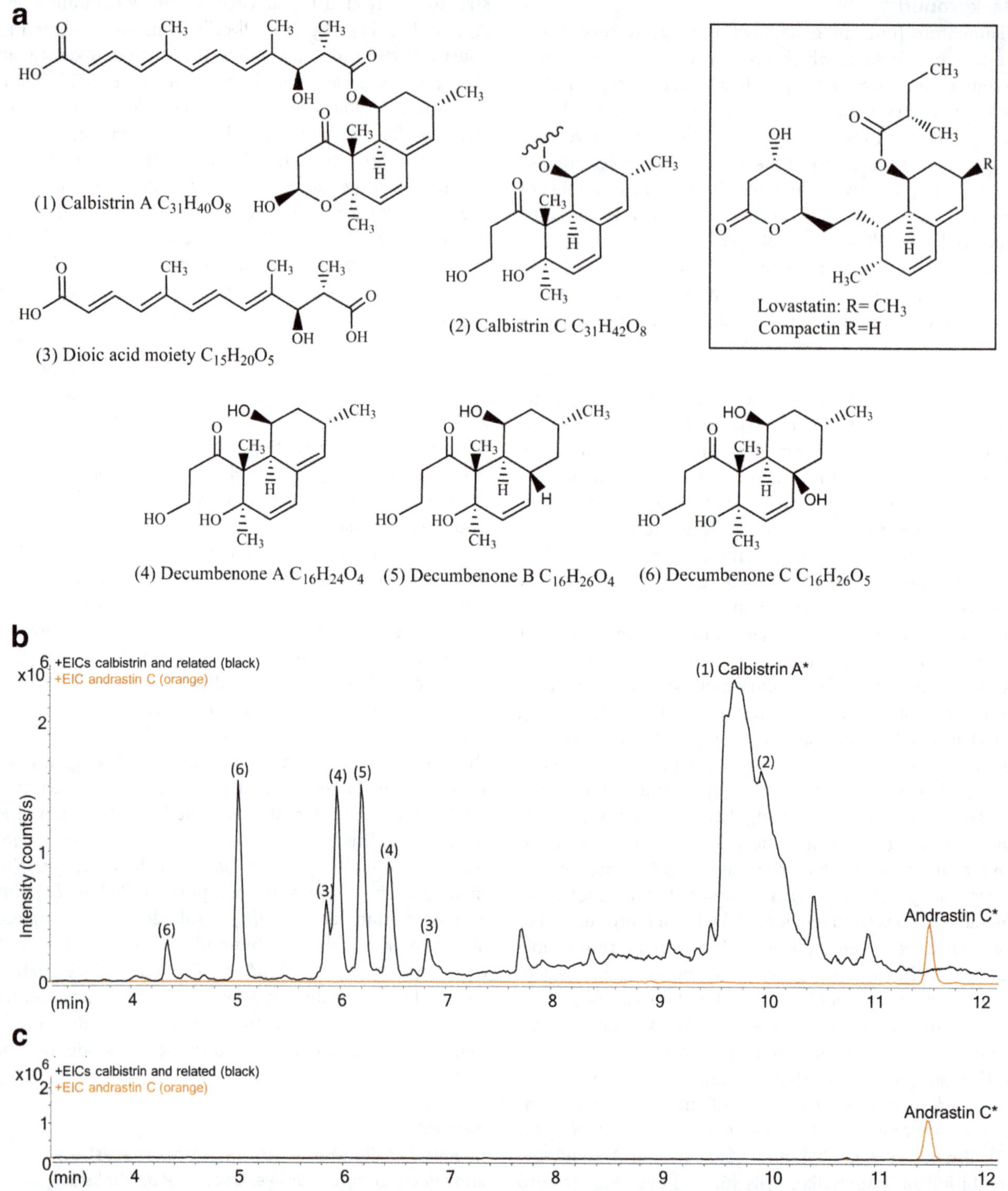


**Fig. 1** Chemical structures of calbistrin and related metabolites and UHPLC-HRMS analysis of *P. decumbens* wild type and PKS mutant strains. **a** Chemical structures of (1) calbistrin A, (2) calbistrin C, (3) putative linear moiety, (4) decumbenone A, (5) decumbenone B, (6) decumbenone C, and in the box compactin and lovastatin. **b** UHPLC-HRMS analysis of the wild type *P. decumbens* culture extract. Merged extracted ion chromatograms (EICs), ± m/z 0.005 of molecular features detected for compounds 1–6: 263.1642; 281.1742; 321.1670; 337.1401; 245.1177; 303.1204; 247.1697; 265.1806; 305.1720; 245.1538; 303.1575; 319.1331; 505.2591; 523.2705; 563.2622; 525.2847; 565.2776; and 285.1463. Additionally, the EIC of andrastin C (m/z 473.2898) is shown in orange. Calbistrin A and andrastin C were confirmed with a reference standard (marked with *), the other compounds were tentative identified based on UV-spectra and MS/HRMS fragmentation patterns. **c** UHPLC-HRMS results of *P. decumbens* ΔPKS culture extract. Merged EICs of molecular features detected for compounds 1-7 and EIC of andrastin C as in B

not observed in the mass spectra due to extensive water losses [7, 8] and we therefore searched for the presence of the sodium ion adducts, $[M+Na]^+$, for the two compounds. Inspection of the chromatograms for the WT revealed the presence of the calbistrin A $[M+Na]^+$ m/z of 563.2623 (calculated 563.2621, mass error of 0.355 ppm) eluting at 9.7 min, and fragment ions corresponding to neutral losses of one, two and three water molecules for calbistrin A (Additional file 1: Additional Information 1A), and the calbistrin C $[M+Na]^+$ m/z of 565.2776 (calculated 565.2777, mass error of 0.177 ppm) eluting at 9.9 min (Additional file 1: Additional Information 1B). These adduct- and fragmentation patterns assisted the establishment of monoisotopic masses and indicated molecular formulas of $C_{31}H_{40}O_8$ and $C_{31}H_{42}O_8$, corresponding to calbistrin A and C, respectively. The identity of calbistrin A was confirmed by comparison of the UV spectrum and the MS/HRMS fragmentation pattern to that of an in-house reference standard for calbistrin A (Additional file 1: Additional Information 1C-D). Tentative identification of calbistrin C was based on comparison of it MS/HRMS fragmentation pattern to that of calbistrin A (Additional file 1: Additional Information 3F-I).

The UHPLC-HRMS analysis of the wild type grown on CM (Fig. 1b) also revealed $[M+Na]^+$ parent ions that corresponded to the three compounds decumbenone A (two isomers eluting at 6.02 and 6.50 min), decumbenone B (eluting at 6.25 min), and decumbenone C (two isomers eluting at 4.4 and 5.05 min). As the decumbenones all have the same polyketide backbone length and decalin moiety as the calbistrins (Fig. 1a), we hypothesized that they are intermediates in, or byproducts of, calbistrin biosynthesis. The identity of these compounds could not be definitively confirmed due to the lack of reference standards, however, the fragmentation patterns for the putative decumbenone A-C compounds were in good agreement with the fragmentation patterns of calbistrin A and C (Additional file 1: Additional Information 3).

Further inspection of the WT chromatogram revealed the presence of two peaks (eluting at 5.9 and 6.7 min in Fig. 1b) that had a composition of $C_{15}H_{20}O_5$, based on HRMS, which corresponds to the composition of the linear dioic acid moiety of calbistrins and therefore also could be related to calbistrin biosynthesis. This hypothesis was further strengthened by the finding that MS/HRMS fragments of these compounds were identical to several MS/HRMS fragments observed upon fragmentation of calbistrin A and C (Additional file 1: Additional Information 4). Furthermore inspection of the MS/HRMS data of the putative dioic acid moieties showed neutral losses of CO (at RT 5.8 min: fragment ions of m/z 199.1112 and m/z 171.1161 give a difference of 27.9951, at RT 6.9 min: fragment ions of m/z 199.1120 and m/z 171.1166 give a difference of 27.9954; theoretical mass CO = 27.9949) and sequential losses of 1C fragments, supporting the predicted molecular features (Additional file 1: Additional Information 2 and 4). Finally, the most abundant peak (5.9 min) had the same distinct UV spectrum as the calbistrins with absorption maxima at 345 nm (Additional file 1: Additional Information 4) (the peak at 6.9 min was too small for detection of UV spectrum). One should note that calbistrins are known to feature several different cis–trans isomers of the linear dioic acid moiety, e.g. calbistrin A constist excluselively of trans conformations while calbistrin B and D include a single cis conformation at various positions [8]. These cis–trans transitions were shown to be induced by light exposure which also occurred during extraction [18].

### Comparative genomics of *P. decumbens* identifies a PKS putatively involved in calbistrin biosynthesis

The genome of *P. decumbens* (IBT11843), a member of the *Penicillium* subgenus *Aspergilloides* clade, was recently sequenced [10]. To narrow down the candidates for the calbistrin PKSs, a comparative genomics analysis with two distantly related known calbistrin producers was conducted. *A. aculeatus* has been reported to produce calbistrin A and C [19], and *A. versicolor* has been reported to produce versiol [20], which has a related structure to the decalin part of calbistrin A. Putative PKSs in *A. aculeatus and A. versicolor* were identified similar as described for *P. decumbens*, yielding 26 and 27 putative PKSs respectively.

Additionally, several further fungal PKS and PKS-NRPS-like biosynthetic systems have been reported to produce decalin containing metabolites, e.g. lovastatin in *A. terreus*, compactin in *Penicillium brevicompactum*, solanapyrone in *Alternaria solani* [21], equisetin/fusarisetin A in *Fusarium heterosporum* and *Fusarium sp.* FN080326 [22, 23] and myceliothermophin in *Myceliophthora thermophile* [24]. The enzymatic basis for decalin formation in these systems is however not identical and falls into at least three distinct groups: (1) PKS/PKS-NRPS based cycloadditions as seen in LovB and MlcA [25], (2) post-PKS bifunctional oxidases/alderases, such as Sol5 in *Alternatria solani* [21], (3) post-PKS monofunctional alderases of diverse evolutionary origin such as the Fsa2 from the fusaristatin/equisetin pathways [23], and MycB (AEO57198) from the myceliothermophin E pathway. Nonetheless, to test how putative orthologous PKSs could be related to the known decalin forming PKSs, we decided to include the KS-AT domains of MlcA, LovB, EqxS, Sol1, Fsa1 and MycA in the phylogenetic analysis.

Subsequently, the KS domains were aligned using the Smith-Waterman algorithm and a neighbour joining tree was constructed to identify putative orthologous enzymes across the three species (Fig. 2). The analysis showed that five of the six known decalin-forming PKSs (highlighted with blue in Fig. 2) clustered within a single well supported clade (bootstrap of 85%) of PKS-NRPS hybrids. This clade includes true PKS-NRPS hybrids, and hybrids where part or the whole NRPS portion has been lost. PdecPKS10 proved to be closest related

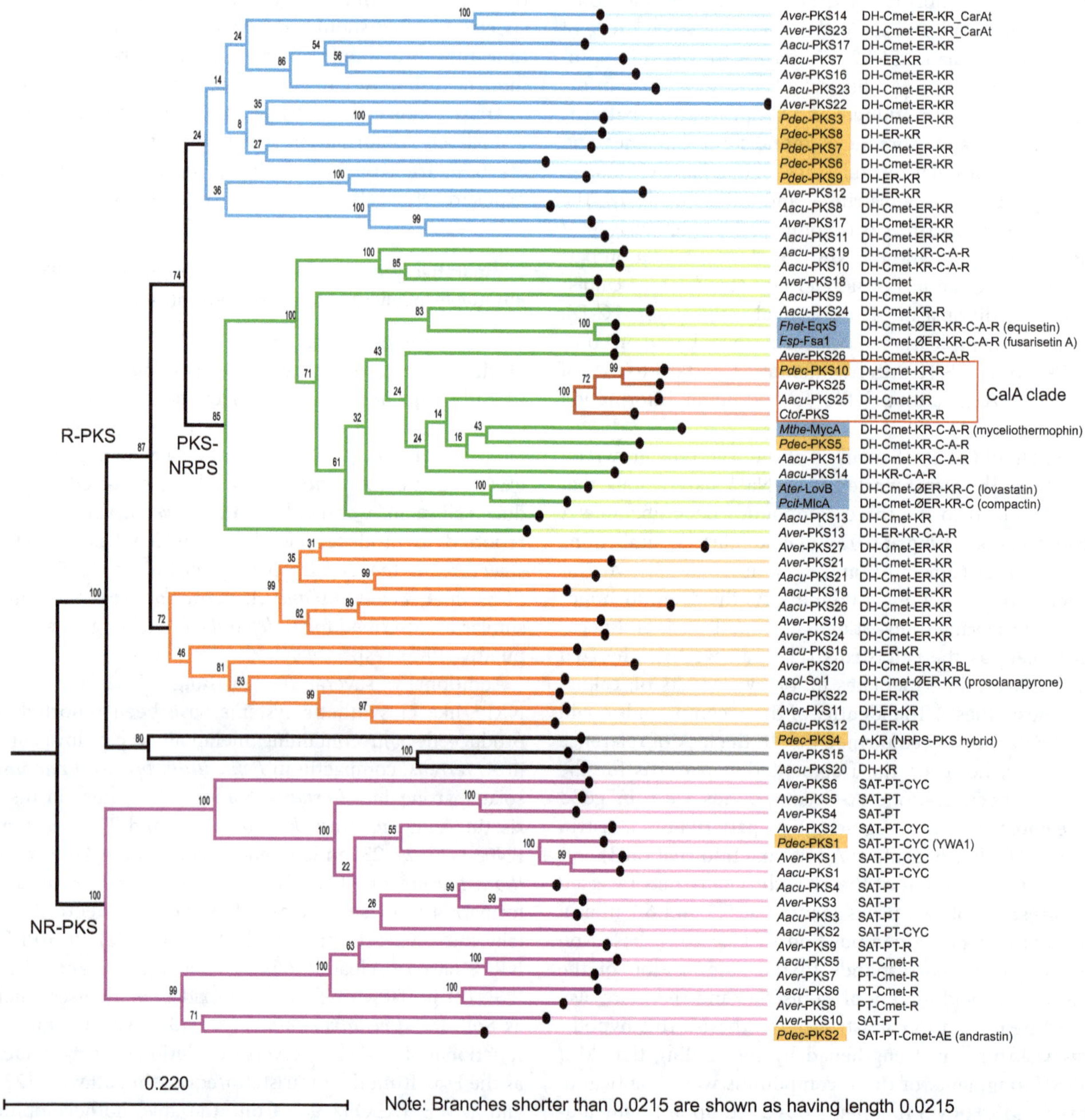


**Fig. 2** Neighbour joining tree of KS-AT domains from *P. decumbens*, *A. aculeatus* and *A. versicolor* PKSs. The four-membered clade with putative calbistrin-forming PKSs is highlighted with a red square. Known decalin forming PKSs are highlighted with blue background. Abbreviations: Species: *Pdec*: *P. decumbens* (highlighted in orange); *Aacu*: *A. aculeatus*; *Aver*: *A. versicolor*; *Fhet*: *F. heterosporum*; *Fsp*: *Fusarium* sp. FN080326; *Mthe*: *Myceliophthora thermophile*; *Ater*: *A. terreus*; *Pcit*: *P. citrinum*. Enzymatic domain: DH: dehydratase; Cmet: C-methyl transferase; ER: enoylreductase; ØER: dysfunctional ER; KR: ketoreductase; C: condensation; A: Adenylation; R: terminal reductase; TE: thioesterase; CarAt: carnetine acyltransferase; BL: beta lactamase; AE: acetylesterase; PT: product template; SAT: Starter acyltransferase, CYC: cyclase

to the myceliothermophin forming PKS-NRPS MycA from *M. thermophile*, then the equisetin forming PKS-NRPSs from *Fusarium* sp. and lastly the statin forming PKS-NRPSs LovB and MlcA. The close association with known decalin forming PKSs supports the hypothesis that PdecPKS10 is responsible for formation of the decalin portion of calbistrin. This hypothesis was further supported by the fact that KS domains from the partially reducing PKSs AspacPKS25 and AspvePKS25 clustered also with PdecPKS10, having an average identity of 76%. These three PKSs were all predicted to include a β-ketosynthase (KS), an acyltransferase (AT), a dehydratase (DH), a methyltransferase (MT), a ketoreductase (KR), an acyl carrier protein (ACP), and a terminal reductase (R) domain.

Further analysis of the neighbour joining tree showed that only one additional clade included members of all three species, suggesting orthologous PKSs (Fig. 2). This clade included KS domains of three non-reducing PKSs (PdecPKS1, AspacPKS1 and AspvePKS1) which showed very high sequence similarities (average of 76%) to the wA PKS from *P. rubens* and therefore likely are responsible for producing YWA-based pigments in the respective species.

The comparative genomics analysis of PKSs in the three known calbistrin producers did not reveal any obvious candidates for the second PKS predicted to be responsible for synthesizing the linear dioic acid portion of calbistrin.

### Deletion of the *pdecPKS10* gene demonstrates involvement of the PKS in calbistrin production

To test the proposed association between PdecPKS10 activity and calbistrin formation we adapted a transformation and targeted genetic engineering system recently developed for *P. chrysogenum* (also known as *Penicillium rubens* Wisconsin 54-1255) [26] for use in *P. decumbens* to delete the PKS encoding gene PENDEC_c013G00595.

This protocol resulted in sufficiently high gene editing efficiencies to generate several clones for characterization of *calA* and also *calB* and *calC* in *P. decumbens* (Additional file 1: Additional Information 5). Initial screening of the generated transformants at the gene locus by colony PCR and sequencing of clones displaying a size shift of the PCR product, indicating that excision of the entire genomic DNA region framed by the used protospacers was the most prevalent recombination event (Additional file 1: Additional Information 13 to 15), followed by some cases where parts from the AMA-plasmid had integrated albeit no microhomology sequences were detected between the inserts and genomic locus. Surprisingly, neither of the analyzed transformants contained simpler short indel mutations as would be expected following incorrect repair of a single cut event by the NHEJ pathway (Additional file 1: Additional Information 13 to 15). Based on our observations, the success rate of future experiments can perhaps be increased by adding a target-specific donor DNA repair templates although this would increase the experimental preparation effort we sought to reduce here.

Analysis of the *PdecPKS10* mutant (ΔPKS) showed that the production of calbistrin was completely abolished whereas production of unrelated compounds, such as andrastin C, remained unaffected (Figs. 1b, 6). These results confirmed that PdecPKS10 is essential for the biosynthesis of the calbistrins. Interestingly, the masses of the putative related metabolites, decumbenones and the putative linear moiety, also disappeared in the PKS deletion strains. This shows that these metabolites are involved in calbistrin biosynthesis, as hypothesized.

### Defining the putative gene clusters boundaries by gene synteny analysis and transcriptomics data

A more detailed bioinformatic analysis of the *PdecPKS10* locus revealed that several of the adjacent genes encoded proteins with putative tailoring enzyme functions were presumably relevant for the biosynthesis of calbistrins. To determine the boundaries of this putative gene cluster, we performed a synteny analysis of the respective contigs containing *PdecPKS10*, *AspacPKS25* and *AspvePKS25*. The analysis clearly showed conserved regions around the predicted PKS genes (Fig. 3a, trimmed to clusters for clarity) covering 10 predicted genes in *P. decumbens* (spanning a region of 35 kb) that displayed sequence similarity with a region containing 14 predicted genes upstream of the PKS in *A. versicolor* and 14 predicted genes in *A. aculeatus* downstream of the PKS. The identified conserved region in *P. decumbens* was continuous, while in *A. versicolor* the syntenic region was disrupted by a single gene that did not show homology with regions in the two other species. The putative cluster in *A. aculeatus* included two regions with no homology to regions in the two other species consisting of one region of seven adjacent genes and a second region of four adjacent genes.

The *P. decumbens* gene cluster included several regions that displayed high sequence similarity to the other two species but which lacked predicted genes, suggesting a less successful gene calling in *P. decumbens*. Guided by the detected homology, we used FGENESH as an alternative gene prediction and predicted three additional genes, which resulted in a total of 13 putative genes in the *P. decumbens* conserved region, named *calA-calM*. The proteins encoded by these genes all showed identities of >75% and >50% at amino acid level with the enzymes encoded the conserved regions in *A. versicolor* and *A.*

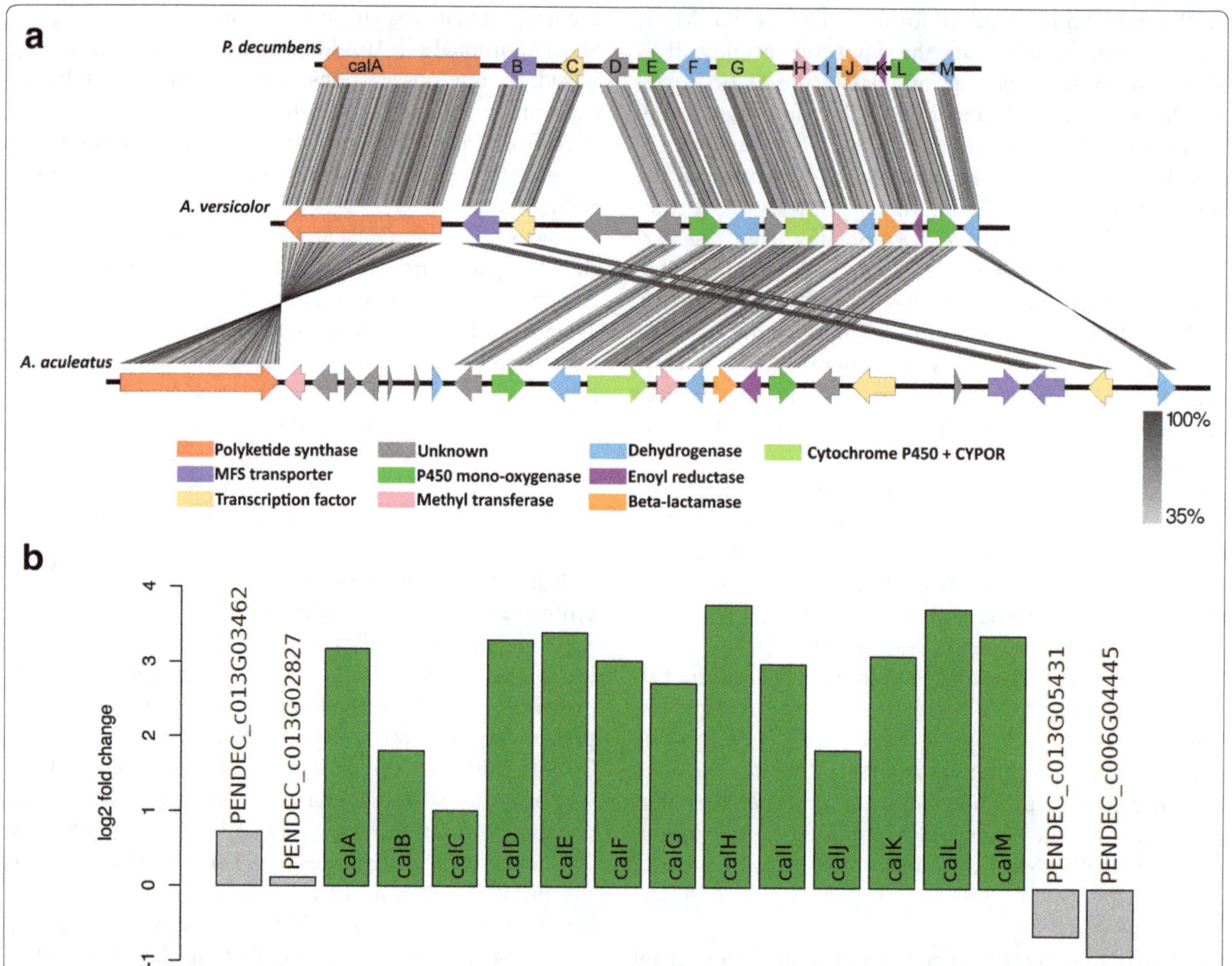


**Fig. 3** Expression and gene synteny in calbistrin cluster. **a** Synteny analysis of putative gene clusters in *P. decumbens*, *A. aculeatus* and *A. versicolor*. Figure made with EasyFig [46]. **b** Transcription analysis of calbistrin cluster in wild-type strain under calbistrin producing vs non-producing conditions. Log2 fold change for read counts in complex medium (inductive) over synthetic medium (non-inductive)

*aculeatus*, respectively (Table 1). At least one conserved functional domain was found in 12 out of the 13 predicted proteins, while none was found in CalD (Table 1). Ten of the proteins included predicted enzymatic functionality which would support a function as tailoring enzymes in secondary metabolite biosynthesis. The predicted enzymes were two cytochrome P450 monooxygenases (CalE and CalL), a bifunctional CYP-P450 monooxygease fused with a CYP-P450 reductase domain (CalG), three dehydrogenases (CalF, CalI and CalM), a methyltransferase (CalH), an enoyl reductase (CalK) and a beta lactamase (CalJ). In addition, two of the proteins included domains indicative of a MFS transporter (CalB) and a GAL4-like Zn(II)2Cys6 transcription factor (CalC), respectively. Analysis of the proteins encoded upstream of the PKS in *A. aculeatus* revealed two proteins (a putative methyl transferase and a short-chain dehydrogenase/reductase) that could be part of a biosynthetic gene cluster. However, these genes are present in multiple copies in the genomes of the two other species and hence likely not involved in calbistrin biosynthesis (Additional file 1: Additional information 6).

Moreover, a BLASTP analysis with the *P. decumbens* CalA-CalM proteins revealed that CalA-CalM showed high identities not only with proteins from *A. versicolor* and *A. aculeatus*, but also with several proteins from *Colletotrichum tofieldiae* and *Colletotrichum chlorophyti*. An additional gene synteny analysis with scaffold 170 (accession LFIV01000170.1) of *C. tofieldiae* revealed the presence of a similar cluster in *C. tofieldiae*, but several rearrangements in the order of the genes (Additional file 1: Additional information 7). All predicted proteins in the calbistrin cluster, except for CalJ, were found to have a homologue in the *C. tofieldiae* cluster.

**Table 1 Putative proteins within the calbistrin cluster in *P. decumbens***

| Name | *P. decumbens locus* | Size (aa) | BLASTP *Aver* | %I | BLASTP *Aacu* | %I | Conserved domain and notes | E-value |
|---|---|---|---|---|---|---|---|---|
| CalA | PENDEC_c013G00595 | 2910 | OJJ08178.1 | 85.3 | XP_020058113.1 | 78.1 | PKS: KS, AT, ACP, DH, Cmet, KR, R<br>Note: similar to MlcA PKS | 0.0 |
| CalB | PENDEC_c013G07044 | 562 | OJJ08177.1 | 84.9 | XP_020058136.1 | 79.4 | TIGR00711, drug resistance transporter,<br>Note: similar to MlcE MFS pump | 4.1E−40 |
| CalC | PENDEC_c013G06298 | 426 | OJJ08176.1 | 74.6 | XP_020058137.1 | 51.9 | smart00066, GAL4-like Zn(II)2Cys6 DNA-binding domain | 3.3E−05 |
| CalD | PENDEC_c013G04601 | 494 | OJJ08174.1 | 89.6 | XP_020058121.1 | 76.1 | No putative conserved domains detected. | |
| CalE | PENDEC_c013G04259 | 494 | OJJ08173.1 | 77.2 | XP_020058122.1 | 49.8 | pfam00067, Cytochrome P450<br>Note: similarity to MlcC monooxygenase | 1.8E−36 |
| CalF | PENDEC_c013G03789 | 575 | OJJ08172.1 | 85.4 | XP_020058123.1 | 75.4 | COG0277, FAD/FMN-containing dehydrogenase<br>Note: similar to the bifunctional Sol5 flavin-dependent oxidase and alderase from *Alternaria solani* | 2.7E−22 |
| CalG | n/a | 1056 | OJJ08171.1 | 84.9 | XP_020058124.1 | 72.8 | pfam00067, Cytochrome P450, + CYPOR<br>Bifunctional: N-term cytochrome P450 and C-term cytochrome P450 reductase domains | 2.5E−78 |
| CalH | PENDEC_c013G02261 | 273 | OJJ08170.1 | 82.6 | XP_020058125.1 | 61.2 | pfam08242, SAM dependent methyltransferase<br>Note: similarity to C-MET domain found in HR-PKSs: FUM1, EasB, LepA, ApdA and AzaB | 1.6E−20 |
| CalI | PENDEC_c013G00477 | 383 | OJJ08169.1 | 82.7 | XP_020058126.1 | 72.4 | PRK06196, oxidoreductase (dehydrogenase) | 1.1E−75 |
| CalJ | n/a | 418 | OJJ08168.1 | 82.1 | XP_020058127.1 | 68.5 | pfam00144, Beta-lactamase (putative acyltransferase)<br>Note: similar to MlcH acyltransferase | 1.7E−33 |
| CalK | PENDEC_c013G03312 | 194 | OJJ08167.1 | 83.0 | XP_020058128.1 | 67.0 | cd08249, enoyl reductase like<br>Note: similar to MlcG ER | 3.5E−110 |
| CalL | PENDEC_c013G00617 | 568 | OJJ08166.1 | 88.1 | XP_020058129.1 | 78.2 | pfam00067, Cytochrome P450 | 5.7E−23 |
| CalM | n/a | 304 | OJJ08165.1 | 89.8 | XP_020058138.1 | 78.3 | PRK06180, short chain dehydrogenase | 1.5E−67 |

The gene names *calA–calM* were defined in this study. The PENDEC_XXXXX accession numbers are as in the original publication of the genome, except for *calG*, *calJ* and *calM*. These new gene models were constructed using Softberry FGENESH supported with homologous genes in *A. versicolor* (*Aver*) and *A. aculeatus* (*Aacu*) (see Additional file 1: additional information 16 for protein sequences of *P. decumbens* CalG, CalJ and CalM proteins). Putative homologues of each of the *P. decumbens* CAL protein in *A. aculeatus* and *A. versicolor* were identify by BLASTP are here presented with accession umber, along with % identity at amino acid level (%I) along with the predicted conserved domains found in the protein and E-value for this prediction

The putative calbistrin cluster was further analysed for co-expression with the aim of identifying the boundaries of the cluster. Transcriptomics data (RNA-seq) of *P. decumbens* grown in liquid CM, supporting calbistrin production, was compared with that of *P. decumbens* grown in liquid DM where calbistrin is not produced (unpublished). The resulting log2 fold change plot showed that all 13 predicted genes in the putative cluster were upregulated in CM compared to DM (Fig. 3b and Additional file 1: additional information 8), while neighbouring genes did not show differential expression. This further strengthened the hypothesis of the proposed boundaries of the cluster.

### The transcription factor CalC is required for calbistrin production

One of the encoded proteins in the *P. decumbens* cluster, CalC, was predicted to include an N-terminal located GAL4-like Zn(II)2Cys6 binuclear zinc cluster DNA-binding domain and a C-terminally located fungal specific transcription factor domain, a domain architecture typically found in secondary metabolite gene cluster specific transcription factor (TF) [27]. Targeted deletion of the gene *calC*, using CRISPR/Cas9, and metabolic profiling of the resulting mutant (Δ*calC*) revealed a similar chemical profile to that of the PKS deletion mutant: a complete disappearance of calbistrins and related compounds (Fig. 4). The deletion did not affect the production of non-related compounds, such as andrastin C, suggesting that the CalC TF is only regulating the transcription of a limited number of genes rather than secondary metabolism in general, as observed for other PKS cluster specific TFs. The function of CalC as an activating transcription factor controlling the calbistrin cluster was further supported by a qPCR based expression analysis of the *calA*, *calB*, and *calF* genes, which showed that deletion of CalC resulted in a significant downregulation of the tree analysed genes (*calA*, *calB* and *calF*) in the cluster (Fig. 5).

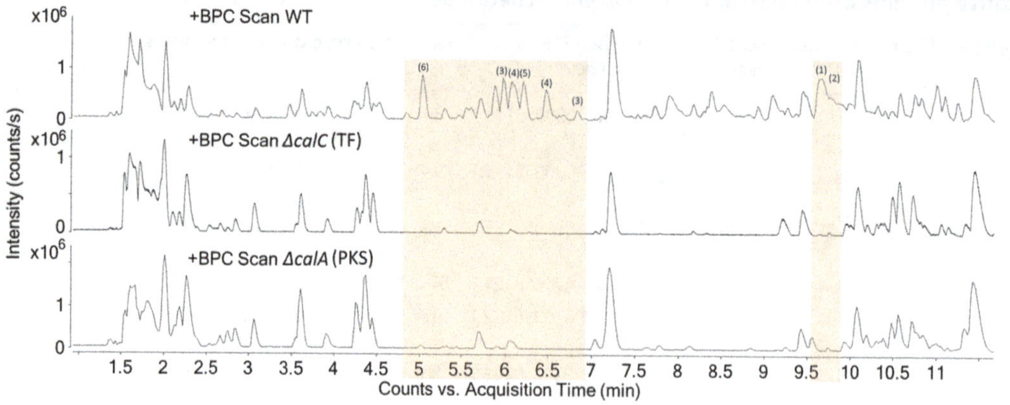


**Fig. 4** Comparison of UHPLC-HRMS results of *P. decumbens* ΔTF and *P. decumbens* ΔPKS compared to WT. Base peak chromatograms (BPCs) of *P. decumbens* WT, *P. decumbens* ΔTF and ΔPKS

**The MFS transporter CalB is involved in calbistrin export**

Targeted deletion of *calB*, encoding a predicted major facilitator superfamily transporter and HPLC-HRMS based profiling of the extracellular secondary metabolites produced in CM broth after day 5 and 7 showed an almost complete absence of calbistrin A and calbistrin C, a decreased abundancy of decumbenone A, B and C to 20–60% of the wild type levels and increased amounts of the linear moiety (Fig. 6). This suggests that CalB is involved in export of calbistrin A and related metabolites containing a decalin moiety. Analysis of the transcriptional response of *calA*, *calC* and *calF* in the Δ*calB* background indicated an earlier decrease in transcription for *calA* and a moderate log2 fold change (log2FC) in expression of 1 for *calC* and *calF* (Fig. 5b), suggesting that the lack of calbistrin export and consequently a putative intracellular increase did not strongly impact the expression of these 2 genes. The need for active transport is likely due to the dioic acid moiety that increases the molecule size of calbistrin and causes changes in surface charge distribution, reducing the likelihood of a partial non calB-dependent transport or passive leakage out of the cells across the membrane as observed with remaining amounts of decumbenones and the linear moiety in the broth after transporter deletion.

**Search for the second PKS required for calbistrin production**

Calbistrin is predicted to consist of two individually formed polyketide chains [8] that differ both in their length and decoration pattern, requiring the activity of two independent polyketide synthases as seen in statin biosynthesis. The high similarity between the KS domain of CalA and other known decalin producing PKS systems strongly indicate that CalA is responsible for biosynthesis of the decalin moiety, while the linear moiety must be produced by a second unknown PKS encoded by a gene located elsewhere in the genome. However, surprisingly deletion of *calA* did not only result in the inability to produce the decalin containing metabolites (calbistrin A, B, decumbenone A, B, C), but also hampered production of the linear dioic acid moiety, suggesting an inaction of the unknown PKS. Similar shutdown of entire biosynthetic pathways has been observed for other secondary metabolite cluster and pathways, e.g. bikaverin biosynthesis in several *Fusarium* species [28], where deletion of structural genes can result in the transcriptional down regulation of the remaining genes in the cluster. The molecular basis for such down regulations is currently unknown, but may be utilized to identify unknown components of a biosynthetic system. Therefore, we performed a qPCR expression analysis of the three PKS candidates (*PdecPKS3*, *PdecPKS6*, *PdecPKS7*) for the unknown dioic acid forming activity in the TF deletion strain (Δ*calC*) and in the MFS deletion strain (Δ*calB*) which was still able to produce all intermediates but performed poorly in export of calbistrin A and B. The analysis showed that the expression of the three PKS encoding genes did not change dramatically, less than two fold, in neither of the two strains (Fig. 7) suggesting that they are most likely are not responsible for forming the linear moiety. Targeted deletion of *pdecPKS6* and chemical analysis of the mycelium and agar-plug extracts confirmed this conclusion for this gene as no change in calbistrin-associated secondary metabolites were detected (data not shown). However, it cannot be conclusively excluded that *PdecPKS7* and *PdecPKS3* based on the presented data and it is possible that formation of the dioic acid occurs in an alternative fashion independent of PKSs.

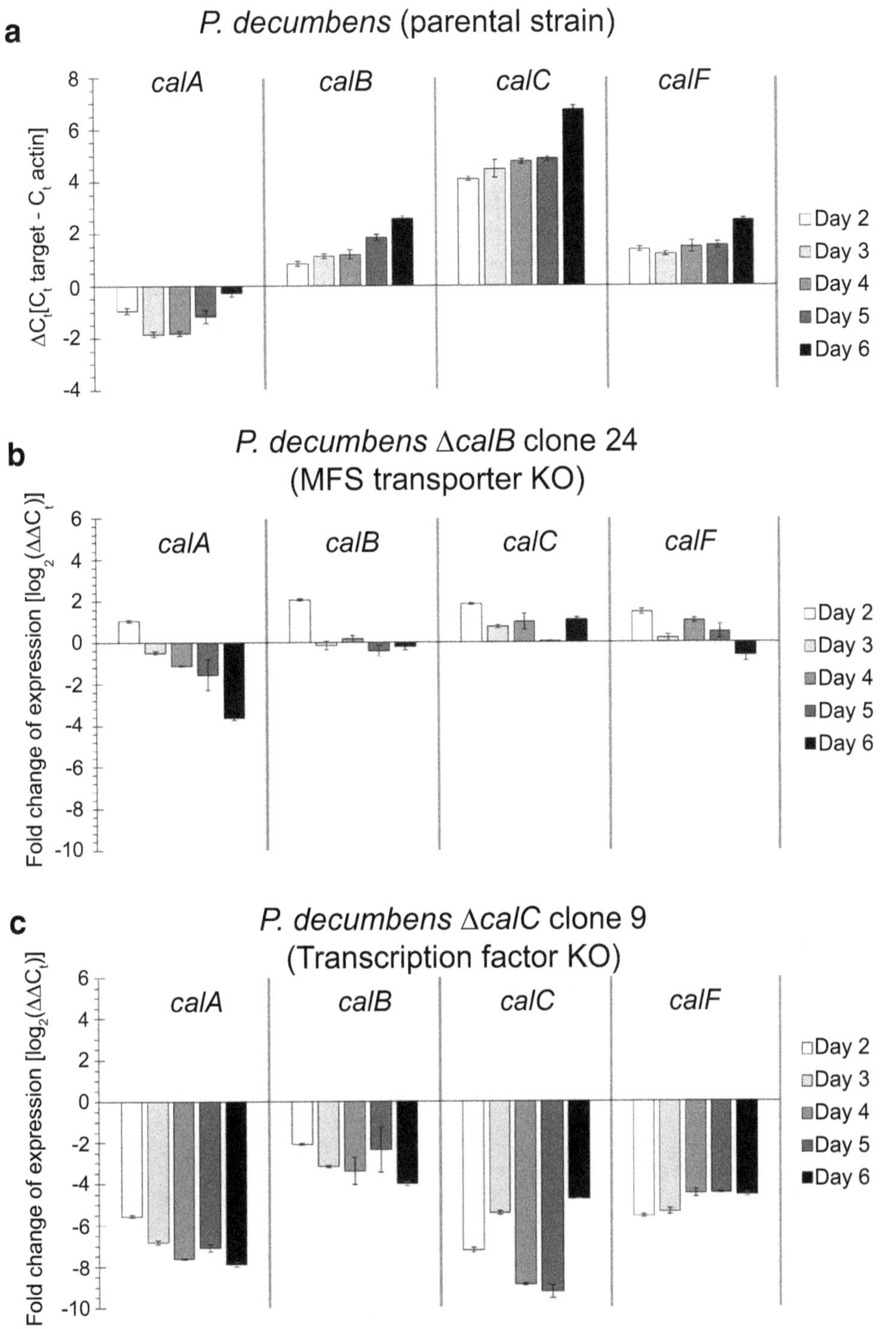


**Fig. 5** Gene expression profiles of *P. decumbens* parental and loss-of-function strains grown in liquid CM. **a** Gene expression of *calA*, *calB*, *calC* and *calF* in wild type *P. decumbens* strain relative to actin. Data are averages from two independent grown flasks analyzed in two technical duplicates. **b** Gene expression profile of *calA*, *calB*, *calC* and *calF* in *P. decumbens* Δ*calB*—loss-of-function strain relative to the wild type strain. **c** Gene expression profile of *calA*, *calB*, *calC* and *calF* in *P. decumbens* Δ*calC*—loss-of-function strain relative to the parental strain

| Metabolite | | Linear moiety | | Decumbenone A | Decumbenone B | | Decumbenone C | | | | Calbistrin A | | Calbistrin C | | Versiol |
|---|---|---|---|---|---|---|---|---|---|---|---|---|---|---|---|
| Adduct | | [M-($H_2O$)+H] | [M-2($H_2O$)+H] | [M+Na] | [M-($H_2O$)+H] | [M+Na] | [M+Na] | [M+K] | [M-($H_2O$)+H] | [M-2($H_2O$)+H] | [M+Na] | [M-($H_2O$)+H] | [M-($H_2O$)+H] | [M+H] | [M+N] |
| m/z | | 263.127 | 245.117 | 303.158 | 265.180 | 305.172 | 321.168 | 337.142 | 281.174 | 263.165 | 263.262 | 523.270 | 525.284 | 263.160 | 285.146 |
| Wild type | Day 3 | 227,940 | 83,067 | 266,850 | 882,043 | 257,323 | 100,909 | 38,326 | 61,696 | 537,718 | 357,616 | 776,395 | 24,323 | 21,369 | 408 |
| | Day 5 | 121,468 | 445,184 | 756,560 | 2,698,623 | 734,737 | 202,760 | 71,142 | 145,875 | 1,073,454 | 661,205 | 2,854,453 | 222,723 | 146,301 | 1,000 |
| | Day 7 | 1,968,779 | 714,996 | 899,452 | 1,395,778 | 332,906 | 197,165 | 67,854 | 155,566 | 1,208,333 | 528,964 | 2,010,466 | 196,002 | 115,686 | 1,128 |
| Δ*calB* clone24 (MFS) | Day 3 | 1,493,686 | 551,187 | 672,048 | 668,002 | 227,921 | 96,061 | 38,282 | 56,552 | 333,701 | 53,901 | 197,962 | n.d. | 6,991 | 661 |
| | Day 5 | 2,584,870 | 925,101 | 797,106 | 1,009,247 | 319,248 | 104,486 | 40,072 | 68,105 | 438,985 | 16,553 | 66,774 | 965 | 4,184 | 1,244 |
| | Day 7 | 3,947,011 | 1,314,112 | 475,997 | 243,076 | 68,626 | 135,785 | 47,542 | 94,718 | 737,908 | 2,606 | 14,544 | n.d. | 1,550 | 2,044 |
| Log2 Δ*calB*/wt | Day 3 | 2.7 | 2.7 | 1.3 | -0.4 | -0.2 | -0.1 | 0.0 | -0.1 | -0.7 | -2.7 | -2.0 | n/a | -1.6 | 0.7 |
| | Day 5 | 1.1 | 1.1 | 0.1 | -1.4 | -1.2 | -1.0 | -0.8 | -1.1 | -1.3 | -5.3 | -5.4 | -7.9 | -5.1 | 0.3 |
| | Day 7 | 1.0 | 0.9 | -0.9 | -2.5 | -2.3 | -0.5 | -0.5 | -0.7 | -0.7 | -7.7 | -7.1 | n/a | 6.2 | 0.9 |

Low High

**Fig. 6** Heat map of tracked masses in MS analysis of *P. decumbens* strains grown in liquid CM. To account for growth differences between strains, peak areas were corrected by CDW. Changes in corrected peak areas of calbistrin A and related compounds in the KO strain of *calB* were compared to the wild type strain. Reduced abundancy of calbistrin A suggests that calB is required for efficient excretion of calbistrins

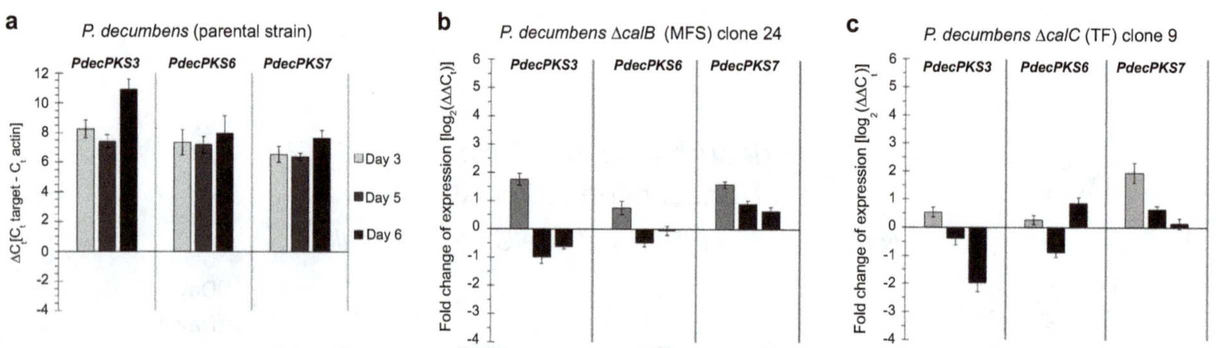


**Fig. 7** Expression analysis of putative candidate PKSs for production of the linear moiety of calbistrins. **a** Gene expression relative to actin for 3 putative PKS capable of performing a C-methylation in the *P. decumbens* wild type strain. **b** and **c** Gene expression changes in loss-of-function mutants of Δ*calB* and Δ*calC*, respectively, compared to the wild type strain. No complete absence of expression was detected in either of the deletion strains, suggesting that none of the PKSs are transcriptionally controlled by molecules from the calbistrin pathway

## Discussion

Comparative genomics analysis of three species producing the bioactive secondary metabolite calbistrin led to the identification of a partly reducing PKS (Fig. 2), that proved to be involved in calbistrin production in *P. decumbens* (Fig. 1). Further comparative analysis identified a region consisting of 13 genes that was shared between the three species. In *P. decumbens* this was a continuous region, while the syntenic region was disrupted in *A. versicolor* by a single gene and in *A. aculeatus* by two regions of seven and four genes, respectively (Fig. 3). In all cases, antiSMASH predicted larger clusters than what was predicted via the synteny based comparative analysis (34 vs. 13 genes in P. *decumbens*, 19 vs. 14 in *A. versicolor* and 33 vs. 23 in *A. aculeatus*). However, the smaller cluster predicted by the synteny analysis was supported by RNA-seq data in *P. decumbens* which showed co-expression of the 13 genes.

Deletion of the PKS encoding gene *pdecPKS10* in *P. decumbens* eliminated calbistrin production proving its involvement in the biosynthesis of calbistrin. However, calbistrin consists of two polyketides, one decalin containing 14 carbon backbone and one linear 12 carbon backbone, and is therefore predicted to be synthesized by two polyketide synthases [8]. Besides the absence of the calbistrins and the putative decalin containing precursors decumbenone A-C, formation of the putative dioic acid moiety was also absent in the *PdecPKS10* deletion strain, producing a situation that made it impossible to conclusively determine whether PdecPKS10 is responsible for synthesis of the decalin or the dioic acid moiety of calbistrin.

However, based on the high sequence identity of the PdecPKS10 KS-AT domains to that of other known decalin forming PKSs, such as MlcA, LovB, EqxS, Sol1 and Fsa1, we suggest that CalA is responsible for forming the decalin moiety (Fig. 2). This hypothesis is strengthened by the reductase (R) domain predicted at the C-terminal end of CalA. The decalin containing decumbenones have a terminal aldehyde instead of the carboxylic acid usually obtained from a classical thioesterase (TE) based release mechanism, and the R domain in CalA could be responsible for reducing the thioester bond to release the product as the observed aldehyde. Resembling the product release mechanism reported for the PKS-NRPS hybrids MycB and EqxS/Fsa1 that both includes terminal reductase domains resulting in the formation of terminal aldehyde groups in the products [23, 24]. A situation that differs markedly from the LovB PKS that does not include TE or R domains, but instead dependent on the trans-acting thioesterase LovG for product release [29].

Calbistrin includes fully reduced ketide units and one would hence expect the involved PKS to include an enoylreductase (ER) domain, however, the identified CalA lacks this domain. Nonetheless, one gene within the calbistrin cluster, *calK*, is predicted to have an ER conserved domain. The involvement of a trans-acting ER is also seen in the lovastatin/compactin, myceliotheramophin and equisetin biosynthesis, where the PKSs contains an inactive ER domain and reduction of the backbone is catalysed by an trans-acting accessory enzyme, LovC in lovastatin biosynthesis [30, 31]. As CalK belongs to the same family of enoylreductases as LovC (conserved protein domain family accession cd08249: enoyl_reductase_like) it could potentially be responsible for carrying out this reductive step on the growing calbistrin polyketide chain.

The enzymatic basis for [4+2] cycloaddition that leads to formation of decalin structures differs significantly between fungal systems and while the statin-forming PKS have been shown to catalyse the reaction themselves [31], other systems depend on trans-acting alderases that act on the polyketide chains following release from the PKS. A search for homologs of the monofunctional alderases Fsa2 and MycB in *P. decumbens* did not return any significant hits, however a search for the bifunctional Sol5 revealed CalF as a significant hit. Sol5 from *A. solani* is a bifunctional flavin-dependent oxidase and Diels-Alderase responsible for catalysing the cycloaddition in solanapyrone [21]. Based on the high level of similarity between CalA and CalF to the enzymes in the salanapyrone pathways and we hence hypothesise that the decalin part of calbistrins is formed via a similar mechanism. The decalin polyketide backbone includes two C-methyl groups, at C7 and C11 in backbone, of which the C7 positions is similarly to what is seen in compactin, where it is known to be added by the PKSs C-methyltransferase domain. A candidate for adding the methyl group at C11, if not done by CalA, is CalH that resembles the C-methyltransferase domains found in the FUM1 (fumonisin), EasB (Emericellamide), LepA (leporins), ApdA (Aspyridones) and AzaB (Azaphilone) PKSs (Table 1).

The genes found upstream of the PKS *calA* gene encodes several tailoring enzymes that potentially could be involved in the modification of the decalin polyketide product (Table 1). This includes three P450 monooxygenases (CalE, CalG and CalL), of which one might be responsible for the introduction of the extra hydroxyl group attached to the backbone of the decalin moiety, at position C9 in the backbone, that allows for attachment of the linear moiety. One tailoring enzyme activity that is expected to be involved in biosynthesis of calbistrin is an acyltransferase for connecting the two polyketide synthase products, such as seen in lovastatin biosynthesis, where the acyltransferase LovD is involved in transferring the polyketide chain from the PKS LovF to the finished polyketide product from the PKS LovB [32]. Blasting of the LovD protein sequence against the predicted *P. decumbens* proteins resulted in the identification of four proteins (E-value below $1 \times 10E{-}38$), of which CalJ had the highest level of sequence identity, of 33%, to LovD of the four hits. CalJ was initially predicted to be an acyltransferase, as the conserved domain with the highest score was a beta lactamase domain. However, this was also the case for LovD which previously has been experimentally proved to act as an acyltransferase. Similarly, it has been demonstrated that EstB, a protein related to beta-lactamases, lacked ß-lactamase activity but instead act as a acyltransferase in the bacteria *Burkholderia gladioli* [33].

The calbistrin cluster identified in this study potentially encodes many of the enzymatic activities predicted to be required for de novo synthesis of calbistrin. However, explaining synthesis of the linear moiety remains a challenge. The first obvious hypothesis for a second PKS responsible for the biosynthesis of the linear moiety would be the presence of another PKS in the close genomic vicinity of *calA* (*PdecPKS10*), similarly to the situation described for the PKSs involved in lovastatin and compactin formation. However, no other PKS was predicted on *P. decumbens* scaffold 13, which suggest that the calbistrin pathway may be encoded by several different loci in the genome. The *P. decumbens* genome is only predicted to encode a total of ten PKSs, of which one was predicted to be responsible for YWA synthesis (Pdec-PKS19), one for andrastin A synthesis (PdecPKS2), and one was found to be involved in calbistrin synthesis in this study (PdecPKS10). The structure of both the decalin and the linear moieties suggest that they undergo a C-methylation of the backbone chain during synthesis, with the decalin possessing two methyl groups and three are present in the linear structure. The *P. decumbens* genome included other than PdecPKS10, three putative PKSs with a predicted C-methylation domain: PdecPKS6, PdecPKS7 and PdecPKS3. Another option is that the C-methylation is catalysed by a post-PKS tailoring enzyme. The gene located on the genome next to *PdecPKS4* was annotated as a putative methyl transferase, and thus could possibly perform a post-PKS C-methylation reaction. Based on our data, we can exclude PdecPKS6 as being responsible for formation of the linear moiety based on targeted deletion. Further investigation of calbistrin biosynthesis could therefore focus on the deletion of *PdecPKS3*, *PdecPKS4* and PdecPKS7, and evaluate their role in biosynthesis of the linear moiety.

Deletion of the predicted transcription factor encoding gene *calC* resulted in the abolishment of the production of calbistrin and its related metabolites, proving involvement of CalC in calbistrin formation (Fig. 4) by regulating expression of *PdecPKS10*. Comparison of the *calC* mutant metabolite profile revealed that it was very similar to that of the PKS deletion strain, suggesting that the transcription factor regulates the cluster, and does not act as a global regulator. Indeed, GAL4-like type of transcription factors are the most common type of in-cluster pathway regulators in fungi [27]. To further investigate the influence of the transcription factor on the calbistrin cluster, expression of several genes in the cluster was compared between the wild-type and the ΔTF strains (Fig. 5). The observation that the final product calbistrin, the decalin intermediates as well as the linear dioic acid intermediates disappeared similarly upon deletion of the PKS and the transcription factor is interesting.

One speculation could be the existence of a negative feedback mechanism triggered by the absence of the decalin intermediates results in the shut-down of the biosynthetic pathway of the linear intermediate, either at enzymatic or gene expression level. Alternatively the lack of the decalin metabolite in the cell results in a situation where the activity of the PKS that forms the linear polyketide is inhibited as it is unable to unload its formed product due to a lack of the decalin reaction partner. Another possibility for the synthesis of the dioic (dicarboxylic) acid would be the oxydation of a free long-chain fatty acid to a ω-hydroxy acid via a cytochrome P450 monoxygenases [34] and subsequent oxidation via alcohol and aldehyde dehydrogenases [35]. Indeed, the putative cluster contains three genes with a putative P450 monoxygenase function (CalE, CalG and CalL), however, this scenario is very speculative as the dioic acid moiety of calbistrin is branched and desaturated, requiring intensive enzymatic activity to source this molecule via the free fatty acid biosynthetic pathway. In contrast, one could argument that the premature release product of CalA could be the starter unit for a P450 monoxygenase and does not undergo cyclysation in this case.

Deletion of the predicted MFS transporter gene *calB* resulted in a strong decrease (log2FC of −6.2 to −7.7 on day 5) of extracellular calbistrin A and C levels and a modereate decrease (log2FC of −0.5 to −2.3 on day 5) of extracellular decumbenone A, B and C, suggesting that calB is involved in export of both decumbenones and calbistrin, however the latter molecules do seem to get exported to some extent via other less specific transport routes, as their level did not decrease as strongly as the level of calbistrin A and C. Similar observations were made for export of andrastatin A in *Penicillium roqueforti* [36] and bikaverin in *Fusarium fujikuroi* [28] when their respective transporter was downregulated or deleted, respectively. Although we did not perform analysis of intracellular accumulation of calbistrin A and C, we can conclude from our expression data that a possible feedback on the transcript level due to accumulation of calbistrin A, C or its pathway intermediates did not occur for *calA*, *calC* and *calF* as the expression levels only changed modestly, making it unlikely that complete abolishment of expression takes place. However, it still remains a possibility that enzymes not analyzed for their expression in this study show a stronger response or the inhibition takes place on the enzyme level.

Future studies could also look beyond calbistrins as molecule class and investigate the mode of action and benefits for the producers of the related decumbenones which were shown to inhibit melanisation of *Magnaporthe grisea* [37] and stimulating the germination of agricultural plants [38]. As our study also identified a potential calbistrin cluster in the root endophyte *Colletotrichum tofieldiae* which supports growth of *Arabidopsis thalina* under low phosphate conditions [39], it might be worth to investigate the production and role of decumbenones and calbistrins in these interactions and wether they are valuable for fending off other soil-thriving fungi or promoting growth of the host plant.

## Conclusions

This study identified a 13-membered gene cluster in *P. decumbens* required for biosynthesis of the structurally interesting polyketide calbistrin. Targeted deletion of three of the identified genes, namely *PdecPKS12* (*calA*), *calB* and *calC*, proved their involvement in the formation of calbistrins, as a polyketide synthase, a pump and a positively acting pathway specific transcription factor, respectively. The identified Cal-cluster encode many of the required enzyme types predicted to be essential for de novo calbistrin biosynthesis, however, the enzyme(s) responsible for formation of the linear moiety remains elusive and further work is hence needed to allow for the future construction of a high yielding calbistrin cell factory.

## Methods

### Strains and media

*Penicillium decumbens* strain IBT11843 was obtained from and is available at the IBT culture collection (Department of Biotechnology and Biomedicine, Technical University of Denmark).

For chemical analysis, strains were grown either on liquid or solidified Czapek yeast autolysate medium (CM) containing (30 g/l sucrose, 5 g/l yeast extract, 3 g/l $NaNO_3$, 1.0 g/L $K_2HPO_4$, 0.5 g/l $MgSO_4 \cdot 7\ H_2O$, 0.5 g/l KCl, 0.01 g/l $FeSO_4$,·7 $H_2O$, 20 g/l agar and 1.0 mL trace metal solution containing 0.1 g/l $ZnSO_4 \cdot 7H2O$ and 0.05 g/L $CuSO_4 \cdot 5\ H_2O$, the pH was adjusted to 6.2 with NaOH.

For transcriptome data referred to in this study, cultivations were performed in CM and defined medium (DM) as described previously [40]. For preparing protopalasts for transformation of *P. decumbens*, YGG medium was used for cultivation as described previously [41].

### Bioinformatic analysis

Genome sequences from *P. decumbens* IBT 11843 (accession MDYL00000000) [10], *A. aculeatus* ATCC 16872 (accession MRCK00000000.1) [42] and *A. versicolor* CBS 583.65 (accession MRBN00000000) [42] were obtained from GenBank.

To identify biosynthetic gene clusters (BGCs) in *P. decumbens*, the genome was analysed via the AntiSMASH (v.3.0.4) server, resulting in the prediction of in total 22 putative BGCs, of which nine included PKS encoding genes. A previous analysis of 24 genome sequenced *Penicillium* species, showed that these in average encoded 17.2 PKS BGCs [10]. The low number of identified PKS encoding genes in *P. decumbens* prompted us to perform an additional BLAST based search for PKS encoding genes that may have been missed in the first round of automated analysis. The manual analysis was performed using the β-ketosynthase (KS) domain from the YWA producing PKS (accession XP_002568608) from *Penicillium rubens* Wisconsin 54-1255 as query in a TBLASTN search against a database containing the translations of the *P. decumbens* whole genome sequence in all six open reading frames and a BLASTP search against a database containing all predicted proteins in the *P. decumbens* genome. Full length protein sequences for hits with an e-value below 1e-6 in the BLASTP analysis were retrieved and annotated using the NCBI Conserved Domain Database [43]. This resulted in the identification of one additional highly reducing PKS, bringing the total to five highly reducing PKSs (HR-PKSs), one partially reducing PKS (PR-PKS), two non-reducing PKSs (NR-PKSs), and two partially reducing PKS-nonribosomal peptide synthetase hybrids (PR-PKS-NRPS).

CLC main Workbench version 7 (QIAGEN Bioinformatics) was used for local BLAST analysis, protein alignment and neighbor joining tree creation. The amino acid sequences of the PKSs for all organisms were trimmed to the KS-AT domains, which are the only universal domains of PKSs and have previously been shown to be a good evolutionary determinant [10, 44]. Phylogenetic trees were exported to the iTOL v3 tool for manual annotation and visualization [45]. Gene predictions in *P. decumbens* were performed using FGENESH (Softberry). Functional conserved domains in the translated protein sequences were predicted using Conserved Domain Search (NCBI). Analysis of syntenic regions was done using the python application Easyfig [46].

RNA-seq data were obtained from Jens Nielsen's lab at Chalmers University (Nielsen et al., unpublished). Raw reads were mapped to the *P. decumbens* reference genome (accession MDYL00000000) using TopHat2 (v. 2.0.9) [47], and gene read counts were quantified using FeatureCount [48], both with default parameters. Differential expression analysis was computed for complex medium relative to defined medium using DESeq 2 [49].

### Fungal transformation and gene disruption in *P. decumbens*

Protoplasts of *P. decumbens* were prepared 48 h after inoculation of $5 \times 10^5$ spores/ml in YGG medium using the methods and media described previously [41] with the following modifications: we reduced the incubation time in glucanex solution (30 mg/ml in KC Buffer) to 75 min, as longer incubation reduced the number of recovered colonies (For an overview of conducted transformations for this publication, see Additional file 1: Additional information 5).

To establish which dominant selection markers can be used for *P. decumbens*, protoplast were initially plated on [41] containing either 0.1% acetamide (Sigma Aldrich, NL) as the sole nitrogen source or 40 mM sodium nitrate and one of the following selection agents: 50 µg/ml phleomycin (Invivogen, USA) or 1.2 µg/ml terbinafine (Terbinafine Hydrochloride, Sigma Aldrich, NL).

In contrast to a lack of inhibition on acetamide plates (due to activity of host acetamidase genes), robust inhibition of growth was observed on plates with phleomycin and terbinafine. Low inhibitory concentration of terbinafine have been previously reported by Sigl et al. [50] for *Penicillium chrysogenum*. As terbinafine acts as an inhibitor of squalene epoxidase in a broad range of fungi and is also convenient from an economic point of view, we used the MoClo modular cloning system [51] to construct an *ergA* overexpression cassette utilizing the widely used pgpdA promoter from *Aspergillus nidulans* (Additional file 1: Additional information 9) and the squalene epoxidase *ergA* from *Penicillium chrysogenum* to build pCP-AMA-ergA, which was utilized when deleting *calC* and *calB*.

For protospacer selection, sgRNA synthesis and RNP delivery we used the methods described in [26] with an additional filtering for highly active protospacers using sgRNA scorer 2.0 [52]. For selection of protoplasts competent in taking up DNA (and presumably other macromolecules such as RNPs), either 3.0 µg pJAK-109 [26] or pCP-AMA-ergA were co-transformed along with RNPs and protoplasts were plated on protoplast recovery plates supplemented with phleomycin or terbinafine and 40 mM sodium nitrate. Plates were incubated for up to 7 days at 25 °C to allow recovery of transformants and formation of colonies.

Colonies were screened by colony PCR using Phire Green 2x Mastermix (Thermo Scientific, The Netherlands) and initial anaylsis of band size shifts on 1% agarose gels. To determine length and location of insertions or deletions (Additional file 1: Additional information 12–15) Sanger sequencing (Macrogen, The Netherlands) of PCR products was performed.

To loose AMA-plasmids obtained during transformation, spores were harvested and diluted out on nonselective R-Agar [41] followed by colony PCR. This procedure was repeated twice. A list of all sgRNAs and primers used in this study can be found in Additional file 1: Additional information 10 and 11, respectively.

#### qPCR analysis of calA, B, C and M in *P. decumbens*

For qPCR analysis of the calbistrin cluster genes in *P. decumbens*, we choose a single Δ*calB* and Δ*calC* clone and 3 biological replicates of the parental strain. 1 ml of a spore solution (1x10$^6$ spores/ml) was used for inoculation of 25 ml liquid CM in 100 ml shake flasks. Cultures were grown for 7 days at 25 °C and 200 rpm. Mycelium for RNA extraction was separated from 5 ml broth by filtration, washed once with 2 volumes of ice-cold $H_2O$ and 100–200 mg wet biomass were mixed with 1 ml Trizol reagent (Thermo Fisher Scientific, The Netherlands), transferred to screw-cap tubes containing glass beads (diameter 0.75–1 mm) and stored at − 80 °C until RNA isolation. Mycelium was disrupted with a FastPrep FP120 system (Qbiogene, France) and total RNA was isolated using the Direct-zol RNA MiniPrep Kit (Zymo Research, USA). For cDNA synthesis, 1500 ng total RNA were reverse transcribed using the Maxima H Minus cDNA Synthesis Master Mix (Life Technologies, The Netherlands) in a volume of 20 µl. Samples were diluted with 80 µl MQ-$H_2O$ and 4 µl of this cDNA were used as input for qPCR in a final volume of 25 µl. As master mix for qPCR, the SensiMix SYBR Hi-ROX (Bioline Reagents, England) was used. All runs were performed on a MiniOpticon system (Bio-Rad). The following conditions were employed for amplification: 95 °C for 10 min, followed by 40 cycles of 95 °C for 15 s, 60 °C for 30 s and 72 °C for 30 s, following an acquisition step. Raw ct data were exported and analysis of relative gene expression was performed with the $2-\Delta\Delta CT$ method [53]. The expression analysis was performed with two technical duplicate per biological sample. The γ-actin gene (PENDEC_c001G04327) was used as internal standard for data normalization. The primers used for qPCR of *calA* (PENDEC_c013G00595), *calB* (PENDEC_c013G07044), *calC* (PENDEC_c013G06298), *calF* (PENDEC_c013G03789) and γ-actin are listed in (Additional file 1: Additional information 11).

#### Chemical analysis

For solid cultures, three agar plugs were sampled from one colony and 1.0 ml of extraction solvent, isopropanol:ethylacetate (1:3) containing 1% formic acid, was added. After ultra-sonication for 1 h the extract was transferred to a clean vial, evaporated to dryness and dissolved in 100 µl methanol. After centrifugation for 5 min the supernatant was directly used for chemical analysis.

Secondary metabolite analysis of solid culture samples was achieved by ultra-high performance liquid chromatography-diode array detection-quadrupole time of flight mass spectrometry (UHPLC-DAD-QTOFMS) on an Agilent 1290 UHPLC system (Agilent Technologies, Torrance, CA) coupled to an Agilent 6545 QTOF equipped with an electrospray ionization (ESI) source. True tandem MS/HRMS spectra were obtained at fixed collision-induced dissociation (CID) energies of 10, 20, and 40 eV

[54] and matched to the available reference standards of calbistrin A and andrastin C.

For analysis of cultures grown in liquid CM, broth and mycelium were separated by centrifugation for 10 min at 14,000 g, followed by filtration of the clarified broth over 0.2 μm PTFE syringe filters (VRW, The Netherlands). The obtained filtrate was directly used for analysis or frozen at − 20 °C. For analysis of liquid culture samples, high performance liquid chromatography electrospray-ionization high–resolution mass spectrometry (HPLC-ESI-HRMS) was conducted on an Accella1250 UPLC system coupled to an Orbitrap Exactive (Thermo Fisher Scientific, The Netherlands) with a scan range of m/z 100–1600. A sample of 10 μL was injected onto a Shim-pack XR-ODS C18 column (75 mm × 3.0 ID, 2.2 μm) (Shimadzu, Japan) kept at 40 °C and operated at a flow rate of 300 μL/min. Separation of compounds was achieved with the following solvents (A: 100% MQ-$H_2O$, B: 100% Acetonitrile, and C: 2% formic in MQ-$H_2O$ being constantly added at 5% to protonate molecules). After injection of sample, column was run for 5 min with isocratic flow at 5% B, following a linear gradient for 25 min reaching 95% B, remaining constant at 95% B for 5 min and equilibrating the column with initial conditions of 5% B for 5 min before injection of the next sample. Each sample was analyzed in technical duplicate. Total ion chromatograms and areas of m/z of interest were generated and processed using MassHunter (Agilent) and XCalibur (ThermoFisher) with a m/z error below 1 ppm for all molecules referred to in this study.

**Authors' contributions**

MW and RF conceived the study. SG performed the bioinformatics analysis under supervision of RF. CP performed the gene deletions in *P. decumbens*, shake-flask cultivations and expression analysis under supervision of YN and AJMD. SG performed the chemical analysis and received help of KFN, JCF and TOL. JCN performed the RNA-seq experiments. SG and CP wrote the manuscript under supervision of RF. All authors read and approved the final manuscript.

**Author details**

[1] Department of Biotechnology and Biomedicine, Technical University of Denmark, 2800 Kgs. Lyngby, Denmark. [2] Molecular Microbiology, Groningen Biomolecular Sciences and Biotechnology Institute, University of Groningen, 9747 AG Groningen, The Netherlands. [3] Department of Biology and Biological Engineering, Chalmers University of Technology, 412 96 Gothenburg, Sweden. [4] Novo Nordisk Foundation Center for Biosustainability, Technical University of Denmark, 2800 Kgs. Lyngby, Denmark.

**Acknowledgements**

The authors acknowledge Christopher Phippen, Aaron John Christian Andersen and Maaike de Vries for assistance with analytics.

**Competing interests**

The authors declare that they have no competing interests.

**Funding**

This work was supported by the European Commission Marie Curie Initial Training Network Quantfung (FP7-People-2013-ITN, Grant 607332) and the Novo Nordisk Foundation. Agilent Technologies is acknowledged for the Thought Leader Donation of the 6545UHPLC-QTOF.

**References**


1. van den Berg M, Albang R, Albermann K, Badger JH, Daran JM, Driessen AJM, et al. Genome sequencing and analysis of the filamentous fungus *Penicillium chrysogenum*. Nat Biotechnol. 2008;26(10):1161–8.
2. Pickens LB, Tang Y, Chooi Y-H. Metabolic engineering for the production of natural products. Annu Rev Chem Biomol Eng. 2011;2:211–36.
3. Nielsen JC, Nielsen J. Development of fungal cell factories for the production of secondary metabolites: linking genomics and metabolism. Synth Syst Biotechnol. 2017;2(1):5–12.
4. O'Connor SE. Engineering of secondary metabolism. Annu Rev Genet. 2015;49:71–94.
5. Jackson M, Karwowski JP, Humphrey PE, Kohl WL, Barlow GJ, Tanaka SK. Calbistrins, novel antifungal agents produced by *Penicillium restrictum*. I. Production, taxonomy of the producing organism and biological activity. J Antibiot (Tokyo). 1993;46(1):34–8.
6. Bertizal FK, Dombrowski AW, Helms GL, Horn WS, Jones ETT, Koupal L, et al. JPH08134059 (A)—Cholesterol lowering agent. 1991.
7. Bladt TT, Durr C, Knudsen PB, Kildgaard S, Frisvad JC, Gotfredsen CH, et al. Bio-activity and dereplication-based discovery of ophiobolins and other fungal secondary metabolites targeting leukemia cells. Molecules. 2013;18(12):14629–50.
8. Brill GM, Chen RH, Rasmussen RR, Whittern DN, McAlpine JB. Calbistrins, novel antifungal agents produced by *Penicillium restrictum* II. Isolation and elucidation of structure. J Antibiot (Tokyo). 1993;46(1):39–47.
9. Stewart M, Capon RJ, Lacey E, Tennant S, Gill JH. Calbistrin E and two other new metabolites from an Australian isolate of *Penicillium striatisporum*. J Nat Prod. 2005;68(4):581–4.
10. Nielsen JC, Grijseels S, Prigent S, Ji B, Dainat J, Nielsen KF, et al. Global analysis of biosynthetic gene clusters reveals vast potential of secondary metabolite production in *Penicillium* species. Nat Microbiol. 2017;2:17044.
11. Fujii Y, Asahara M, Ichinoe M, Nakajima H. Fungal melanin inhibitor and related compounds from *Penicillium decumbens*. Phytochemistry. 2002;60(7):703–8.
12. Endo A, Monacolin K. new hypocholesteroleic agent produced by a *Monascus species*. J Antibiot (Tokyo). 1979;32(8):852–4.
13. Alberts AW, Chen J, Kuron G, Hunt V, Huff J, Hoffman C, et al. Mevinolin: a highly potent competitive inhibitor of hydroxymethylglutaryl-coenzyme A reductase and a cholesterol-lowering agent. Proc Natl Acad Sci U S A. 1980;77(7):3957–61.
14. Brown BAG, Srnale TC, Pharmaceuticals B, Park B, King TJ, Hasenkamp R, et al. Crystal and molecular structure of compactin, a new antifungal metabolite from *Penicillium brevicompactum*. JCS Perkin. 1976;1:1165–8.
15. Frisvad JC, Filtenborg O. Terverticillate *Penicillia*: chemotaxonomy and mycotoxin production. Mycologia. 1989;81(6):837–61.
16. Endo A, Kuroda M, Tsujita Y. ML-236A, ML-236B, and ML-236C, new inhibitors of cholesterogensis produced by *Penicillium citrinum*. J Antibiot (Tokyo). 1976;29(12):1346–8.
17. Auclair K, Sutherland A, Kennedy J, Witter DJ, Van Den Heever JP, Hutchinson CR, et al. Lovastatin nonaketide synthase catalyzes an intramolecular diels—alder reaction of a substrate analogue. J Am Chem Soc. 2000;1(12):11519–20.
18. Horn WS, Bierilo KK, Bills GF, Dombrowski AW, Helms GL, Jones ET, et al. Characterization of the light- and base-mediated instability of calbistrin A. J Nat Prod. 1993;56(10):1779–85.
19. Petersen LM, Hoeck C, Frisvad JC, Gotfredsen CH, Larsen TO. Dereplication guided discovery of secondary metabolites of mixed biosynthetic origin from *Aspergillus aculeatus*. Molecules. 2014;19(8):10898–921.
20. Fukuyama K, Hamasaki T, Hatsuda Y, Hamasaki T, Nakagomi T, Fukuyama K, et al. Structure and absolute configuration of versiol, a metabolite from *Aspergillus versicolor*. J Chem Soc Perkin Trans 2. 1978. https://doi.org/10.1039/P29780000683.

21. Kasahara K, Miyamoto T, Fujimoto T, Oguri H, Tokiwano T, Oikawa H, et al. Solanapyrone synthase, a possible Diels–Alderase and iterative type I polyketide synthase encoded in a biosynthetic gene cluster from *Alternaria solani*. ChemBioChem. 2010;11(9):1245–52.
22. Kakule TB, Sardar D, Lin Z, Schmidt EW. Two related pyrrolidinedione synthetase loci in *Fusarium heterosporum* ATCC 74349 produce divergent metabolites. ACS Chem Biol. 2013;8(7):1549–57.
23. Kato N, Nogawa T, Hirota H, Jang J-H, Takahashi S, Ahn JS, et al. A new enzyme involved in the control of the stereochemistry in the decalin formation during equisetin biosynthesis. Biochem Biophys Res Commun. 2015;460(2):210–5.
24. Li L, Yu P, Tang M-C, Zou Y, Gao S-S, Hung Y-S, et al. Biochemical characterization of a eukaryotic decalin-forming Diels–Alderase. J Am Chem Soc. 2016;138(49):15837–40.
25. Campbell CD, Vederas JC. Biosynthesis of lovastatin and related metabolites formed by fungal iterative PKS enzymes. Biopolymers. 2010;93(9):755–63.
26. Pohl C, Kiel JAKW, Driessen AJM, Bovenberg RAL, Nygård Y. CRISPR/Cas9 based genome editing of *Penicillium chrysogenum*. ACS Synth Biol. 2016;5:754–64.
27. Yin W, Keller NP. Transcriptional regulatory elements in fungal secondary metabolism. J Microbiol. 2011;49(3):329–39.
28. Wiemann P, Willmann A, Straeten M, Kleigrewe K, Beyer M, Humpf HU, et al. Biosynthesis of the red pigment bikaverin in *Fusarium fujikuroi*: genes, their function and regulation. Mol Microbiol. 2009;72:931–46.
29. Xu W, Chooi YH, Choi JW, Li S, Vederas JC, Da Silva NA, et al. LovG: the thioesterase required for dihydromonacolin L release and lovastatin nonaketide synthase turnover in lovastatin biosynthesis. Angew Chemie Int Ed. 2013;52(25):6472–5.
30. Auclair K, Kennedy J, Hutchinson CR, Vederas JC. Conversion of cyclic nonaketides to lovastatin and compactin by a lovC deficient mutant of *Aspergillus terreus*. Bioorganic Med Chem Lett. 2001;11(12):1527–31.
31. Kennedy J, Auclair K, Kendrew SG, Park C, Vederas JC, Hutchinson CR. Modulation of polyketide synthase activity by accessory proteins during lovastatin biosynthesis. Science (80-). 1999;284(5418):1368–72.
32. Xie X, Meehan MJ, Xu W, Dorrestein PC, Tang Y. Acyltransferase mediated polyketide release from a fungal megasynthase. J Am Chem Soc. 2009;131(24):8388–9.
33. Petersen EI, Valinger G, Solkner B, Stubenrauch G, Schwab H. A novel esterase from *Burkholderia gladioli* which shows high deacetylation activity on cephalosporins is related to beta-lactamases and DD-peptidases. J Biotechnol. 2001;89(1):11–25.
34. Durairaj P, Malla S, Nadarajan SP, Lee P-G, Jung E, Park HH, Kim B-G, Yun H. Fungal cytochrome P450 monooxygenases of *Fusarium oxysporum* for the synthesis of ω-hydroxy fatty acids in engineered *Saccharomyces cerevisiae*. Microb Cell Fact. 2015;14:45.
35. Bowen CH, Bonin J, Kogler A, Barba-Ostria C, Zhang F. Engineering *Escherichia coli* for conversion of glucose to medium- chain ω-hydroxy fatty acids and α, ω-dicarboxylic acids. ACS Synth Biol. 2016;5:200–6.
36. Rojas-Aedo JF, Gil-Durán C, Del-Cid A, Valdés N, Álamos P, Vaca I, García-Rico RO, Levicán G, Tello M, Chávez R. The biosynthetic gene cluster for andrastin A in *Penicillium roqueforti*. Front Microbiol. 2017;8:813.
37. Fujiia Y, Asahara M, Ichinoe M, Nakajima H. Fungal melanin inhibitor and related compounds from *Penicillium decumbens*. Phytochemistry. 2002;60(7):703–8.
38. Anisimov MM, Chaikina EL, Afiyatullov SS, Zhuravleva OI, Klykov AG, Kraskovskaja NA, Aminin DL. Decumbenones A–C from marine fungus *Aspergillus sulphureus* as stimulators of the initial stages of development of agricultural plants. Agric Sci. 2012;3(8):1019–22.
39. Hiruma K, Gerlach N, Sacristán S, Nakano RT, Hacquard S, Kracher B, Neumann U, Ramírez D, Bucher M, O'Connell RJ, Schulze-Lefert P. Root endophyte *Colletotrichum tofieldiae* confers plant fitness benefits that are phosphate status dependent. Cell. 2016;165(2):464–74.
40. Grijseels S, Nielsen JC, Nielsen J, Larsen TO, Frisvad JC, Fog Nielsen K, et al. Physiological characterization of secondary metabolite producing *Penicillium* cell factories. Fungal Biol Biotechnol. 2017;4(8):1–12.
41. Kovalchuk A, Weber SS, Nijland JG, Bovenberg RAL, Driessen AJM. Fungal ABC transporter deletion and localization analysis. In: Bolton M, Thomma B, editors. Plant fungal pathogens methods in molecular biology, vol. 835. New York: Humana Press; 2012. p. 1–16.
42. de Vries RP, Riley R, Wiebenga A, Aguilar-Osorio G, Amillis S, Uchima CA, et al. Comparative genomics reveals high biological diversity and specific adaptations in the industrially and medically important fungal genus *Aspergillus*. Genome Biol. 2017;18(1):28.
43. Marchler-Bauer A, Derbyshire MK, Gonzales NR, Lu S, Chitsaz F, Geer LY, et al. CDD: NCBI's conserved domain database. Nucleic Acids Res. 2015;43(D1):D222–6.
44. Kroken S, Glass NL, Taylor JW, Yoder OC, Turgeon BG. Phylogenomic analysis of type I polyketide synthase genes in pathogenic and saprobic ascomycetes. Proc Natl Acad Sci. 2003;100(26):15670–5.
45. Letunic I, Bork P. Interactive tree of life (iTOL) v3: an online tool for the display and annotation of phylogenetic and other trees. Nucleic Acids Res. 2016;44(W1):W242–5.
46. Sullivan MJ, Petty NK, Beatson SA. Easyfig: a genome comparison visualizer. Bioinformatics. 2011;27(7):1009–10.
47. Kim D, Pertea G, Trapnell C, Pimentel H, Kelley R, Salzberg SL. TopHat2: accurate alignment of transcriptomes in the presence of insertions, deletions and gene fusions. Genome Biol. 2013;14(4):R36.
48. Liao Y, Smyth GK, Shi W. FeatureCounts: an efficient general purpose program for assigning sequence reads to genomic features. Bioinformatics. 2014;30(7):923–30.
49. Love MI, Huber W, Anders S. Moderated estimation of fold change and dispersion for RNA-seq data with DESeq2. Genome Biol. 2014;15(12):550.
50. Sigl C, Handler M, Sprenger G, Kurnsteiner H, Zadra I. A novel homologous dominant selection marker for genetic transformation of *Penicillium chrysogenum*: overexpression of squalene epoxidase-encoding ergA. J Biotechnol. 2010;150(3):307–11.
51. Weber E, Engler C, Gruetzner R, Werner S, Marillonnet S. A modular cloning system for standardized assembly of multigene constructs. PLoS One. 2011;6(2):e16765.
52. Chari R, Yeo NC, Chavez A, Church GM. SgRNA scorer 2.0: a species-independent model to predict CRISPR/Cas9 activity. ACS Synth Biol. 2017;6(5):902–4.
53. Livak KJ, Schmittgen TD. Analysis of relative gene expression data using real-time quantitative PCR and the 2(-Delta Delta C(T)) method. Methods. 2001;25(4):402–8.
54. Kildgaard S, Mansson M, Dosen I, Klitgaard A, Frisvad JC, Larsen TO, et al. Accurate dereplication of bioactive secondary metabolites from marine-derived fungi by UHPLC-DAD-QTOFMS and a MS/HRMS library. Mar Drugs. 2014;12(6):3681–705.

12

# A community-driven reconstruction of the *Aspergillus niger* metabolic network

Julian Brandl[1], Maria Victoria Aguilar-Pontes[2], Paul Schäpe[6], Anders Noerregaard[1], Mikko Arvas[3,7], Arthur F. J. Ram[5], Vera Meyer[6], Adrian Tsang[4], Ronald P. de Vries[2] and Mikael R. Andersen[1*]

**Abstract**

**Background:** *Aspergillus niger* is an important fungus used in industrial applications for enzyme and acid production. To enable rational metabolic engineering of the species, available information can be collected and integrated in a genome-scale model to devise strategies for improving its performance as a host organism.

**Results:** In this paper, we update an existing model of *A. niger* metabolism to include the information collected from 876 publications, thereby expanding the coverage of the model by 940 reactions, 777 metabolites and 454 genes. In the presented consensus genome-scale model of *A. niger* iJB1325 , we integrated experimental data from publications and patents, as well as our own experiments, into a consistent network. This information has been included in a standardized way, allowing for automated testing and continuous improvements in the future. This repository of experimental data allowed the definition of 471 individual test cases, of which the model complies with 373 of them. We further re-analyzed existing transcriptomics and quantitative physiology data to gain new insights on metabolism. Additionally, the model contains 3482 checks on the model structure, thereby representing the best validated genome-scale model on *A. niger* developed until now. Strain-specific model versions for strains ATCC 1015 and CBS 513.88 have been created containing all data used for model building, thereby allowing users to adopt the models and check the updated version against the experimental data. The resulting model is compliant with the SBML standard and therefore enables users to easily simulate it using their preferred software solution.

**Conclusion:** Experimental data on most organisms are scattered across hundreds of publications and several repositories.To allow for a systems level understanding of metabolism, the data must be integrated in a consistent knowledge network. The *A. niger* iJB1325 model presented here integrates the available data into a highly curated genome-scale model to facilitate the simulation of flux distributions, as well as the interpretation of other genome-scale data by providing the metabolic context.

**Keywords:** *Aspergillus niger*, Primary metabolism, Secondary metabolism, Genome-scale model

## Background

Genome-scale metabolic models have been successfully used as tools for guiding metabolic engineering, analyzing cellular phenotypes and contextualizing omics data [1–3]. For all of these tasks, high quality reconstructions are needed to minimize problems introduced by errors in the model. Today, there are multiple approaches for model generation spanning from classic manual model building, semi-automated and fully-automated generation of genome-scale models [1, 4, 5]. The latter approaches are especially valuable for new or undercharacterized species as the genome sequence can form the basis for the construction of a draft genome-scale model. For well-characterized species, the classic model building approach provides the opportunity of integrating available experimental knowledge into a structured framework, allowing for consistency checking and identification of knowledge gaps. The probably best curated genome-scale models are available for widely used model species such as *E. coli* and *S. cerevisiae*. The consensus

*Correspondence: mr@bio.dtu.dk
[1] Technical University of Denmark, Soeltofts Plads, Building 223, 2800 Kongens Lyngby, Denmark
Full list of author information is available at the end of the article

reconstruction of *S. cerevisiae* has been curated in a community-driven effort for several years and is able to simulate gene deletions and growth performance with high accuracy [6]. In this paper we aimed at establishing a community consensus genome-scale model of *A. niger* that enables researchers to run constraint-based analyses like prediction of gene knockout phenotypes or maximum yields under different conditions. While particular areas of metabolism in Aspergilli have attracted significant attention [7], there are still big gaps [8] in our understanding to be addressed.

*Aspergillus niger* described in 1867 by Van Tiegham, sparked considerable interest due to the observation of citric acid overproduction in the beginning of the last century [9]. Besides being an industrial work-horse in citric acid production, *A. niger* and its close relatives are also widely used hosts for enzyme production [10, 11]. Owing to the commercial interest in *A. niger* and its metabolic flexibility with respect to utilizable substrates, there has been sustained research to elucidate the metabolism of this organism. The first genome-scale model of *A. niger* was published by one of the authors [12], which has been built on a former reconstruction of central carbon metabolism of *A. niger*[13]. The original model has been widely used for a variety of applications, e.g. for modeling acid production [14] and predicting protein yields [15]. However, these modeling efforts are based on the state of the art in 2008. During the last decade, substantial amounts of research have been conducted on *A. niger* metabolism as well as on the metabolism of closely related fungi, which can be used to greatly improve the predictions and metabolic network of *A. niger*. Additionally, the organization of biological data has changed tremendously in the last decade. Standards for structuring models (e.g. SBML), identifying chemical reactions (E.C. numbers and KEGG identifiers), referencing literature (e.g. DOIs, Pubmed IDs, and PMC IDs), and identification of chemical compounds (ChEBI's and InChI's) have been established and/or updated, enabling a much higher degree of cross-referencing of information and establishing interlinked data structures.

Recently, a different update of the original genome-scale model has been published by Lu et al. [16]. The authors used a partially overlapping updating strategy. First, they updated the annotations of metabolites and reactions and balanced all unbalanced reactions. The authors then used the information from four databases to add new reactions to the model and update existing gene-protein-reaction associations (GPRs). While also making use of the structured information provided by those databases, we aimed for a systematic storage of the primary data in the model thereby ensuring long-term evolution of the model. In this work, we present an updated genome-scale model of *A. niger* that has undergone major revisions with respect to the metabolic coverage as well as the quality of gene assignments, and is in compliance with state-of-the-art data standards. The end result is a gold-standard curated and validated genome-scale model, incorporating the information of 876 publications.

## Results

### Update methodology and statistics

The aim of the update was to improve the original genome-scale model of *A. niger* [12], both with respect to coverage as well as with respect to the annotations included in the model, in particular the assignment of genes to reactions. Furthermore, specific interests have been to include modeling of secreted secondary metabolites and the hundreds of proteins, which *A. niger* is known to produce. In a first step, metabolites have been annotated with their respective ChEBI identifier [17] to enable their unequivocal identification in relevant databases. The undissociated form of the individual metabolites has been used to avoid problems caused by the largely unknown proton stoichiometry of transport reactions. Reactions have been checked for mass balance accordingly and have been adapted where needed, in particular to include protons in all reactions where they are known to be present.

In a second step, the model has been updated based on primary literature, patents, as well as on information contained in the AspGD [18], BRENDA [19] and AMIGO2 [20] databases. The total number of publications in PubMed on *A. niger* has roughly doubled since the publication of the *i*MA871 model, and the abstracts of all these publications have been examined manually and complemented by additional searches. Manual searches for literature have been further complemented by comparing the genome sequence using the BLAST algorithm against the non-redundant patent sequence database [21] as well as the UniProtKB/Swiss-Prot database [22]. The experimental information included in those resources have been tracked back to the primary source and added in a structured way to the model with a reference to its origin.

As an integrated part of the model update process, we have implemented the model in fully functional SBML. Genome-scale metabolic models are usually shared in the standardized SBML format [23]. This format is routinely read by popular modeling software packages thereby preventing the error prone process of custom model parsing [24]. While the standard focuses on the safe distribution of models, the XML basis of the format allows for the introduction of additional information without breaking the format. We have therefore further improved the association of literature and model reactions by integrating the references and their level of support directly in the SBML file. We separated the experimental data into two classes: Evidence items (see panels B and C in Fig. 1) and Test cases (see panel A in Fig. 1). Tests can be viewed as simple Input/Output tests and consist of a list of test conditions i.e. medium composition and gene knockouts and a list of reported outcomes that are tested for. With this setup, the differential growth of strains of *A. niger* on combinations of C- and N-sources can be saved in a structured way that can later be tested through simulation in an automated manner. Evidence items are used for the storage of information containing the presence/absence of a specific reaction or metabolite, as well as for the presence/absence of a connection between a gene and reaction. Additionally localization of a gene product to a specific compartment can be represented as well. Relevant information contained in the literature used for building the model has been saved as corresponding test case or evidence information.

As a third step, and in order to further improve the amount of available experimental information for *A. niger*, we employed phenotype arrays to screen the capability of *A. niger* spores to germinate and grow on 190 different C-sources as well as 95 N-sources using phenotype screening plates from Biolog Inc. If *A. niger* showed ability to grow on the substrate (Additional files 1 and 2), the model was updated to include the relevant catabolic pathway if possible. This led to a further addition of 34 pathways to the model. The absence of growth has not been used as information in the modelling process as this might be either caused by absence of transport, a missing catabolic route or lack of expression of the former two and might therefore represent a false negative result.

Fourthly, two other expanded models for *A. niger* have been published during the development of this one, one *de novo*-generated based on an advanced automated method [5], and one by Lu et al. [16], expanding our previous iMA871 model. We have analyzed the content of these models and integrated information from these where appropriate, thus generating a consensus-type model.

The update strategy described led to the expansion of the model with respect to several pathways (Table 1) as well as the update of pathways already included in the model. The final version presented here was named *A. niger* iJB1325. The comparison of the key statistics of the different models is shown in Table 2. The current update of the model includes 1325 genes and therefore adds 454 genes to the original model, which is on par with the model published by Lu et al. [16]. The number of metabolites has been increased by 773–1818 while including 1130 additional reactions. Overall experimental information from 876 sources has been included in the model, which represents an addition of 505 publications in comparison to our first model. The experimental information has been broken up into evidence items and test cases thereby allowing for the easy backtracking of the experimental information as well as for the automatic validation of the model structure, growth/production capabilities against the knowledge of the 876 publications used for building the model.

### Strain-specific model implementations

In order to provide strain-specific models for the most commonly used strains of *A. niger* we generated individual models for the sequenced strains ATCC 1015,

**Table 1 Newly introduced pathways**

| Degradation | Biosynthesis | Secondary metabolites |
|---|---|---|
| Agmatine degradation | CoQ biosynthesis | Azanigerones |
| Amide degradation | Coprogen biosynthesis | Malformins |
| Aromatics degradation | Storage compounds | Nigragillin |
| Peroxisomal beta oxidation | Ferrichrome | Kotanin |
| Cyanide degradation | Iron assimilation | Funalenone |
| Galacturonic acid degradation | Lipoic acid biosynthesis | Pyranonigrin |
| Detoxification of compounds | Metabolite repair | Aurasperone |
| Glucuronate degradation | NAD biosynthesis | Tetraacetic acid lactone |
| Isoleucine degradation | Thiamin biosynthesis | |
| Leucine degradation | Vitamin metabolism | |
| Lipid degradation | Molybdenum cofactor | |
| Plant biomass degradation | Riboflavin biosynthesis | |
| Purine degradation | | |
| Valine degradation | | |
| L-Rhamnose metabolism | | |

**Table 2 Table depicting key statistics of the different models**

| Model | iMA873 | iJB1325 | CoReCo | iHL1210 |
|---|---|---|---|---|
| *Reactions* | | | | |
| Total | 1380 | 2320 | 4917 | 1764 |
| Transport | 189 | 447 | 0 | 285 |
| Boundary | 0 | 385 | 148 | 189 |
| Unbalanced | 40 | 68 | 148 | – |
| Annotated | 1013 | 1239 | 0 | – |
| No genes | 340 | 604 | 3049 | – |
| Evidence for presence | – | 767 | – | – |
| Known gene | – | 654 | – | – |
| *Metabolites* | | | | |
| Total | 1084 | 1818 | 4025 | 1254 |
| Annotated | 0 | 1533 | 0 | |
| Dead-End | 270 | 295 | 1739 | |
| *Genes* | | | | |
| Total | 871 | 1325 | 4533 | 1210 |
| Verified location | – | 296 | – | – |
| Predicted location | – | 107 | | |
| Known function | – | 707 | – | – |
| *Evidences* | | | | |
| Total | – | 3482 | – | – |
| Gene-reaction | – | 1677 | – | – |
| Metabolite presence | – | 333 | – | – |
| Reaction presence | – | 539 | – | – |
| Gene-compartment | – | 907 | – | – |
| *Test cases* | | | | |
| Total | – | 471 | – | 99 |
| Passing | – | 373 | – | 83 |
| Failing | – | 98 | – | 16 |
| *References* | | | | |
| Total | 371 | 876 | – | – |

and CBS 513.88 (see Additional files 2 and 3). Reciprocal best blast hits have been used to transfer reaction assignments and the experimental evidence from the development model to the individual models.There are 18 genes in ATCC1015 that have no hit in the CBS513.88 genome (see Additional file 4: Table 4).

### Validation, iterative improvements, and test cases

In the iJB1325 model, we included 471 test cases that can be run when updating the model in order to ensure consistency with the information that has been used for building the model. Those tests mainly comprise absence or presence of growth on different combinations of carbon and nitrogen sources (392 cases). A smaller set consists of tests for gene deletion phenotype i.e. absence or presence of growth of deletion mutants (73 cases). An even smaller number of tests comprise overall system checks i.e. possibility to produce biomass precursors, no growth in the absence of known essential medium components, and a check for the possibility to oxidize fatty acids in the peroxisome (6 cases). Running all tests with the current version of the model leads to 373 (79%) passing and 98 (21%) failing test cases. The failing tests consist of 75 cases failing due to an unknown metabolic pathway, 15 failing due to inconsistencies between experiments, and 8 tests failing due to reasons we have not been able to determine (see Additional file 5 for more information).

Having a collection of test conditions enables the identification of missing reactions in the model. One such example is the utilization of L-histidine as single N-source in combination with D-galactose. This was reported by Hayer et al. [25] as well as in combination with glycerol by Steinberg [26]. Growth on L-histidine in combination with glucose can also be observed in our own phenotype screening arrays. The previous version of the model was incapable of simulating growth on L-histidine as single N-source. The metabolic reactions involved in the utilization of L-histidine have not been reported to our knowledge. In *A. nidulans*, the presence of a histidase (histidine ammonia-lyase, E.C. 4.3.1.3) has been reported, but the corresponding gene remains to be identified. Using the reviewed entries in Uniprot, we could identify three candidate genes for histidase in *A. niger* (JGI *A. niger* ATCC 1015 (Aspni7) ProteinIDs 1129557, 1126350 and 1081533). Sequence comparison with the characterized enzyme from *P. putida* [27] showed high conservation for the active site residues for the three candidates (see Additional file 6); therefore, all three candidate genes have been added to the model.

Another example for the identification of missing reactions in the model is the growth on L-methionine or L-cysteine as single N-source. Growth of *A. niger* on both nitrogen sources has been observed by Hayer et al. 2014 [25] as well as in our Biolog experiments. Failing tests for growth of *A. niger* on those two N-sources hinted to the absence of the corresponding pathway. Evidence shown in *A. nidulans* by Sienko et al. [28] hints towards the degradation of L-methionine towards L-cysteine by a reverse transsulfuration reaction involving the genes *mecA* and *mecB*. Putative orthologs for these genes are present in *A. niger*. The metabolic fate of L-cysteine as single N-source has to our knowledge not been demonstrated conclusively in *A. niger* or any related fungus. We therefore did not include a L-cysteine degradation pathway and left this gap for future improvements.

Another example of a failing test is the no growth phenotype for the Δ*gaaA* strain on D-galacturonate. Besides the *gaaA* gene, the putative *H. jecorina* ortholog (JGI *A. niger* ATCC 1015 (Aspni7) transcriptID 1109007) has been included due to sequence similarity

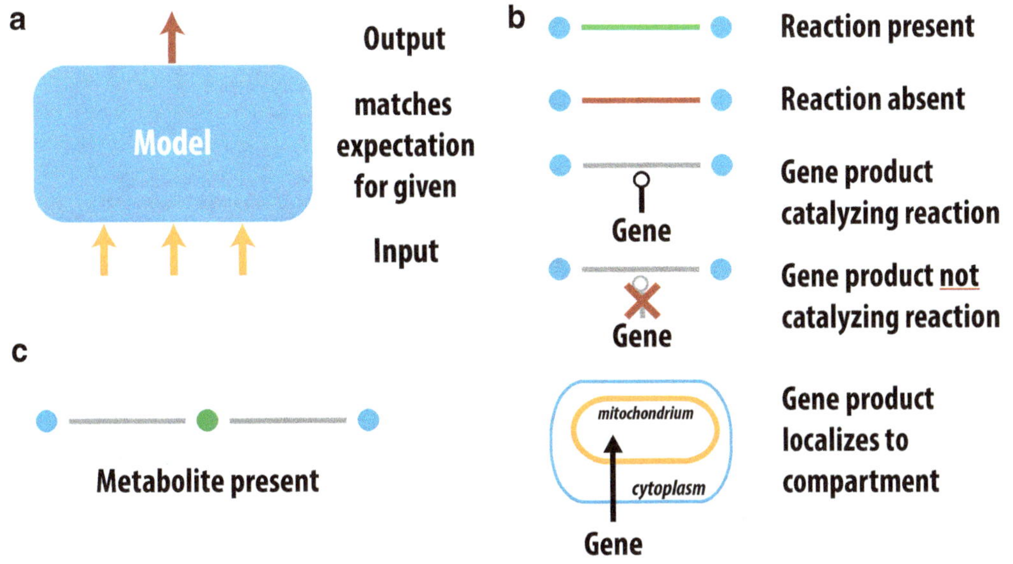


**Fig. 1** Types of experimental evidence included in the current version of the model. Panel A) depicts the concept of model test cases included in iJB1325. The test cases consist of a set of active input reactions, inactive genes as well as conditions for the test to pass. Inputs therefore refer to metabolites present in the growth medium or the knockout of a specific gene. Test conditions can be the ability to produce biomass or other compounds. Panel B) shows the different pieces of evidence stored in the model for genes and reactions while Panel C) depicts the information stored about metabolites

as a D-galacturonic acid reductase. The experimental data reported by Mojzita et al. [29] however demonstrate that a strain with a defect in *gaaA* is not able to grow on D-galacturonate, indicating that the alternative gene either is not a D-galacturonate reductase, or is not expressed under the experimental condition. However, Alazi et al. [30] found only a reduction of growth in the Δ*gaaA* strain indicating partial redundancy of the pathway. The putative ortholog has therefore been kept in the model.

#### Evidence-based support for reactions

In the iJB1325 model, we included several levels of evidence from the literature to enable to continuous testing of model connectivity during future improvements. The different types of evidence included in the model are depicted in Fig. 1. For a format description of the evidence items, see Additional file 7. For the assignment of genes to individual metabolic reactions, primary literature for *A. niger* as well as related species has been used. Making use of the evidence code ontology [31], we included a measure of certainty for the individual connections. The localization of the individual proteins has been predicted using Mitofates [32] for mitochondrial proteins as well as PTS1 predictor [33] for peroxisomal localization predictions. In order to identify secreted proteins, we compiled a list of published proteomics experiments on *A. niger* [34–44]. Extracellular localization has been considered a true positive if the presence in the extracellular space has been reported by three independent publications. The prediction of localizations is complicated by the fact that some proteins are localized in multiple compartments by alternative translational start sites [45], differential splicing [46, 47] as well as stop-codon readthrough [48]. However due to the lack of experimental validation of the localization of most proteins, predictions for mitochondrial and peroxisomal localization have been included as best guess.

Additionally, evidence items for the presence of individual metabolites in *A. niger* have been included where reported in the literature. The presence and absence of reactions have also been included when reported. In total, we have been able to include 3482 pieces of evidence that link individual components inside the model and are associated with an evidence code as a measure of certainty, a small description where appropriate describing the underlying experiment, and a link to the resource the conclusion has been drawn from. This small summary allows for a quick assessment of the quality of the gene assignment and thereby easing the interpretation.

The distribution of the experimental support of the individual reactions is shown in Fig. 2 panel A. About 19% of the reactions included in the model have direct experimental support by either being measured *in vitro* or having a gene with a confirmed activity assigned. Another 17% of the reactions have a gene assigned

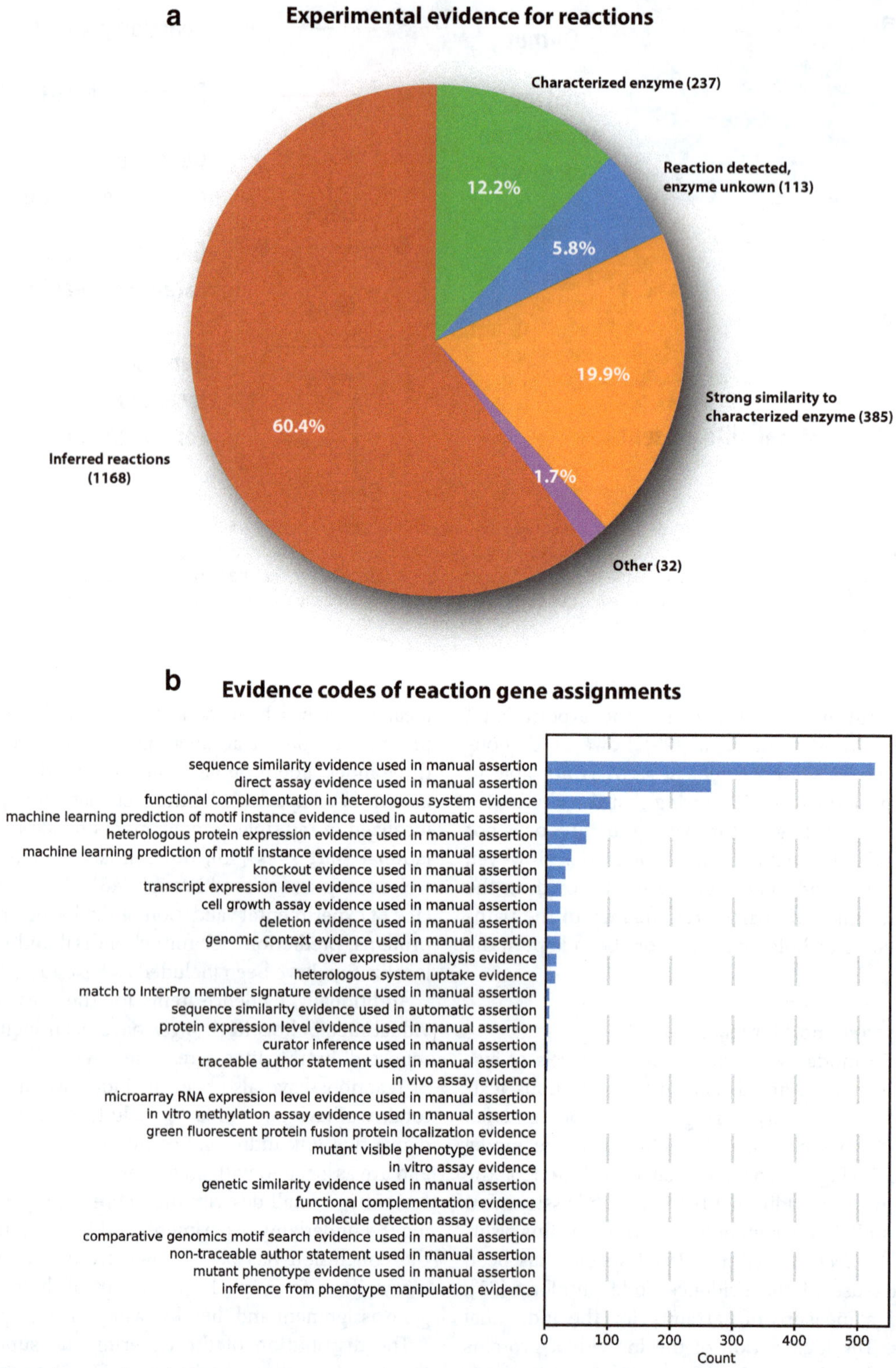


**Fig. 2** Experimental support for individual reactions. **a** Depicts the strongest experimental support for the presence of individual reactions in the model. The categories according to decreasing experimental support are: "Characterized enzyme", "Measured, but unknown enzyme", "Strong similarity to characterized enzyme", "Other" and "No experimental evidence". **b** Shows the evidence codes associated with the individual reaction gene assignments

based on a strong similarity to a characterized enzyme in a closely related species. These gene reactions associations are mainly derived from characterized genes of the related fungal model organisms *A. nidulans* and *A. fumigatus*. The remaining reactions either have no genes assigned or are inherited from the iMA871 model as the best candidates for a given reaction based on sequence similarity to characterized enzymes or domain predictions. The distribution of the individual evidence codes of all reaction-gene-assignments is depicted in Fig. 2 panel B. The experimental support for individual metabolites are depicted in Fig. 3. About 12% of the metabolites included in the model have been measured experimentally leaving 88% of the metabolites that were inferred during the reconstruction process.

### Application of the model network for transcriptomics data analysis

In addition to our 471 modeling test cases, we wanted to acknowledge that a significant application of genome-scale models is the use of the underlying metabolic networks for data analysis and interpretation [2, 49]. Using the current version of our genome-scale model, we analyzed an compendium of transcriptomics data based on published studies using a microarray developed for the *A. niger* ATCC 1015 strain [14, 15, 50–53], as well as a compendium based on an *A. niger* CBS 513.88 array data [54].

One possible application for a genome-scale model is the analysis of expression data using the gene-protein-reaction (GPR) associations included in the model. The grouping of reactions into pathways, thereby also grouping the associated genes, allows for overall contextualization of expression changes in transcriptomics data as depicted in Fig. 4 for the ATCC 1015 transcriptome data (see Additional file 8: Figure 8 for the corresponding analysis for the CBS 513.88 dataset). In both cases, plotting the expression level of different subnetworks demonstrates the expected up-regulation of the genes associated with D-xylose and L-arabinose catabolism on those two carbon sources while there appears to be no overall change of genes involved in glycolysis or the TCA cycle. Some of the genes associated with plant biomass degradation which consists mainly of polysaccharide degrading enzymes are upregulated, but the majority of these genes does not show a change in expression under these conditions. Using the correlation of gene expression levels to individual pathways the transporter with the JGI Aspni7 TranscriptID 1178899 appeared to be co-regulated with

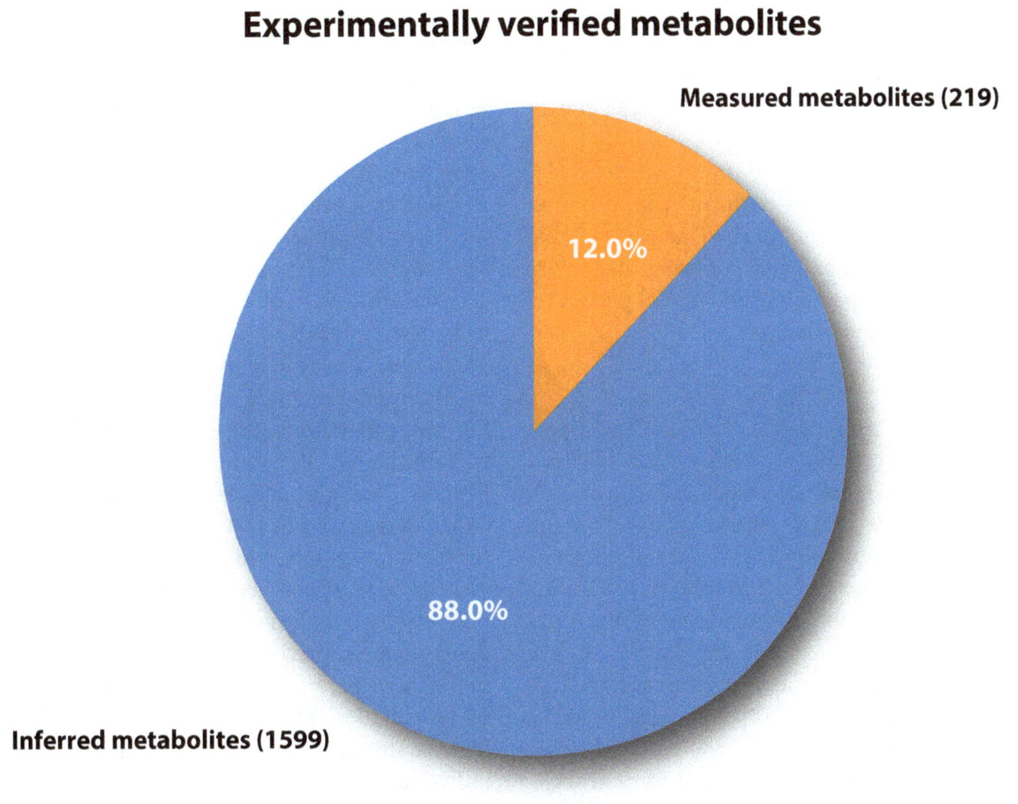


**Fig. 3** Experimental support for individual metabolites

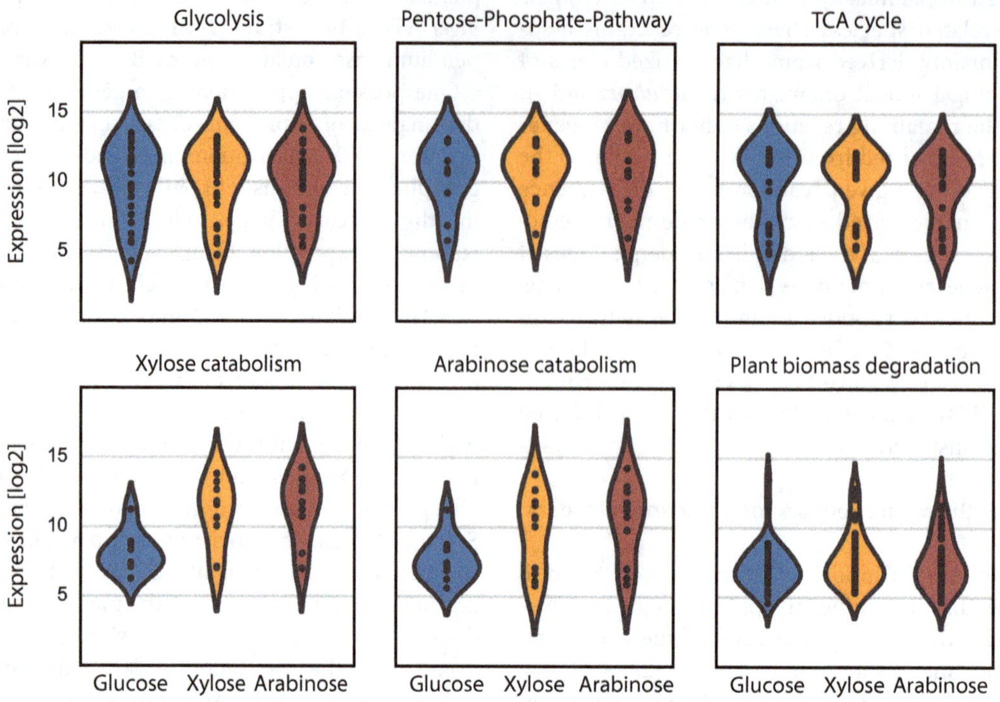


**Fig. 4** Change in transcription level of the genes assigned to metabolic pathways under different conditions. Violin plots showing the change of expression of genes involved in different metabolic pathways depending on the carbon source used

the known members of the *ada* cluster that has been characterized by Li et al. [55]. The transporter is located next to the *adaA* gene and therefore we named this gene *adaE* and included it as a putative TAN-1612 transporter in the model.

Transcriptomics data can also be used for the validation of genome-scale models as performed by Lu et al. [16]. Unfortunately only the part of the data set covering the genes in their model has been published, therefore not allowing a direct comparison of the model presented here to their transcriptomics data set.

## Discussion

*Aspergillus niger* has been used as a biotechnological workhorse for about a century producing citric acid and several enzymes in high amounts. During this time much research has been dedicated to shedding light on the underlying metabolic network. The resulting information has been published in several hundreds to thousands of papers containing the individual pieces of the puzzle. With the advent of the genomic era, databases have been developed that try to catalog the literature information on the individual genes [18]. However, in order to be able to analyze and understand the metabolism at a systems level, integration into a coherent framework is needed. Genome-scale models can provide such a framework in which the knowledge about metabolism can be integrated, tested for logical consistency, and predictions made.

In this study we present an update of the genome-scale model that has been developed in our group almost a decade ago [12]. The updated model integrates the knowledge from 876 publications into a consistent framework, thereby representing the experimentally best supported model currently available. As the model update process represents an iterative process spanning years to decades, a sustainable way of keeping track of the information used in the curation process is needed. With this in mind, we created a new structure to store all information used for the model building in the model file, thereby allowing users to modify the model and check the changed version against all literature information in an automated fashion. This approach led to the inclusion of 3482 evidence items and 471 test cases. Whereas test cases allow for the representation of both quantitative and qualitative macroscopic observations representing overall model predictions (e.g. growth or production capacities), evidence items represent the experimental support for the model structure (e.g. presence/absence of a compound or biochemical reaction). In order to allow backtracking, we stored the primary literature references with the individual evidence items and test cases. As the information used for model building is currently not shared in a standardized form, we aimed at extending the SBML format [23] in a manner compliant with the specification, making it usable by other researchers. To our knowledge this is the first time a model has been presented, which allows for efficient continuous improvement making use of automated testing. While this strategy increases the complexity of the reconstruction, it allows backtracking of the experimental information used for the model building. This feature makes the model a true knowledgebase for *A. niger* metabolism that is not only valuable for future improvements, but also provides a structured way for the search of existing experimental knowledge. The usefulness of having this information has been demonstrated by the ability for identifying missing reactions, as well as in checking the connectivity of the network.

The model development has been focused on the update of the information and gene assignments for the *A. niger* ATCC1015 strain. However, as the strains are very similar and in order to allow the utilization of the model by more users, we used reciprocal best blast hits to translate the genes in the model to the identifier of the *A. niger* CBS513.88 strain.

Recently a model has been published by Lu et al. [16] that already included improvements to the metabolite annotation and reaction quality that are presented in the current publication. The authors also validated their model using 99 growth tests for different carbon and nitrogen sources where the model performed successfully in 83 of 99 tests. We also included tests for those growth conditions if not already present in the model. While 373 passing tests out of 471 seems worse than the relation presented by Lu et al. the number of failing tests is explained by the wider coverage thereby including substrates with unknown metabolic pathways that could not be included in the model. Inclusion of these test cases leads to presence of a substantial number of dead-end metabolites in the current version of the model which are frequently removed from genome-scale models as reactions associated with those metabolites are guaranteed to carry no flux. We decided to keep those metabolites in the model as points for future model improvement and to contain the experimental data that led to the inclusion of those reactions.

One important aspect of eukaryotic metabolism is the compartmentalization of reactions into different organelles. Unfortunately, information about the subcellular localization of individual enzymes is only scarcely available in *Aspergillus niger*. Another existing challenge for genome-scale modelling in filamentous fungi is the presence of multiple seemingly orthologous genes for many metabolic functions. The existence of these multiple copies poses the challenge to assign the individual contribution of those genes to the metabolic activity. One way of identifying the best candidate for a function is comparing the expression values of a gene to verified genes upstream or downstream in the same pathway. If known constituents of the pathway are expressed at a very high level (e.g. see Glycolysis or TCA cycle in Fig. 4) missing members are expected to have an expression level in the same range as the known members. We tried using transcriptomics data in such a manner as proxy for the individual contribution. This proved to be challenging as for most reactions the expected activity level is unknown, as well as for many genes the correct assignment is not evident. Due to the mentioned difficulties we did not include the transcript level as experimental evidence in the majority of the reactions. With the development of the CRISPR/Cas9 system in filamentous fungi, large-scale genetic manipulations become increasingly feasible thereby allowing the assessment of the contribution of individual genes to a specific phenotype on a genome-scale model. The development of a large-scale knockout library would be an interesting project for the validation of the genome-scale model presented here.

## Conclusion

Here we presented the largest and most thoroughly curated genome-scale model of *A. niger* metabolism to date. The model has been built on an extensive body of primary literature which has been structured and saved along with the model. We therefore extended the SBML format of the model to include the literature information about reactions and gene functions resulting in 1677 evidence items for gene reaction links, 539 items for the presence or absence of reactions, and 907 items for subcellular protein localization. At the same time we integrated the growth capabilities of *A. niger* as reported in the literature with our own experimental data, leading to the validation of the model against 471 test cases. From this validated model, strain specific models have been generated for *A. niger* ATCC1015, and CBS513.88. Finally, the model has been demonstrated to be useful for the interpretation of -omics data providing the metabolic context of the individual genes .

## Methods

### Software

During the process of updating the model we used a newly developed software for the editing of the model. This software will be published separately and is based on existing open source software packages COBRApy [56] and Escher [57] for simulation and visualization, respectively. The software is already available at https://github.com/JuBra/GEMEditor. Users interested in testing, updating and viewing the experimental information included in the model are referred to the GEMEditor wiki for instructions about installation and analysis of the model.

### Bibliomic data

In order to identify extracellular proteins, available extracellular proteomics data has been collected [34–44]. Characterized proteins have been identified from primary literature, individual patents by searching for patents on *Aspergillus niger* as well as by blasting against the non-redundant patent sequence database [21]. Proteins not accounted for in the model have additionally been blasted against the UniProtKB/Swiss-Prot database [22] to identify proteins with known functions.

### Integration of public experimental data

Experimental data have been gathered from several literature resources. We combined all available proteomics information for *A. niger* for assessing the sub-cellular localization of different gene products. Additionally, subcellular prediction of proteins has been performed using MitoFates [32] for mitochondrial and PTS1 predictor [33] for peroxisomal localization prediction. The analysis of the transcriptomics data has been performed on a dataset for *A. niger* ATCC1015 collected from [14, 15, 50–53] and for *A. niger* CBS513.88 (GEO accession number GSE98572, Samples: GSM2600962, GSM2600963, GSM2600941, GSM2600942, GSM2600992 and GSM2600993) collected from Gruben et al. [58].

### Phenotype arrays

Screening for growth on nitrogen and carbon-sources has been performed using the phenotype plates PM1, PM2A and PM3B from Biolog Inc. The plates have been prepared according to the manufacturers manual with the spore density being adjusted to $10^7$ spores per ml. The plates have been incubated at 28 °C for up to 10 days. Growth has been assessed by inspecting the plates visually for sporulation, see Additional files 1 and 9.

### Test cases

The test cases introduced in the current version of the model consist of a list of settings representing the simulation conditions, a list of deactivated genes and a list of outcomes the resulting solution is checked against. All boundary reactions that are not specified in the settings list are set to only being able to consume the metabolite. Outcomes are a combination of a specific reaction, a greater than or less than modifier and a numerical value (see Additional file 10).

### Model simulation

Test simulations have been run using the FBA or pFBA method as implemented in COBRApy [56] (see Additional file 11).

## Additional file

**Additional file 1.** Image of Biolog PM3B screening plate. The image shows a picture of a Biolog PM3B screening plate inoculated with *A. niger* ATCC1015 after 4 days of incubation at 32 °C. Nitrogen sources on plate: Control; Ammonia; Nitrite; Nitrate; Urea; Biuret; L-Alanine; L-Arginine; L-Asparagine; L-Aspartic Acid; L-Cysteine; L-Glutamic Acid; L-Glutamine; Glycine; L-Histidine; L-Isoleucine; L-Leucine; L-Lysine; L-Methionine; L-Phenylalanine; L-Proline; L-Serine; L-Threonine; L-Tryptophan; L-Tyrosine; L-Valine; D-Alanine; D-Asparagine; D-Aspartic Acid; D-Glutamic Acid; D-Lysine; D-Serine; D-Valine; L-Citrulline; L-Homoserine; L-Ornithine; N-Acetyl- L-Glutamic Acid; N-Phthaloyl- L-Glutamic Acid; L-Pyroglutamic Acid; Hydroxylamine; Methylamine; N-Amylamine; N-Butylamine; Ethylamine; Ethanolamine; Ethylenediamine; Putrescine; Agmatine; Histamine; beta-Phenylethyl- amine; Tyramine; Acetamide; Formamide; Glucuronamide; D, L-Lactamide; D-Glucosamine; D-Galactosamine; D-Mannosamine; N-Acetyl- D-Glucosamine; N-Acetyl- D-Galactosamine; N-Acetyl- D-Mannosamine; Adenine; Adenosine; Cytidine; Cytosine; Guanine; Guanosine; Thymine; Thymidine; Uracil; Uridine; Inosine; Xanthine; Xanthosine; Uric Acid; Alloxan; Allantoin; Parabanic Acid; D, L-alpha-Amino-N-Butyric Acid; gamma-Amino-N-Butyric Acid; epsilon-Amino-N-Caproic Acid; D, L-alpha-Amino- Caprylic Acid; delta-Amino-N-Valeric Acid; alpha-Amino-N-Valeric Acid; Ala-Asp; Ala-Gln; Ala-Glu; Ala-Gly; Ala-His; Ala-Leu; Ala-Thr; Gly-Asn; Gly-Gln; Gly-Glu; Gly-Met; Met-Ala.

**Additional file 2.** Genome-scale model for Aspergillus niger ATCC1015. Model file in SBML level 3 version 1.

**Additional file 3.** Genome-scale model for Aspergillus niger CBS513.88. Model file in SBML level 3 version 1.

**Additional file 4.** Gene differences between strain ATCC1015 and CBS513.88. The list shows the genes for which no reciprocal best blast hit could be identified in *A. niger* CBS513.88.

**Additional file 5.** Failing test cases. The file consists of the currently failing test cases of the model with an explanation why they are failing.

**Additional file 6.** Sequence comparison of the candidates for histidase. Multiple sequence alignment of the candidates for the histidase activity.

**Additional file 7.** Example of the format for storing experimental data. Example entries for the experimental data as saved in the current version of the model.

**Additional file 8.** Changes in expression of genes belonging to different pathways. The change of gene expression in the CBS513.88 model.

**Additional file 9.** Image of Biolog PM1 screening plate. The image shows a picture of a Biolog PM1 screening plate inoculated with *A. niger* ATCC1015 after 9 days of incubation at 32 °C. Carbon sources on plate: Control; L-Arabinose; N-Acetyl-D-Glucosamine; D-Saccharic Acid; Succinic Acid; D-Galactose; L-Aspartic Acid; L-Proline; D-Alanine; D-Trehalose; D-Mannose; Dulcitol; D-Serine; D-Sorbitol; Glycerol; L-Fucose; D-Glucuronic Acid; D-Gluconic Acid; D, L-alpha-Glycerol- Phosphate; D-Xylose; L-Lactic Acid; Formic Acid; D-Mannitol; L-Glutamic Acid; D-Glucose-6-Phosphate; D-Galactonic Acid-gamma-Lactone; D, L-Malic Acid; D-Ribose; Tween 20; L-Rhamnose; D-Fructose; Acetic Acid; alpha-D-Glucose; Maltose; D-Melibiose; Thymidine; L-Asparagine; D-Aspartic Acid; D-Glucosaminic Acid; 1,2-Propanediol; Tween 40; alpha-Keto-Glutaric Acid; alpha-Keto-Butyric Acid; alpha-Methyl-D-Galactoside; alpha-D-Lactose; Lactulose; Sucrose; Uridine; L-Glutamine; m-Tartaric Acid; D-Glucose-1-Phosphate; D-Fructose-6-Phosphate; Tween 80; alpha-Hydroxy Glutaric Acid-gamma-Lactone; alpha-Hydroxy Butyric Acid;

beta-Methyl-D-Glucoside; Adonitol; Maltotriose; 2-Deoxy Adenosine; Adenosine; Glycyl- L-Aspartic Acid; Citric Acid; m-Inositol; D-Threonine; Fumaric Acid; Bromo Succinic Acid; Propionic Acid; Mucic Acid; Glycolic Acid; Glyoxylic Acid; D-Cellobiose; Inosine; Glycyl- L-Glutamic Acid; Tricarballylic Acid; L-Serine; L-Threonine; L-Alanine; L-Alanyl-Glycine; Acetoacetic Acid; N-Acetyl-beta-D-Mannosamine; Mono Methyl Succinate; Methyl Pyruvate; D-Malic Acid; L-Malic Acid; Glycyl- L-Proline; p-Hydroxy Phenyl Acetic Acid; m-Hydroxy Phenyl Acetic Acid; Tyramine; D-Psicose; L-Lyxose; Glucuronamide; Pyruvic Acid; L-Galactonic Acid-gamma-Lactone; D-Galacturonic Acid; Phenylethyl-amine; 2-Aminoethanol.

**Additional file 10.** Correlation of members of the Anthracenone cluster. The file shows the expression values for individual genes that have been identified to be part of the ada cluster. The adaD gene has not been included in the source microarray of the data.

**Additional file 11.** Export of the evidence items contained in the ATCC1015 model. Gene IDs correspond to ATCC1015 Aspni7 transcriptIds.

### Authors' contributions

JB built the model, curated the literature, modeled metabolism, analysed data, and wrote the manuscript. MVAP provided analysis of metabolic pathways and read and commented on the manuscript. AFJR, PS and VM contributed and analyzed transcriptome data, and read and commented on the manuscript. AN performed preliminary work on the Biolog phenotype arrays and established a protocol for this. MA provided analysis of the CoReCo models and read and commented on the manuscript. AT provided analysis of *A. niger* genomes and read and commented on the manuscript. RPdV provided analysis of the pathway data, and read and commented on the manuscript. MRA designed the study, supervised the work, and read and commented on the manuscript. All authors read and approved the final manuscript.

### Author details

[1] Technical University of Denmark, Soeltofts Plads, Building 223, 2800 Kongens Lyngby, Denmark. [2] Fungal Physiology, Westerdijk Fungal Biodiversity Institute and Fungal Molecular Physiology, Utrecht University, Uppsalalaan 8, 3584 CT Utrecht, The Netherlands. [3] VTT Technical Research Centre of Finland, Tietotie 2, 02044 Espoo, Finland. [4] Concordia University, 7141 Sherbrooke Street West, H4B1R6 Montreal, Québec, Canada. [5] Leiden University, Sylviusweg 72, 2333 BE Leiden, The Netherlands. [6] Berlin University of Technology, Gustav-Meyer-Allee 25, 13355 Berlin, Germany. [7] Present Address: Finnish Red Cross Blood Service, Helsinki, Finland.

### Competing interests

The authors declare that they have no competing interests.

### Funding

JB and MR gratefully acknowledge funding from the Novo Nordisk Foundation, Grant NNF13OC0004831. MR gratefully acknowledges funding from the Villum Foundation, grant VKR023427. MVAP was supported by a Grant of the Dutch Technology Foundation STW, Applied Science Division of NWO, and the Technology Program of the Ministry of Economic Affairs 016.130.609 to RPdV.

### References

1. Brandl J, Andersen MR. Current state of genome-scale modeling in filamentous fungi. Biotechnol Lett. 2015;37(6):1131–9.
2. Hyduke DR, Lewis NE, Palsson BØ. Analysis of omics data with genome-scale models of metabolism. Mol BioSyst. 2013;9(2):167–74.
3. Zhang C, Hua Q. Applications of genome-scale metabolic models in biotechnology and systems medicine. Front Physiol. 2015;6:413.
4. Devoid S, Overbeek R, DeJongh M, Vonstein V, Best AA, Henry C. Automated genome annotation and metabolic model reconstruction in the SEED and Model SEED. Methods Mol Biol (Clifton, N.J.). 2013;985:17–45.
5. Pitkänen E, Jouhten P, Hou J, Syed MF, Blomberg P, Kludas J, Oja M, Holm L, Penttilä M, Rousu J, Arvas M. Comparative genome-scale reconstruction of gapless metabolic networks for present and ancestral species. PLoS Comput Biol. 2014;10(2):1003465.
6. Heavner BD, Smallbone K, Price ND, Walker LP. Version 6 of the consensus yeast metabolic network refines biochemical coverage and improves model performance. Database J Biol Databases Curation. 2013;2013:059.
7. Brandl J, Andersen MR. Aspergilli: models for systems biology in filamentous fungi. Curr Opin Syst Biol. 2017;6:67–73.
8. Meyer V, Andersen MR, Brakhage AA, Braus GH, Caddick MX, Cairns TC, de Vries RP, Haarmann T, Hansen K, Hertz-Fowler C, Krappmann S, Mortensen UH, Peñalva MA, Ram AFJ, Head RM. Current challenges of research on filamentous fungi in relation to human welfare and a sustainable bio-economy: a white paper. Fungal Biol Biotechnol. 2016;3:6.
9. Currie JN. The citric acid fermentation of *Aspergillus niger*. J Biol Chem. 1917;31:15–37.
10. AMFEP. List of commercial enzymes 2015. http://www.amfep.org/sites/default/files/201505/Amfep List of Enzymes update May 2015.pdf
11. Workman M, Andersen MR, Thykaer J. Integrated approaches for assessment of cellular performance in industrially relevant filamantous fungi. Ind Biotechnol. 2013;9(6):337–44.
12. Andersen MR, Nielsen ML, Nielsen J. Metabolic model integration of the bibliome, genome, metabolome and reactome of *Aspergillus niger*. Mol Syst Biol. 2008;4:178.
13. David H, Akesson M, Nielsen J. Reconstruction of the central carbon metabolism of *Aspergillus niger*. Eur J Biochem. 2003;270(21):4243–53.
14. Andersen MR, Lehmann L, Nielsen J. Systemic analysis of the response of *Aspergillus niger* to ambient pH. Genome Biol. 2009;10(5):47.
15. Andersen MR, Salazar MP, Schaap PJ, van de Vondervoort PJI, Culley D, Thykaer J, Frisvad JC, Nielsen KF, Albang R, Albermann K, Berka RM, Braus GH, Braus-Stromeyer Sa, Corrochano LM, Dai Z, van Dijck PWM, Hofmann G, Lasure LL, Magnuson JK, Menke H, Meijer M, Meijer SL, Nielsen JJB, Nielsen ML, van Ooyen AJJ, Pel HJ, Poulsen L, Samson RA, Stam H, Tsang A, van den Brink JM, Atkins A, Aerts A, Shapiro H, Pangilinan J, Salamov A, Lou Y, Lindquist E, Lucas S, Grimwood J, Grigoriev IV, Kubicek CP, Martinez D, van Peij NNME, Roubos JA, Nielsen JJB, Baker SE. Comparative genomics of citric-acid-producing *Aspergillus niger* ATCC 1015 versus enzyme-producing CBS 513.88. Genome Res. 2011;21(6):885–97.
16. Lu H, Cao W, Ouyang L, Xia J, Huang M, Chu J, Zhuang Y, Zhang S, Noorman H. Comprehensive reconstruction and in silico analysis of *Aspergillus niger* genome-scale metabolic network model that accounts for 1210 ORFs. Biotechnol Bioeng. 2017;114(3):685–95.
17. Hastings J, De Matos P, Dekker A, Ennis M, Harsha B, Kale N, Muthukrishnan V, Owen G, Turner S, Williams M, Steinbeck C. The ChEBI reference database and ontology for biologically relevant chemistry: enhancements for 2013. Nucleic Acids Res. 2013;41(D1):D456–63.
18. Cerqueira GC, Arnaud MB, Inglis DO, Skrzypek MS, Binkley G, Simison M, Miyasato SR, Binkley J, Orvis J, Shah P, Wymore F, Sherlock G, Wortman JR. The Aspergillus genome database: multispecies curation and incorporation of RNA-Seq data to improve structural gene annotations. Nucleic Acids Res. 2014;42(D1):D705–10.
19. Chang A, Schomburg I, Placzek S, Jeske L, Ulbrich M, Xiao M, Sensen CW, Schomburg D. BRENDA in 2015: exciting developments in its 25th year of existence. Nucleic Acids Res. 2015;43(D1):439–46.
20. Balsa-Canto E, Henriques D, Gabor A, Banga JR. AMIGO2, a toolbox for dynamic modeling, optimization and control in systems biology. Bioinformatics. 2016;32(21):3357–9.
21. Li W, Kondratowicz B, McWilliam H, Nauche S, Lopez R. The annotation-enriched non-redundant patent sequence databases. Database J Biol Databases Curation. 2013;2013:005.
22. Consortium TU. UniProt: the universal protein knowledgebase. Nucleic Acids Res. 2017;45(D1):158–69.
23. Hucka M, Finney A, Sauro HM, Bolouri H, Doyle JC, Kitano H, Arkin AP, Bornstein BJ, Bray D, Cornish-Bowden A, Cuellar AA, Dronov S, Gilles ED, Ginkel M, Gor V, Goryanin II, Hedley WJ, Hodgman TC, Hofmeyr J-H, Hunter PJ, Juty NS, Kasberger JL, Kremling A, Kummer U, Le Novère N, Loew LM, Lucio D, Mendes P, Minch E, Mjolsness ED, Nakayama Y, Nelson MR, Nielsen PF, Sakurada T, Schaff JC, Shapiro BE, Shimizu TS, Spence HD, Stelling J, Takahashi K, Tomita M, Wagner J, Wang J. The systems biology markup language (SBML): a medium for representation and exchange

of biochemical network models. Bioinformatics (Oxford, England). 2003;19(4):524–31.

24. Ebrahim A, Almaas E, Bauer E, Bordbar A, Burgard AP, Chang RL, Dräger A, Famili I, Feist AM, Fleming RM, Fong SS, Hatzimanikatis V, Herrgård MJ, Holder A, Hucka M, Hyduke D, Jamshidi N, Lee SY, LeNovère N, Lerman JA, Lewis NE, Ma D, Mahadevan R, Maranas C, Nagarajan H, Navid A, Nielsen J, Nielsen LK, Nogales J, Noronha A, Pal C, Palsson BØ, Papin JA, Patil KR, Price ND, Reed JL, Saunders M, Senger RS, Nikolaus S, Yuekai S, Thiele I. Do genome-scale models need exact solvers or clearer standards? Mol Syst Biol. 2015;11(10):831.
25. Hayer K, Stratford M, Archer DB. Germination of *Aspergillus niger* conidia is triggered by nitrogen compounds related to L-amino acids. Appl Environ Microbiol. 2014;80(19):6046–53.
26. Steinberg RA. Influence of carbon dioxide on response of *Aspergillus niger* to trace elements. Plant Physiol. 1942;17(1):129–1321691942.
27. Schwede TF, Rétey J, Schulz GE. Crystal structure of histidine ammonia-lyase revealing a novel polypeptide modification as the catalytic electrophile. Biochemistry. 1999;38(17):5355–61.
28. Sieńko M, Natorff R, Owczarek S, Olewiecki I, Paszewski A. Aspergillus nidulans genes encoding reverse transsulfuration enzymes belong to homocysteine regulon. Curr Genet. 2009;55(5):561–70.
29. Mojzita D, Wiebe M, Hilditch S, Boer H, Penttilä M, Richard P. Metabolic engineering of fungal strains for conversion of D-galacturonate to meso-galactarate. Appl Environ Microbiol. 2010;76(1):169–75.
30. Alazi E, Khosravi C, Homan TG, du Pré S, Arentshorst M, Di Falco M, Pham TTM, Peng M, Aguilar-Pontes MV, Visser J, Tsang A, de Vries RP, Ram AFJ. The pathway intermediate 2-keto-3-deoxy-L-galactonate mediates the induction of genes involved in D-galacturonic acid utilization in *Aspergillus niger*. FEBS Lett. 2017;591(10):1408–18.
31. Chibucos MC, Mungall CJ, Balakrishnan R, Christie KR, Huntley RP, White O, Blake JA, Lewis SE, Giglio M. Standardized description of scientific evidence using the Evidence Ontology (ECO). Database J Biol Databases Curation. 2014;2014:075.
32. Fukasawa Y, Tsuji J, Fu S-C, Tomii K, Horton P, Imai K. MitoFates: improved prediction of mitochondrial targeting sequences and their cleavage sites. Mol Cell Proteomics. 2015;14(4):1113–26.
33. Neuberger G, Maurer-Stroh S, Eisenhaber B, Hartig A, Eisenhaber F. Motif refinement of the peroxisomal targeting signal 1 and evaluation of taxon-specific differences. J Mol Biol. 2003;328(3):567–79.
34. Tsang A, Butler G, Powlowski J, Panisko EA, Baker SE. Analytical and computational approaches to define the *Aspergillus niger* secretome. Fungal Genet Biol FG & B. 2009;46(Suppl 1):153–60.
35. Lu X, Sun J, Nimtz M, Wissing J, Zeng A-P, Rinas U. The intra- and extracellular proteome of *Aspergillus niger* growing on defined medium with xylose or maltose as carbon substrate. Microb Cell Factories. 2010;9:23.
36. Adav SS, Li AA, Manavalan A, Punt P, Sze SK. Quantitative iTRAQ secretome analysis of *Aspergillus niger* reveals novel hydrolytic enzymes. J Proteome Res. 2010;9(8):3932–40.
37. Braaksma M, Martens-Uzunova ES, Punt PJ, Schaap PJ. An inventory of the *Aspergillus niger* secretome by combining in silico predictions with shotgun proteomics data. BMC Genomics. 2010;11:584.
38. de Oliveira JMPF, van Passel MWJ, Schaap PJ, de Graaff LH. Proteomic analysis of the secretory response of *Aspergillus niger* to D-maltose and D-xylose. PloS ONE. 2011;6(6):20865.
39. Wang L, Aryal UK, Dai Z, Mason AC, Monroe ME, Tian Z-X, Zhou J-Y, Su D, Weitz KK, Liu T, Camp DG, Smith RD, Baker SE, Qian W-J. Mapping N-linked glycosylation sites in the secretome and whole cells of *Aspergillus niger* using hydrazide chemistry and mass spectrometry. J Proteome Res. 2012;11(1):143–56.
40. Krijgsheld P, Altelaar AFM, Post H, Ringrose JH, Müller WH, Heck AJR, Wösten HAB. Spatially resolving the secretome within the mycelium of the cell factory *Aspergillus niger*. J Proteome Res. 2012;11(5):2807–18.
41. Nitsche BM, Jørgensen TR, Akeroyd M, Meyer V, Ram AFJ. The carbon starvation response of *Aspergillus niger* during submerged cultivation: insights from the transcriptome and secretome. BMC Genomics. 2012;13:380.
42. Krijgsheld P, Bleichrodt R, van Veluw GJ, Wang F, Müller WH, Dijksterhuis J, Wösten HAB. Development in Aspergillus. Stud Mycol. 2013;74:1–29.
43. Dai Z, Aryal UK, Shukla A, Qian W-J, Smith RD, Magnuson JK, Adney WS, Beckham GT, Brunecky R, Himmel ME, Decker SR, Ju X, Zhang X, Baker SE. Impact of alg3 gene deletion on growth, development, pigment production, protein secretion, and functions of recombinant *Trichoderma reesei* cellobiohydrolases in *Aspergillus niger*. Fungal Genet Biol FG & B. 2013;61:120–32.
44. Klaubauf S, Narang HM, Post H, Zhou M, Brunner K, Mach-Aigner AR, Mach RL, Heck AJR, Altelaar AFM, de Vries RP. Similar is not the same: differences in the function of the (hemi-)cellulolytic regulator XlnR (Xlr1/Xyr1) in filamentous fungi. Fungal Genet Biol FG & B. 2014;72:73–81.
45. Sato I, Shimizu M, Hoshino T, Takaya N. The glutathione system of *Aspergillus nidulans* involves a fungus-specific glutathione S-transferase. J Biol Chem. 2009;284(12):8042–53.
46. Kirimura K, Yoda M, Kumatani M, Ishii Y, Kino K, Usami S. Cloning and expression of *Aspergillus niger* icdA gene encoding mitochondrial NADP+-specific isocitrate dehydrogenase. J Biosci Bioeng. 2002;93(2):136–44.
47. Meixner-Monori B, Kubicek CP, Harrer W, Schreferl G, Rohr M. NADP-specific isocitrate dehydrogenase from the citric acid-accumulating fungus *Aspergillus niger*. Biochem J. 1986;236(2):549–57.
48. Freitag J, Ast J, Bölker M. Cryptic peroxisomal targeting via alternative splicing and stop codon read-through in fungi. Nature. 2012;485(7399):522–5.
49. Ebrahim A, Brunk E, Tan J, O'Brien EJ, Kim D, Szubin R, Lerman JA, Lechner A, Sastry A, Bordbar A, Feist AM, Palsson BO. Multi-omic data integration enables discovery of hidden biological regularities. Nature Commun. 2016;7:13091.
50. Andersen MR, Vongsangnak W, Panagiotou G, Salazar MP, Lehmann L, Nielsen J. A trispecies Aspergillus microarray: comparative transcriptomics of three Aspergillus species. Proc Natl Acad Sci USA. 2008;105(11):4387–92.
51. Meijer S, Otero J, Olivares R, Andersen MR, Olsson L, Nielsen J. Overexpression of isocitrate lyase-glyoxylate bypass influence on metabolism in *Aspergillus niger*. Metab Eng. 2009;11(2):107–16.
52. Andersen MR, Giese M, de Vries RP, Nielsen J. Mapping the polysaccharide degradation potential of *Aspergillus niger*. BMC Genomics. 2012;13:313.
53. Poulsen L, Andersen MR, Lantz AE, Thykaer J. Identification of a transcription factor controlling pH-dependent organic acid response in *Aspergillus niger*. PloS ONE. 2012;7(12):50596.
54. Paege N, Jung S, Schäpe P, Müller-Hagen D, Ouedraogo J-P, Heiderich C, Jedamzick J, Nitsche BM, van den Hondel CA, Ram AF, Meyer V. A transcriptome meta-analysis proposes novel biological roles for the antifungal protein AnAFP in *Aspergillus niger*. PloS ONE. 2016;11(11):0165755.
55. Li Y, Chooi YH, Sheng Y, Valentine JS, Tang Y. Comparative characterization of fungal anthracenone and naphthacenedione biosynthetic pathways reveals an $\alpha$-hydroxylation-dependent claisen-like cyclization catalyzed by a dimanganese thioesterase. J Am Chem Soc. 2011;133(39):15773–85.
56. Ebrahim A, Lerman JA, Palsson BO, Hyduke DR. COBRApy: COnstraints-based reconstruction and analysis for python. BMC Syst Biol. 2013;7(1):74.
57. King ZA, Dräger A, Ebrahim A, Sonnenschein N, Lewis NE, Palsson BO. Escher: a web application for building, sharing, and embedding data-rich visualizations of biological pathways. PLoS Comput Biol. 2015;11(8):e1004321.
58. Gruben BS, Mäkelä MR, Kowalczyk JE, Zhou M, Benoit-Gelber I, De Vries RP. Expression-based clustering of CAZyme-encoding genes of *Aspergillus niger*. BMC Genomics. 2017;18(1):900.

# 13

# Emergence and loss of spliceosomal twin introns

Michel Flipphi[1], Norbert Ág[1], Levente Karaffa[1], Napsugár Kavalecz[1], Gustavo Cerqueira[2], Claudio Scazzocchio[3,4] and Erzsébet Fekete[1*]

**Abstract**

**Background:** In the primary transcript of nuclear genes, coding sequences—exons—usually alternate with non-coding sequences—introns. In the evolution of spliceosomal intron–exon structure, extant intron positions can be abandoned and new intron positions can be occupied. Spliceosomal twin introns ("stwintrons") are unconventional intervening sequences where a standard "internal" intron interrupts a canonical splicing motif of a second, "external" intron. The availability of genome sequences of more than a thousand species of fungi provides a unique opportunity to study spliceosomal intron evolution throughout a whole kingdom by means of molecular phylogenetics.

**Results:** A new stwintron was encountered in *Aspergillus nidulans* and *Aspergillus niger*. It is present across three classes of Leotiomyceta in the transcript of a well-conserved gene encoding a putative lipase (*lipS*). It occupies the same position as a standard intron in the orthologue gene in species of the early divergent classes of the Pezizomycetes and the Orbiliomycetes, suggesting that an internal intron has appeared within a pre-extant intron. On the other hand, the stwintron has been lost from certain taxa in Leotiomycetes and Eurotiomycetes at several occasions, most likely by a mechanism involving reverse transcription and homologous recombination. Another ancient stwintron present across whole Pezizomycotina orders—in the transcript of the bifunctional biotin biosynthesis gene *bioDA*—occurs at the same position as a standard intron in many species of non-Dikarya. Nevertheless, also the *bioDA* stwintron has disappeared from certain lineages within the taxa where it occurs, i.e., Sordariomycetes and Botryosphaeriales. Intriguingly, only the internal intron was lost from the Sordariomycetes *bioDA* stwintron at all but one occasion, leaving a standard intron in the same position, while where the putative lipase stwintron was lost, no intronic sequences remain.

**Conclusions:** Molecular phylogeny of the peptide product was used to monitor the existence and fate of a stwintron in the transcripts of two neatly defined fungal genes, encoding well conserved proteins. Both defining events—stwintron emergence and loss—can be explained with extant models for intron insertion and loss. We thus demonstrate that stwintrons can serve as model systems to study spliceosomal intron evolution.

**Keywords:** Spliceosomal twin introns, Spliceosomal intron evolution, Intron gain, Intron loss, Molecular phylogenetics, Pezizomycotina, *Aspergillus nidulans*

## Background

In the primary transcript of nuclear genes, coding sequences—exons—usually alternate with non-coding sequences—introns. The latter are removed and former are joined by means of splicing to create the mRNA ORF that translates into the functional peptide product (for a review, [1]). In the evolution of intron–exon structures of transcripts of nuclear genes, extant intron positions can be abandoned and new intron positions can be occupied. Loss and gain of introns are processes that can take place at different rates leading to three distinct modes of intron dynamics [2]. In most present day species, intron loss and -gain balance each other overall. The availability of complete genome sequences of more than a thousand

*Correspondence: kicsizsoka@yahoo.com
[1] Department of Biochemical Engineering, University of Debrecen, Debrecen 4032, Hungary

species of fungi provides a unique opportunity to study intron evolution in conserved genes throughout a whole kingdom.

Spliceosomal twin introns ("stwintrons") are complex intervening sequences that consist of a canonical U2 intron within another canonical U2 intron, arranged in such a way that one of these (the "internal" intron) interrupts one of the conserved domain sequences of the second one (the "external" intron)—i.e., the donor at the 5′ splice site, the acceptor at the 3′ splice site or the sequence around the lariat branch point adenosine. A consequence of this intron nestling is that the external intron is only functional *after* excision of the internal intron and, hence, consecutive splicing reactions are necessary to generate a mature mRNA. We have characterised stwintrons in different fungal species where the internal intron interrupts the donor of the external intron between the first and the second, or between the second and third nucleotides—[D1,2] and [D2,3] stwintrons, respectively [3–5]. The two-step splicing process resulting in stwintron excision demonstrated that splice sites pair via intron definition (cf. [6]) in filamentous ascomycota (Pezizomycotina). For the [D1,2] stwintron we found in the alternative oxidase transcript of *Helminthosporium solani*, we demonstrated a complex structure that could undergo two alternative pathways of sequential splicing. An internal intron could be envisaged as interrupting the donor sequence *or* the acceptor sequence of an external intron. In every case where a [D1,2] stwintron is extant (which implies that the complex intervening sequence starts with two successive Gs at its 5′ terminus), a stwintron in which the acceptor sequence of the external intron is disrupted between the penultimate and ultimate nucleotides—an [A2,3] stwintron—will occur every time that the 5′ base of the downstream exon is also a G. This complex arrangement results in two mutually exclusive, alternative ways to obtain the same mature mRNA via distinct splicing intermediates. Such an alternatively spliced [D1,2]/[A2,3] stwintron was also found recently in the gene encoding a putative multidrug efflux pump in some taxa of Pezizomycotina [5].

Molecular phylogeny of orthologue proteins encoded by genes that harbour stwintrons and the evolutionary fate of that stwintron may contribute to our understanding of the mechanisms by which spliceosomal introns come into existence and by which they cease to exist. In previous publications, we identified and characterised fungal stwintrons that we had serendipitously encountered while pursuing other work. Unfortunately, all but one of these occur exclusively in one or in a few closely related species. Only the [D2,3] stwintron in the biotin-biosynthetic bifunctional gene *bioDA* was found across a complete class of fungi, the Sordariomycetes. This stwintron thus appears evolutionary stable: loss of any of the internal splice sites will inevitably result in the inability to properly remove the complete intervening sequence from the primary transcript and hence, in biotin auxotrophy. Nevertheless in the *bioDA* gene of *Nectria haematococca*, a standard U2 intron is present at the position occupied by a [D2,3] stwintron in other Sordariomycetes.

To facilitate molecular and mutational study of mechanisms involved in the emergence and disappearance of spliceosomal introns using stwintrons as model systems, we initiated a search for *bona fide* stwintrons in the amenable species *Aspergillus nidulans*, in which molecular- and classical genetic tools are readily developed [7–13]. We devised a crude informatics tool that enabled us to identify putative stwintrons in whole genome sequences, the principle and basics of which were described in detail in [5]. Here, we describe a stwintron we have uncovered in a putative lipase gene after a preliminary screen of the *A. nidulans* genome sequences for [D1,2] stwintrons. We also experimentally verified its presence in *Aspergillus niger*. This stwintron is present across three classes of Pezizomycotina in the transcript of a well-conserved gene without paralogues of note, sufficiently ancient to monitor its emergence and evolution.

## Results

### A new [D1,2] stwintron in *Aspergillus nidulans* and *Aspergillus niger*

An informatics method to detect putative [D1,2] stwintrons has been detailed elsewhere as auxiliary methodology [5]. We screened the genome sequences of *A. nidulans* (whole genome shotgun master accession AACD00000000: annotated scaffolds CH236920–CH236936) and detected, amongst dozens of others, a candidate [D1,2] stwintron at locus AN7524. This locus is predicted by auto-annotation to harbour two intervening sequences. At 69 nt from the ATG, there would be a small canonical intron of 52 nt (5′-GUACGU—33-nt—ACUGAC—4-nt—CAG). 497 nt further downstream, a 289-nt-long, phase 2 intron (5′-GUAUGC—268-nt—GCUAAU—6-nt—CAG) would split a CGA (Arg) codon, while the mature messenger would code for a peptide of 824 amino acids. However, our stwintron search tool suggested the presence of a [D1,2] stwintron at the 5′ end of this large auto-annotated second intron, of which the proposed donor sequence would actually serve as the donor of the 53-nt long internal intron (5′-GUAUGC—33-nt—GCUAAC—5-nt—UAG). This internal intron would disrupt the donor of a 46-nt long external intron between the first and second nt (5′-G|UGAGU—25-nt—GCUGAC—6-nt—AAG). As shown in the splicing scheme depicted in Fig. 1, the 99-nt long [D1,2] stwintron

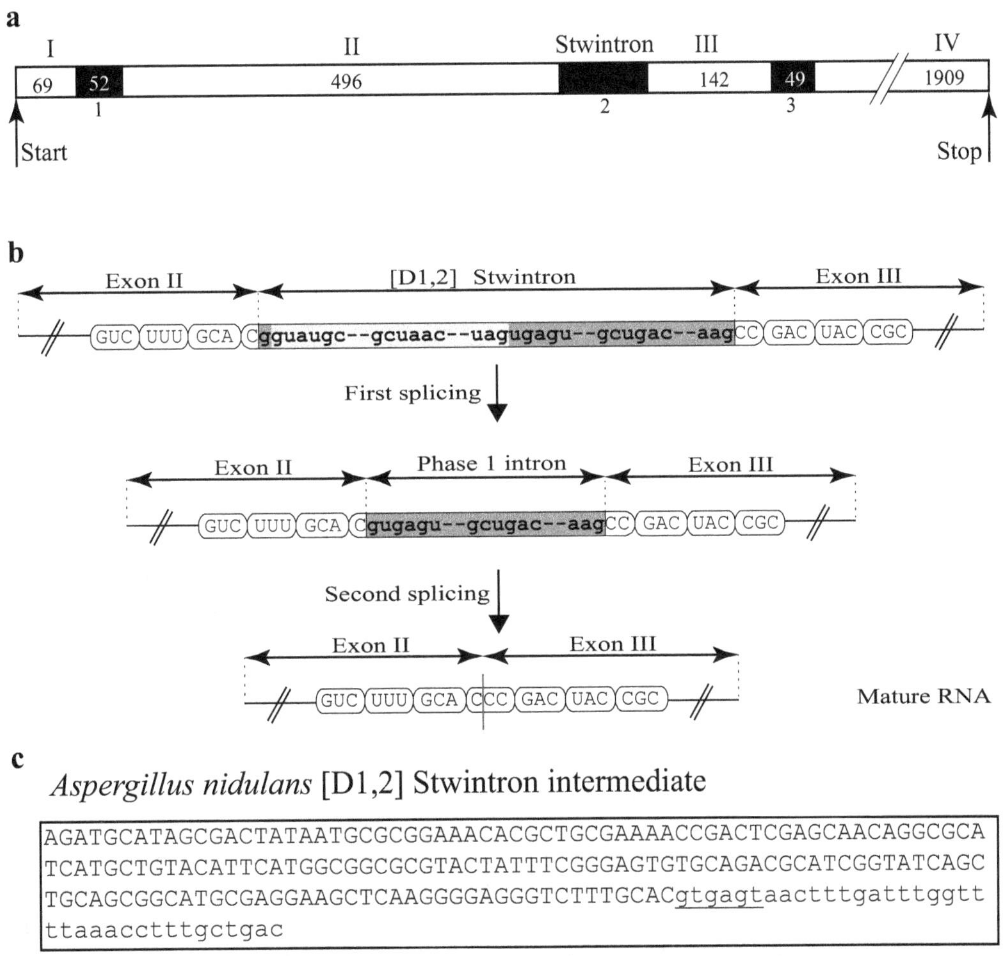


**Fig. 1** The [D1,2] stwintron of the putative lipase encoding gene of *A. nidulans*. **a** Depicts the determined *A. nidulans lipS* intron–exon structure with the exact sizes of the alternating exons and intervening sequences given. The stwintron is the second intervening sequence. **b** Schematically the structure of the phase-1 stwintron that splits the CCC codon of Pro189, and the two consecutive splicing events necessary to remove it. Exonic sequences are printed in capitals, grouped as the consecutive codons (Val186–Phe187–Ala188–Pro189–Asp190–Tyr191–Arg192). The internal intron is marked by the lighter grey bar; its 5′-donor-, lariat branch point domain- and 3′-acceptor sequences are printed in blue lettering. The external intron is marked by the darker grey bar; its 5′-donor-, lariat branch point domain- and 3′-acceptor sequences are printed in red lettering. **c** The existence of the splicing intermediate from which the internal intron has been removed by the first excision, has been confirmed experimentally (GenBank MF612152). The newly formed donor splice site sequence of the retained external intron, (5′-gugagu), is underlined

would be excised by consecutive U2 splicing reactions and would split a CCC (Pro) codon behind the first nt. At the end of the large theoretical auto-annotated intron (289-nt), we predicted the presence of a small U2 intron of 49 nt (5′-GUGAGU—28-nt—GCUAAU—6-nt—CAG) using the closest canonical donor sequence available upstream of the lariat branchpoint domain/acceptor couple. Strict application of intron definition (cf. [6]) for the stwintron at the second intron position and the downstream third intron leaves an exon of 142 nt in between, unrecognised by auto-annotation. According to our predicted intron–exon structure of locus AN7524 (Fig. 1a), there would thus be four exons.

The *A. nidulans* gene is best expressed growing on complete medium as compared with minimal medium conditions used for high-throughput RNA sequencing (JBrowse module at AspGD; [8]). Nevertheless, there are no RNASeq reads from which the complete stwintron was absent nor reads from which the large auto-annotated intron sequence was removed (data not shown). We therefore sought to confirm excision of all three intervening sequences we predicted by sequencing of cDNA generated from total RNA isolated from a 16-h cultivation in complete medium (see "Methods" section for details). The cDNA (GenBank Accession number MF612150) specifies a reading frame of 2616 nt that codes for a peptide of 871 amino acids. The first intron separates the codons for His23 and Tyr24, the [D1,2] stwintron splits the codon for Pro189 while the third intron splits the codon for Ser236 (Fig. 1a). Next, we confirmed the existence of the predicted splicing intermediate of the [D1,2] stwintron (Fig. 1b) with our previously published RT-PCR strategy using a reverse primer that terminally overlaps the 3′ distal junction with the predicted external intron to avoid amplification off fully spliced RNA. Upon cloning and sequencing, the smaller amplified fragment was shown to lack the predicted internal intron (53 nt) with the 5′-GUGAGU donor of the external intron reconstructed (Fig. 1c). We deposited the determined sequence of this splicing intermediate at GenBank (MF612152).

The four exon gene model is supported by Expressed Sequence Tag (EST) clone Asn_02874 from the related fungus *A. niger* (Accession DR703192), covering the stwintron position and the last intron position, 142 nt downstream. We independently verified the positions of the [D1,2] stwintron and the two other extant introns in the orthologue gene (miscalled locus ASPNIDRAFT_53020) by sequencing cDNA (GenBank MF612151) including the complete coding region encoding a peptide of 909 amino acids in *A. niger* ATCC 1015 (see Fig. 2a for the gene model). The 109-nt long [D1,2] stwintron splits the CCU codon for Pro184 in phase 1 (Fig. 2b). A 52-nt long internal intron (5′-GUAAGA—31-nt—GCUAAC—6-nt—CAG) would be nestled in the donor of a 57-nt long external intron between the first and second nt (5′-G|UGAGU—35-nt—GCUAAU—7-nt—CAG). We experimentally confirmed the expected splicing intermediate of the *A. niger* stwintron (GenBank MF612153) from which the predicted 52-nt internal intron was indeed removed (Fig. 2c).

#### Identification of orthologues of the stwintron-containing gene

The peptide product of the stwintron-containing gene at locus AN7524 locally bears considerable similarity (46% identity, 56% similarity in a 182-residue subsequence) with a putative lipase/esterase (EC 3.1.1—carboxylic ester hydrolase) of 316 amino acids found in bacteria from the Burkholderiaceae family (RefSeq protein accession WP_054929686). This bacterial enzyme harbours a characteristic *alpha/beta* hydrolase fold 3 domain (Pfam07859) and is classified as belonging to the Hormone-sensitive_lipase_like_1 family of the *alpha/beta* hydrolase superfamily [14]. We therefore named the gene at locus AN7524, *lipS*, for putative lipase gene with stwintron. Note that the fungal protein is considerably larger than the bacterial *alpha/beta* hydrolases, with the Pfam07859 domain residing in the N-terminal half. Preliminary BLAST searches suggest the existence of a single, orthologue protein of substantial similarity beyond the *alpha/beta* hydrolase fold 3 domain in Pezizomycotina, Taphrinomycotina and all principal lineages of Basidiomycota albeit it is notably absent from Saccharomycotina, the third Ascomycota subphylum. In Ascomycota, there do not appear to be paralogues of high similarity like is the case for the family one Drug/H+ antiporter (DHA1), where we recently identified a genuine stwintron in some encoding genes in 73 species of Pezizomycotina from different classes [5].

46 amino acids from the central part of the sequences corresponding with the (bacterial) *alpha/beta* hydrolase fold 3 domain are encoded in the third exon between the [D1,2] stwintron and the downstream intron. The stwintron thus interrups the DNA encoding a structurally well-defined conserved domain, facilitating in depth analysis of its distribution. We collected hundreds of *lipS* orthologue genes in both phyla of Dikarya upon TBLASTN screening of the DNA databases with the *A. nidulans* protein as the query (data not shown). We noticed that in 52 Basidiomycota orthologue genes—a selection of genome-sequenced species representing all three subphyla—the stwintron position was not occupied despite the wealth of introns in certain taxa: *Basidioascus undulatus*, for instance, has 22 introns in its *lipS* orthologue gene and *Mrakia frigida*, 18. In the Taphrinomycotina subphylum (Ascomycota), the orthologue genes in *Taphrina* species have one intron at a unique position, 21 nt behind the start codon, while the gene is intronless in five other genome-sequenced species (including *Saitoella complicata*). On the other hand, both the stwintron position and the downstream third intron position in the DNA coding for the Pfam07859 domain were occupied in certain taxa of Pezizomycotina (as specified further below): the stwintron position thus appears to be specific to that subphylum. This was confirmed after manually verifying the intron–exon structure of some 150 *lipS* homologue genes in non-Dikarya species where some species harbour more than a dozen *lipS* paralogues (up to 27 paralogues in *Conidiobolus incongruus*).

Figure 3 shows a maximum likelihood tree of the putative lipase LipS protein in 292 species of Ascomycota with Taphrinomycotina proteins as the outgroup. The evolutionary relationships between the proteins in this analysis largely reflect fungal taxonomy. In species belonging to the early divergent classes of the Pezizomycetes and the Orbiliomycetes, the intron position where the [D1,2] stwintron resides in *Aspergillus*, is occupied by a standard U2 intron (the relevant species names are printed in light blue color in Fig. 3). In *Tuber melanosporum*, the U2 intron at the stwintron position is confirmed comparing the genomic sequence with that of EST clone SY0AAB55YD16 (Accession FP429675). Exceptionally,

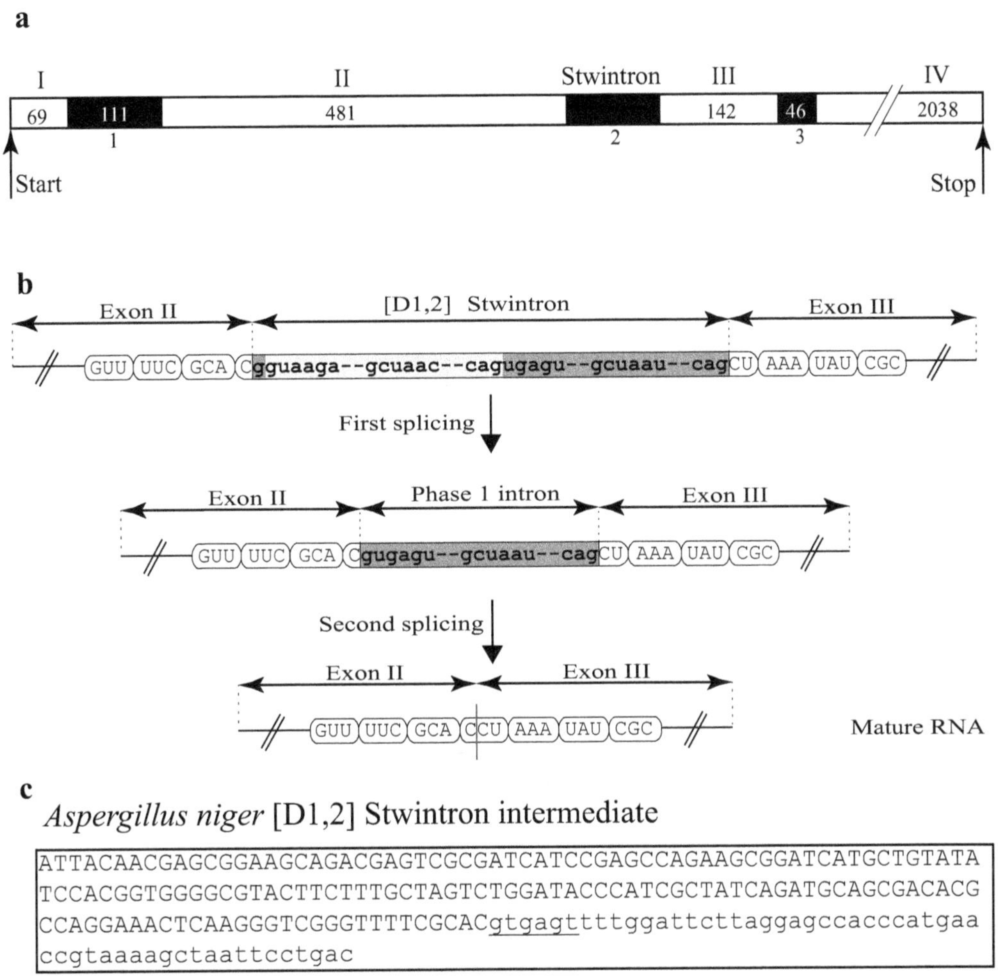


**Fig. 2** The [D1,2] stwintron of the putative lipase encoding gene of *A. niger* ATCC 1015. **a** Depicts the determined *A. niger lipS* intron–exon structure with the exact sizes of the exons and intervening sequences given: The stwintron is the second intervening sequence. **b** Schematically the structure of the stwintron that splits the CCU codon of Pro184, and the two consecutive splicing events necessary to remove it. Exonic sequences (in capitals) are grouped as the consecutive codons (Val181–Phe182–Ala183–Pro184–Lys185–Tyr186–Arg187). The internal intron is marked by the lighter grey bar; its 5′-donor-, lariat branch point domain- and 3′-acceptor sequences are printed in blue lettering. The external intron is marked by the darker grey bar; its 5′-donor-, lariat branch point domain- and 3′-acceptor sequences are printed in red lettering. **c** The existence of the splicing intermediate from which the internal intron has been removed by the first excision, has been confirmed experimentally (GenBank MF612153). The newly formed donor splice site sequence of the retained external intron, (5′-gugagu), is underlined

the *lipS* gene is duplicated in *Dactylellina haptotyla*; both genes have more than ten introns and include two very small exons of 3 nt. On the other hand, *Pyronema omphalodes* appears to have lost all but one intron, and lacks an intron at the stwintron position. Remarkably, the third intron position in *Aspergillus* (142 nt downstream of the stwintron position) is never occupied in species of the two early divergent classes of Pezizomycotina, suggesting it is exclusive to the Leotiomyceta superclass (see, e.g., [15], for a phylogeny-based classification of the Ascomycota). On the other hand, the first *Aspergillus* intron position (69 nt behind the ATG, splitting a His from a Tyr/Phe codon) is occupied in *T. melanosporum* and thus appears to be older.

The *lipS* [D1,2] stwintron is present across the classes of the Eurotiomycetes, Lecanoromycetes and Leotiomycetes (underlined in Fig. 3) and would thus be older than most stwintrons we have uncovered to date. In these Leotiomyceta, all three intron positions that occur in the *A. nidulans* gene (see above) are conserved although certain lineages appear to have lost intronic sequences (as specified further below). On the contrary, in the classes of the Sordariomycetes and the Dothideomycetes, the positions of stwintron and the third *Aspergillus* intron are

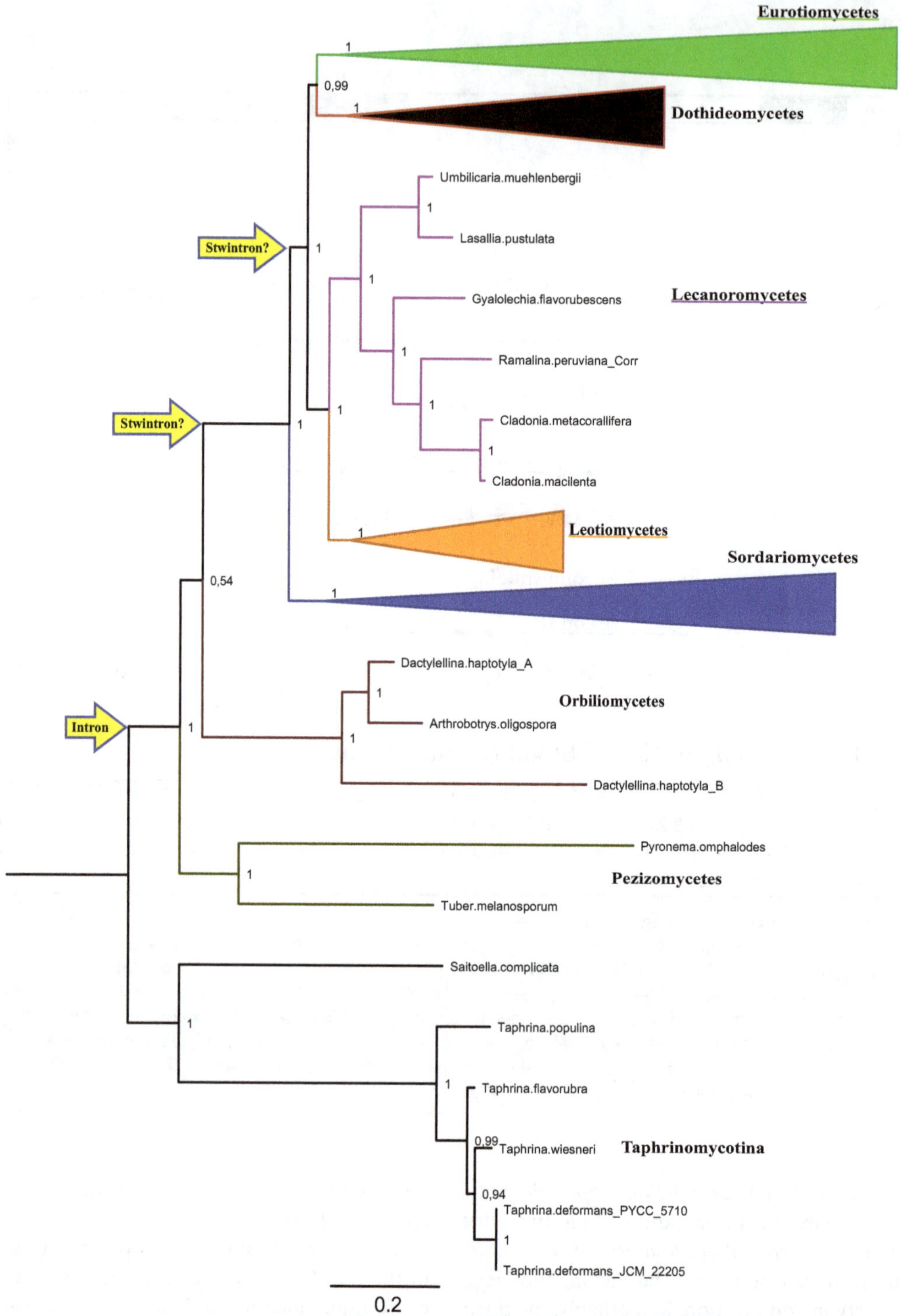


**Fig. 3** Maximum likelihood phylogeny of the LipS orthologue in the Ascomycota: Emergence of the [D1,2] stwintron. Taphrinomycotina proteins constitute the outgroup. Branch statistics are given as Approximate Likelihood Ratio Test values (0–1) at each node. The branches are color coded to distinguish the classes of Pezizomycotina; olive green: Pezizomycetes, auburn: Orbiliomycetes, blue: Sordariomycetes, green: Eurotiomycetes (NB. Including *A. nidulans*), red: Dothideomycetes, orange: Leotiomycetes, violet: Lecanoromycetes. The latter five classes belong to the Leotiomyceta super class. For the sake of simplicity, the clades for the four well-represented classes—Sordariomycetes, Eurotiomycetes, Dothideomycetes and Leotiomycetes—are collapsed. Classes in which the stwintron occurs are underlined. Species of Pezizomycetes and Orbiliomycetes, in which the stwintron position (second intron position in *Aspergillus*) is occupied by a standard intron in the orthologue gene, have their names printed in light blue. The two defining events in the formation of the stwintron are indicated by the yellow arrows at the left. The data do not allow to determine whether the stwintron emerged in Leotiomyceta before or after the divergence of the Sordariomycetes

not occupied. Sordariomycetes and some taxa of Dothideomycetes (Botryosphaeriales and Dothideales, e.g.) still harbor an intron at the first *Aspergillus* intron position but in Pleosporales, the *lipS* intron–exon structure has been reset completely with three introns appearing at new positions. In most Capnodiales, the *lipS* gene is intronless. Importantly, we could not find Leotiomyceta species in which the *lipS* stwintron position is occupied by a standard canonical intron, hence, this latter situation is unique to Pezizomycetes and Orbiliomycetes species. The LipS phylogeny thus strongly suggests that the [D1,2] stwintron has emerged at the position of a pre-extant canonical "host" intron, currently serving as the stwintron's external intron.

Figures 4 and 5 show the clades of species that include the stwintron in more detail, to highlight instances of stwintron loss in Eurotiomycetes and Leotiomycetes, respectively (fungi that lost the *lipS* stwintron are marked in red lettering in both figures). In the Eurotiomycetes (Fig. 4), the stwintron is completely absent from genome-sequenced species in the orders of the Chaetothyriales, Verrucariales and Phaeomoniellales. In the Eurotiales order, the stwintron is completely absent from the genus of *Penicillium*, the sequenced *Monascus* and *Xeromyces* species as well as from *Aspergillus terreus* while in the Onygenales, *Ascosphaera apis* lacks it. Thus the stwintron has been lost with the bordering exons fused at at least five independent occasions in the Eurotiomycetes. Meanwhile, the stwintron is absent in eight species of Leotiomycetes (Fig. 5). The latter include all species from the family of the Sclerotiniaceae in our analysis, along with *Pseudogymnoascus pannorum*, *Glarea lozoyensis*, *Cairneyella variabilis* and "*Geotrichum candidum* 3C" (NB. The *G. candidum* 3C genome sequences (WGS Master Accession JMRO01000000) strongly suggest that they are from a species of Leotiomycetes rather than from a species of Saccharomycetales: All Saccharomycetales lack a *lipS* orthologue, see above). It is remarkable that in the majority of the above cases of stwintron loss, the intron at the third *Aspergillus* position has also disappeared, suggesting simultaneous loss. Only in *A. terreus*, *A. apis* and "*G. candidum* 3C", the downstream intron is retained, while we found the reverse situation (i.e., only the intron at the third *Aspergillus* position lost) in *Marssonina brunnea* and *Calycina herbarum*.

#### Reappraisal of the *bioDA* stwintron

In a previous stwintron study, we had identified another potentially ancient stwintron in *bioDA*, a structural gene of D-biotin biosynthesis encoding a bifunctional protein exhibiting dethiobiotin synthetase (EC 6.3.3.3—BioD) and *S*-adenosyl-L-methionine:8-amino-7-oxononanoate aminotransferase (EC 2.6.1.62—BioA) activities (cf. [16]). We showed that a [D2,3] stwintron occurs at the position of the first (standard) intron in *A. nidulans bioDA* in most (31) species of the Sordariomycetes class and the two species of the Botryosphaeriales order of the Dothideomycetes class for which genome sequences were available at the time. In a phylogenetic analysis ([3]; see Supplementary Fig. S4 thereof), the two Botryosphaeriales proteins clustered with those from the Sordariomycetes and not with those from the other Dothideomycetes (24 species). Remarkably, in *Nectria haemotococca* (formerly known as *Fusarium solani f. sp. pisi*) a standard intron occupies the position of the stwintron in the other Sordariomycetes, including in other *Fusarium* species. Since the spring of 2013, hundreds of new Pezizomycotina genome sequences have been added to publicly accessible databases. We took the opportunity to update our analysis of the *bioDA* stwintron, to compare defining aspects of its evolution with those of the (new) stwintron in the putative lipase *lipS* gene (see above).

We screened the databases with TBLASTN using the *A. nidulans* protein (protein accession number ACR44943) as the query and collected hundreds of fungal *bioDA* genes for which we manually deduced the intron–exon structure and the cognate protein (results not shown). From the updated collection from non-Pezizomycotina taxa, we can now confirm that the intron position at which the stwintron occurs (in Sordariomycetes and Botryosphaeriales, see below) is deeply rooted in fungal evolution (see, e.g., [17], for the taxonomy of the fungal kingdom). This intron position is occupied by a standard U2 intron in species of Mucoromycotina, Mortierellomycotina, Entomophthoromycotina and Chytridiomycota albeit absent from *Rhizophagus irregularis* (Glomeromycota) and Blastocladiomycota (results not shown). Remarkably, in both sequenced strains of *Batrachochytrium dendrobatidis*, the donor of this intron is a non-canonical 5′-GAAAGA. (NB. *bioDA* is absent from Microsporidia, Cryptomycota, Neocallimastigomycota and Kickxellomycotina). Within Dikarya, this intron position is conserved across the three subphyla of the Basidiomycota. We also found the stwintron position occupied by a standard intron in three species of Saccharomycetales (*Saprochaete clavata*, *Lipomyces starkeyi* and *Sugiyamaella xylanicola*) amongst those that (still) have the *bioDA* gene. On the other hand, there is no intron present at that position in the seven genome-sequenced species of Taphrinomycotina that have the *bioDA* gene (results not shown).

We carried out a phylogenetic analysis of 298 Ascomycete BioDA proteins which we had deduced manually from our *bioDA* gene collection. Taphrinomycotina serve as the outgroup and the Saccharomycotina proteins constitute a sister clade to the Pezizomycotina (Fig. 6).

**Fig. 4** Maximum likelihood phylogeny of LipS: Instances of stwintron loss in Eurotiomycetes. A subtree of the phylogenetic analysis depicted in Fig. 3, is shown in detail to highlight the loss of the stwintron from taxa of Eurotiomycetes. Class-specific color coding is the same as in Fig. 3. Species that have lost the stwintron from their *lipS* gene have their name printed in red lettering. For the sake of simplicity, we have collapsed groups of related fungi that behave identically with respect to stwintron presence/absence. All the species in the collapsed clades for the Chaetothyriales order and the *Penicillium* genus have no intronic sequences at the stwintron position (second intron position in *Aspergillus*) and the taxon names are therefore marked with red boxes to highlight stwintron loss. Independent events of stwintron loss are also indicated by red triangles on the directly preceding branches (NB. The position of the triangles does not correspond with the exact time point at which stwintron loss has taken place)

**Fig. 5** Maximum likelihood phylogeny of LipS: Instances of stwintron loss in Leotiomycetes. A subtree of the phylogenetic analysis depicted in Fig. 3, is shown in detail to highlight the loss of the stwintron from taxa of Leotiomycetes. Class-specific color coding is the same as in Fig. 3. Species that have lost the stwintron from their *lipS* gene have their name printed in red lettering. Independent events of stwintron loss are also indicated by red triangles on the directly preceding branches (NB. The position of the triangles does not correspond with the exact time point at which stwintron loss has taken place)

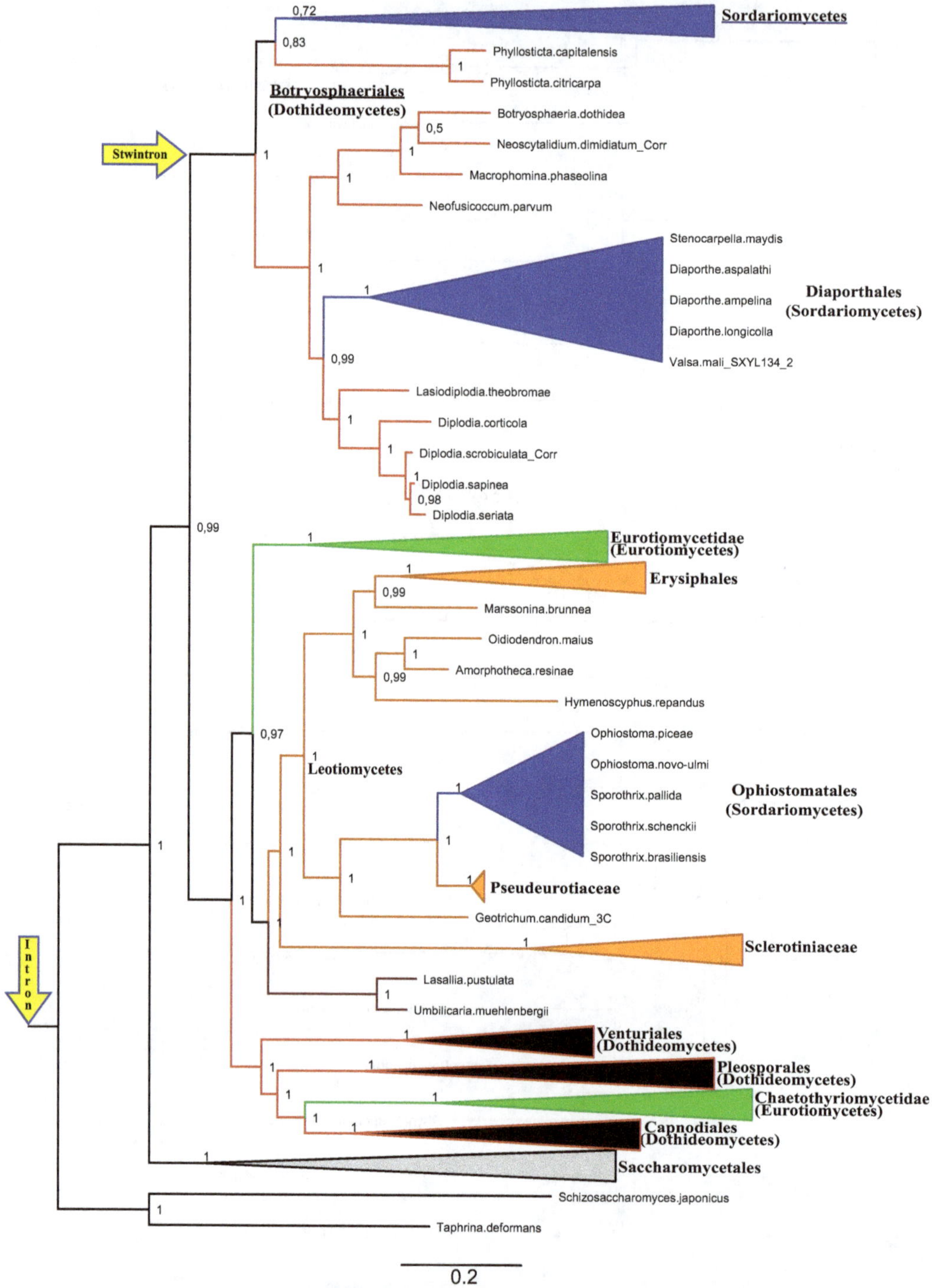


**Fig. 6** Maximum likelihood phylogeny of the BioDA protein in the Ascomycota: Emergence of the [D2,3] stwintron. Taphrinomycotina proteins constitute the outgroup. Branch statistics are given as Approximate Likelihood Ratio Test values (0–1) at each node. The branches are color coded to distinguish classes of Pezizomycotina, as described in the legend to Fig. 3. Some fungal taxa are collapsed, to the level of whole classes for most of the Pezizomycotina. Taxa in which the stwintron occurs—Sordariomycetes and Botryosphaeriales—are underlined. The two groups of Sordariomycetes that stand out for the absence of intronic sequences at the *bioDA* stwintron position—members of two families of Diaporthales and five species of the order of the Ophiostomatales, respectively—are cartooned (blue) rather than collapsed. The branch where the stwintron has emerged at the position of an ancient intron is indicated by the upper yellow arrow. The latter "host" intron occurs in many species of non-Dikarya (see "Results" section): To indicate its kingdom-wide existence, the lower yellow arrow points at the origin of the tree. Note that a Maximum Likelihood phylogeny based on an alignment calculated using more stringent parameters (similarity matrix BLOSUM62) suggested that the Botryosphaeriales clade is paraphyletic to the (main) Sordariomycetes clade

Unlike for the putative lipase LipS, the evolutionary relations between Pezizomycotina BioDA proteins do not conform the standard fungal taxonomy as, for instance, the classes of the Sordariomycetes (cf. [18]), Dothideomycetes (cf. [19]) and Eurotiomycetes (cf. [20]) do not behave as monophyletic.

However, the fate of the relevant intron position in Pezizomycotina *bioDA* appeared more conventional. The [D2,3] stwintron occurs in all orders and in most species of Sordariomycetes and in Botryosphaeriales (Dothideomycetes), which cluster together in one clade (Fig. 6) albeit not all species in this clade have the *bioDA* stwintron. In the other Dothideomycetes (including, the Pleosporales and Capnodiales orders) as well as in the Eurotiomycetes class, a standard U2 intron is present at the *bioDA* stwintron position. On the other hand, in the Leotiomycetes class and in the Umbilicariomycetidae family (NB. *Lasallia pustulata* and *Umbilicaria muehlenbergii*; the other genome-sequenced Lecanoromycetes do not have *bioDA*) that ancient intron position is lost.

The *bioDA* genes from some of the species of the Ophiostomatales order of the Sordariomycetes appear to originate from a Leotiomycetes ancestor, and consequently do not harbour an intron at the stwintron position. This is probably the strongest indication to lateral transmission of *bioDA* between not directly related taxons of Pezizomycotina. This particular *bioDA* gene transfer is likely to have occurred recently as Ophiostomatales species *Rafaelea quercivora* and *Leptographium procerum* do harbour the *bioDA* stwintron and their BioDA proteins cluster together with those of all other Sordariomycetes orders. Such an event is not exceptional for structural genes of biotin biosynthesis: The acquisition of *bioA* and *bioD* genes from an unspecified bacterial source is well documented for *Saccharomyces cerevisiae* [21].

Figure 7 shows the clade of species that harbour the *bioDA* stwintron in more detail. Unlike the situation with the putative lipase stwintron, most instances of stwintron loss from the *bioDA* gene result in the presence of a standard intron at the same position (relevant species names in green in Fig. 7). Loss of the internal intron has happened at at least five independent instances in the clade that constitute the large majority of the Sordariomycetes species, in three different orders—Hypocreales, Xylariales and Microascales—including the instance previously observed in *N. haematococca bioDA*. In Botryosphaeriales, we observe the loss of the internal intron in five of the eleven species investigated, corresponding to two independent events. Furthermore, in two Botryosphaeriales species, *Phyllosticta capitalensis* and *Neoscytalidium dimidiatum*, there are no intronic sequences left at the stwintron position (species names in red in Fig. 7). The current phylogenetic analysis suggests that the loss of the complete stwintron in *N. dimidiatum* happened in one event, like in all cases of stwintron loss from the *lipS* gene (see above; Figs. 4, 5). On the contrary, the stwintron sequences may have been lost from *P. capitalensis bioDA* in two consecutive intron loss events, since its close relative *P. citricarpa* (same genus) has retained a standard intron at the stwintron position.

A number of species from the Diaporthales order—more specifically, the members of the Diaporthaceae and Valsaceae families—also have a *bioDA* gene from which the stwintron is completely absent (relevant species names in red in Fig. 7). The genes encoding the five clustered Diaporthales proteins appear to derive from a Botryosphaeriales taxon diverged from the main Sordariomycetes branch, the latter including species of another Diaporthales family, the Cryphonectriaceae (*Chrysoporthe cubensis*, *Chrysoporthe deuterocubensis* and *Cryphonectria radicalis*). Interestingly, *Valsa mali* (WGS Master accessions JUIY00000000 and JUIZ00000000) has two *bioDA* genes: The gene that encodes the protein in the separate Diaporthales clade has lost all stwintron sequences while the second gene, which carries a complete stwintron, is more related with Xylariales genes. Our current phylogenetic analysis suggests that the absence of all *bioDA* stwintron sequences from the five species of Diaporthales is due to their loss in one event (like in *N. dimidiatum*, see above) since one of the species in the divergent branch of Botryosphaeriales, *Lasiodiplodia theobromae*, (still) harbours a stwintron.

(See figure on next page.)
**Fig. 7** Maximum likelihood phylogeny of the BioDA protein: Instances of stwintron loss. A subtree of the phylogenetic analysis depicted in Fig. 6, is shown in detail to highlight the loss of the stwintron from taxa of Sordariomycetes and Botryosphaeriales. Class-specific color coding is the same as in Fig. 3. Species that have lost the internal intron from the stwintron in their *bioDA* gene (retaining a standard intron at the stwintron position) have their name printed in green lettering. For the sake of simplicity, we have collapsed groups of related fungi that behave the same with respect to the stwintron. Independent events of internal intron loss are also indicated by the green triangles on the directly preceding branches. (NB. The position of the triangles does not correspond with the exact time point at which intron loss has taken place). The complete absence of intronic sequences at the stwintron position from the genes encoding the BioDA proteins in the separate Diaporthales clade and in two of the Botryosphaeriales (7 proteins in red lettering) may have occurred either in one or in two consecutive events of intron loss. Complete stwintron loss was indicated by a red triangle (as in the legend to Fig. 4). We annotated the tree for the two-step process in the case of *P. capitalensis*, first losing the internal intron in an ancestor shared with *P.citricarpa* (green triangle). The subsequent loss of the standard intron that remained after the first event (red triangle) resulted in the complete absence of intronic sequences from *P. capitalensis bioDA*

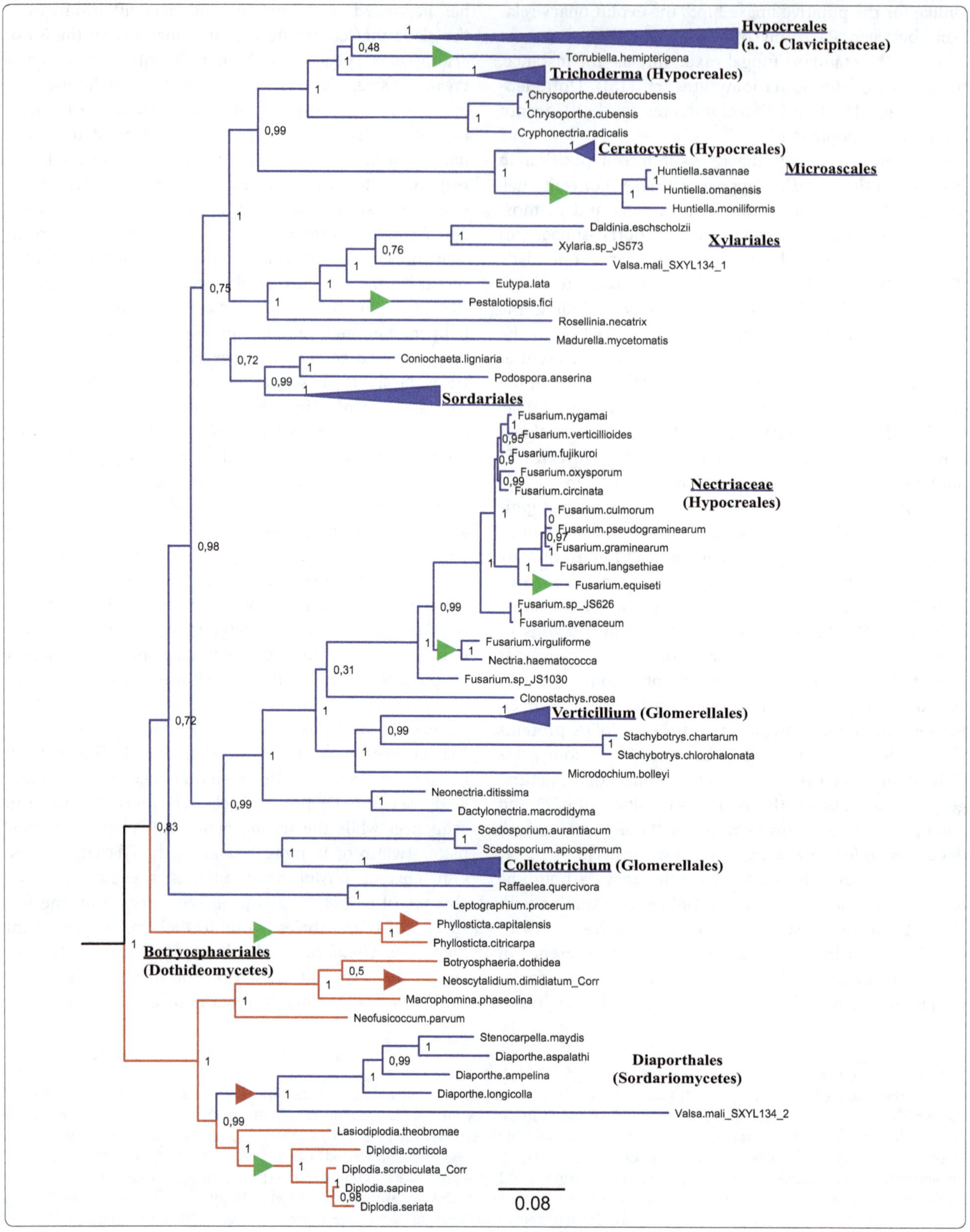
Hypocreales
(a. o. Clavicipitaceae)
Torrubiella.hemipterigena
Trichoderma (Hypocreales)
Chrysoporthe.deuterocubensis
Chrysoporthe.cubensis
Cryphonectria.radicalis
Ceratocystis (Hypocreales)
Microascales
Huntiella.savannae
Huntiella.omanensis
Huntiella.moniliformis
Daldinia.eschscholzii
Xylaria.sp_JS573
Xylariales
Valsa.mali_SXYL134_1
Eutypa.lata
Pestalotiopsis.fici
Rosellinia.necatrix
Madurella.mycetomatis
Coniochaeta.ligniaria
Podospora.anserina
Sordariales
Fusarium.nygamai
Fusarium.verticillioides
Fusarium.fujikuroi
Fusarium.oxysporum
Fusarium.circinata
Nectriaceae
(Hypocreales)
Fusarium.culmorum
Fusarium.pseudograminearum
Fusarium.graminearum
Fusarium.langsethiae
Fusarium.equiseti
Fusarium.sp_JS626
Fusarium.avenaceum
Fusarium.virguliforme
Nectria.haematococca
Fusarium.sp_JS1030
Clonostachys.rosea
Verticillium (Glomerellales)
Stachybotrys.chartarum
Stachybotrys.chlorohalonata
Microdochium.bolleyi
Neonectria.ditissima
Dactylonectria.macrodidyma
Scedosporium.aurantiacum
Scedosporium.apiospermum
Colletotrichum (Glomerellales)
Raffaelea.quercivora
Leptographium.procerum
Phyllosticta.capitalensis
Phyllosticta.citricarpa
Botryosphaeriales
(Dothideomycetes)
Botryosphaeria.dothidea
Neoscytalidium.dimidiatum_Corr
Macrophomina.phaseolina
Neofusicoccum.parvum
Stenocarpella.maydis
Diaporthe.aspalathi
Diaporthales
(Sordariomycetes)
Diaporthe.ampelina
Diaporthe.longicolla
Valsa.mali_SXYL134_2
Lasiodiplodia.theobromae
Diplodia.corticola
Diplodia.scrobiculata_Corr
Diplodia.sapinea
Diplodia.seriata
0.08

## Discussion

In this paper, we evidence a new [D1,2] stwintron in a putative lipase gene in *A. nidulans* and *A. niger*. It resides within DNA that encodes a well conserved *alpha/beta* hydrolase fold domain allowing deep phylogenetic analysis to trace the emergence and evolution of this stwintron. In parallel, we monitored the fate of the ancient stwintron in the bifunctional biotin biosynthesis gene *bioDA*, previously characterised in *Trichoderma reesei* (cf. [3]).

The fungal BioDA phylogeny does not conform to the accepted taxonomy, however, the groups of fungi wherein the [D2,3] stwintron occurs cluster together to form a clearly defined clade (Fig. 6). The [D2,3] stwintron in most Sordariomycetes and in various Botryosphaeriales occupies the same position as a standard U2 intron in other Dothideomycetes orders and in the Eurotiomycetes. This is a very ancient intron position which appears to predate the emergence of Dikarya. Therefore, our results leave little doubt that the [D2,3] stwintron emerged by the appearance or insertion of an "internal" intron between the second and third nt of the donor sequence of a pre-extant "host" intron in the *bioDA* gene. It is possible that the internal intron has "matured" within the host intron, a process we have previously termed "stwintronisation" [3] as it resembles the "intronisation" of exonic sequences [22]. Alternatively, intron insertion may have taken place abruptly in one event, during the repair of double strand DNA breaks by the non-homologous end joining machinery [23]). These two modes of stwintron formation by the appearance of a new intron in one of the terminal splice site sequences of a pre-extant intron, are consistent with two proposed mechanisms of intron gain from an endogenous origin (reviewed by [24]).

Phylogenetics of the putative lipase LipS (Fig. 3) suggests that the [D1,2] stwintron present in certain taxa of the Leotiomyceta superclass was formed in essentially the same way, at the position of an older, canonical U2 intron. This latter has survived as a standard intron in the early divergent classes of Pezizomycetes and Orbiliomycetes while the external intron of the stwintron in Leotiomyceta derives from it. From our data collection, it is not clear whether the *lipS* stwintron appeared before or after the divergence of the Sordariomycetes from the other Leotiomyceta lineages. Sordariomycetes lack intervening sequences at the *lipS* stwintron position. This situation could result from either the loss of the predecessor standard intron before the first appearance of the stwintron, or the class-specific loss of the complete stwintron after its formation in a common ancestor of all Leotiomyceta.

Such a "stwintron loss" event appears to have occurred at several occasions in the evolution of the *lipS* orthologous gene in the Leotiomyceta superclass, including other events that have led to the absence of the stwintron from all species from the class of the Dothideomycetes, all species from the superorder of the Chaetothyriomycetidae (Eurotiomycetes class) and all species from the *Penicillium* genus (Eurotiales order of the Eurotiomycetes class). Each of these instances of stwintron loss appears to coincide with the absence or the loss of the intron at the third position in the *Aspergillus lipS* gene, 142 bp downstream of the stwintron. Indeed, the simultaneous loss of separate intervening sequences may have occurred in all but four instances (all four apparently involving just one species—*A. terreus*, *A. apis*, *C. variabilis* and "*G. candidum* 3C") (Figs. 4, 5). On the other hand, in Sordariomycetes, the loss of the stwintron from the *bioDA* gene has resulted in the retention of a standard U2 intron in all but one instance (i.e., that involving the separate clade of five Diaportales, see above) (Fig. 7). It would thus appear that in most Sordariomycetes *bioDA*, the event(s) leading to the emergence of an internal intron was/were reversed to return to the primordial situation.

Despite this rather striking difference between the two ancient stwintrons, all observed instances of accurate intron loss are compatible with one single mechanism, involving reverse transcription of spliced transcripts and homologous recombination of cDNA at the genome locus (e.g., [25]; reviewed by [26]). This mechanism allows the simultaneous disappearance of two introns (bordering the same exon) but also the removal of the internal intron from a stwintron, if the stwintron splicing intermediate rather than the mature messenger served as reverse transcriptase template. It is tempting to speculate that the longevity of the stwintron splicing intermediate in the nucleus—dependent of the relative excision rates of the internal- and the external introns and the export of the messenger from the nucleus—is an important factor in the reversal of a stwintron into a standard intron at the same position. We have proven recently that alternative removal of terminally overlapping internal introns from an alternatively spliced [D1,2]/[A2,3] stwintron leads to discordant introns in orthologue genes [5], suggesting that the events we observed in the current study of the *bioDA* stwintron are not exceptional.

## Conclusions

Molecular phylogeny of the peptide product was used to monitor the emergence and disappearance of an ancient spliceosomal twin intron in the transcripts of two neatly defined fungal genes, encoding well conserved proteins. These stwintrons occur across complete fungal orders and classes and both defining events can be explained with extant models for intron insertion and -loss. We thus demonstrated that stwintrons can serve as model systems to study spliceosomal intron evolution.

## Methods

### Nucleic acid isolation

*Aspergillus nidulans* ATCC 48756 (R21) and *A. niger* ATCC 1015 were used to confirm the existence of a stwintron in the transcript of their orthologue genes for a putative lipase characterised by the well-conserved *alpha/beta* hydrolase fold 3 domain (Pfam07859). Standardised medium compositions are described elsewhere [10]. Fungal biomass was generated in 500-mL Erlenmeyer flasks with 100 mL of complete medium seeded with vegetative spore inoculum, in a rotary shaker (Infors HT Multitron) at 200 rotations per min. 16 h after inoculation, mycelia were harvested by filtration, thoroughly washed with distilled water, and subsequently frozen and ground to powder under liquid nitrogen. For the extraction of genomic DNA and total RNA from the mycelial powder, Macherey–Nagel NucleoSpin kits (NucleoSpin Plant II and NucleoSpin RNA Plant) were used.

### Reverse transcription PCR (RT-PCR)

Reverse transcription was performed with 1 μg of total RNA as the template and Oligo(dT) as a primer in a 20 μL reaction volume using the RevertAid First Strand cDNA Synthesis Kit (Thermo Scientific). Subsequent PCR reactions were done in a 25 μL volume containing 4 μL of the single strand cDNA, using gene-specific oligonucleotides (Additional file 1: Table S1) as primers and DreamTaq DNA Polymerase (Thermo Scientific). Cycling conditions after initial denaturation at 95 °C for 2 min were: 35 cycles of 95 °C for 30 s, 60 °C for 1 min, and 72 °C for 1 min, followed by one post-cyclic elongation at 72 °C for 5 min. Amplified fragments were resolved in native agarose gels.

To confirm the existence of the predicted stwintron splicing intermediates, we used the same approach as previously with PCR primer pairs that do not amplify off cDNA template from fully spliced mRNA. This strategy usually yields two fragments of defined sizes of which the smaller one corresponds to the splicing intermediate and the bigger one, to primary transcript. All experiments were done in duplicate, starting with biomass from two independent liquid cultures.

### cDNA sequencing

Double strand cDNA was gel-purified (NucleoSpin Gel and PCR Clean-up, Macherey–Nagel) and cloned (pGEM-T Easy Vector System I, Promega). Plasmid DNA was isolated using the NucleoSpin Plasmid EasyPure kit (Macherey–Nagel). Three independent clones were sequenced over both strands using universal primers hybridising to the vector (MWG-Biotech AG, Ebersberg, Germany) or gene-specific oligonucleotide primers where appropriate. cDNA sequences were deposited at GenBank under accession numbers MF612150–MF612153.

### Phylogenetic analyses

For the analysis of the putative lipase orthologue, 292 proteins were aligned with Multiple Alignment using Fast Fourier Transform (MAFFT; version 7) [27], applying the E-INS-i algorithm and a BLOSUM 62 similarity matrix. The MAFFT alignment was trimmed with Block Mapping and Gathering with Entropy software (BMGE version 1.12) [28] using BLOSUM 30 and a block size of 5. A Maximum Likelihood tree was then calculated online from the curated alignment by PhyML 3.0 [29] using the WAG substitution model.

For the study of the bifunctional BioDA protein, 298 proteins were aligned using a BLOSUM 45 similarity matrix in the MAFFT E-INS-i module. The alignment was BMGE trimmed using BLOSUM 55 and a block size of 5, before a Maximum Likelihood phylogeny was inferred using the WAG substitution model.

For each tree, approximate Likelihood Ratio Tests provide statistical branch support [30]. The trees were drawn from the Newick output of PhyML with the FigTree program (http://tree.bio.ed.ac.uk/software/figtree) and further annotated with Adobe Illustrator.

#### Abbreviation

nt: nucleotides.

#### Authors' contributions

MF, LK and EF conceived this study. MF, GC and EF discovered the new *lipS* stwintron. NK and EF provided the experimental evidence. MF, NÁ and EF mined the DNA databases. MF and CS performed phylogenetic analyses. MF and EF wrote the paper, with contributions from NÁ and CS. All Authors read and approved the final manuscript.


#### Author details

[1] Department of Biochemical Engineering, University of Debrecen, Debrecen 4032, Hungary. [2] Broad Institute of MIT and Harvard, Cambridge, MA, USA. [3] Department of Microbiology, Imperial College London, London, UK. [4] Institut de Biologie Intégrative de la Cellule, CEA/CNRS, Université Paris-Saclay UMR, 9198 Orsay, France.


#### Competing interests

The authors declare that they have no competing interests.


#### Funding

This work was supported by the EU and co-financed by the European Regional Development Fund [GINOP-2.3.2-15-2016-00008], the Hungarian Scientific Research Fund [OTKA NN116519 to LK], and a Bólyai János Research Scholarship [BO/00548/14 to LK].

**References**

1. Irimia M, Roy SW. Origin of spliceosomal introns and alternative splicing. Cold Spring Harb Perspect Biol. 2014. doi:10.1101/cshperspect.a016071.
2. Carmel L, Wolf YI, Rogozin IB, Koonin EV. Three distinct modes of intron dynamics in the evolution of eukaryotes. Genome Res. 2007;17:1034–44.
3. Flipphi M, Fekete E, Ág N, Scazzocchio C, Karaffa L. Spliceosome twin introns in fungal nuclear transcripts. Fungal Genet Biol. 2013;57:48–57.
4. Ág N, Flipphi M, Karaffa L, Scazzocchio C, Fekete E. Alternatively spliced, spliceosomal twin introns in *Helminthosporium solani*. Fungal Genet Biol. 2015;85:7–13.
5. Fekete E, Flipphi M, Ág N, Kavalecz N, Cerqueira G, Scazzocchio C, et al. A mechanism for a single nucleotide intron shift. Nucleic Acids Res. 2017;45:9085–92.
6. Berget SM. Exon recognition in vertebrate splicing. J Biol Chem. 1995;270:2411–4.
7. Casselton L, Zolan M. The art and design of genetic screens: filamentous fungi. Nat Rev Genet. 2002;3:683–97.
8. Cerqueira GC, Arnaud MB, Inglis DO, Skrzypek MS, Binkley G, Simison M, et al. The Aspergillus Genome Database: multispecies curation and incorporation of RNA-Seq data to improve structural gene annotations. Nucleic Acids Res. 2014;42:D705–10.
9. Galagan JE, Calvo SE, Cuomo C, Ma LJ, Wortman JR, Batzoglou S, et al. Sequencing of *Aspergillus nidulans* and comparative analysis with *A. fumigatus* and *A. oryzae*. Nature. 2005;438:1105–15.
10. Pontecorvo G, Roper JA, Hemmons LM, MacDonald KD, Bufton AWJ. The genetics of *Aspergillus nidulans*. Adv Genet. 1953;5:141–238.
11. Sibthorp C, Wu H, Cowley G, Wong PWH, Palaima P, Morozov IY, et al. Transcriptome analysis of the filamentous fungus *Aspergillus nidulans* directed to the global identification of promoters. BMC Genom. 2013;14:847.
12. Tilburn J, Scazzocchio C, Taylor GG, Zabicky-Zissman JH, Lockington RA, Davies RW. Transformation by integration in *Aspergillus nidulans*. Gene. 1983;26:205–21.
13. Wortman JR, Gilsenan JM, Joardar V, Deegan J, Clutterbuck J, Andersen MR, et al. The 2008 update of the *Aspergillus nidulans* genome annotation: a community effort. Fungal Genet Biol. 2009;46(Suppl 1):S2–13.
14. Lenfant N, Hotelier T, Velluet E, Bourne Y, Marchot P, Chatonnet A. ESTHER, the database of the *alpha/beta*-hydrolase fold superfamily of proteins: tools to explore diversity of functions. Nucleic Acids Res. 2013;41:D423–9.
15. Schoch CL, Sung GH, López-Giráldez F, Townsend JP, Miadlikowska J, Hofstetter V, et al. The Ascomycota tree of life: a phylum-wide phylogeny clarifies the origin and evolution of fundamental reproductive and ecological traits. Syst Biol. 2009;58:224–39.
16. Magliano P, Flipphi M, Sanglard D, Poirier Y. Characterization of the *Aspergillus nidulans* biotin biosynthetic gene cluster and use of the *bioDA* gene as a new transformation marker. Fungal Genet Biol. 2011;48:208–15.
17. Hibbett DS, Binder M, Bischoff JF, Blackwell M, Cannon PF, Eriksson OE, et al. A higher-level phylogenetic classification of the Fungi. Mycol Res. 2007;111:509–47.
18. Maharachchikumbura SSN, Hyde KD, Jones EBG, McKenzie EHC, Huang S-K, Abdel-Wahab MA, et al. Towards a natural classification and backbone tree for Sordariomycetes. Fungal Divers. 2015;72:199–301.
19. Hyde KD, Jones EBG, Liu J-K, Ariyawansa H, Boehm E, Boonmee S, et al. Families of dothideomycetes. Fungal Divers. 2013;63:1–313.
20. Geiser DM, Gueidan C, Miadlikowska J, Lutzoni F, Kauff F, Hofstetter V, et al. Eurotiomycetes: eurotiomycetidae and chaetothyriomycetidae. Mycologia. 2006;98:1053–64.
21. Hall C, Dietrich FS. The reacquisition of biotin prototrophy in *Saccharomyces cerevisiae* involved horizontal gene transfer, gene duplication and gene clustering. Genetics. 2007;177:2293–307.
22. Irimia M, Rukov JL, Penny D, Vinther J, Garcia-Fernandez J, Roy SW. Origin of introns by 'intronization' of exonic sequences. Trends Genet. 2008;24:378–81.
23. Farlow A, Meduri E, Schlotterer C. DNA double-strand break repair and the evolution of intron density. Trends Genet. 2011;27:1–6.
24. Yenerall P, Zhou L. Identifying the mechanisms of intron gain: progress and trends. Biol Direct. 2012;7:29.
25. Derr LK. The involvement of cellular recombination and repair genes in RNA-mediated recombination in *Saccharomyces cerevisiae*. Genetics. 1998;148:937–45.
26. Roy SW, Gilbert W. The evolution of spliceosomal introns: patterns, puzzles and progress. Nat Rev Genet. 2006;7:211–21.
27. Katoh K, Standley DM. MAFFT multiple sequence alignment software version 7: improvements in performance and usability. Mol Biol Evol. 2013;30:772–80.
28. Criscuolo A, Gribaldo S. BMGE (Block Mapping and Gathering with Entropy): a new software for selection of phylogenetic informative regions from multiple sequence alignments. BMC Evol Biol. 2010;10:210.
29. Guindon S, Dufayard JF, Lefort V, Anisimova M, Hordijk W, Gascuel O. New algorithms and methods to estimate maximum-likelihood phylogenies: assessing the performance of PhyML 3.0. Syst Biol. 2010;59:307–21.
30. Anisimova M, Gascuel O. Approximate likelihood-ratio test for branches: a fast, accurate, and powerful alternative. Syst Biol. 2006;55:539–52.

14

# A silver bullet in a golden age of functional genomics: the impact of *Agrobacterium*-mediated transformation of fungi

Alexander Idnurm[1*], Andy M. Bailey[2], Timothy C. Cairns[3], Candace E. Elliott[1], Gary D. Foster[2], Giuseppe Ianiri[4] and Junhyun Jeon[5]

**Abstract**

The implementation of *Agrobacterium tumefaciens* as a transformation tool revolutionized approaches to discover and understand gene functions in a large number of fungal species. *A. tumefaciens* mediated transformation (*At*MT) is one of the most transformative technologies for research on fungi developed in the last 20 years, a development arguably only surpassed by the impact of genomics. *At*MT has been widely applied in forward genetics, whereby generation of strain libraries using random T-DNA insertional mutagenesis, combined with phenotypic screening, has enabled the genetic basis of many processes to be elucidated. Alternatively, *At*MT has been fundamental for reverse genetics, where mutant isolates are generated with targeted gene deletions or disruptions, enabling gene functional roles to be determined. When combined with concomitant advances in genomics, both forward and reverse approaches using *At*MT have enabled complex fungal phenotypes to be dissected at the molecular and genetic level. Additionally, in several cases *At*MT has paved the way for the development of new species to act as models for specific areas of fungal biology, particularly in plant pathogenic ascomycetes and in a number of basidiomycete species. Despite its impact, the implementation of *At*MT has been uneven in the fungi. This review provides insight into the dynamics of expansion of new research tools into a large research community and across multiple organisms. As such, *At*MT in the fungi, beyond the demonstrated and continuing power for gene discovery and as a facile transformation tool, provides a model to understand how other technologies that are just being pioneered, e.g. CRISPR/Cas, may play roles in fungi and other eukaryotic species.

**Keywords:** Functional genomics, Mycota, Pathogenicity genes, *Rhizobium radiobacter*, Transfer DNA

## Background

In 1998, Dunn-Coleman and Wang published a commentary on a newly described system for the transformation of foreign DNA into filamentous fungi using *Agrobacterium tumefaciens*: the catchy term in their article's title was that this method was potentially a "silver bullet" [1]. We think of a "silver bullet" as a missile to combat werewolves or other fantasy monsters, yet here there is an additional metaphor; DNA is shot into a fungal genome to cause damage to a key gene, and thereby that can provide information on the strengths and weaknesses of the fungus. This remarkable use of a plant pathogenic bacterium, *A. tumefaciens*, to transform fungi had first been demonstrated in the model yeast *Saccharomyces cerevisiae* just a few years earlier [2, 3], and then extended in 1998 to seven species of filamentous fungi in both the Ascomycota and Basidiomycota lineages [4]. Within a decade from its first reported use in *S. cerevisiae*, by 2005 over 50 fungal species had been transformed with *A. tumefaciens* [5]. In the decade since then, the use of *A. tumefaciens* mediated transformation (*At*MT) continued

*Correspondence: alexander.idnurm@unimelb.edu.au
[1] School of BioSciences, University of Melbourne, Melbourne, VIC 3010, Australia
Full list of author information is available at the end of the article

to expand to become a standard experimental technique within the tool-box for gene manipulation in many fungal species. For some species it became the easiest or even the only method by which to introduce foreign DNA. In other species, it emerged as a powerful technique for forward genetics, for use in the creation of large collections of strains carrying random T-DNA insertions and their analysis, for reverse genetics to create specific targeted gene replacements, or for manipulation of gene expression for biotechnological benefits.

In this review we describe the rise and influence of *At*MT on the understanding of fundamental aspects of fungi. We describe species or groups of fungi in which *At*MT has had greatest impact, some of the limitations that have subsequently emerged in applications, and areas of research or fungal species in which this transformation technology did not have as great an impact. Understanding how this technology was implemented can guide or anticipate the benefits of future technologies for advancing research on fungi.

It is not possible to include specific details from all the publications reporting the use of *At*MT on fungi, even if covering those since the review by Michielse et al. [5]. A PubMed search of "Agrobacterium AND fungus" returns more than 900 papers, and as an example in the *Cryptococcus neoformans* species complex alone *At*MT has been used in more than 30 studies. Other comprehensive and insightful reviews address specific aspects of this technique, e.g. different vectors that are available [6], or the proteins encoded by *A. tumefaciens* that are required to transform organisms [7, 8], which this review aims to complement. Finally, "impact" is relative in that what may appear important to one set of researchers may not to another set, while individual people may have personal favorite experiments or discoveries made using the technique.

## *Agrobacterium tumefaciens* and how it transforms species

*Agrobacterium tumefaciens* is a plant pathogen in the class α-Proteobacterium that is best known as one of nature's natural agents in creating genetically modified organisms. In this process, the bacterium inserts a piece of a plasmid into the nucleus of the plant host cell, and in the wild that bacterial DNA encodes proteins that modify the plant growth in favor of the bacterium. In most circumstances, this results in the formation of a non-proliferative gall or tumor-like growth on a plant, with alterations in the recipient genome that are not normally carried on into subsequent plant generations. However, analysis of the genome of sweet potato indicates that in rare cases these transformation events can be integrated more permanently into the genome [9]. The *Agrobacterium* genus is within the family Rhizobiaceae and as such it is closely related to the genus *Rhizobium*, members of which also form intimate associations with plants to fix atmospheric nitrogen. *A. tumefaciens* was renamed *Rhizobium radiobacter* [10], although the community using this species as a transformation technology for fungi continues to use the name *A. tumefaciens*.

Before the development of genome sequencing projects, the only known example of horizontal gene transfer from bacteria to eukaryotes was the trans-conjugation mediated by *A. tumefaciens* [11]. *Agrobacterium* naturally exists in an environment where it encounters numerous hosts, including fungi that are likely to be present at the plant wounds, which induce T-DNA transfer. Knight et al. [12] demonstrated that it is entirely feasible that such transformation events happen in a natural environment. They co-cultivated the plant pathogenic fungus *Verticillium albo-atrum* on plant material alongside an *Agrobacterium* strain containing a plasmid that could potentially transform fungi, and observed transformation of the fungus under these *in planta* conditions [12]. Of course, in the wild such an event is unlikely to deliver any beneficial DNA sequence into the fungus, so may well not confer any selectable advantage, but it is interesting to speculate on the frequency of such events over an evolutionary timescale: indeed, in addition to plants [9], genome sequencing projects have identified *Agrobacterium*-like DNA in the genomes of some fungi such as *Aspergillus oryzae* [13].

Plant molecular biologists altered wild strains of *A. tumefaciens* to their own advantage. The bacterial strains and genetic material were modified to prevent gall formation, and to establish systems in which DNA for transformation into a plant can be placed between two direct repeats of 25 bp (the left and right borders of the transfer or T-DNA) (Fig. 1). From the perspective of bacterial genetics, rather than transformation it is more accurate to describe the movement of the T-DNA from the bacterium into the eukaryotic host as a trans-conjugation method of gene transfer, in this case a conjugation mechanism that is capable of occurring across different species. The promiscuous nature of *A. tumefaciens*, to target a wide diversity of hosts, enabled it to be applied to numerous other eukaryotic species, with members of the fungi being the best examples beyond the model plant *Arabidopsis thaliana* [11].

One of the initial downsides of transformation of fungi with *Agrobacterium* was the perceived inflexibility with the vector systems. Some of the *Agrobacterium* vectors were rather unwieldy, with limited restriction sites for conventional cloning and, having been developed for

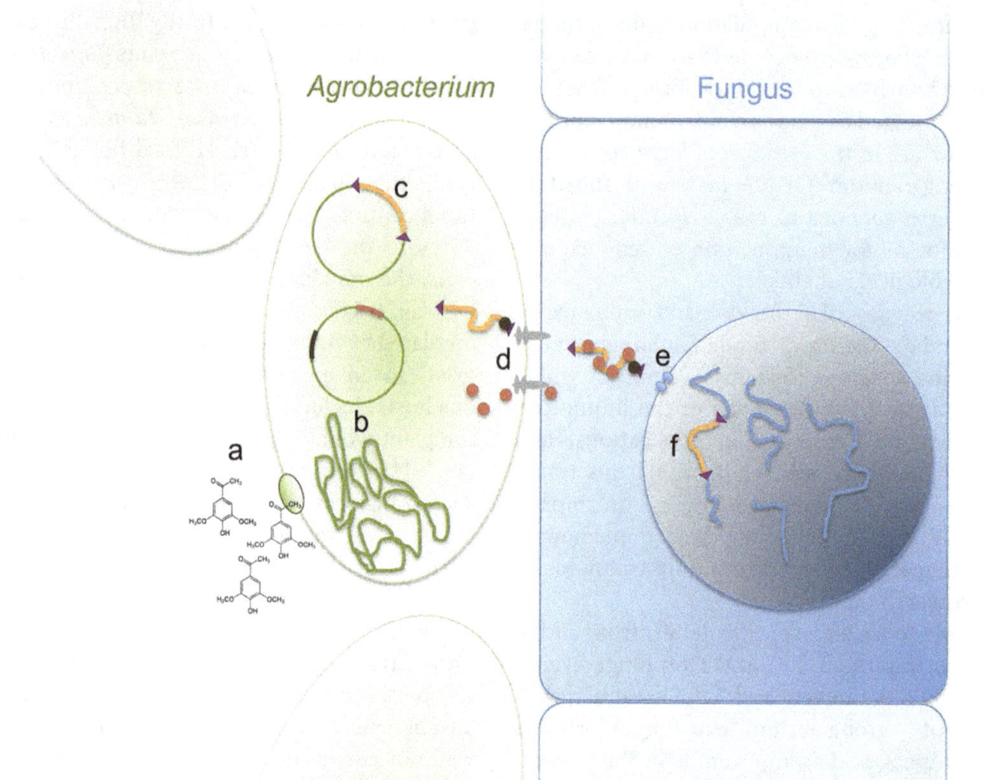


**Fig. 1** A summary of the transformation process of fungi by *Agrobacterium tumefaciens*. At step *a*, acetosyringone present in the transformation media is recognized by the bacterium, triggering *b* the expression of genes within the chromosomal DNA and from the helper plasmid. At stage *c*, the T-DNA is excised from the T-plasmid. The linear DNA is bound by VirD2, which is recognised by the type IV secretion system for export from the cell, along with other Vir proteins (*d*). At step *e*, the DNA enters the fungal nucleus guided by nuclear localization signals on the bacterial-encoded proteins. In the final event *f*, the T-DNA inserts into the chromosomal DNA of the fungus. The T-DNA should encode a selectable marker such that the integration events can be selected

plants, were prefabricated with selection cassettes and regulatory elements specifically for plant transformation. Fortunately some of the suites of vectors made for plants (e.g. pGreen, pCambia, etc.) also included small vectors that had empty T-DNA regions that were amenable to easy manipulation. These have been updated to allow vector construction by means such as yeast recombination in *S. cerevisiae* [14–16], the Gateway system [17] or Golden Gate assembly [18], making vector construction and deployment simple and amenable to high throughput approaches.

Another limit of the early vectors was the length of DNA that could be inserted, which was a significant technical restriction for complementation of large genes. One solution to this problem was the development of the BIBAC system, which enabled the modification of any bacterial artificial chromosome (BAC) containing a large piece of fungal DNA to contain left and right border sequences, thus enabling direct fungal transformation. Large fragments (up to 75 kb) of DNA were successfully transferred into the *Fusarium oxysporum* f. sp. *lycopersici* genome using this method [19]. In a similar manner, a system to convert BACs into vectors suitable for *At*MT in the *Ustilago maydis* has also been developed [20].

## Advantages of *At*MT over other transformation techniques

Transformation tools existed for fungi prior to the development of *At*MT, for example using the protoplast/polyethylene glycol or cation/polyethylene glycol approaches [21, 22]. However, *At*MT provided improvements over many of these methods, explaining why it became a transformation tool of choice in many fungal species.

Firstly, *At*MT eliminates the need to remove the fungal cell wall to make protoplasts. While protoplasting is an established method in some species, in others it is difficult and variable in success. Fungi have a suite of cell wall types that differ between species and that change at different stages during growth and development. These differences likely explain why the ease and success of protoplasting vary between species; this is not helped by the difficulties in obtaining suitable cell wall degrading enzymes. In contrast, although *Agrobacterium* does show cell type preferences, as discussed later in the section on the mushroom-forming Agaricomycotina, *Agrobacterium* can transform species across a wide spectrum of evolution, including mammalian cells [23] and oomycetes [24], and many different tissue or cell types in fungi.

A second significant advantage to *At*MT over other approaches is that the T-DNA can integrate randomly into the genome. Consequently, much of the impact of *At*MT comes from the perspective of random mutagenesis as a resource for forward genetic screens. At the time of development of *At*MT, the process of restriction enzyme mediated integration (REMI) was the insertional mutagenesis method of choice. This method includes restriction enzymes in the stage when DNA is transformed into protoplasted cells. A number of problems arose with this method such as mutations not linked to the inserted DNA, which were proposed to be caused by the restriction enzymes causing damage to the DNA. Other insertional mutagenesis tools include transposon insertions, although these usually require the design of specific constructs for each species. *At*MT largely superseded REMI as the insertional mutagenesis tool in fungi [5]. Usually the T-DNA inserts as a single copy into the genome, so any change in phenotype is likely caused by the insertion. After screening libraries of T-DNA insertion transformants for phenotypes, either side of the T-DNA insert are then obtained by difference methods, which are most often PCR-based, in order to identify the affected gene. Typically, the function of the genes identified using T-DNA mutagenesis is confirmed through (a) linkage analysis of the progeny obtained from crosses between a strain of opposite mating type and the T-DNA mutant, if the mutation does not affect the sexual cycle; (b) generation of a targeted replacement allele by means of transformation techniques suitable for the studied organism, or (c) complementation with a wild type copy of the gene.

Thirdly, *At*MT is amenable for use in reverse genetic approaches for targeted gene deletion or disruption. This differs from random insertional mutagenesis, as transformation vectors are supplemented with DNA sequences that mediate homologous recombination of the exogenous cassette with specific loci of the recipient genome. Thus, *At*MT can be used for targeted replacement at desired genomic regions, most obviously a putative open reading frame. *At*MT was often developed in conjuncture with the isolation of mutants in the non-homologous end joining DNA repair process [25]. Mutation of this pathway helps increase the proportion of transformants that have gene replacements. Over the past decade, there has been a rapid increase in publicly available genome sequences of fungi [26], which has enabled facile identification of individual genes or gene families that can be analyzed by targeted gene deletion. This has occurred in parallel with development of numerous molecular tools, including inducible promoter systems [27], recyclable markers [25] and most recently CRISPR-Cas genome editing [28]. These techniques now promise functional genomic analyses at a high throughput level, and systems-level insight into industrial tractability, processes essential for disease, and putative drug targets of many fungi [29]. Consequently, targeted manipulation of fungal genomes using *At*MT is a critical technique that will facilitate the implementation of more recent breakthroughs.

Finally, having a method for easy transformation "leveled the playing field" for discovering gene function in what had up until then been dominated by the model species for molecular biology experiments. This particularly became the case for non-conventional species as soon as their genome sequence became available. Examples are given later in this review.

## Trends in the research of fungi using *At*MT

In their 2005 review, Michielse et al. [5] described specific features about *At*MT, and then some research trends many that continued over the following decade. A large focus has been on the efficiency of transformation as influenced by co-culture conditions, e.g. bacterial and fungal cell concentrations, temperature, length of co-incubation, and concentration of acetosyringone, which is a plant metabolite released from wounded roots that enhances *A. tumefaciens* transformation (Fig. 1). The direction of this focus on transformation efficiency relates to the prior challenges in obtaining large numbers of transformants from protoplasts or other methods, and hence the efforts to optimize conditions to maximize the number of transformants obtained per experiment. However, this is not crucial because *At*MT is technically easy, and if more transformants are needed they can be obtained just by increasing the number of transformation experiments to be performed. Conversely, it is ideal to generate a library of single T-DNA insertion mutants quickly and unequivocally link the phenotype of interest with the genetic mutation, but little work has addressed if the numbers of transformants obtained correlate with the number of integration events per transformant. There is a tendency for more publications reporting the first use of *At*MT in a species, and fewer on the full implementation of the technique for making mutations in genes, or other purposes. Given that the following examples represent only a small proportion of all species successfully transformed using *Agrobacterium*, there remains a large untapped resource waiting for gene discovery in a wide diversity of fungi.

## Fungal species and biological questions in which *At*MT made greatest contributions

Application of *At*MT depends on the question being asked, and this versatile tool has suited answering such questions in different fungi. Approaches include individual gene deletion or disruption experiments, analyses of gene classes resulting in several dozen mutant strains, to the generation of large libraries consisting of thousands of strains. Nevertheless, it is notable that some species, or indeed fungal lineages, have more widely adapted *At*MT as a common research tool. The following sections are divided based on the evolution of the fungi (Fig. 2): examples from Ascomycetes with the focus on the plant pathogens, Basidiomycetes, and a brief section on the earlier, paraphyletic fungal lineages.

### Phylum Ascomycota and the role of *At*MT on understanding gene functions in plant pathogenicity

The impact of *At*MT in the ascomycetes has been of greatest relevance to the plant pathogens, with little impact on model species like *S. cerevisiae*, *Neurospora crassa* or *Aspergillus nidulans* wherein a long history of research and efficient methods for transformation, classical genetics, and gene identification were already in place. There are thousands of ascomycete species that infect plants. One way to measure the impact of *At*MT is to examine its role in understanding gene functions in particularly problematic species. In a proposed "top ten" list of plant pathogens (Table 1, [30]), many species benefited from this technique. For those species that did not benefit, this was either because efficient methods for transformation and gene discovery were already available for them (e.g. the basidiomycete *U. maydis*) or because they are obligate pathogens and therefore difficult to co-culture with *Agrobacterium* (e.g. *Puccinia* species, *Blumeria graminis* and *Melampsora lini*). In the diverse species in which *At*MT was adopted, this technology opened the opportunity for high throughput mutant screens or construction of mutant libraries. Thus, *At*MT has been applied to a number of plant pathogenic fungi including many of the economically important pathogens. Although T-DNA integration varies depending on the system, *At*MT has been consistently an efficient tool for the genetic study of fungal pathogenesis. The following sections describe ascomycete plant pathogenic species or genera in which *At*MT has been widely used, and finishes with one human pathogen example.

#### *Pyricularia oryzae*

*Pyricularia oryzae* (Sordariomycetes) is the causal agent of rice blast, the most serious disease of cultivated rice. The species is also referred to as *Magnaporthe oryzae* or *M. grisea* in older literature [31]. Rice blast disease

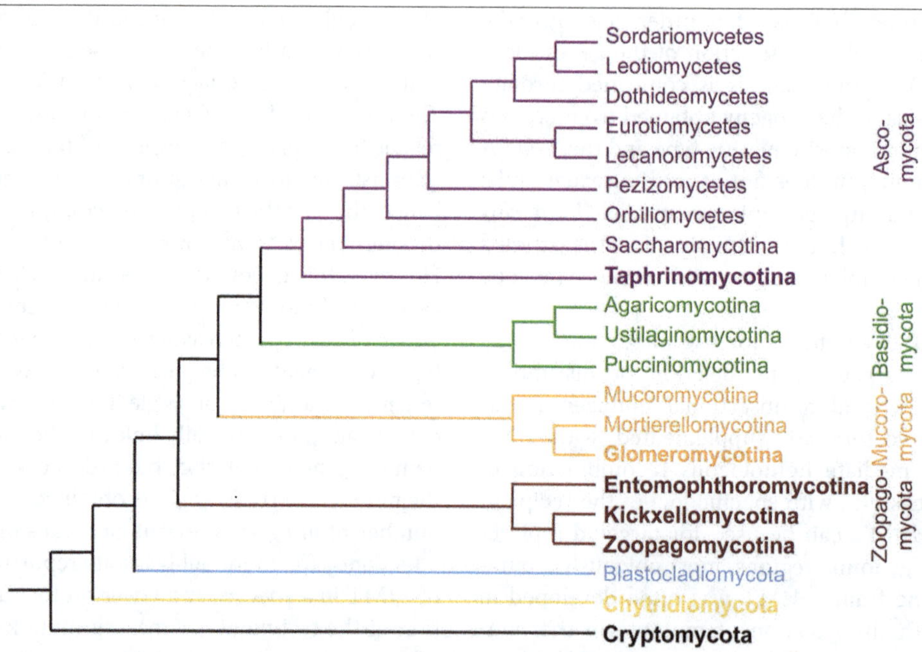


**Fig. 2** Phylogeny of major lineages within the fungi. The relationships are based on references [204, 243]; note that some nodes remain poorly defined. For hierarchy: phylum -mycota, subphylum -mycotina, class -mycetes. The Pezizomycotina subphylum is split into seven classes in the phylogeny. Groups in which species have not yet been transformed successfully by *Agrobacterium* are in bold

**Table 1 A "top ten" list of fungal plant pathogens for research at the molecular biology level [30], and the impact of *At*MT on these species**

| Rank | Name | Growth capabilities | *At*MT success | Impact | Key references |
|---|---|---|---|---|---|
| 1 | *Pyricularia oryzae* (*Magnaporthe oryzae*) | In vitro | Yes | Major | [15, 38–42] |
| 2 | *Botrytis cinerea* | In vitro | Yes | Modest | [244] |
| 3 | *Puccinia* spp. | Obligate pathogen | No | N/A | N/A |
| 4 | *Fusarium graminearum* | In vitro | Yes | Minor | [65] |
| 5 | *Fusarium oxysporum* | In vitro | Yes | Major | [75, 76] |
| 6 | *Blumeria graminis* | Obligate pathogen | No | N/A | N/A |
| 7 | *Zymoseptoria tritici* (*Mycosphaerella graminicola*) | In vitro | Yes | Major | [52–54, 245] |
| 8 | *Colletotrichum* spp. | In vitro | Yes | Major | [84, 86] |
| 9 | *Ustilago maydis* | In vitro | Yes | Minor | [198] |
| 10 | *Melampsora lini* | Obligate pathogen | Yes | Minor | [176] |

destroys enough rice to feed 60 million people every year [32]. Considering that rice is a staple food accounting for major caloric and protein intake in many countries (http://www.irri.org/), the disease is one of the major threats to global food security. Due to the experimental tractability of *P. oryzae* and the socioeconomic impact of rice blast, the fungus has served as an important model to understand the biology of fungal plant pathogens [33, 34]. Although genome sequence information had offered great opportunities to discern possible genetic attributes that confer pathogenicity on the fungus [35], low efficiency of gene knockout hampered translating genome sequences into meaningful biological information. To overcome this bottleneck, insertional mutagenesis techniques such as restriction enzyme-mediated integration (REMI) and transposon-arrayed gene knockout (TAGKO) were developed to generate mutants and examine the function of disrupted genes [36, 37].

The first demonstration of *At*MT for large-scale analysis of gene functions in any plant pathogenic fungus came from a study in which over 20,000 *P. oryzae* insertional transformants were generated [38]. Southern blot analysis revealed that >80% of transformants had a single T-DNA copy within their genome. In parallel, laboratories in the USA [39] and China produced more than 150,000 *At*MT mutants, establishing the most extensive insertional mutant libraries in any plant pathogenic fungus.

Analyses of T-DNA insertion patterns in the *P. oryzae* transformants showed that T-DNA integration favored promoter regions of genes that have an AT-rich base composition [40–42]. In addition, direct or inverted repeats of T-DNA, chromosomal rearrangements and inclusion of additional plasmid vector were also observed. Despite these biases, T-DNA insertions are relatively evenly distributed throughout all of the chromosomes, suggesting the potential of *At*MT as a tool for forward genetics. A high throughput phenotype screening system was developed to identify and characterize the transformants that are affected in key developmental steps of the life cycle, including pathogenicity [38]. The screens yielded more than 180,000 data points, which are archived and analyzed by a relational database (http://atmt.riceblast.snu.ac.kr/) (Fig. 3). Such high throughput phenotype screening in combination with identification of genes tagged by T-DNA in individual mutants led to the identification of 203 independent loci implicated in fungal pathogenicity. This represents the largest, unbiased set of putative pathogenicity genes for a single fungal species. The majority of putative pathogenicity genes tagged by T-DNA in the study included novel genes, although the list contained known pathogenicity genes, such as *NTH1*, which encodes neutral trehalase [43]. The value of the *At*MT approach is exemplified by many subsequent discoveries of genes that control aspects of fungal pathogenicity. One example amongst many is the detailed analysis of a novel gene, *DES1* (plant defense suppression 1), required for the suppression of the basal defense responses in the host plant [44]. *DES1* was identified from a T-DNA insertion site that was 750 bp from the closest predicted gene, and analysis of progeny of sexual crosses confirmed the segregation of antibiotic resistance with this interesting phenotype of aberrant conidial morphogenesis. This is an example of a gene that would not have been prioritized for analysis via a reverse genetics approach as it encodes a serine rich protein with no obvious functional domains or characterized homologues in other ascomycetes.

*At*MT can also be used to make gene replacement mutant strains in *P. oryzae*. For example, binary plasmids have been generated that are compatible with replication in *Agrobacterium* and *S. cerevisiae*, thus enabling assembly of transformation vectors using yeast recombinational cloning [15]. Such vectors can be used to create targeted gene knockouts in *P. oryzae* by homologous

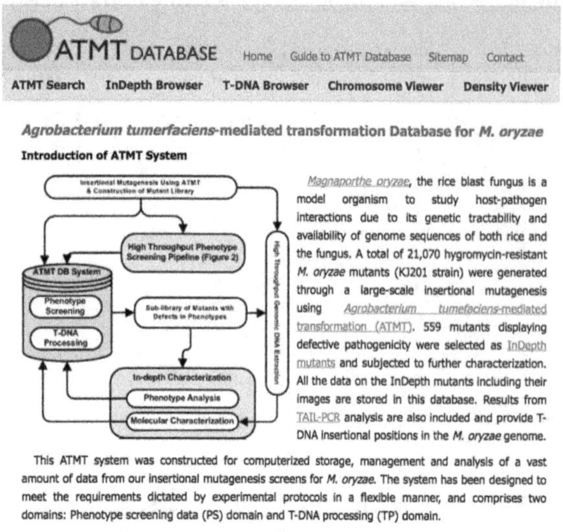


**Fig. 3** Databases of T-DNA insertion sites, mutant phenotypes, and strain availability have been developed for some fungal species. A screen shot of the front page of the ATMT Database (http://atmt.snu.ac.kr) that features information on T-DNA mutants of *P. oryzae*, primarily based on information from [38]. Limited funding for maintaining such databases and strain resources after projects end may undermine the potential impact of large-scale mutant screens

integration: for example, a library of 102 deletion strains in genes encoding $Zn_2Cys_6$-type transcription factors was created, with phenotypic analysis defining 61 regulators of development and eight required for infection of rice and barley [15].

Thus, for *P. oryzae*, *At*MT has had two directions of impact for gene function discovery. One was through forward genetics by screening tens of thousands of T-DNA insertional mutants. The second was in creating gene replacement strains, which were then tested for pathogenicity on plants.

*Zymoseptoria tritici*

*Zymoseptoria tritici* (previously named *Mycosphaerella graminicola*; Dothideomycetes) causes the most important foliar disease of wheat, Septoria blotch, with average yield losses of 5–10% annually [45]. Throughout Europe, the crop destruction and deployment of antifungals to combat *Z. tritici* are estimated to run into billions of Euros, with losses in Germany, for example, estimated to cost 500 and 310 million per year, respectively [45]. Worryingly, resistance of *Z. tritici* to every class of antifungal compounds is increasing, yet the number of new compounds developed to combat this pathogen is in decline [46]. Consequently, validation of new pathogen targets for rational fungicide development, concomitant with development of highly resistant wheat varieties, are pressing objectives for food security. *At*MT has been essential for improved understanding of the molecular basis of *Z. tritici* infection, which may ultimately lead to durable disease control strategies.

The first transformation of *Z. tritici* utilized a classical protoplast-polyethylene glycol (PEG) approach, whereby the cell walls of in vitro grown conidia were digested and a DNA cassette encoding hygromycin or carbendazim resistance was introduced into recipient genomes [47]. From this study, several limitations to protoplast-based transformation became apparent. Firstly, there were significant variations in successful protoplast generation between *Z. tritici* isolates. Secondly, this experimental challenge was further complicated by the commercial discontinuation of Novozyme 234, which was the enzyme mix used to digest conidial cell walls [47]. While other enzyme preparations exist for fungal protoplasting [48], variations in availability of validated transformation reagents introduced an additional challenge for the widespread adoption of this approach by the *Z. tritici* research community.

Consequently, with few exceptions [47, 49, 50], molecular studies of *Z. tritici* rely on *At*MT [51]. As one example, a T-DNA mutagenesis forward genetic screen generated an insertional library of 615 *Z. tritici* transformants [52]. Virulence analysis of this mutant library revealed one isolate, 5-29H, which was avirulent in a detached leaf infection assay. Mapping of the T-DNA insertion locus revealed disruption of a putative mannosyltransferase-encoding gene, and subsequent phenotypic screening and proteomic analyses demonstrated that protein *N*-glycosylation was essential for a switch from yeast-like conidia to infectious hyphal growth, and ultimately successful disease initiation [52]. Using *At*MT in a targeted approach, Cousin and co-workers deleted a mitogen-activated protein kinase (MAPK) encoding gene that had high sequence homology to the *FUS3* gene of *S. cerevisiae* [53]. This MAPK plays a critical role in mating and growth in this yeast, and deletion in *Z. tritici* resulted in aberrant polarized growth and deficiencies in host penetration during infection. In a similar approach using *At*MT, deletion of a gene encoding another MAPK, termed MgSlt2, demonstrated a critical role of this gene in resistance to several fungicides, and hyphal branching following leaf penetration [54]. Taken together, these studies demonstrate that impaired hyphal development, deficient stomatal penetration [53], or an inability to proliferate after disease initiation [54] all result in reduced pathogenicity, and ultimately suggest interference with the normal *Z. tritici* developmental program offers an opportunity to inhibit disease. Thus, *At*MT represents a critical tool in the researchers' repertoire for identifying the targets for the rational development of novel fungicides. With regards to future *Z. tritici* experiments, and the molecular analyses of phytopathogenic fungi in general, it is clear that *At*MT will continue to be utilized even in face of major paradigm shifts in the field of fungal pathogenicity.

#### *Fusarium* species

Members of the *Fusarium* genus (Sordariomycetes) include agronomically important plant pathogens some of which are known for producing mycotoxins and also as opportunistic human pathogens [55]. One recent phylogenetic study suggests that the *Fusarium* genus could be subdivided into 20 species complexes with the terminal *Fusarium* clade originating in the middle Cretaceous period [56]. The most intensely studied *Fusarium* species include members of the species complexes Sambucinum (*F. graminearum*, *F. pseudograminearum* and *F. culmorum* causing blights and rots of wheat and barley), Fujikuroi (*F. fujikuroi* causing bakane disease of rice and *F. verticillioides* causing ear rot of maize), and Oxysporum (a species complex of soil-borne filamentous fungi whose members include more than 12 forma speciales causing vascular wilts of many different plant species). The Oxysporum complex has been a focus of interest to evolutionary biologists due to the range of life strategies of its members. Although all members are soil-borne, not all isolates are plant pathogens, a feature associated with the elevated genome plasticity of *Fusarium* and that allow the acquisition of mobile pathogenesis chromosomes [57]. Moreover, the finding that an isolate pathogenic on tomato can cause disease in immunocompromised mice raised the profile of this species complex as a model for cross-kingdom pathogenesis studies [58]. Given the size, diversity and economic damage caused by members of the genus it is not surprising that *At*MT has made an impact on some species of *Fusarium* more than others, and therefore the main focus of this section will be on a limited number of species such as *F. graminearum* and *F. oxysporum*.

In the first comprehensive report of transformation of a *Fusarium* species (*F. oxysporum* f. sp. *raphani* strain 699), the efficiency ranged from 1 to 40 transformants per $10^7$ protoplasts [59]. Although transformation efficiency using protoplasting methods was low, homologous recombination was easily achieved in *F. oxysporum* and *F. graminearum* with 20–50% homologous integration events [60, 61], opening the way for reverse genetic experiments. The development of *At*MT for filamentous fungi including *F. venenatum*, the source of the edible mycoprotein Quorn™, promised to improve transformation efficiency for the genus *Fusarium* [4]. Two more *Fusarium* species, *F. circinatum* and *F. oxysporum*, were transformed in 2001, providing new vectors for fungal transformation as well as a plasmid rescue cassette to enable the easy retrieval of DNA sequences flanking the T-DNA insertion in the fungal genome [62, 63]. These publications reported improved transformation efficiencies to 300–500 transformants per $10^6$ conidia thus paving the way for further forward genetic approaches in *Fusarium* [64]. Since 2001, *At*MT has been reported in at least nine *Fusarium* species including *F. culmorum*, *F. graminearum*, *F. pseudograminearum* [65], *F. verticillioides* [66], *F. virguliforme* [67], and *F. avenaceum* [68].

Although *F. graminearum* can be transformed by *A. tumefaciens*, the efficiency of *At*MT using non-homologous DNA is extremely low. Malz et al. [65] compared the efficiency of *At*MT for the random integration of a hygromycin resistance cassette into three species of *Fusarium* and found that in a single transformation experiment *F. pseudograminearum* yielded 409 transformants, whereas *F. culmorum* and *F. graminearum* yielded only 13 and 9 transformants, respectively. Interestingly, *F. graminearum* can be transformed to high efficiency (up to 2000 transformants per $10^7$ conidia) with *At*MT as long as the vector used contains some homologous *F. graminearum* DNA within it, suggesting that homologous recombination may be the dominant type of integration event in *F. graminearum* as it is in *S. cerevisiae* [69]. Consequently REMI, and not *At*MT, is still used for forward genetics approaches in *F. graminearum* [70].

In contrast, *At*MT is clearly the method of choice for large-scale random mutagenesis approaches in *F. oxysporum*. The first random mutagenesis studies published were done using REMI [71, 72] or transposon tagging [73, 74] with numbers of mutants generated in the range of 182–1129. *At*MT has enabled studies of a much larger scale. In 2009, two large-scale studies were published using *At*MT. Screening of a mutant collection of 10,290 transformants of *F. oxysporum* f. sp. *lycopersici* identified 106 isolates with reduced pathogenicity on tomato and 111 potential pathogenicity genes [75]. Similarly, Li et al. [76] created a bank of 20,000 mutants of *F. oxysporum* f. sp. *cubense* race 4, and screened them over a 6-year period for reduced pathogenicity on Cavendish banana plantlets. This revealed 27 reduced or loss of pathogenicity isolates, one of which had a T-DNA insertion in the gene *FoOCH1* that encodes an α-1-6-mannosyltransferase. Interestingly, further characterization of *FoOCH1* via targeted deletion and complementation experiments was achieved by transforming protoplasts, illustrating that in contrast with fungal systems covered in other sections of this review, there is a choice of efficient transformation methods available to researchers working on *F. oxysporum*.

### *Colletotrichum* species

The genus *Colletotrichum* (Sordariomycetes) contains a large number of plant pathogens causing diseases in most crops, including grains and fruit, as well as resulting in post-harvest losses [30, 77, 78]. Many species are hemibiotrophs (i.e. growing first as a biotroph without causing disease symptoms before switching to a necrotrophic mode of damage), although some purely nectrotophic species are also known.

Early insertional mutagenesis approaches in *Colletotrichum* spp. yielded discoveries into genes required for pathogenicity, indicating that insertional mutagenesis would be an effective approach for gene discovery. Examples of successful gene identification include those encoding a class V chitin synthase in *C. graminicola* [79] and a serine/threonine protein kinase in *C. lindemuthianum* [80]. However, large numbers of transformants have been made using *At*MT and screened on plants.

The current "record holders" in terms of strain numbers are *C. higginsianum*, a pathogen of Cruciferae species, and *C. gloeosporioides*, a pathogen with a wide host range that includes both monocot and dicot plants. For *C. higginsianum* three sets of mutants that total more than 21,000 T-DNA mutants have been isolated and screened on *A. thaliana* [81–84]. These studies both traced the stages during the infection cycle in which mutants were blocked in causing disease, and went on to identify 17 new pathogenicity genes for this species. For *C. gloeosporioides* two screens on more than 14,000 T-DNA strains in total have been conducted [85, 86]. The recent study by Wu et al. used an in vitro screen to identify genes required for the production of asexual spores that are required to establish disease to identify 11 candidates for genes required for conidiation [85]. Cai et al. generated more than 4000 insertional strains and screened them on detacted rubber tree leaves, to identify 16 genes required for pathogenicity [86]. In addition to random insertional mutagenesis, *At*MT can be used in *Colletotrichum* species to make targeted gene replacements, and this has been enhanced by the isolation of mutants in the Ku genes for the non-homologous end joining DNA repair pathway [82]. This pathway is involved in the ectopic insertion of DNA into fungal genomes, and thus pathway mutants have a high proportion of targeted gene replacement events after transformation.

An example of convergent discoveries by using *At*MT in *Colletotrichum* species has been the recent independent identification of T-DNA insertional mutants in the same signaling pathway in *C. higginsianum* [87] and *C. orbiculare* [88]. Both studies found insertions in components of the Regulation of Ace2 and Cellular Morphogenesis (RAM) pathway, a complex of two kinases and associated proteins that controls cell morphology in fungi. With six components, collectively this provides a large target for T-DNA insertions, and as discussed under the section on the basidiomycete *C. neoformans*, the same pathway was also first identified as impacting multiple functions in this human pathogenic yeast from T-DNA insertional mutants [89].

### *Leptosphaeria maculans*

A plant pathogen that is not included in the "top 10" [30] list is *Leptosphaeria maculans* (Dothideomycetes), in which substantial use of *At*MT has been made as an insertional mutagenesis tool. *L. maculans* is a phytopathogen capable of attacking cultivated Brassicas such as *B. napus, B. rapa, B. juncea, B. oleracea* as well as numerous wild Cruciferae species [90].

*Leptosphaeria maculans* was among the first ascomycetes genetically transformed, by PEG-mediated transformation of protoplasts [91]. The first forward genetic screen in *L. maculans* was undertaken using REMI, a method that has been discussed above in sections "Advantages of *At*MT over other transformation techniques" and "*Pyricularia oryzae*". A screen of 516 transformants identified 12 loss of pathogenicity mutants [92]. An evaluation of 47 randomly selected insertional mutants revealed complex patterns of insertions with 31 containing insertions at multiple loci, 12 with single loci insertions of multiple copies of plasmid and only 4 had single copies of the insertion plasmid, aptly illustrating the drawbacks of REMI discussed previously. PEG-mediated transformation was also used successfully to deliver gene disruption cassettes to *L. maculans* resulting in the disruption of genes, although at frequencies as low as less that 0.25% of transformants [93].

*At*MT was first shown to be effective in *L. maculans* with the successful knockout of genes encoding an ATP binding cassette transporter and two-component histidine kinase [94]. The frequency of gene deletions achieved in this study was still less than 1 in 140, therefore a negative selectable marker was developed whereby two copies of a thymidine kinase gene from the herpes simplex virus were amended to the ends of the deletion vector. Transformants arising through ectopic integration of plasmid DNA would still contain the thymidine kinase negative selectable marker and should not grow in the presence of thymidine analogues fluorodeoxyuridine or trifluorothymidine. This strategy increased the rate of homologous integration up to 1 in 30.

The greatest impacts of *At*MT in *L. maculans* are its use as a forward genetics tool for random mutagenesis and a transformation tool to deliver gene constructs that manipulate the expression of endogenous fungal genes. In comparison with other fungi discussed in this review, the forward genetics studies are smaller in scale and highlight some of the unusual events that can occur when a T-DNA is inserted into a host genome. For example, in the first forward genetic approach published, 91 transformants were screened to identify one reduced pathogenicity mutant with a T-DNA insertion in the promoter region of two divergently described genes that resulted in increased expression of both genes and rendered the transformed strain "uncomplementable" [95]. To circumvent this problem *At*MT was used to recreate this state of overexpression in new mutant strains, overexpressing both genes individually or together, and this led to the finding that overexpression of a maleylacetate reductase results in loss of pathogenicity in *L. maculans*. Interestingly this effect of T-DNA insertion altering the gene expression or transcript stability in unexpected ways has since been reported on several occasions [96, 97]. In two independent studies, the T-DNA inserted into the 3′ regulatory sequences of genes thereby altering the length and/or stability of transcripts, and resulted in the discovery of the *IFRD* gene, which is important for cell wall integrity, conidial germination and pathogenicity, and the *cpcA* gene, which is responsible for regulating the production of amino acids during starvation growth. In both of these studies where the insertion of T-DNA resulted in a complex phenotype, *At*MT was also used to deliver RNA interference constructs to create isolates with reduced expression of the gene of interest.

Blaise et al. [98] and Bourras et al. [99] published the largest and most extensively characterized forward genetic screens conducted in *L. maculans* to date. Blaise et al. found that 53 transformants out of 1388 tested had altered but reproducible pathogenicity phenotypes, ranging from lost, reduced, delayed and growth condition dependent defects. By genetic crossing of 12 mutants they could show that the T-DNA insertion was linked to the loss of pathogenicity in only 50% of the cases, thus highlighting a limitation of *At*MT. They retrieved left border sequences from 135 randomly selected transformants and observed a trend towards integration into gene rich regions with a possible bias towards regulatory or intergenic regions. These findings were substantiated in much greater detail by Bourras et al. [99] whereby 400 border sequences were obtained through thermal asymmetric interlaced (TAIL)-PCR and primer walking, thus identifying 318 single locus T-DNA integration events. With the backing of an annotated genome, the authors were able to confirm that 97% of T-DNA integrations were mapped into GC-rich and transcriptionally active regions of the genome. There was also some evidence of chromosomal bias with statistically more T-DNA insertions in chromosomes 5 and 10 and less in chromosome 18 than would have been predicted for completely random insertions. A detailed examination of insertions into the gene-rich areas showed that there were more insertions into gene regulatory regions and introns than would have been expected under a random integration hypothesis, and less insertions than expected in intergenic regions and exons. Futhermore, a comparison of the promoter sequences of targeted genes with the T-DNA left border flanking sequence revealed 5 bp long consecutive stretches of homologous sequences, termed microhomology domains, consistent with the integration of T-DNA into the *L. maculans* genome via a microhomology-mediated end-joining pathway. What is readily apparent from these studies is that T-DNA insertion into the *L. maculans* genome does not necessarily result in gene disruption or loss of gene function, and extreme truncations and chromosomal translocations can occur [100].

Given that all known genes encoding effectors, which are small secreted proteins involved in plant pathogen-host recognition, characterized in *L. maculans* to date are located in AT-rich regions of the genome and if one would assume that genes required for infection are lowly expressed in culture [101], the efficiency of *At*MT to target pathogenicity genes is questionable. However, the ease with which *L. maculans* can be transformed via *At*MT offers realms of opportunities for other forward genetics screens and the delivery of genome tailoring enzymes in the future. Furthermore, *At*MT is used to deliver RNA silencing constructs into *L. maculans*, and can be used to confirm the functions of effector genes [102].

#### *Histoplasma capsulatum*

The ascomycetes also include a number of human pathogenic species, such as *Histoplasma capsulatum* (Eurotiomycetes), the etiologic agent responsible for histoplasmosis or "cave disease". It is an infection of the lungs normally arising through inhalation of fungal spores and is especially common in immunocompromised patients. *H. capsulatum* is a thermally dimorphic fungus characterized by two different growth forms. The saprophytic form of the fungus grows as a filamentous mold with aerial hyphae in the environment (especially in soil that contains bird or bat droppings) whence spores can become airborne and inhaled by people; the person's body temperature allows the pathogen to grow into the next stage of its life cycle that consists of a yeast that can infect lungs, or it can travel to lymph nodes and spread through the bloodstream to other parts of the body, such as the central nervous system [103]. *H. capsulatum* can also cause significant mortality and morbidity in healthy hosts with approximately 25,000 estimated life-threatening infections per year in countries where the fungus is endemic [104].

Molecular research on *H. caspulatum* has been performed for many years, and technological developments were reviewed more than a decade ago [105]. Electroporation was the most successful technique for transformation compared to biolistic and lithium acetate/ PEG-mediated transformation for efficiency and reliability, although episomal plasmids and multiple random insertions of heterologous DNA limited the exploitation for its use in functional genetics studies despite several optimization attempts [106]. As a further step toward the development of reliable and efficient molecular tools, a protocol based on *At*MT was developed for *H. capsulatum* and the related dimorphic species *Blastomyces dermatitidis* [107]; since then, *At*MT protocols have been developed for other dimorphic fungi such as *Coccidioides* spp., *Sporothrix schenkii, Paracoccidioides brasiliensis* and *Talaromyces* (*Penicillium*) *marneffei* (see review [108]). We focus attention on *At*MT of *H. capsulatum* and the main discoveries arising from use of this technique.

Sullivan and colleagues [107] compared the feasibility of *At*MT using two different selection markers, the native *H. capsulatum URA5* gene, and the *hph* gene (for hygromycin resistance) placed under the control of the *A. nidulans gpd* promoter and *trpC* terminator. In both *H. capsulatum* and *B. dermatitidis*, a 5- to 10-fold higher transformation efficiency was achieved using the selection for uracil prototrophs. Moreover, T-DNA insertions were always found in the host genome with more than 80% of transformants obtained bearing single T-DNA insertions; however, a small percentage of multiple copies of T-DNA, small rearrangements or deletions, and integration of plasmids regions beyond the T-DNA borders were also observed. As an insertional mutagenesis tool, *At*MT works most effectively when the DNA is transformed into uninucleate cells, which are more easily obtained for *H. capsulatum* than for *B. dermatitidis.*

Following this first report of *At*MT in these dimorphic species, other promoters (i.e. from *TEF1*), selection markers or reporter genes (i.e. *GFP, BLE*), and other transformation parameters were optimized [109]. For example, Marion et al. [110] optimized *At*MT and went on to identify *H. capsulatum* loci that impact the production of cell wall α-(1,3)-glucan, based on a simple and effective visual screening: wild-type strains of *H. capsulatum* have a visibly "rough" colony morphology on culture plates, while mutants that lack α-(1,3)-glucan have a "smooth" colony appearance. Beside *AGS1*, which was already characterized as important for cell wall construction in *H. capsulatum*, two novel genes (*AMY1* and *UGP1*) required for α-(1,3)-glucan biosynthesis were identifed, of which *AMY1* was also found to be required to kill macrophages and to colonize murine lungs [110].

The dimorphic transition is key for pathogenicity, but until 2008 little was known about what genes regulated the transition. Sil and colleagues identified *At*MT mutants that were unable to make the transition from the filamentous (fuzzy colonies) to the pathogenic yeast form (smooth colonies) under temperature shift from room temperature to 37 °C [111, 112]; the mutated genes were named *RYP1, RYP2* and *RYP3* from "required for yeast phase growth". Subsequent studies demonstrated that these genes encode a connected network of transcription factors that regulate each other and target common genes to activate a transcriptional program that is required for cell shape changes and expression of virulence genes in response to host temperature in *H. capsulatum* [113].

*At*MT was also used to identify genes of *H. capsulatum* required for intracellular growth and virulence by assessing the survival rate of T-DNA mutants within macrophages, which led to the identification of the genes *VMA1* and *HSP82,* both crucial for virulence in an pulmonary murine model for histoplasmosis [114, 115]. More recently, in another *At*MT screen, Isaac and colleagues revealed a mechanism of evasion of *H. capsulatum* from macrophages that involves the protein calcium-binding protein Cbp1, which had been previously characterized [116] and also identified by another group using *At*MT coupled with reverse genetics and PCR screening [117]. In their screen the authors identified three independent *cbp1* mutants that grew at wild type level within macrophages but failed to elicit host-cell death; *cbp1* mutants also showed attenuated virulence in an animal model, thus suggesting a key role for Cbp1 in favoring dissemination of the fungus in the host through a mechanism that seems to be specific for *H. capsulatum* and related dimorphic fungi.

#### *Other pathogenic ascomycetes*

Whilst *AtMT* has had considerable success in studying gene function in plant pathogens, it has also been deployed in other pathogenic fungi, in some cases to investigate the wider functional applicability of the virulence factors first characterized in other fungal species.

Amongst insect pathogenic fungi, *Metarhizium* spp. (e.g. [118]), *Beauveria* spp. (e.g. [119]) and *Lecanicillium lecanii* [120] (all three in the Sordariomycetes) have been transformed by this method, with a sizable *At*MT T-DNA mutant collection generated in *B. bassiana* [121]. In *Metarhizium* spp, targeted gene disruption has been reported using this approach to characterize genes such as the non-ribosomal peptide synthase needed for serinocyclin synthesis [122] and further developed for high throughput gene disruption [123].

Amongst mycopathogenic fungi, *Coniothyrium minitans* (Dothideomycetes), a fungal parasite of the plant pathogen *Sclerotinia sclerotiorum*, has been successfully transformed by *At*MT [124], and in the mushroom pathogen *Lecanicillium fungicola* (Sordariomycetes), the method has been used for targeted disruption of cell wall degrading β-1-6 glucanase [125] and the Pmk1-like MAP kinase [126], with mutation of the latter gene somewhat surprisingly not having any impact on virulence.

### Phylum Basidiomycota

Basidiomycetes are distinguished morphologically by their sexual spore formation, produced on the ends of club-shaped cells (basidia) in which meiosis has taken place. The phylum is divided into three major subphyla, the Agaricomycotina, Pucciniomycotina and Ustilaginomycotina (Fig. 2). While some species of basidiomycetes (e.g. *U. maydis*, *Coprinopsis cinerea* or *C. neoformans*) have served as models for aspects of plant pathology, medical mycology, fungal or general biology, compared to ascomycetes relatively less was known about gene functions in the phylum prior to the advent of *At*MT.

It was fortuitous that the two main model basidiomycetes that had been preferred for classical Mendelian genetics, the inkcap toadstool *C. cinerea* (Agaricomycotina) and maize smut *U. maydis* (Ustilaginomycotina), both proved to be readily amenable to protoplast-based transformation methods. These species provided reliable and reproducible starting material for protoplasting, in the form of asexual ooidia for *C. cinerea* and yeast-like sporidial growth for *U. maydis*. Both species gave good yields of transformants, and *U. maydis* had the additional benefit of having both integrative and autonomous transformation vectors, and a very efficient homologous recombination system allowing easy gene targeting. The early progress achieved in these species encouraged researchers to investigate the tractability of other basidiomycetes, but difficulties were often encountered when attempting to transfer the methods developed in these models to other species. In particular the absence of asexual spores in the majority of species in the Agaricomycotina meant that protoplasting had to be performed on highly variable mycelial cultures, and the obligate pathogens such as the Pucciniomycotina species causing rusts were largely ignored due to the inherent problems in only being able to work with such species in planta.

#### *The mushroom-forming species of Agaricomycetes*

Historically the transformations of species in the Agaricomycetes (Agaricomycotina) were often of very low efficiency, variable in terms of success, and few suitable vectors had ever been developed. Indeed, for the cultivated button mushroom *Agaricus bisporus*, only one lab was successful in protoplast-based transformation. The report in 1998 of transformation of seven fungi [4], including *A. bisporus*, was therefore met with great excitement by the basidiomycete community and was followed by a flurry of papers on different species, although often without the high transformation frequencies seen in ascomycetes. A number of confounding factors then became apparent that help explain why the initial transformation attempts were often without success.

The key breakthrough came when Chen et al. [127] demonstrated that whilst most tissues of *A. bisporus* would only yield low transformation efficiencies, the use of gill tissue excised from fruiting bodies immediately prior to veil-break gave high efficiencies. A similar situation occurs in the important forestry pathogen *Armillaria mellea*, where the most amenable tissue for transformation is basidiospores collected from either wild-grown fruiting bodies [128] or laboratory-raised fruiting bodies [129]. These studies flag the importance of selecting the appropriate developmental stage of fungal material, since not all stages are equally amenable to transformation.

Another breakthrough in the transformation of the Agaricomycetes came with the observation that transgenes often needed to include an intron, ideally at the 5′ end of the gene. This proved to be important whether the selection or antibiotic resistance cassette was introduced via protoplasts or via *At*MT, and impacted both on choice of reporter genes and on some of the selectable markers [130]. This requirement varies from species to species, and indeed from gene to gene. For instance it is fortuitous that it is not normally required for the function of the hygromycin resistance gene typically used in initial selection of transformants, but is needed in some cases for successful deployment of the gene conferring resistance to phleomycin [131].

The choice of promoters to drive transgene expression is important because this determines when, where and if the transgene is expressed. Not all fungal promoters are active when transferred into the genome of a related species and this had to be assessed on a case-by-case basis, which added constraints to the wider utility of some of the vectors [132]. Examples from the Agaricomycetes illustrate this point. Whilst the *A. bisporus gpd* promoter showed a reasonable spectrum of activity in other fungi, there were instances where it was not very successful (e.g. [133, 134]) and there was no readily apparent pattern to explain this. This is in contrast to many of the ascomycete vectors in common use where promoters such as *trpC* or *gpdA* from *A. nidulans* have been used in other ascomycete species over wide evolutionary distances. Effective promoters that function a cross phyla are less common in Agaricomycotina, but it is perhaps noteworthy that the DNA immediately to the 5′ of the start codon of the *C. neoformans* actin gene was successful in driving hygromycin resistance in *Hypsizigus marmoreus, Flammulina velutipes,* and *Grifola frondosa,* suggesting that this promoter may have broad utility [135]. Curiously, RNA-seq data indicate that this DNA is a combination of promoter and the 5′ untranslated region, with that region containing an intron that is spliced in *C. neoformans* [136].

To add further complexity to the deployment of *At*MT in basidiomycetes, Kilaru et al. [131] highlighted that two slightly different forms of the hygromycin resistance cassette were in common usage, and that these gave very differing transformation efficiencies in a species-specific manner. Once all these factors—cell material, introns in markers, promoters and cassettes—were fully appreciated and factored into planned investigations, it has become far easier to transform basidiomycetes. This has allowed a wide set of studies to be undertaken on diverse aspects of basidiomycete biology.

To date, the transformation experiments on the Agaricomycetes have primarily focused on species with either edible fruiting bodies or where there is a biotechnological application, and publications have focussed on methodological development. This methodology has since been deployed to help assess expression patterns using reporter genes and now mutant screens in other species such as *Laccaria bicolor* [137]. One interesting approach has been in modifying the stress tolerance in fruiting bodies, for example in conferring cold-tolerance to the paddy straw mushroom *Volvariella volvacea* [138]. In other cases, the yield of pharmaceutically relevant compounds, such as ganodermic acid in *Ganoderma lucidum* [139], clavaric acid in *Hypholoma sublateritium* [140] or various triterpenes in *Antrodia cinnamomea* [141], has been enhanced as a result of *At*MT by overexpression of a core biosynthetic gene. One drawback to using *At*MT for overexpression is that it usually only delivers a single copy of the transformation construct and may not achieve as high a titre of the desired compound. In contrast, protoplast-mediated events often result in multicopy integrations, delivering a wider range of expression levels, with some transformed strains having very high titre, which can be beneficial to create high expression strains.

Perhaps the most powerful application of *At*MT is in delivering constructs to effect gene silencing. This is of particular interest in basidiomycetes as the hyphae, which are often the starting material for transformation, are often maintained in a dikaryotic state, precluding the easy use of gene disruption (which requires nuclear integration of a construct) to assess gene functionality. Because post-transcriptional gene silencing operates within the cytoplasm, it has a dominant effect and thereby can cause a phenotype in the mutant lines despite their heterokaryotic state. Effective gene silencing has been deployed in a number of basidiomycetes, including the important mycorrhizal symbiont *L. bicolor* [142, 143] and in *A. bisporus* where it has been deployed to identify core synthetic genes and also the proteases involved in nutrient acquisition [144, 145].

Transformation, however, is still often challenging. While some species are naturally amenable to gene targeting/deletion via homologous recombination, and others can be made amenable by use of mutants in the non-homologous end joining pathway such as *KU70* mutants (e.g. *C. cinerea* [146]), many still have no reports of successful gene targeting. The recent high profile report of successful deployment of CRISPR/Cas in *A. bisporus* [147] may serve to overcome these issues, and we would expect the Cas proteins and guide RNA construct(s) to be deployed using *Agrobacterium*-vectors given how successful they have been in the Agaricomycetes in broadening the range of species amenable for transformation.

#### *The C. neoformans* ***species complex***

*Cryptococcus* species are major fungal pathogens of humans within the Agaricomycotina [104], divided into serotypes, varieties, two species *C. neoformans* and *C. gattii,* and the most recent classification splitting *C. neoformans* into two species and *C. gattii* into five species [148]. All species cause disease in humans and animals, and among them, *C. neoformans* (sensu stricto) is the one most commonly isolated in clinical settings [149]. In immunocompromised individuals the fungus infects the lungs, crosses the blood–brain barrier and invades the cerebrospinal fluid, causing fatal meningitis if untreated [150]. The disease causes hundreds of thousands of deaths globally each year [151].

Molecular studies of *C. neoformans* benefit from effective tools for random and targeted mutagenesis, conditional gene expression, gene editing and protein localization. Moreover, approaches of functional genetics were dramatically streamlined by the availability of genome sequences for several *Cryptococcus* species since the early 2000s (see review [152]). Electroporation and biolistic methods were the first transformation methods developed for *C. neoformans* [153, 154] and were employed for delivering episomal plasmids into the fungus, heterologous gene expression and gene-targeted mutagenesis. As tools for random insertional mutagenesis they have been used to study several biological processes in *C. neoformans*: examples are the identification of the essential gene *CAM1* encoding calmodulin through fortuitous insertion of a marker in the 3′ UTR of the gene yielding a temperature-sensitive mutant [155], and an insertion in a chloride transporter required to balance ions for the synthesis of the virulence factor melanin [156]. Although effective as transformation tools, both electroporation and biolistics are characterized by the high rate of genetic instability of transformants probably due to transgenes not integrating into the host genome, with reports ranging from 70 to 85% of such transformant being unstable [149]. Therefore electroporation and biolistics are mainly used for the generation of targeted mutants through homologous recombination [153, 154].

*At*MT was first used for the *C. neoformans*/*C. gattii* complexes in the early 2000s [157, 158]. The selective markers used included those that confer resistance to nourseothricin, G418 and hygromycin [157, 159], which have been extensively used for *At*MT functional genetics studies. Other plasmids for *At*MT of *C. neoformans* include those enabling the fusion with genes to assess protein localization and conditional promoters. From these early reports, the high potential of *At*MT became clear when compared to electroporation and biolistic transformation, with the advantages of having a higher rate of transformation and stability of the T-DNA insertion (close to 100%). One relevant limitation or feature of *At*MT for *Cryptococcus* is that it has not yet been successfully deployed for targeted gene replacement [158], making it different from most other fungi.

*At*MT has featured in more than 30 studies on *Cryptococcus* species, many of which have used the T-DNA insertions as a mutagenic tool in forward genetics. As such, *At*MT has been valuable for the identification of *Cryptococcus* genes that are not conserved in *S. cerevisiae*, which represents a reference organism in fungi and thus a starting point for the identification of *C. neoformans* orthologs whose specific function is in general assessed by targeted mutagenesis. However, some of the most significant discoveries in *Cryptococcus* species have been made by extending the capabilities of *At*MT beyond just an insertional mutagenesis tool for wild type strains. Hence, an additional focus is placed on these species in the following sections.

To gain insight about *C. neoformans* pathogenesis, T-DNA mutant screens have been performed using surrogate markers for virulence, such as changes in the production of the pathogenicity factors melanin and capsule, and the ability to grow at human body temperature (37 °C). Screens to identify the molecular basis of melanin biosynthesis exploit the presence of visible dark pigments produced by *C. neoformans* on media containing L-DOPA or other phenolic precursors on which melanin-deficient mutants are easily identified as white or pale colonies. This screen was performed as a proof-of-principle in the development of *C. neoformans* *At*MT with the identification of the genes *LAC1* and *CLC1*, encoding the main laccase involved in melanin biosynthesis and a putative voltage-gated chloride channel, respectively [157]. In a subsequent screen the same authors identified three more independent *lac1* mutants (Fig. 4a). Other mutants with melanin defects as well as mutants unable to grow at human body temperature were also identified [159, 160]. Successful identification of T-DNA mutants impaired in capsule production was performed by visual analysis of colony morphology or by selecting mutants unable to use heme, leading to the identification of the *CAP60*, *ARF1* and *VPS23* genes [89, 161].

Beside in vitro studies, a method to identify *C. neoformans* genes required for virulence in vivo used signature tags that were incorporated into the T-DNA molecules. Pools of signature-tagged T-DNA mutants were used for murine pulmonary infection experiments, and once the disease developed and the mice sacrificed, the reduction of the signature tag signal in the population of strains recovered from the mouse lungs indicated strains with possible reduction in virulence. This approach allowed the identification of a T-DNA mutant with reduced numbers of cells from the lungs. The impaired gene, *ENA1*, that encodes a putative ion transporter was then shown to be required for virulence in conventional virulence assays with an independently created gene deletion strain [160]. Remarkably, this application of *At*MT was the first forward genetic screen in a eukaryotic human pathogen that used virulence in animals as the phenotype. In contrast, such loss-of-virulence/pathogenicity screens have been extensively exploited with plant pathogenic fungi, as discussed above.

The application of *At*MT has also greatly contributed to decipher the mechanisms of light responses in *C. neoformans* and more broadly the fungi. While the *C. neoformans* *BWC1* gene encoding the blue light sensor was identified through searches of orthologs, other signaling components could not be identified through bioinformatic searches. Thus, a haploid strain was modified to become self-filamentous by the introduction of a construct to express the opposite mating type homedomain protein, and then used in a T-DNA insertional mutagenesis screen to find mutants with impaired sexual filamentation. One strain whose filamentation was not repressed when exposed to either light or dark had an insertion in the *BWC2* gene: the Bwc2 protein was then shown to directly interact with Bwc1 and be involved in the light responses during mating, and UV resistance. Furthermore, it was required for virulence [162]. The *BWC2* gene was independently identified by other investigators in subsequent *At*MT screens [163, 164], indicating a level of saturation in T-DNA mutant screens in this organism. Last, the downstream factor responsible for the UV sensitivity phenotype in *bwcl* or *bwc2* mutants was also

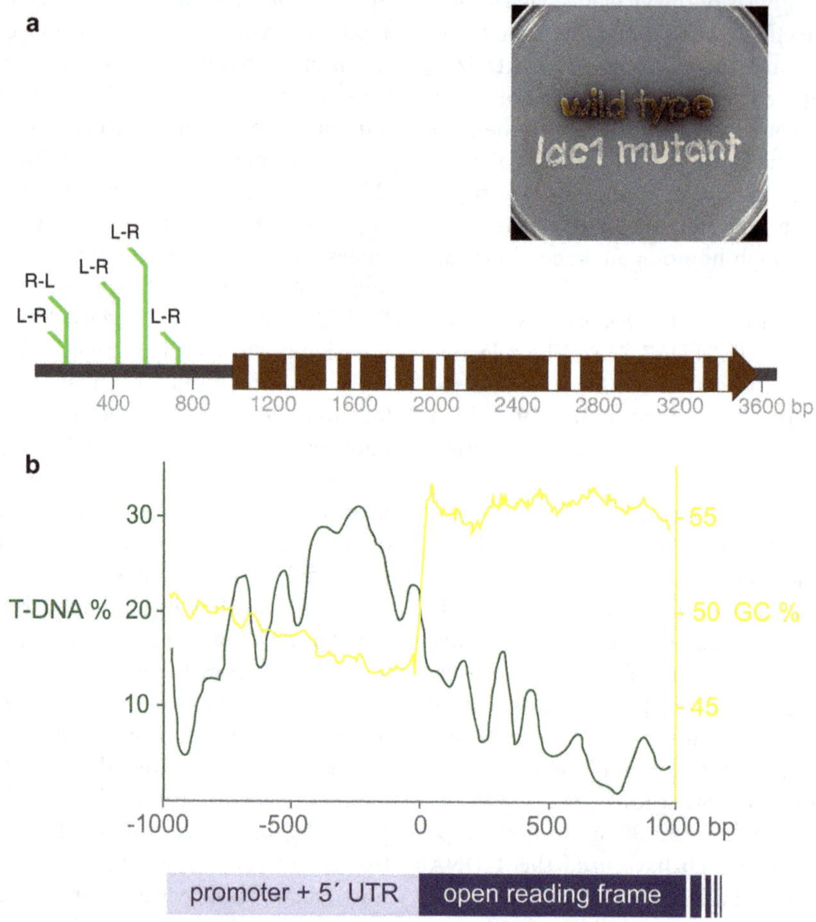


**Fig. 4** Two examples of T-DNA insertion bias into the 5′ non-coding (or flanking) regions of the genes in *Cryptococcus neoformans* (a basidiomycete yeast and human pathogen) and *Pyricularia oryzae* (a filamentous ascomycete and plant pathogen). **a** Example of a single gene in *C. neoformans* targeted by T-DNA insertion on five occasions [157, 159, 160] causing loss of pigmentation. A schematic drawing of the *LAC1* gene represented by an arrow shows coding regions as brown boxes, introns as white boxes and upstream and downstream non-coding sequences as a grey line. T-DNA insertions from five independent transformation events each causing loss of pigmentation all lie within the upstream region, with none to date in the coding region. R and L refer to the left and right border and the relative positions of these when the T-DNA inserted; note that the same site is targeted by two independent insertion events in opposite orientation. The plate shows *C. neoformans* wild type and one of the T-DNA insertional mutants growing on medium containing the substrate for laccase, L-DOPA. **b** Example of insertional bias of T-DNA into the genome of *P. oryzae* transformants. Results and figure are modified from Ref. [40]. An analysis of the distribution of 799 insertion sites mapped into 50 bp windows illustrates a twofold higher insertion frequency in promoter or untranslated regions compared to coding regions. This is even more striking in that such regions have a higher proportion of AT nucleotides while in *P. oryzae* T-DNA insertions are preferentially into higher GC content DNA

identified as a UV sensitive T-DNA insertional mutant, in the gene encoding the Uve1 endonuclease required for repair of UV damage [165].

Using similar assays, *At*MT has been extensively used to identify genes required for mating and the transition from the yeast to hyphal form. A forward genetic screen performed on a hyperfilamentous strain derived from *Cryptococcus deneoformans* (previously *C. neformans* var. *neoformans*) isolated seven mutants unable to produce filaments during the α–α mating reproduction process (a homothallic mating system observed within the *Cryptococcus* clade). One mutant had a T-DNA insertion in the transcription factor gene *MAT2*, a key regulator of the sexual reproduction pathway essential for pheromone sensing, responding and cell fusion [166]. Two other independent screens using the same hyperfilamentous strain identified a new transcription factor, Znf3, that governs sexual reproduction though a *MAT2*-independent pathway, and Spo11 and Ubc5 proteins that are essential for sporulation [164], and the long non-coding RNA *RZE1*, which controls yeast-to-hypha transition through regulating the key morphogenesis regulator Znf2 [167]. Filamentation can still occur in the *mat2Δ* deletion mutant, and recently additional components required for filamentation were identified by a screen of 77,000 insertion mutants in a *mat2Δ* background [168]. *STE50* was identified as being required for all steps of monokaryotic fruiting and sexual reproduction, i.e. from response to pheromone to production of hyphae [169]. In these studies many other mutants with impaired mating and filamentation defects have been identified, and they include both previously characterized genes (e.g. *ZNF2, MAT2, STE7* and *BWC2*) and those of unknown function that at the present are subject to further studies.

*At*MT has also been used for new drug target discovery. In two studies *At*MT insertional mutagenesis was used to identify genes essential for viability. In the first it was performed on a diploid strain derived from *C. neoformans*, using Mendelian genetic segregation analysis on the haploid spores arising after meiosis [170]. In the second, a haploid strain was mutagenized by the introduction of a regulatable promoter within the T-DNA [16]. The essential genes identified are potential targets for new antifungal drugs.

Additional important discoveries in *C. neoformans* based on *At*MT include the identification of the *CTR2* gene required for evasion of macrophages [171], the RAM pathway components that confer altered colony morphology and control cell polarity [89], genes required for growth under hypoxic conditions [172], and the basidiomycetous-specific gene *RRA1* involved in the *RIM101*-mediated alkaline responses [173].

#### *Pucciniomycotina*

*The Pucciniales that cause rust diseases* The rust-disease fungi represent the largest group within the Pucciniomycotina with about 7000 species that are obligate pathogens of many crop plants and trees, and for this reason they represent the most economically important group in this clade [174]. It is extremely difficult to conduct molecular studies and functional characterizations of genes in Pucciniales fungi for several reasons. First, as obligate biotrophic pathogens they cannot be cultured on artificial media; second, most of their life stages are dikaryotic, including the urediniospores that are commonly used in laboratory experiments; third, the need of a reliable transformation system that allows stable ectopic integration of the exogenous DNA; last, the lack of gene markers that force the selection of clear phenotypes only when the rust fungus is inoculated inside the host. Hence, the first and few reports on functional genetics consist of transient transgene expression achieved through biolistic transformation. However, transformation efficiencies were low and the transformants obtained were unstable, with two rare exceptions in *Puccinia triticina* [175].

The only successful and stable transformation experiment was in the flax rust fungus *Melamposora lini* using *At*MT [176]. The approach was exceptionally innovative. The genetic marker for selection was developed based on a previous finding that an avirulence gene of *M. lini* (*AvrL567*) when mutated can lead to disease in flax cultivars with a specific resistance gene (*L6*). Lawrence et al. used a hairpin antisense structure to silence the *AvrL567* avirulence gene and performed *At*MT of *M. lini* within the plant itself, i.e. within stems of a plant cultivar with no resistance genes that had been inoculated 5 days before with a *M. lini* isolate bearing homozygous copies of the *AvrL567* gene [176]. Candidate silenced urediospores were collected at different time points from these plants and inoculated into a cultivar containing the *L6* resistance gene, thus allowing the transformants to be selected by their ability to cause lesions. *M. lini* isolates that were able to cause disease were obtained, and molecular analyses confirmed stable integration of the T-DNA and robust silencing of the native *AvrL567* gene. Despite this elegant strategy, functional genetics through *A. tumefaciens* transformation-mediated gene silencing has not taken place yet in rust fungi and the work of Lawrence and colleagues is still the only report published [176].

*Microbotryum lychnidis-dioicae* The Pucciniomycotina includes other plant pathogens, such as *M. lychnidis-dioicae* (*M. violaceum* sensu lato) that causes anther smut of plants in the Caryophyllaceae family. *M. lychnidis-dioicae* is a dimorphic non-obligate biotrophic fungus that has been intensively studied both at the genomics and genetics levels (see review [177]). It is considered a model system in non-agricultural settings and ecological studies, offering alternatives to study host–pathogen interactions in diverse host environments. Further, *M. lychnidis-dioicae* has been used as a model for studying the evolution of sex chromosomes in fungi, and it was the first fungus in which heteromorphic mating type chromosomes were described [178].

The potential of the resources that have been generated has not been fully exploited at the level of gene functions due to the lack of a reliable transformation system, despite early and apparently successful attempts. In 1989 Bej and Perlin reported the first transformation of *M. lychnidis-dioicae* [179] where they used lithium acetate and PEG to deliver into both protoplasts and intact cells a plasmid containing the *hygB* gene as the selective marker. Transformation efficiency was high and the exogenous DNA was integrated stably into the nuclear genome. Subsequently, the same authors reported the successful transformation of *M. lychnidis-dioicae* using bacterial DNA conferring resistance to neomycin [180]. Despite the positive outcome achieved, these techniques were not reproducible by other researchers, and transformation attempts using biolistics were unsuccessful [177].

By exploiting the newly acquired genomic and transcriptomic data for *M. lychnidis-dioicae*, a robust transformation system based on *At*MT has recently been developed [181]. The selection markers delivered through *At*MT consist of endogenous promoters of the most highly expressed genes under different phases of the fungal lifecycle, as assessed by previous transcriptomic data [182], fused with the *HYG2* gene alone or in combination with *eGFP.* Stable and random integration of the T-DNA in the *M. lychnidis-dioicae* genome was achieved, and also expression/over-expression of inserted genes, corroborating transcriptomic data. Although this is the only report of stable transformation in *M. lychnidis-dioicae,* the authors showed its potential as an insertional mutagen, thus opening a new field of functional genetics in this fungus. Currently further molecular tools, such as the overexpression of a heterologous marker using a native promoter as the driver and a targeted knockout system, are under development [177].

*Red yeasts* The red yeasts in the Pucciniomycotina are a polyphyletic group that included the four genera *Sporobolomyces, Sporidiobolus, Rhodotorula* and *Rhodosporidium,* but in a recent reclassification most of the *Sporidiobolus* and *Rhodosporidium* teleomorphic species were grouped with their anamorphic counterparts *Sporobolomyces* and *Rhodotorula,* respectively, and a new genus (*Rhodosporidiobolus*) was created [183]. Compared to ascomycetous yeasts, whose importance in biotechnology has been known since ancient times for fermented beverages and food, basidiomycetous red yeasts have been relatively understudied in terms of their potential importance in biotechnology, agriculture, food processing, and environmental impact. The last few decades have revealed that these yeasts have a multitude of unique beneficial attributes, which include the production of secondary metabolites such as carotenoids and fragrances, sources of enzymes important in pharmaceutical production and chemical syntheses, biodegradation of pollutants and mycotoxins, antagonistic activity against plant pathogenic fungi, and high levels of lipid synthesis for biofuel production [184].

Gene functions in red yeasts were unexplored, with a single report of transformation in 1985 and nothing afterwards until 2010 [185]. In 1985 *Rhodosporidium toruloides* was transformed with a protoplast-lithium acetate/PEG approach: although transformation was successful, it yielded a very low number of transformants (~$10^3$/μg DNA) and the majority of which were unable to retain the introduced exogenous DNA. Since 2010 there has been a drive to generate transgenic strains by two methods, biolistic and *Agrobacterium* T-DNA delivery.

Initial *At*MT attempts in red yeasts were unsuccessful due to the lack of Pucciniomycotina-specific gene markers and the capability of some species to become spontaneously resistant to the common drugs used in transformation protocols in the Agaricomycotina and Ustilagomycotina yeasts. The first successful *At*MT of red yeasts was for *Sporobolomyces* sp. IAM 13481 [186], whose draft genome sequence and annotation had been released by the Joint Genome Institute (JGI). The strategy employed was based on the use of the endogenous *Sporobolomyces* sp. IAM 13481 *URA5* and *URA3* genes as selective markers to restore prototrophy in *ura5* or *ura3* auxotrophs that were isolated as spontaneous mutants resistant to 5-fluoroorotic acid [186]. However, plasmids developed for *Sporobolomyces* were not effective for other Pucciniomycotina species, and to fill this gap several binary vectors were generated with other selective markers. These include the *URA5* and *URA3* genes of *Rhodotorula graminis* strain WP1, the naturally high G+C content gene encoding nourseothricin acetyltransferase placed under the control of the promoter and terminator of the tubulin-encoding *TUB2* gene [187], and a codon-optimized hygromycin phosphotransferase gene alone or fused with a codon-optimized enhanced green fluorescent protein gene, both placed under the control of endogenous promoter of the *GPD1* gene. These latter markers were used successfully to transform *R. toruloides* [188].

The crucial factors for successful transformation of red yeasts were selectable markers that were native copies of genes or including native regulatory elements, appropriate G+C content, or recoding the DNA for optimal expression [187, 188]. Following these key findings, vectors based on other antibiotic-resistance marker genes (bleomycin) and/or other promoters to drive gene expression were also evaluated. These experiments demonstrated multiple stable T-DNA integrations in the genome of *R. toruloides* [189], and showed that strongest heterologous expression was achieved using the constitutive glucose 6-phosphate isomerase promoter [190]. To date *At*MT is effective in transforming all the Pucciniomycotina red yeasts tested, which include species within the genera *Sporobolomyces* and *Rhodotorula,* and *Cystobasidium slooffiae* which belongs to a different class (i.e. Cystobasidiomycetes) [187]. The *At*MT tools have yielded new insights into genes required for biological processes in the Pucciniomycotina, as illustrated by the following examples.

A T-DNA forward genetic screen was developed with the aim to elucidate the genetic mechanisms behind the ability of red yeasts to resist and degrade the mycotoxin patulin, using *Sporobolomyces* sp. IAM 13481 as a model

[191]. Although mutants defective in patulin degradation were not identified (chemical analysis subsequently revealed two independent pathways for degradation in this strain), genes involved in resistance to diverse stresses were identified. The putative functions of the mutated genes as well as the mutant phenotypes correlated with the cytotoxic effects of patulin, allowing the formulation of a hypothesis, which was then substantiated by transcriptomic analyses [192], that *Sporobolomyces* sp. IAM 13481 activates stress responsive genes to overcome the initial insult caused by the mycotoxin.

In another study, *At*MT insertional mutants of *Sporobolomyces* sp. IAM 13481 were used to investigate the molecular basis of the intriguing mechanism of ballistospore formation and dispersal that is unique to basidiomycete fungi. This property can be evaluated in the laboratory by the production of a mirror image of a culture from one Petri dish onto an uninoculated second dish [193]. The screen of more than 5000 transformants led to identification of 18 *mirror* mutants unable to shoot spores to produce the mirror image. One mutant bearing a T-DNA insertion in gene *PHS1*, which encodes 3-hydroxyacyl-CoA dehydratase required for the third step in very long chain fatty acid (VLCFA) biosynthesis, was further subjected to molecular, phenotypic and biochemical characterizations that unequivocally confirmed the involvement of Phs1 and VLCFAs in basidiomycete dispersal through a delay in ballistospore formation. Strikingly, *PHS1* is an essential gene in other fungi, and its identification in a *Sporobolomyces* sp. mutant was fortuitous as the T-DNA inserted within the last intron of the gene and past the essential amino acid residues required for function, which resulted in the production of an altered transcript yet still viable phenotype [193].

An industrial and biotechnological application of gene manipulation by *At*MT has been recently reported for *R. toruloides* [194]. In order to achieve a higher rate of lipid production, *At*MT was used to generate metabolically engineered strains through a "push–pull" approach for two genes involved in triacylglycerol biosynthesis. A plasmid for *At*MT was manipulated to overexpress simultaneously the *ACC1* and the *DGA1* genes, which encode an acetyl-CoA carboxylase and a diacylglycerol acyltransferase, with the glyceraldehyde-3-phosphate dehydrogenase and the ATP-citrate lyase promoters, respectively. Due to the random integration of the T-DNA into the genome, the transformants showed variability in the production of lipids. However, one transformant had a 2-fold increase of lipids during growth on glucose, compared to wild type.

Beside the random integration of the T-DNA into the host genome, *At*MT also can generate targeted gene replacements through homologous recombination in the Pucciniomycotina. This was first demonstrated using a strategy that exploited the change in pigmentation of *Sporobolomyces* sp. IAM 13481 from red to white when mutating the *CAR2* gene [187]. Based on the proportion of white transformants obtained, at least 1000 bp of flanking DNA is necessary to achieve a good percentage (~6%) of homologous recombination events. This percentage was about half of that achieved using biolistics transformation, which was included as a comparison since it was known to be effective for target gene replacement in *Sporobolomyces* sp. IAM 13481 [186, 187]. However, one advantage of *At*MT compared to biolistic transformation is that it does not require expensive equipment but the basic resources that are common in a molecular biology or microbiology laboratory. A *ku70Δ* mutant of *R. toruloides* was generated that showed a drastic improvement of target gene replacement efficiency through homologous recombination, with values ranging from 20% of deletion mutants obtained using just 100-bp of flanking regions to ~91% when 1000 bp or 1500 bp were used [195].

A last point worth raising on the role of *At*MT in the Pucciniomycotina is that Liu and colleagues used the reliability of the *At*MT of *R. toruloides* to develop an inducible promoter system for red yeasts [196]. The *DAO1* gene promoter was fused to a codon-optimized luciferase gene, which was then inserted by *At*MT at the *CAR2* locus of the *R. toruloides ku70Δ* strain, allowing the monitoring of the luciferase expression driven by the *pDAO1* in the albino mutants. An optimized *DAO1* promoter, which contains an intronic enhancer sequence and an artificially created ATG start codon, tightly induces luciferase expression in the presence of D-alanine; conversely, its expression was repressed by L-alanine or glucose and ammonium sulfate. More recently, to further exploit the biotecnological potential of *R. toruloides*, the same authors characterized the promoters of six genes involved in lipid biosynthesis, and identified that the promoter of the perilipin/lipid droplet protein 1 gene (*LDP1*) displays 4- to 11-fold stronger activity than that of the glyceraldehyde-3-phosphate dehydrogenase gene (*GPD1*), one of the strongest promoters known in yeasts [197].

### *Ustilaginomycotina*

The third subphylum in the Basidiomycota with a large number of species is the Ustilaginomycotina. These are best known for the species that cause smut diseases of plants. Deployment of *At*MT in these basidiomycetes has been patchy. For *U. maydis* and close relatives, where protoplast-based methods have been routine, and gene disruption was already easy, there was no immediate advantage in deploying *At*MT. However Ji et al. [198] used *At*MT in *U. maydis* for efficient transformation and gene disruption including on previously frozen cells and obviating any need for protoplasting, and Ali et al. [20] used *At*MT to deliver large-insert vectors into *U. hordei*. *At*MT has also been used to transform *Pseudozyma antarctica* (*Moesziomyces antarcticus*), which is an extremophile due to its growth under low temperatures and hence of interest for the potential cold-adapted enzymes it may produce [199].

Recently, *At*MT was found to be the only transformation method effective in species of *Malassezia* [200, 201]. This genus is associated with skin and hair of animals, and represents the main fungal component of the human skin microflora [202]. Despite this high prevalence and links to common human skin diseases, the genus had been poorly studied at the level of gene functions in part due to challenges in culturing the species, all of which are fatty acid auxotrophs. Building on recently developed genomic resources [203], *At*MT was optimized for *M. furfur*, *M. sympodialis* and *M. pachydermatis*, with co-cultivation conditions modified to support higher levels of transformation. A noteworthy aspect of these experiments was high frequencies of targeted gene disruptions (>60% of transformants being knockouts) for two genes in *M. furfur* [200].

## Fungi outside the Dikarya

The Ascomycetes and Basidiomycetes form a monophyletic lineage in the fungi (Fig. 2), termed the Dikarya due to the presence of two-nucleus cells during the sexual stages of their lifecycle. However, there are at least nine additional lineages that at one point were classified into two groups, the zygomycetes and chytrids, many with limited molecular biology. Given the promiscuity of *A. tumefaciens*—it can transform plants, fungi, oomycetes and human cells [23, 24]—one application of *At*MT that is relatively limited is in the fungal species outside of the Dikarya lineage (Fig. 2).

The Mucoromycota are a large group of species and gene function has been studied through isolation of mutants by chemical mutagenesis screens, and through targeted gene disruption in a few species [204]. *At*MT has been applied to a number of these species, and it was first reported in the Mucoromycotina species *Rhizopus oryzae* [205]. A common limitation is the loss of the transgenes in species such as *Backusella lamprospora* or *Mucor* spp. [206–209], which may reflect transient transformation, gene silencing or a foreign DNA surveillance system, or loss over time during passaging of these species that have coenocytic hyphae (without septa) and multinucleate spores. *Umbelopsis isabellina* is an interesting case, wherein standard *A. tumefaciens* strains used to transform fungi were compared with a strain of *A. rhizogenes*, which is another *Agrobacterium* species able to transform plants [210]. In this case, the latter bacterial species yielded higher numbers of transformants.

To date, the most successful applications of *At*MT in a biotechnology capacity have been on the Mortierellomycotina subphylum, in the species *Mortierella alpina*, which is a source of polyunsaturated fatty acids that may be beneficial for health. Some fatty acids derived from *M. alpina* are added to infant formula, while others like eicosapentaenoic acid are only otherwise available from fish oil. The initial transformation approaches aimed to modify or increase the fatty acid composition of this fungus [211, 212]. Subsequent studies altered gene expression to change the lipid profiles by overexpressing an ω3-desaturase enzyme or pathways that alter NADPH levels [213–215].

Other than the Mucoromycota, a chytrid species in the Blastocladiomycota, *Blastocladiella emersonii*, has been reported as transformed using *Agrobacterium* [216]. The Glomeromycota are obligate symbionts with the roots of plant species, and are challenging to work with for this reason. For instance, use of drugs for selection would likely also impede growth of the host plant. *Rhizophagus irregularis* (previously *Glomus intraradices*) has also been subjected to *At*MT, using as a selection system the delivery of constructs that express a nuclear-localized GFP [217]. The method was inefficient, hindered by the natural levels of autofluoresence of the fungus, and transformation success could not be confirmed with other approaches.

## All that shines is not silver: problems and limitations with the *At*MT technique

The previous sections are highlights in which the use of *At*MT has led to new and major advances in fungi. However, not all is perfect with the tool, and researchers should consider some of these limitations when designing forward genetic experiments, planning to make targeted gene replacements, or interpreting data from strains generated with *At*MT.

Given that generating hundreds or thousands of mutants is a prerequisite for functional genomic studies, several adaptations to the *At*MT pipeline have been made to overcome significant experimental limitations and ensure this technique is sufficiently high-throughput. In some instances, technical challenges cannot be obviated. For example, while 96-well high throughput protocols exist for lithium acetate or protoplast-PEG mediated transformation of model yeasts [218], similar experiments in 96-well format are not possible using *At*MT. This is likely due to the technical challenges of miniaturising *Agrobacterium* cultures, which must be sufficiently aerated to reach an optimised growth phase prior to transformation. Nevertheless, much progress has been made in reducing the burden of experimentally intensive cloning for *At*MT, which relies on generation of *Agrobacterium* compatible plasmids. Such cloning is more challenging when compared to lithium acetate or PEG mediated transformation, where linear DNA cassettes can be assembled by simple PCR steps [219]. Investigators have generated *Agrobacterium* plasmids that are compatible with Gateway® cloning technology from Invitrogen, e.g. enabling ultra-high throughput generation of *Z. tritici* over-expression strains [220]. Elsewhere, *Agrobacterium* compatible vector construction using yeast recombination in *S. cerevisiae* [14–16], or Golden Gate assembly [18], provide comparable improvements in high throughput vector construction. These studies highlight how some important technical challenges have been addressed, and their recent utilization demonstrate *At*MT is applicable for high throughput functional genomic analyses of fungi [220]. More generally, they highlight that *At*MT is a robust technique that will continue to be essential for analysis of fungi in the "big data" era of fungal functional genomics, as long as *At*MT is used taking into consideration both the advantages and the limitations of the technique. Four points to consider are as follows.

First, an ideal insertional mutagenesis protocol should provide the ability to "hit" any and every gene, but T-DNA insertions have non-random integration patterns into fungal genomes. A single gene example of this bias is illustrated by the isolation of T-DNA insertional mutants into the *LAC1* gene encoding laccase for melanin biosynthesis in *C. neoformans*. Five insertions have been isolated within a 1 kb promoter region and none in the 2.8 kb coding region [157, 159, 160] (Fig. 4a). This single gene example is consistent with insertional bias from the analysis of collectively thousands of T-DNA insertion strains in the plant pathogenic ascomycetes (see examples given above, and illustrated from *P. oryzae* in Fig. 4b). This is perhaps the biggest limitation to the use of *At*MT as a tool for creating a library of mutants with insertions in every gene.

Despite this limitation, the skew to insert the T-DNA outside of genes has been recently converted into an advantage for the identification of essential genes [16]. The regulatory sequences for the *C. neoformans GAL7* gene were cloned adjacent to the right border, and transformed into the basidiomycete *C. neoformans*. Transformation and selection was conducted on media containing galactose to ensure any gene near that insertion would be expressed, and then strains tested for growth on glucose, which represses *GAL7* expression. Approximately 1% of transformants did not grow on glucose media. Analysis of the positions of the T-DNA insertions in the *C. neoformans* genome revealed they were in genes with essential functions in ascomycete species.

Second, in ideal cases the T-DNA inserts perfectly from right to left border, the subsequent mutant (or set of mutants) with an interesting phenotype is isolated, followed by rapid identification of the gene of interest. Identification of the deleted or disrupted gene might be achieved by TAIL-PCR, inverse PCR, or some other method. However, this gene identification step can become a bottleneck or even end point for several reasons. The T-DNA may insert in ways which make subsequent mapping challenging, for example following signficiant cassette truncation, insertion of additional plasmid DNA beyond the border sequences, or with multiple copies either inserted in tandem or dispersed throughout the genome. The insertion events can be associated with deletions and chromosomal rearrangements of the genome. The tendency for insertions to fall within intergenic regions can cause issues establishing which of the two genes may be affected by the T-DNA, and sometimes qPCR of both genes can help identifying which one is affected by the insertion. In many cases the assumption is that the insertion will reduce expression of the adjacent genes. However, there are also examples such as from *L. maculans* in which the insertion event causes an increase in gene expression of adjacent genes [95].

Third, *At*MT may not be able to provide valuable insighs compared to other transformation methods in some species. For a number of fungi, *At*MT did not enhance our understanding of gene function in fungi. This includes the first fungus transformed using *A. tumefaciens*, *S. cerevisiae* [2]. This makes sense given the other transformation techniques available for *S. cerevisiae*, the ability to use chemical mutagenesis and then clone by complementation, construction and access to sets of strains with all the genes deleted, and other genomic-level resources. Hence, with so many tools to isolate mutants and identify the affected gene, or screening whole genome deletion sets for phenotypes, there was little additional benefit of having *At*MT as another tool. As mentioned above, a similar situation exists for the filamentous fungus model *N. crassa*: although one of the first filamentous species transformed with *Agrobacterium* [4] the method was not used since that time. A third example is *Schizosaccharomyces pombe*, a model used to uncover regulators of cell cycle control that was the topic of the 2001 Nobel Prize in Physiology or Medicine; indeed, this species and it seems no member of the Taphrinomycotina have been transformed with this method.

A fourth limitation with *At*MT is that transformation with T-DNA will always be insertional, whereas in some fungi, such as *U. maydis* [221] or with the use of the

AMA1 sequence in plasmids for *Aspergillus* spp. [222], plasmids that replicate autonomously are available. Such plasmids can be rescued back into *E. coli*, easing the way for cloning by complementation, or for deliberate loss by counter-selection.

## The future and further maximizing the impact of *AtMT* in fungi

*At*MT is a tool that is applicable to plant and fungal molecular biology, but the host side of the transformation is little understood. A set of 129 *A. thaliana* mutants has been identified that are resistant to transformation [223]. However, the fungi surpass all other eukaryotes in terms of the molecular tools and resources available for their study at the genome-wide scale and as such could be ideal hosts to define the eukaryotic side of the host cell–bacterial cell interaction. The process of T-DNA transfer and integration has been recently reviewed [7]. Using fungi as the host could provide insights into ways of improving efficiency of the method and the mechanisms by which the T-DNA is integrated. *S. cerevisiae* mutants have been used to identify those that have altered efficiency of transformation [224–229], and a discovery of the role of purine concentrations and biosynthesis was then also linked to transformation efficiency in plants [230]. Hence, testing other deletion sets, such as for *S. pombe* or the ongoing projects for *N. crassa* or *C. neoformans*, to find strains recalcitrant to transformation. The rationale for exploring the eukaryotic genes required for transformation in multiple species is because there are clear differences between fungi, e.g. gene replacement using *At*MT is not possible in *C. neoformans* but it is in other basidiomycetes or in *S. cerevisiae* the influence of purine during transformation on efficiency is dependent on both genes for purine synthesis and the yeast strain [230]. This research direction could have wide impact, if that information could be transferred across to explain why some plant species are difficult to transform with the method. The other side of the transformation interaction is *A. tumefaciens* itself. There is evidence that different strains of *A. tumefaciens* behave differently. Few studies have investigated strains side-by-side for bias in insert preference or gene targeting efficiency.

*At*MT can be a powerful teaching tool. "It is in human nature to value any novelty, however slight, in one's own possession" [231]. Those who have discovered a new mutant or mutant phenotype will appreciate the insight of Charles Darwin's comment, particularly for engaging students or others at many education levels with science. It is not known how many people have used *At*MT as a teaching tool on the philosophy underlying molecular genetics, i.e. the phenotype of a mutant reflects the inverse of its function in the cell. In an undergraduate practical class setting, it is an effective method for demonstrating this principle, as well as exposing students to fundamental aspects of biology. The method has been taught at the Marine Biological Laboratory Molecular Mycology course (Woods Hole, MA, USA) since 2004 where students have performed large-scale mutant screens as part of research projects [166, 173, 232].

Several of the authors of this review have used *At*MT as a teaching tool for undergraduates or even high school students, in cases resulting in the students contributing to research publications [89, 159, 193, 233]. For implementing *At*MT as a teaching tool, it is wise to select phenotype screens whereby multiple genes can control that particular phenotype to increase the chances of students isolating mutants.

It is essential to make any new method or resources available, such as through the deposition of plasmids or strains. One drawback of the use of *A. tumefaciens* is that some countries classify laboratory strains of *A. tumefaciens* as plant pathogens and have import restrictions due to this classification. An invaluable role has been and continues to be played by public repositories, such as the Fungal Genetics Stock Center in the USA. However, lack of funding for such organizations both threatens these and undermines decades of research and limits the potential to perform experiments into the future [234]. In addition to the physical resources, online databases of genomes, T-DNA insertions and phenotypes (Fig. 3) also require ongoing support to avoid the loss of this hard-gained information.

At present, the identification of T-DNA junctions can still be a rate-limiting step. Sequencing a genome can be more cost effective than PCRs or other methods to identify junctions. Indeed, next generation sequencing has been applied for the identification of the mutated genes. In the case of *L. maculans*, there was difficulty in identifying junctions and four strains were sequenced separately to identify the T-DNA insertion sites [100]. Another approach was used for *C. neoformans* in which pools of DNA from mutants were sequenced simultaneously, and then the affected genes in the individual strains identified by specific PCRs [168, 232]. In future, incorporating signature tags, bar codes or using asymmetric restriction enzymes will make identification of flanking regions easier with the next-generation sequencing capabilities that are available.

In plants there have been many reports of using co-transformation to introduce multiple T-DNA cassettes at the same time (e.g. [235]); however, this is not a common approach in fungi [236]. Such an approach has the potential to allow *At*MT delivery of separate partial selection cassettes, in a similar approach to that used very successfully in the bipartite or split-marker approach to enhancing targeted gene disruption in fungi [219]. Future implementation may further enhance gene targeting abilities in fungi [237, 238].

An exciting tool in biological research is the CRISPR/Cas genome editing systems and likely, as *At*MT did in the last 2 decades, this methodology could revolutionize fungal biology. As with *At*MT in *S. cerevisiae*, it is likely that for the well-established fungi CRISPR/Cas will be less ground breaking than for those species with challenges, especially species with low frequencies of targeted gene replacements. It is also worth pointing out that for many fungi *At*MT will still be an instrumental tool to transform in the CRISPR/Cas constructs. Another feature of *At*MT yet to be exploited in fungi is transient transformation. In plants this has been a routine method for introduction of constructs such as viral infectious clones [239], and given the difficulty currently faced with manipulation of mycoviruses, it is likely that such deployment will occur in the future. There is also interest in exploiting the transient transformation systems for other purposes like CRISPR/Cas that would circumvent the need to have this system integrated into the genome and the potential that causes for off-target mutations. In Ascomycetes this has been achieved by exploiting unstable autonomous vectors delivered by protoplast transformation [240, 241], whilst in *U. maydis* an unstable autonomous plasmid was successfully deployed into protoplasts to deliver an efficient CRISPR/Cas system [242], but clearly there is also the possibility of exploiting the benefits of *At*MT for such cases in the future, potentially by modification of some of the components required for integration of the T-DNA into the fungal genome.

## Concluding remarks

In closing, nearly 20 years on we have a chance to answer Dunn-Coleman and Wang's question about *At*MT, "a silver bullet for filamentous fungi?" This "silver bullet" conjures a mixture of concepts, ranging from the use of silver against werewolves, Paul Ehrlich's "magische Kugel", or the lesser metal status of silver over gold. The perfect agent to comprehensively determine gene function in fungi, which ultimately provide these species with their remarkable capabilities, is likely elusive. *At*MT has, however, achieved many milestones in defining the genetic basic behind numerous and diverse traits, and promises to continue to play an important role for our understanding of fungal biology.

**Authors' contributions**

All authors contributed to writing the manuscript. All authors read and approved the final manuscript.

**Author details**

[1] School of BioSciences, University of Melbourne, Melbourne, VIC 3010, Australia. [2] School of Biological Sciences, University of Bristol, Bristol, UK. [3] Department of Applied and Molecular Microbiology, Technische Universität Berlin, Berlin, Germany. [4] Department of Molecular Genetics and Microbiology, Duke University Medical Center, Durham, USA. [5] College of Life and Applied Sciences, Yeungnam University, Gyeongsan, South Korea.

**Acknowledgements**

We thank Barbara Howlett for comments and edits on the manuscript.

**Competing interests**

The authors declare that they have no competing interests.

**Funding**

Research that has used *AtMT* by the authors has been supported by the following organizations: United States National Institutes of Health (AI094364 AI and GI), Australian Research Council (FT130100146 AI), Procter and Gamble (AI and GI), Australian Grains Research and Development Corporation (UM00050 AI and CEE), Technische Universität Berlin Postdoctoral Research Funding (TCC), and the Biotechnology and Biological Sciences Research Council, UK (D19266 AMB).

**References**

1. Dunn-Coleman N, Wang H. *Agrobacterium* T-DNA: a silver bullet for filamentous fungi? Nat Biotechnol. 1998;16:817–8.
2. Bundock P, den Dulk-Ras A, Beijersbergen A, Hooykaas PJJ. Trans-kingdom T-DNA transfer from *Agrobacterium tumefaciens* to *Saccharomyces cerevisiae*. EMBO J. 1995;14:3206–14.
3. Piers KL, Heath JD, Liang X, Stephens KM, Nester EW. *Agrobacterium tumefaciens*-mediated transformation of yeast. Proc Natl Acad Sci USA. 1996;93:1613–8.
4. de Groot MJA, Bundock P, Hooykaas PJJ, Beijersbergen AGM. *Agrobacterium tumefaciens*-mediated transformation of filamentous fungi. Nat Biotechnol. 1998;16:839–42.
5. Michielse CB, Hooykaas PJJ, van den Hondel CAMJJ, Ram AFJ. *Agrobacterium*-mediated transformation as a tool for functional genomics in fungi. Curr Genet. 2005;48:1–17.
6. Frandsen RJN. A guide to binary vectors and strategies for targeted genome modification in fungi using *Agrobacterium tumefaciens*-mediated transformation. J Microbiol Methods. 2011;87:247–62.
7. Bourras S, Rouxel T, Meyer M. *Agrobacterium tumefaciens* gene transfer: how a plant pathogen hacks the nuclei of plant and nonplant organisms. Phytopathology. 2015;105:1288–301.
8. Lacroix B, Citovsky V. The roles of bacterial and host plant factors in *Agrobacterium*-mediated genetic transformation. Int J Dev Biol. 2013;57:467–81.
9. Kyndt T, Quispe D, Zhai H, Jarret R, Ghislain M, Liu Q, Gheysen G, Kreuze JF. The genome of cultivated sweet potato contains *Agrobacterium* T-DNAs with expressed genes: an example of a naturally transgenic food crop. Proc Natl Acad Sci USA. 2015;112:5844–9.
10. Young JM, Kuykendall LD, Martínez-Romero E, Kerr A, Sawada HA, et al. A revision of *Rhizobium* Frank 1889, with an emended description of the genus, and the inclusion of all species of *Agrobacterium* Conn 1942 and *Allorhizobium undicola* de Lajudie et al. 1998 as new combinations: *Rhizobium radiobacter*, *R. rhizogenes*, *R. rubi*, *R. undicola* and *R. vitis*. Int J Syst Evol Microbiol. 2001;51:89–103.
11. Lacroix B, Citovsky V. Transfer of DNA from bacteria to eukaryotes. mBio. 2016;7:e00863-00816.
12. Knight CJ, Bailey AM, Foster GD. Investigating *Agrobacterium*-mediated transformation of *Verticillium albo-atrum* on plant surfaces. PLoS ONE. 2010;5:e13684.

13. Machida M, Asai K, Sano M, Tanaka T, Kumagai T, Terai G, Kusumoto K, Arima T, Akita O, Kashiwagi Y, et al. Genome sequencing and analysis of *Aspergillus oryzae*. Nature. 2005;438:1157–61.
14. Kilaru S, Steinberg G. Yeast recombination-based cloning as an efficient way of constructing vectors for *Zymoseptoria tritici*. Fungal Genet Biol. 2015;79:76–83.
15. Lu J, Cao H, Zhang L, Huang P, Lin F. Systematic analysis of $Zn_2Cys_6$ transcription factors required for development and pathogenicity by high-throughput gene knockout in the rice blast fungus. PLoS Pathog. 2014;10:e1004432.
16. Ianiri G, Boyce KJ, Idnurm A. Isolation of conditional mutations in genes essential for viability of *Cryptococcus neoformans*. Curr Genet. 2017;63:519–30.
17. Sidhu YS, Chaudhari YK, Usher J, Cairns TC, Csukai M, Haynes K. A suite of Gateway® compatible ternary expression vectors for functional analysis in *Zymoseptoria tritici*. Fungal Genet Biol. 2015;79:180–5.
18. Engler C, Youles M, Gruetzner R, Ehnert T-M, Werner S, Jones JDG, Patron NJ, Marillonnet S. A Golden Gate modular cloning toolbox for plants. ACS Synth Biol. 2014;3:839–43.
19. Takken FLW, van Wijk R, Michielse CB, Houterman PM, Ram AFJ, Cornelissen BJC. A one-step method to convert vectors into binary vectors suited for *Agrobacterium*-mediated transformation. Curr Genet. 2004;45:242–8.
20. Ali S, Bakkeren G. Introduction of large DNA inserts into the barley pathogenic fungus, *Ustilago hordei*, via recombined binary BAC vectors and *Agrobacterium*-mediated transformation. Curr Genet. 2011;57:63–73.
21. van den Berg MA, Maruthachalam K. Genetic transformation systems in fungi, vol. I & II. Switzerland: Springer; 2015.
22. Ito H, Fukuda Y, Murata K, Kimura A. Transformation of intact yeast cells treated with alkali cations. J Bacteriol. 1983;153:163–8.
23. Kunik T, Tzfira T, Kapulnik Y, Gafni Y, Dingwall C, Citovsky V. Genetic transformation of HeLa cells by *Agrobacterium*. Proc Natl Acad Sci USA. 2001;98:1871–6.
24. Vijn I, Govers F. *Agrobacterium tumefaciens* mediated transformation of the oomycete plant pathogen *Phytophthora infestans*. Mol Plant Pathol. 2003;4:459–68.
25. Khrunyk Y, Münch K, Schipper K, Lupas AN, Kahmann R. The use of FLP-mediated recombination for the functional analysis of an effector gene family in the biotrophic smut fungus *Ustilago maydis*. New Phytol. 2010;187:957–68.
26. Grigoriev IV, Nikitin R, Haridas S, Kuo A, Ohm R, Otillar R, Riley R, Salamov A, Zhao X, Korzeniewski F, et al. MycoCosm portal: gearing up for 1000 fungal genomes. Nucleic Acids Res. 2014;42:D699–704.
27. Gamboa-Meléndez H, Judelson HS. Development of a bipartite ecdysone-responsive gene switch for the oomycete *Phytophthora infestans* and its use to manipulate transcription during axenic culture and plant infection. Mol Plant Pathol. 2015;16:83–91.
28. Arazoe T, Miyoshi K, Yamato T, Ogawa T, Ohsato S, Arie T, Kuwata S. Tailor-made CRISPR/Cas system for highly efficient targeted gene replacement in the rice blast fungus. Biotechnol Bioeng. 2015;112:2543–9.
29. Cairns TC, Studholme DJ, Talbot NJ, Haynes K. New and improved techniques for the study of pathogenic fungi. Trends Microbiol. 2016;24:35–50.
30. Dean R, Van Kan JAL, Pretorius ZA, Hammond-Kosack KE, Di Pietro A, Spanu PD, Rudd JJ, Dickman M, Kahmann R, Ellis J, et al. The top 10 fungal pathogens in molecular plant pathology. Mol Plant Pathol. 2012;13:414–30.
31. Klaubauf S, Tharreau D, Fournier E, Groenewald JZ, Crous PW, de Vries RP, Lebrun M-H. Resolving the polyphyletic nature of *Pyricularia* (*Pyriculariaceae*). Stud Mycol. 2014;79:85–120.
32. Zeigler RS, Leong SA, Teng PS. Rice blast disease. Wallingford: C.A.B. International, International Rice Research Institute; 1994.
33. Talbot NJ. On the trail of a cereal killer: exploring the biology of *Magnaporthe grisea*. Annu Rev Microbiol. 2003;57:177–202.
34. Martin-Urdiroz M, Oses-Ruiz M, Ryder LS, Talbot NJ. Investigating the biology of plant infection by the rice blast fungus *Magnaporthe oryzae*. Fungal Genet Biol. 2016;90:61–8.
35. Dean RA, Talbot NJ, Ebbole DJ, Farman ML, Mitchell TK, Orbach MJ, Thon M, Kulkarni R, Xu J-R, Pan H, et al. The genome sequence of the rice blast fungus *Magnaporthe grisea*. Nature. 2005;434:980–6.
36. Sweigard JA, Carroll AM, Farrall L, Chumley FG, Valent B. *Magnaporthe grisea* pathogenicity genes obtained through insertional mutagenesis. Mol Plant Microbe Interact. 1998;11:404–12.
37. Hamer L, Adachi K, Montenegro-Chamorro MV, Tanzer MM, Mahanty SK, Lo C, Tarpey RW, Skalchunes AR, Heiniger RW, Frank SA, et al. Gene discovery and gene function assignment in filamentous fungi. Proc Natl Acad Sci USA. 2001;98:5110–5.
38. Jeon J, Park S-Y, Chi M-H, Choi J, Park J, Rho H-S, Kim S, Goh J, Yoo S, Choi J, et al. Genome-wide functional analysis of pathogenicity genes in the rice blast fungus. Nat Genet. 2007;39:561–5.
39. Betts MF, Tucker SL, Galadima N, Meng Y, Patel G, Li L, Donofrio N, Floyd A, Nolin S, Brown D, et al. Development of a high throughput transformation system for insertional mutagenesis in *Magnaporthe oryzae*. Fungal Genet Biol. 2007;44:1035–49.
40. Choi J, Park J, Jeon J, Chi M-H, Goh J, Yoo S-Y, Park J, Jung K, Kim H, Park S-Y, et al. Genome-wide analysis of T-DNA integration into the chromosomes of *Magnaporthe oryzae*. Mol Microbiol. 2007;66:371–82.
41. Li G, Zhou Z, Liu G, Zheng F, He C. Characterization of T-DNA insertion patterns in the genome of rice blast fungus *Magnaporthe oryzae*. Curr Genet. 2007;51:233–43.
42. Meng Y, Patel G, Heist M, Betts MF, Tucker SL, Galadima N, Donofrio NM, Brown D, Mitchell TK, Li L, et al. A systematic analysis of T-DNA insertion events in *Magnaporthe oryzae*. Fungal Genet Biol. 2007;44:1050–64.
43. Foster AJ, Jenkinson JM, Talbot NJ. Trehalose synthesis and metabolism are required at different stages of plant infection by *Magnaporthe grisea*. EMBO J. 2003;22:225–35.
44. Chi MH, Park SY, Kim S, Lee YH. A novel pathogenicity gene is required in the rice blast fungus to suppress the basal defenses of the host. PLoS Pathog. 2009;5:e1000401.
45. Fones H, Gurr S. The impact of Septoria tritici Blotch disease on wheat: An EU perspective. Fungal Genet Biol. 2015;79:3–7.
46. Torriani SFF, Melichar JPE, Mills C, Pain N, Sierotzki H, Courbot M. *Zymoseptoria tritici*: a major threat to wheat production, integrated approaches to control. Fungal Genet Biol. 2015;79:8–12.
47. Payne AC, Grosjean-Cournoyer M-C, Hollomon DW. Transformation of the phytopathogen *Mycosphaerella graminicola* to carbendazim and hygromycin B resistance. Curr Genet. 1998;34:100–4.
48. Szewczyk E, Nayak T, Oakley CE, Edgerton H, Xiong Y, Taheri-Talesh N, Osmani SA, Oakley BR. Fusion PCR and gene targeting in *Aspergillus nidulans*. Nat Protoc. 2006;1:3111–20.
49. Adachi K, Nelson GH, Peoples KA, Frank SA, Montenegro-Chamorro MV, DeZwaan TM, Ramamurthy L, Shuster JR, Hamer L, Tanzer MM. Efficient gene identification and targeted gene disruption in the wheat blotch fungus *Mycosphaerella graminicola* using TAGKO. Curr Genet. 2002;42:123–7.
50. Skinner W, Bailey A, Renwick A, Keon J, Gurr S, Hargreaves J. A single amino-acid substitution in the iron-sulphur protein subunit of succinate dehydrogenase determines resistance to carboxin in *Mycosphaerella graminicola*. Curr Genet. 1998;34:393–8.
51. Talbot NJ. Taming a wild beast: developing molecular tools and new methods to understand the biology of *Zymoseptoria tritici*. Fungal Genet Biol. 2015;79:193–5.
52. Motteram J, Lovegrove A, Pirie E, Marsh J, Devonshire J, van de Meene A, Hammond-Kosack K, Rudd JJ. Aberrant protein *N*-glycosylation impacts upon infection-related growth transitions of the haploid plant-pathogenic fungus *Mycosphaerella graminicola*. Mol Microbiol. 2011;81:415–33.
53. Cousin A, Mehrabi R, Guilleroux M, Dufresne M, van der Lee T, Waalwijk C, Langin T, Kema GHJ. The MAP kinase-encoding gene *MgFus3* of the non-appressorium phytopathogen *Mycosphaerella graminicola* is required for penetration and in vitro pycnidia formation. Mol Plant Pathol. 2006;7:269–78.
54. Mehrabi R, van der Lee T, Waalwijk C, Kema GHJ. *MgSlt2*, a cellular integrity MAP kinase gene of the fungal wheat pathogen *Mycosphaerella graminicola*, is dispensable for penetration but essential for invasive growth. Mol Plant Microbe Interact. 2006;19:389–98.
55. Michielse CB, Rep M. Pathogen profile update: *Fusarium oxysporum*. Mol Plant Pathol. 2009;10:311–24.
56. O'Donnell K, Rooney AP, Proctor RH, Brown DW, McCormick SP, Ward TJ, Frandsen RJN, Lysøe E, Rehner SA, Aoki T, et al. Phylogenetic analyses of

*RPB1* and *RPB2* support a middle Cretaceous origin for a clade comprising all agriculturally and medically important fusaria. Fungal Genet Biol. 2013;52:20–31.

57. Ma L-J, van der Does HC, Borkovich KA, Coleman JJ, Daboussi M-J, Di Pietro A, Dufresne M, Freitag M, Grabherr M, Henrissat B, et al. Comparative genomics reveals mobile pathogenicity chromosomes in *Fusarium*. Nature. 2010;464:367–73.
58. Ortoneda M, Guarro J, Madrid MP, Caracuel Z, Roncero MIG, Mayayo E, Di Pietro A. *Fusarium oxysporum* as a multihost model for the genetic dissection of fungal virulence in plants and mammals. Infect Immun. 2004;72:1760–6.
59. Kistler HC, Benny UK. Genetic transformation of the fungal plant wilt pathogen, *Fusarium oxysporum*. Curr Genet. 1988;13:145–9.
60. Diolez A, Langin T, Gerlinger C, Brygoo Y, Daboussi M-J. The *nia* gene of *Fusarium oxysporum*: isolation, sequence and development of a homologous transformation system. Gene. 1993;131:61–7.
61. Proctor RH, Hohn TM, McCormick SP. Reduced virulence of *Gibberella zeae* caused by disruption of a trichothecene toxin biosynthetic gene. Mol Plant-Microbe Interact. 1995;8:593–601.
62. Mullins ED, Kang S. Transformation: a tool for studying fungal pathogens of plants. Cell Mol Life Sci. 2001;58:2043–52.
63. Covert SF, Kapoor P, Lee M-H, Briley A, Nairn CJ. *Agrobacterium tumefaciens*-mediated transformation of *Fusarium circinatum*. Mycol Res. 2001;105:259–64.
64. Mullins ED, Chen X, Romaine P, Raina R, Geiser DM, Kang S. *Agrobacterium*-mediated transformation of *Fusarium oxysporum*: an efficient tool for insertional mutagenesis and gene transfer. Phytopathology. 2001;91:173–80.
65. Malz S, Grell MN, Thrane C, Maier FJ, Rosager P, Felk A, Albertsen KS, Salomon S, Bohn L, Schäfer W, et al. Identification of a gene cluster responsible for the biosynthesis of aurofusarin in the *Fusarium graminearum* species complex. Fungal Genet Biol. 2005;42:420–33.
66. Visentin I, Montis V, Döll K, Alabouvette C, Tamietti G, Karlovsky P, Cardinale F. Transcription of genes in the biosynthetic pathway for fumonisin mycotoxins is epigenetically and differentially regulated in the fungal maize pathogen *Fusarium verticillioides*. Eukaryot Cell. 2012;11:252–9.
67. Pudake RN, Swaminathan S, Sahu BB, Leandro LF, Bhattacharyya MK. Investigation of the *Fusarium virguliforme fvtox1* mutants revealed that the FvTox1 toxin is involved in foliar sudden death syndrome development in soybean. Curr Genet. 2013;59:107–17.
68. Sørensen LQ, Lysøe E, Larsen JE, Khorsand-Jamal P, Nielsen KF, Frandsen RJN. Genetic transformation of *Fusarium avenaceum* by *Agrobacterium tumefaciens* mediated transformation and the development of a USER-Brick vector construction system. BMC Mol Biol. 2014;15:15.
69. Frandsen RJN, Frandsen M, Giese H. Targeted gene replacement in fungal pathogens via *Agrobacterium tumefaciens*-mediated transformation. Methods Mol Biol. 2012;835:17–45.
70. Urban M, King R, Hassani-Pak K, Hammond-Kosack KE. Whole-genome analysis of *Fusarium graminearum* insertional mutants identifies virulence associated genes and unmasks untagged chromosomal deletions. BMC Genom. 2015;16:261.
71. Namiki F, Matsunaga M, Okuda M, Inoue I, Nishi K, Fujita Y, Tsuge T. Mutation of an arginine biosynthesis gene causes reduced pathogenicity in *Fusarium oxysporum* f. sp. *melonis*. Mol Plant-Microbe Interact. 2001;14:580–4.
72. Morita K, Kimura S, Saito M, Shinoyama H, Usami T, Amemiya Y, Shishido M. Generation and characterization of reduced virulence *Fusarium oxysporum* f. sp. *lycopersici* mutants through plasmid-vector insertion. Mycopathologia. 2005;160:67–73.
73. López-Berges MS, Di PA, Daboussi M-J, Wahab HA, Vasnier C, Roncero MIG, Dufresne M, Hera C. Identification of virulence genes in *Fusarium oxysporum* f. sp. *lycopersici* by large-scale transposon tagging. *Mol.* Plant Pathol. 2009;10:95–107.
74. Dufresne M, van der Lee T, Ben M'Barek S, Xu X, Zhang X, Liu T, Waalwijk C, Zhang W, Kema GHJ, Daboussi M-J. Transposon-tagging identifies novel pathogenicity genes in *Fusarium graminearum*. Fungal Genet Biol. 2008;45:1552–61.
75. Michielse CB, van Wijk R, Reijnen L, Cornelissen BJC, Rep M. Insight into the molecular requirements for pathogenicity of *Fusarium oxysporum* f. sp. *lycopersici* through large-scale insertional mutagenesis. Genome Biol. 2009;10:R4.
76. Li M-H, Xie X-L, Lin X-F, Shi J-X, Ding Z-J, Ling J-F, Xi P-G, Zhou J-N, Leng Y, Zhong S, et al. Functional characterization of the gene *FoOCH1* encoding a putative α-1,6-mannosyltransferase in *Fusarium oxysporum* f. sp. *cubense*. Fungal Genet Biol. 2014;65:1–13.
77. Münch S, Lingner U, Floss DS, Ludwig N, Sauer N, Deising HB. The hemibiotrophic lifestyle of *Colletotrichum* species. J Plant Physiol. 2008;165:41–51.
78. De Silva DD, Crous PW, Ades PK, Hyde KD, Taylor PWJ. Life styles of *Colletotrichum* species and implications for plant biosecurity. Fungal Biol Rev. 2017;31:155–68.
79. Amnuaykanjanasin A, Epstein L. A class V chitin synthase gene, *chsA* is essential for conidial and hyphal wall strength in the fungus *Colletotrichum graminicola* (*Glomerella graminicola*). Fungal Genet Biol. 2003;38:272–85.
80. Dufresne M, Bailey JA, Dron M, Langin T. *clk1*, a serine/threonine protein kinase-encoding gene, is involved in pathogenicity of *Colletotrichum lindemuthianum* on common bean. Mol Plant-Microbe Interact. 1998;11:99–108.
81. Gu Q, Chen M, Huang J, Wei YD, Hsiang T, Zheng L. Multifaceted roles of the Ras guanine-nucleotide exchange factor ChRgf in development, pathogenesis, and stress responses of *Colletotrichum higginsianum*. Phytopathology. 2017;107:433–43.
82. Korn M, Schmidpeter J, Dahl M, Müller S, Voll LM, Koch C. A genetic screen for pathogenicity genes in the hemibiotrophic fungus *Colletotrichum higginsianum* identifies the plasma membrane proton pump Pma2 required for host penetration. PLoS ONE. 2015;10:e0125960.
83. Liu L, Zhao D, Zheng L, Hsiang T, Wei Y, Fu Y, Huang J. Identification of virulence genes in the crucifer anthracnose fungus *Colletotrichum higginsianum* by insertional mutagenesis. Microb Pathog. 2013;64:6–17.
84. Huser A, Takahara H, Schmalenbach W, O'Connell R. Discovery of pathogenicity genes in the crucifer anthracnose fungus *Colletotrichum higginsianum*, using random insertional mutagenesis. Mol Plant-Microbe Interact. 2009;22:143–56.
85. Wu J, Ji Z, Wang N, Chi F, Xu C, Zhou Z, Zhang J. Identification of conidiogenesis-associated genes in *Colletotrichum gloeosporioides* by *Agrobacterium tumefaciens*-mediated transformation. Curr Microbiol. 2016;73:802–10.
86. Cai Z, Li G, Lin C, Shi T, Zhai L, Chen Y, Huang G. Identifying pathogenicity genes in the rubber tree anthracnose fungus *Colletotrichum gloeosporioides* through random insertional mutagenesis. Microbiol Res. 2013;168:340–50.
87. Schmidpeter J, Dahl M, Hofmann J, Koch C. ChMob2 binds to ChCbk1 and promotes virulence and conidiation of the fungal pathogen *Colletotrichum higginsianum*. BMC Microbiol. 2017;17:22.
88. Kodama S, Ishizuka J, Miyashita I, Ishii T, Nishiuchi T, Miyoshi H, Kubo Y. The morphogenesis-related NDR kinase pathway of *Colletotrichum orbiculare* is required for translating plant surface signals into infection-related morphogenesis and pathogenesis. PLoS Pathog. 2017;13:e1006189.
89. Walton FJ, Heitman J, Idnurm A. Conserved elements of the RAM signaling pathway establish cell polarity in the basidiomycete *Cryptococcus neoformans* in a divergent fashion from other fungi. Mol Biol Cell. 2006;17:3768–80.
90. West JS, Kharbanda PD, Barbetti MJ, Fitt BDL. Epidemiology and management of *Leptosphaeria maculans* (phoma stem canker) on oilseed rape in Australia, Canada and Europe. Plant Pathol. 2001;50:10–27.
91. Farman ML, Oliver RP. The transformation of protoplasts of *Leptosphaeria maculans* to hygromycin B resistance. Curr Genet. 1988;13:327–30.
92. Idnurm A, Howlett BJ. Analysis of loss of pathogenicity mutants reveals that repeat-induced point mutations can occur in the Dothideomycete *Leptosphaeria maculans*. Fungal Genet Biol. 2003;39:31–7.
93. Idnurm A, Taylor JL, Pedras MSC, Howlett BJ. Small scale functional genomics of the blackleg fungus, *Leptosphaeria maculans*: analysis of a 38 kb region. Australas Plant Pathol. 2003;32:511–9.
94. Gardiner DM, Howlett BJ. Negative selection using thymidine kinase increases the efficiency of recovery of transformants with targeted genes in the filamentous fungus *Leptosphaeria maculans*. Curr Genet. 2004;45:249–55.
95. Elliott CE, Howlett BJ. Overexpression of a 3-ketoacyl-CoA thiolase in *Leptosphaeria maculans* causes reduced pathogenicity on *Brassica napus*. Mol Plant Microbe Interact. 2006;19:588–96.

96. Van de Wouw AP, Pettolino FA, Howlett BJ, Elliott CE. Mutations to *LmIFRD* affect cell wall integrity, development and pathogenicity of the ascomycete *Leptosphaeria maculans*. Fungal Genet Biol. 2009;46:695–706.
97. Elliott CE, Fox EM, Jarvis RS, Howlett BJ. The cross-pathway control system regulates production of the secondary metabolite toxin, sirodesmin PL, in the ascomycete, *Leptosphaeria maculans*. BMC Microbiol. 2011;11:169.
98. Blaise F, Rémy E, Meyer M, Zhou L, Narcy J-P, Roux J, Balesdent M-H, Rouxel T. A critical assessment of *Agrobacterium tumefaciens*-mediated transformation as a tool for pathogenicity gene discovery in the phytopathogenic fungus *Leptosphaeria maculans*. Fungal Genet Biol. 2007;44:123–38.
99. Bourras S, Meyer M, Grandaubert J, Lapalu N, Fudal I, Linglin J, Ollivier B, Blaise F, Balesdent M-H, Rouxel T. Incidence of genome structure, DNA asymmetry, and cell physiology on T-DNA integration in chromosomes of the phytopathogenic fungus *Leptosphaeria maculans*. G3. 2012;2:891–904.
100. Chambers K, Lowe RGT, Howlett BJ, Zander M, Batley J, Van de Wouw AP, Elliott CE. Next-generation genome sequencing can be used to rapidly characterise sequences flanking T-DNA insertions in random insertional mutants of *Leptosphaeria maculans*. Fungal Biol Biotechnol. 2014;1:10.
101. Lowe RGT, Cassin A, Grandaubert J, Clark BL, Van de Wouw AP, Rouxel T, Howlett BJ. Genomes and transcriptomes of partners in plant–fungal-interactions between canola (*Brassica napus*) and two *Leptosphaeria* species. PLoS ONE. 2014;9:e103098.
102. Plissonneau C, Daverdin G, Ollivier B, Blaise F, Degrave A, Fudal I, Rouxel T, Balesdent M-H. A game of hide and seek between avirulence genes *AvrLm4-7* and *AvrLm3* in *Leptosphaeria maculans*. New Phytol. 2016;209:1613–24.
103. Assi MA, Sandid MS, Baddour LM, Roberts GD, Walker RC. Systemic histoplasmosis: a 15-year retrospective institutional review of 111 patients. Medicine (Baltimore). 2007;86:162–9.
104. Brown GD, Denning DW, Gow NAR, Levitz SM, Netea MG, White TC. Hidden killers: human fungal infections. Sci Transl Med. 2012;4:165rv113.
105. Magrini V, Goldman WE. Molecular mycology: a genetic toolbox for *Histoplasma capsulatum*. Trends Microbiol. 2001;9:541–6.
106. Woods JP, Heinecke EL, Goldman WE. Electrotransformation and expression of bacterial genes encoding hygromycin phosphotransferase and β-galactosidase in the pathogenic fungus *Histoplasma capsulatum*. Infect Immun. 1998;66:1697–707.
107. Sullivan TD, Rooney PJ, Klein BS. *Agrobacterium tumefaciens* integrates transfer DNA into single chromosomal sites of dimorphic fungi and yields homokaryotic progeny from multinucleate yeast. Eukaryot Cell. 2002;1:895–905.
108. Sil A, Andrianopoulos A. Thermally dimorphic human fungal pathogens–polyphyletic pathogens with a convergent pathogenicity trait. Cold Spring Harb Perspect Med. 2015;5:a019794.
109. Kemski MM, Stevens B, Rappleye CA. Spectrum of T-DNA integrations for insertional mutagenesis of *Histoplasma capsulatum*. Fungal Biol. 2013;117:41–51.
110. Marion CL, Rappleye CA, Engle JT, Goldman WE. An α-(1,4)-amylase is essential for α-(1,3)-glucan production and virulence in *Histoplasma capsulatum*. Mol Microbiol. 2006;62:970–83.
111. Nguyen VQ, Sil A. Temperature-induced switch to the pathogenic yeast form of *Histoplasma capsulatum* requires Ryp1, a conserved transcriptional regulator. Proc Natl Acad Sci USA. 2008;105:4880–5.
112. Webster RH, Sil A. Conserved factors Ryp2 and Ryp3 control cell morphology and infectious spore formation in the fungal pathogen *Histoplasma capsulatum*. Proc Natl Acad Sci USA. 2008;105:14573–8.
113. Beyhan S, Gutierrez M, Voorhies M, Sil A. A temperature-responsive network links cell shape and virulence traits in a primary fungal pathogen. PLoS Biol. 2013;11:e1001614.
114. Hilty J, Smulian AG, Newman SL. The *Histoplasma capsulatum* vacuolar ATPase is required for iron homeostasis, intracellular replication in macrophages and virulence in a murine model of histoplasmosis. Mol Microbiol. 2008;70:127–39.
115. Edwards JA, Zemska O, Rappleye CA. Discovery of a role for Hsp82 in *Histoplasma* virulence through a quantitative screen for macrophage lethality. Infect Immun. 2011;79:3348–57.
116. Sebghati TS, Engle JT, Goldman WE. Intracellular parasitism by *Histoplasma capsulatum*: fungal virulence and calcium dependence. Science. 2000;290:1368–72.
117. Youseff BH, Dougherty JA, Rappleye CA. Reverse genetics through random mutagenesis in *Histoplasma capsulatum*. BMC Microbiol. 2009;9:236.
118. Fang W, Pei Y, Bidochka MJ. Transformation of *Metarhizium anisopliae* mediated by *Agrobacterium tumefaciens*. Can J Microbiol. 2006;52:623–6.
119. Leclerque A, Wan H, Abschütz A, Chen S, Mitina GV, Zimmermann G, Schairer HU. *Agrobacterium*-mediated insertional mutagenesis (AIM) of the entomopathogenic fungus *Beauveria bassiana*. Curr Genet. 2004;45:111–9.
120. Zhang Y-J, Zhao J-J, Xie M, Peng D-L. *Agrobacterium tumefaciens*-mediated transformation in the entomopathogenic fungus *Lecanicillium lecanii* and development of benzimidazole fungicide resistant strains. J Microbiol Methods. 2014;105:168–73.
121. Fan Y, Zhang S, Kruer N, Keyhani NO. High-throughput insertion mutagenesis and functional screening in the entomopathogenic fungus *Beauveria bassiana*. J Invertebr Pathol. 2011;106:274–9.
122. Moon Y-S, Donzelli BGG, Krasnoff SB, McLane H, Griggs MH, Cooke P, Vandenberg JD, Gibson DM, Churchill ACL. *Agrobacterium*-mediated disruption of a nonribosomal peptide synthetase gene in the invertebrate pathogen *Metarhizium anisopliae* reveals a peptide spore factor. Appl Environ Microbiol. 2008;74:4366–80.
123. Xu C, Zhang X, Qian Y, Chen X, Liu R, Zeng G, Zhao H, Fang W. A high-throughput gene disruption methodology for the entomopathogenic fungus *Metarhizium robertsii*. PLoS ONE. 2014;9:e107657.
124. Li M, Gong X, Zheng J, Jiang D, Fu Y, Hou M. Transformation of *Coniothyrium minitans*, a parasite of *Sclerotinia sclerotiorum*, with *Agrobacterium tumefaciens*. FEMS Microbiol Lett. 2005;243:323–9.
125. Amey RC, Mills PR, Bailey A, Foster GD. Investigating the role of a *Verticillium fungicola* β-1,6-glucanase during infection of *Agaricus bisporus* using targeted gene disruption. Fungal Genet Biol. 2003;39:264–75.
126. Collopy PD, Amey RC, Sergeant MJ, Challen MP, Mills PR, Foster GD, Bailey AM. The pmk1-like mitogen-activated protein kinase from *Lecanicillium (Verticillium) fungicola* is not required for virulence on *Agaricus bisporus*. Microbiology. 2010;156:1439–47.
127. Chen X, Stone M, Schlagnhaufer C, Romaine CP. A fruiting body tissue method for efficient *Agrobacterium*-mediated transformation of *Agaricus bisporus*. Appl Environ Microbiol. 2000;66:4510–3.
128. Baumgartner K, Fujiyoshi P, Foster GD, Bailey AM. *Agrobacterium tumefaciens*-mediated transformation for investigation of somatic recombination in the fungal pathogen *Armillaria mellea*. Appl Environ Microbiol. 2010;76:7990–6.
129. Ford KL, Baumgartner K, Henricot B, Bailey AM, Foster GD. A reliable in vitro fruiting system for *Armillaria mellea* for evaluation of *Agrobacterium tumefaciens* transformation vectors. Fungal Biol. 2015;119:859–69.
130. Burns C, Gregory KE, Kirby M, Cheung MK, Riquelme M, Elliott TJ, Challen MP, Bailey A, Foster GD. Efficient GFP expression in the mushrooms *Agaricus bisporus* and *Coprinus cinereus* requires introns. Fungal Genet Biol. 2005;42:191–9.
131. Kilaru S, Collins CM, Hartley AJ, Burns C, Foster GD, Bailey AM. Investigating dominant selection markers for *Coprinopsis cinerea*: a carboxin resistance system and re-evaluation of hygromycin and phleomycin resistance vectors. Curr Genet. 2009;55:543–50.
132. Burns C, Leach KM, Elliott TJ, Challen MP, Foster GD, Bailey A. Evaluation of Agrobacterium-mediated transformation of *Agaricus bisporus* using a range of promoters linked to hygromycin resistance. Mol Biotechnol. 2006;32:129–38.
133. Zhang JJ, Shi L, Chen H, Sun YQ, Zhao MW, Ren A, Chen MJ, Wang H, Feng ZY. An efficient Agrobacterium-mediated transformation method for the edible mushroom *Hypsizygus marmoreus*. Microbiol Res. 2014;169:741–8.

134. Ford KL, Baumgartner K, Henricot B, Bailey AM, Foster GD. A native promoter and inclusion of an intron is necessary for efficient expression of GFP or mRFP in *Armillaria mellea*. Sci Rep. 2016;6:29226.
135. Hatoh K, Izumitsu K, Morita A, Shimizu K, Ohta A, Kawai M, Yamanaka T, Neda H, Ota Y, Tanaka C. Transformation of the mushroom species *Hypsizigus marmoreus*, *Flammulina velutipes*, and *Grifola frondosa* by an *Agrobacterium*-mediated method using a universal transformation plasmid. Mycoscience. 2013;54:8–12.
136. Janbon G, Ormerod KL, Paulet D, Byrnes EJ 3rd, Yadav V, Chatterjee G, Mullapudi N, Hon C-C, Billmyre RB, Brunel F, et al. Analysis of the genome and transcriptome of *Cryptococcus neoformans* var. *grubii* reveals complex RNA expression and microevolution leading to virulence attenuation. PLoS Genet. 2014;10:e1004261.
137. Stephan BI, Alvarez Crespo MC, Kemppainen MJ, Pardo AG. *Agrobacterium*-mediated insertional mutagenesis in the mycorrhizal fungus *Laccaria bicolor*. Curr Genet. 2017;63:215–27.
138. Wang J, Guo L, Zhang K, Wu Q, Lin J. Highly efficient *Agrobacterium*-mediated transformation of *Volvariella volvacea*. Bioresour Technol. 2008;99:8524–7.
139. Xu J-W, Xu Y-N, Zhong J-J. Enhancement of ganoderic acid accumulation by overexpression of an N-terminally truncated 3-hydroxy-3-methylglutaryl coenzyme A reductase gene in the basidiomycete *Ganoderma lucidum*. Appl Environ Microbiol. 2012;78:7968–76.
140. Godio RP, Fouces R, Martín JF. A squalene epoxidase is involved in biosynthesis of both the antitumor compound clavaric acid and sterols in the basidiomycete *H. sublateritium*. Chem Biol. 2007;14:1334–46.
141. Lin Y-L, Lee Y-R, Tsao N-W, Wang S-Y, Shaw J-F, Chu F-H. Characterization of the 2,3-oxidosqualene cyclase gene from *Antrodia cinnamomea* and enhancement of cytotoxic triterpenoid compound production. J Nat Prod. 2015;78:1556–62.
142. Kemppainen MJ, Pardo AG. Gene knockdown by ihpRNA-triggering in the ectomycorrhizal basidiomycete fungus *Laccaria bicolor*. Bioeng Bugs. 2010;1:354–8.
143. Kemppainen MJ, Pardo AG. pHg/pSILBAgamma vector system for efficient gene silencing in homobasidiomycetes: optimization of ihpRNA—triggering in the mycorrhizal fungus *Laccaria bicolor*. Microb Biotechnol. 2010;3:178–200.
144. Costa ASMB, Thomas DJI, Eastwood D, Cutler SB, Bailey AM, Foster GD, Mills PR, Challen MP. Quantifiable downregulation of endogenous genes in *Agaricus bisporus* mediated by expression of RNA hairpins. J Microbiol Biotechnol. 2009;19:271–6.
145. Heneghan MN, Burns C, Costa AMSB, Burton KS, Challen MP, Bailey AM, Foster GD. Functional analysis of *Agaricus bisporus* serine proteinase 1 reveals roles in utilization of humic rich substrates and adaptation to the leaf-litter ecological niche. Environ Microbiol. 2016;18:4687–96.
146. Nakazawa T, Ando Y, Kitaaki K, Nakahori K, Kamada T. Efficient gene targeting in Δ*Cc.ku70* or Δ*Cc.lig4* mutants of the agaricomycete *Coprinopsis cinerea*. Fungal Genet Biol. 2011;48:939–46.
147. Waltz E. Gene-edited CRISPR mushroom escapes US regulation. Nature. 2016;532:293.
148. Hagen F, Khayhan K, Theelen B, Kolecka A, Polacheck I, Sionov E, Falk R, Parnmen S, Lumbsch HT, Boekhout T. Recognition of seven species in the *Cryptococcus gattii/Cryptococcus neoformans* species complex. Fungal Genet Biol. 2015;78:16–48.
149. Heitman J, Kozel TR, Kwon-Chung KJ, Perfect JR, Casadevall A, editors. *Cryptococcus*: from human pathogen to model yeast. Washington: American Society for Microbiology Press; 2011.
150. Idnurm A, Bahn Y-S, Nielsen K, Lin X, Fraser JA, Heitman J. Deciphering the model pathogenic fungus *Cryptococcus neoformans*. Nat Rev Microbiol. 2005;3:753–64.
151. Rajasingham R, Smith RM, Park BJ, Jarvis JN, Govender NP, Chiller TM, Denning DW, Loyse A, Boulware DR. Global burden of disease of HIV-associated cryptococcal meningitis: an updated analysis. Lancet Infect Dis. 2017;17:873–81.
152. Zhang N, Park Y-D, Williamson PR. New technology and resources for cryptococcal research. Fungal Genet Biol. 2015;78:99–107.
153. Edman JC, Kwon-Chung KJ. Isolation of the *URA5* gene from *Cryptococcus neoformans* var. *neoformans* and its use as a selective marker for transformation. Mol Cell Biol. 1990;10:4538–44.
154. Toffaletti DL, Rude TH, Johnston SA, Durack DT, Perfect JR. Gene transfer in *Cryptococcus neoformans* by use of biolistic delivery of DNA. J Bacteriol. 1993;175:1405–11.
155. Kraus PR, Nichols CB, Heitman J. Calcium and calcineurin-independent roles for calmodulin in *Cryptococcus neoformans* morphogenesis and high-temperature growth. Eukaryot Cell. 2005;4:1079–87.
156. Zhu X, Williamson PR. A CLC-type chloride channel gene is required for laccase activity and virulence in *Cryptococcus neoformans*. Mol Microbiol. 2003;50:1271–82.
157. Idnurm A, Reedy JL, Nussbaum JC, Heitman J. *Cryptococcus neoformans* virulence gene discovery through insertional mutagenesis. Eukaryot Cell. 2004;3:420–9.
158. McClelland CM, Chang YC, Kwon-Chung KJ. High frequency transformation of *Cryptococcus neoformans* and *Cryptococcus gattii* by *Agrobacterium tumefaciens*. Fungal Genet Biol. 2005;42:904–13.
159. Walton FJ, Idnurm A, Heitman J. Novel gene functions required for melanization of the human pathogen *Cryptococcus neoformans*. Mol Microbiol. 2005;57:1381–96.
160. Idnurm A, Walton FJ, Floyd A, Reedy JL, Heitman J. Identification of *ENA1* as a virulence gene of the human pathogenic fungus *Cryptococcus neoformans* through signature-tagged insertional mutagenesis. Eukaryot Cell. 2009;8:315–26.
161. Hu G, Caza M, Cadieux B, Chan V, Liu V, Kronstad J. *Cryptococcus neoformans* requires the ESCRT protein Vps23 for iron acquisition from heme, for capsule formation, and for virulence. Infect Immun. 2013;81:292–302.
162. Idnurm A, Heitman J. Light controls growth and development via a conserved pathway in the fungal kingdom. PLoS Biol. 2005;3:615–26.
163. Yeh Y-L, Lin Y-S, Su B-J, Shen W-C. A screening for suppressor mutants reveals components involved in the blue light-inhibited sexual filamentation in *Cryptococcus neoformans*. Fungal Genet Biol. 2009;46:42–54.
164. Feretzaki M, Heitman J. Genetic circuits that govern bisexual and unisexual reproduction in *Cryptococcus neoformans*. PLoS Genet. 2013;9:e1003688.
165. Verma S, Idnurm A. Uve1 endonuclease is regulated by White collar to protect *Cryptococcus neoformans* from UV damage. PLoS Genet. 2013;9:e1003769.
166. Lin X, Jackson JC, Feretzaki M, Xue C, Heitman J. Transcription factors Mat2 and Znf2 operate cellular circuits orchestrating opposite- and same-sex mating in *Cryptococcus neoformans*. PLoS Genet. 2010;6:e1000953.
167. Chacko N, Zhao Y, Yang E, Wang L, Cai JJ, Lin X. The lncRNA *RZE1* controls cryptococcal morphological transition. PLoS Genet. 2015;11:e1005692.
168. Gyawali R, Zhao Y, Lin J, Fan Y, Xu X, Upadhyay S, Lin X. Pheromone independent unisexual development in *Cryptococcus neoformans*. PLoS Genet. 2017;13:e1006772.
169. Fu J, Mares C, Lizcano A, Liu Y, Wickes BL. Insertional mutagenesis combined with an inducible filamentation phenotype reveals a conserved *STE50* homologue in *Cryptococcus neoformans* that is required for monokaryotic fruiting and sexual reproduction. Mol Microbiol. 2011;79:990–1007.
170. Ianiri G, Idnurm A. Essential gene discovery in the basidiomycete *Cryptococcus neoformans* for antifungal drug target prioritization. mBio. 2015;6:e02334-02314.
171. Chun CD, Madhani HD. Ctr2 links copper homeostasis to polysaccharide capsule formation and phagocytosis inhibition in the human fungal pathogen *Cryptococcus neoformans*. PLoS ONE. 2010;5:e12503.
172. Chang YC, Lamichhane AK, Garraffo HM, Walter PJ, Leerkes M, Kwon-Chung KJ. Molecular mechanisms of hypoxic responses via unique roles of Ras1, Cdc24 and Ptp3 in a human fungal pathogen *Cryptococcus neoformans*. PLoS Genet. 2014;10:e1004292.
173. Ost KS, O'Meara TR, Huda N, Esher SK, Alspaugh JA. The *Cryptococcus neoformans* alkaline response pathway: identification of a novel rim pathway activator. PLoS Genet. 2015;11:e1005159.
174. Aime MC, Matheny PB, Henk DA, Frieders EM, Nilsson RH, Piepenbring M, McLaughlin DJ, Szabo LJ, Begerow D, Sampaio JP, et al. An overview of the higher level classification of Pucciniomycotina based on combined analyses of nuclear large and small subunit rDNA sequences. Mycologia. 2006;98:896–905.

175. Webb CA, Szabo LJ, Bakkeren G, Garry C, Staples RC, Eversmeyer M, Fellers JP. Transient expression and insertional mutagenesis of *Puccinia triticina* using biolistics. Funct Integr Genomics. 2006;6:250–60.
176. Lawrence GJ, Dodds PN, Ellis JG. Transformation of the flax rust fungus, *Melampsora lini*: selection via silencing of an avirulence gene. Plant J. 2010;61:364–9.
177. Toh SS, Perlin MH. Resurgence of less-studied smut fungi as models of phytopathogenesis in the omics age. Phytopathology. 2016;106:1244–54.
178. Hood ME. Dimorphic mating-type chromosomes in the fungus *Microbotryum violaceum*. Genetics. 2002;160:457–61.
179. Bej AK, Perlin MH. A high efficiency transformation system for the basidiomycete *Ustilago violacea* employing hygromycin resistance and lithium-acetate treatment. Gene. 1989;80:171–6.
180. Perlin MH, Bej AK, Will OH 3rd, Jacob RJ. Introduction and maintenance of prokaryotic DNA in *Ustilago violacea*. J Ind Microbiol. 1990;5:355–63.
181. Toh SS, Treves DS, Barati MT, Perlin MH. Reliable transformation system for *Microbotryum lychnidis-dioicae* informed by genome and transcriptome project. Arch Microbiol. 2016;198:813–25.
182. Perlin MH, Amselem J, Fontanillas E, Toh SS, Chen Z, Goldberg J, Duplessis S, Henrissat B, Young S, Zeng Q, et al. Sex and parasites: genomic and transcriptomic analysis of *Microbotryum lychnidis-dioicae*, the biotrophic and plant-castrating anther smut fungus. BMC Genom. 2015;16:461.
183. Wang Q-M, Yurkov AM, Göker M, Lumbsch HT, Leavitt SD, Groenewald M, Theelen B, Liu X-Z, Boekhout T, Bai F-Y. Phylogenetic classification of yeasts and related taxa within *Pucciniomycotina*. Stud Mycol. 2015;81:149–89.
184. Johnson EA. Biotechnology of non-*Saccharomyces* yeasts-the basidiomycetes. Appl Microbiol Biotechnol. 2013;97:7563–77.
185. Tully M, Gilbert HJ. Transformation of *Rhodosporidium toruloides*. Gene. 1985;36:235–40.
186. Ianiri G, Wright SAI, Castoria R, Idnurm A. Development of resources for the analysis of gene function in Pucciniomycotina red yeasts. Fungal Genet Biol. 2011;48:685–95.
187. Abbott EP, Ianiri G, Castoria R, Idnurm A. Overcoming recalcitrant transformation and gene manipulation in *Pucciniomycotina* yeasts. Appl Microbiol Biotechnol. 2013;97:283–95.
188. Liu Y, Koh CMJ, Sun L, Hlaing MM, Du M, Peng N, Ji L. Characterization of glyceraldehyde-3-phosphate dehydrogenase gene Rt*GPD1* and development of genetic transformation method by dominant selection in oleaginous yeast *Rhodosporidium toruloides*. Appl Microbiol Biotechnol. 2013;97:719–29.
189. Lin X, Wang Y, Zhang S, Zhu Z, Zhou YJ, Yang F, Sun W, Wang X, Zhao ZK. Functional integration of multiple genes into the genome of the oleaginous yeast *Rhodosporidium toruloides*. FEMS Yeast Res. 2014;14:547–55.
190. Wang Y, Lin X, Zhang S, Sun W, Ma S, Zhao ZK. Cloning and evaluation of different constitutive promoters in the oleaginous yeast *Rhodosporidium toruloides*. Yeast. 2016;33:99–106.
191. Ianiri G, Idnurm A, Wright SAI, Durán-Patrón R, Mannina L, Ferracane R, Ritieni A, Castoria R. Searching for genes responsible for patulin degradation in a biocontrol yeast provides insight into the basis for resistance to this mycotoxin. Appl Environ Microbiol. 2013;79:3101–15.
192. Ianiri G, Idnurm A, Castoria R. Transcriptomic responses of the basidiomycete yeast *Sporobolomyces* sp. to the mycotoxin patulin. BMC Genom. 2016;17:210.
193. Ianiri G, Abhyankar R, Kihara A, Idnurm A. Phs1 and the synthesis of very long chain fatty acids are required for ballistospore formation. PLoS ONE. 2014;9:e105147.
194. Zhang S, Skerker JM, Rutter CD, Maurer MJ, Arkin AP, Rao CV. Engineering *Rhodosporidium toruloides* for increased lipid production. Biotechnol Bioeng. 2016;113:1056–66.
195. Koh CMJ, Liu Y, Du Moehninsi M, Ji L. Molecular characterization of *KU70* and *KU80* homologues and exploitation of a *KU70*-deficient mutant for improving gene deletion frequency in *Rhodosporidium toruloides*. BMC Microbiol. 2014;14:50.
196. Liu Y, Koh CMJ, Ngoh ST, Ji L. Engineering an efficient and tight D-amino acid-inducible gene expression system in *Rhodosporidium/Rhodotorula* species. Microb Cell Fact. 2015;14:170.
197. Liu Y, Yap SA, Kohn CMJ, Ji L. Developing a set of strong intronic promoters for robust metabolic engineering in oleaginous *Rhodotorula* (*Rhodosporidium*) yeast species. Microb Cell Fact. 2016;15:200.
198. Ji L, Jiang Z-D, Liu Y, Koh CM, Zhang L-H. A simplified and efficient method for transformation and gene tagging of *Ustilago maydis* using frozen cells. Fungal Genet Biol. 2010;47:279–87.
199. Marchand G, Fortier E, Neveu B, Bolduc S, Belzile F, Bélanger RR. Alternative methods for genetic transformation of *Pseudozyma antarctica*, a basidiomycetous yeast-like fungus. J Microbiol Methods. 2007;70:519–27.
200. Ianiri G, Averette AF, Kingsbury JM, Heitman J, Idnurm A. Gene function analysis in the ubiquitous human commensal and pathogen *Malassezia* genus. mBio. 2016;7:e01853-01816.
201. Celis AM, Vos AM, Triana S, Medina CA, Escobar N, Restrepo S, Wösten HAB, de Cock H. Highly efficient transformation system for *Malassezia furfur* and *Malassezia pachydermatis* using *Agrobacterium tumefaciens*-mediated transformation. J Microbiol Methods. 2017;134:1–6.
202. Findley K, Oh J, Yang J, Conlan S, Deming C, Meyer JA, Schoenfeld D, Nomicos E, Park M, Program NIHISCCS, et al. Topographic diversity of fungal and bacterial communities in human skin. Nature. 2013;498:367–70.
203. Wu G, Zhao H, Li C, Rajapakse MP, Wong WC, Xu J, Saunders CW, Reeder NL, Reilman RA, Scheynius A, et al. Genus-wide comparative genomics of *Malassezia* delineates its phylogeny, physiology, and niche adaptation on human skin. PLoS Genet. 2015;11:e1005614.
204. Spatafora JW, Chang Y, Benny GL, Lazarus K, Smith ME, Berbee ML, Bonito G, Corradi N, Grigoriev I, Gryganskyi A, et al. A phylum-level phylogenetic classification of zygomycete fungi based on genome-scale data. Mycologia. 2016;108:1028–46.
205. Michielse CB, Salim K, Ragas P, Ram AFJ, Kudla B, Jarry B, Punt PJ, van den Hondel CAMJJ. Development of a system for integrative and stable transformation of the zygomycete *Rhizopus oryzae* by *Agrobacterium*-mediated DNA transfer. Mol Genet Genomics. 2004;271:499–510.
206. Nyilasi I, Papp T, Csernetics A, Vágvölgyi C. *Agrobacterium tumefaciens*-mediated transformation of the zygomycete fungus *Backusella lamprospora*. J Basic Microbiol. 2008;48:59–64.
207. Nyilasi I, Ács K, Papp T, Nagy E, Vágvölgyi C. *Agrobacterium tumefaciens*-mediated transformation of *Mucor circinelloides*. Folia Microbiol. 2005;50:415–20.
208. Papp T, Csernetics A, Nagy G, Bencsik O, Iturriaga EA, Eslava AP, Vágvölgyi C. Canthaxanthin production with modified *Mucor circinelloides* strains. Appl Microbiol Biotechnol. 2013;97:4937–50.
209. Monfort A, Cordero L, Maicas S, Polaina J. Transformation of *Mucor miehei* results in plasmid deletion and phenotypic instability. FEMS Microbiol Lett. 2003;224:101–6.
210. Wei D-S, Zhang Y-H, Xing L-J, Li M-C. *Agrobacterium rhizogenes*-mediated transformation of a high oil-producing filamentous fungus *Umbelopsis isabellina*. J Appl Genet. 2010;51:225–32.
211. Ando A, Sumida Y, Negoro H, Suroto DA, Ogawa J, Sakuradani E, Shimizu S. Establishment of *Agrobacterium tumefaciens*-mediated transformation of an oleaginous fungus, *Mortierella alpina* 1S-4, and its application for eicosapentaenoic acid producer breeding. Appl Environ Microbiol. 2009;75:5529–35.
212. Hao G, Chen H, Wang L, Gu Z, Song Y, Zhang H, Chen W, Chen YQ. Role of malic enzyme during fatty acid synthesis in the oleaginous fungus *Mortierella alpina*. Appl Environ Microbiol. 2014;80:2672–8.
213. Shi H, Chen H, Gu Z, Zhang H, Chen W, Chen YQ. Application of a delta-6 desaturase with α-linolenic acid preference on eicosapentaenoic acid production in *Mortierella alpina*. Microb Cell Fact. 2016;15:117.
214. Hao G, Chen H, Gu Z, Zhang H, Chen W, Chen YQ. Metabolic engineering of *Mortierella alpina* for enhanced arachidonic acid production through the NADPH-supplying strategy. Appl Environ Microbiol. 2016;82:3280–8.
215. Hao G, Chen H, Du K, Huang X, Song Y, Gu Z, Wang L, Zhang H, Chen W, Chen YQ. Increased fatty acid unsaturation and production of arachidonic acid by homologous over-expression of the mitochondrial malic enzyme in *Mortierella alpina*. Biotechnol Lett. 2014;36:1827–34.
216. Vieira ALG, Camilo CM. *Agrobacterium tumefasciens* [*sic*]-mediated transformation of the aquatic fungus *Blastocladiella emersonii*. Fungal Genet Biol. 2011;48:806–11.

217. Helber N, Requena N. Expression of the fluorescence markers DsRed and GFP fused to a nuclear localization signal in the arbuscular mycorrhizal fungus *Glomus intraradices*. New Phytol. 2008;177:537–48.
218. Gietz RD, Schiestl RH. High-efficiency yeast transformation using the LiAc/SS carrier DNA/PEG method. Nat Protoc. 2007;2:31–4.
219. Catlett NL, Lee B-N, Yoder OC, Turgeon BG. Split-marker recombination for efficient targeted deletion of fungal genes. Fungal Genet Newsl. 2003;50:9–11.
220. Cairns TC, Sidhu YS, Chaudhari YK, Talbot NJ, Studholme DJ, Haynes K. Construction and high-throughput phenotypic screening of *Zymoseptoria tritici* over-expression strains. Fungal Genet Biol. 2015;79:110–7.
221. Kojic M, Holloman WK. Shuttle vectors for genetic manipulations in *Ustilago maydis*. Can J Microbiol. 2000;46:333–8.
222. Aleksenko A, Clutterbuck AJ. Autonomous plasmid replication in *Aspergillus nidulans*: AMA1 and MATE elements. Fungal Genet Biol. 1997;21:373–87.
223. Zhu Y, Nam J, Humara JM, Mysore KS, Lee LY, Cao H, Valentine L, Li J, Kaiser AD, Kopecky AL, et al. Identification of Arabidopsis *rat* mutants. Plant Physiol. 2003;132:494–505.
224. Luo Y, Chen Z, Zhu D, Tu H, Pan SQ. Yeast actin-related protein ARP6 negatively regulates *Agrobacterium*-mediated transformation of yeast cell. Biomed Res Int. 2015;2015:275092.
225. Ohmine Y, Satoh Y, Kiyokawa K, Yamamoto S, Moriguchi K, Suzuki K. DNA repair genes *RAD52* and *SRS2*, a cell wall synthesis regulator gene *SMI1*, and the membrane sterol synthesis scaffold gene *ERG28* are important in efficient *Agrobacterium*-mediated yeast transformation with chromosomal T-DNA. BMC Microbiol. 2016;16:58.
226. van Attikum H, Bundock P, Hooykaas PJJ. Non-homologous end-joining proteins are required for *Agrobacterium* T-DNA integration. EMBO J. 2001;20:6550–8.
227. van Attikum H, Hooykaas PJJ. Genetic requirements for the targeted integration of *Agrobacterium* T-DNA in *Saccharomyces cerevisiae*. Nucleic Acids Res. 2003;31:826–32.
228. Rolloos M, Dohmen MHC, Hooykaas PJJ, van der Zaal BJ. Involvement of Rad52 in T-DNA circle formation during *Agrobacterium tumefaciens*-mediated transformation of *Saccharomyces cerevisiae*. Mol Microbiol. 2014;91:1240–51.
229. Soltani J, van Heusden GPH, Hooykaas PJJ. Deletion of host histone acetyltransferases and deacetylases strongly affects *Agrobacterium*-mediated transformation of *Saccharomyces cerevisiae*. FEMS Microbiol Lett. 2009;298:228–33.
230. Roberts RL, Metz M, Monks DE, Mullaney ML, Hall T, Nester EW. Purine synthesis and increased *Agrobacterium tumefaciens* transformation of yeast and plants. Proc Natl Acad Sci USA. 2003;100:6634–9.
231. Darwin C. On the origin of species by means of natural selection. London: John Murray; 1859.
232. Esher SK, Granek JA, Alspaugh JA. Rapid mapping of insertional mutations to probe cell wall regulation in *Cryptococcus neoformans*. Fungal Genet Biol. 2015;82:9–21.
233. Shine AM, Shakya VPS, Idnurm A. Phytochelatin synthase is required for tolerating metal toxicity in a basidiomycete yeast and is a conserved factor involved in metal homeostasis in fungi. Fungal Biol Biotechnol. 2015;2:3.
234. McCluskey K, Boundy-Mills K, Dye G, Ehmke E, Gunnell GF, Kiaris H, Polihronakis Richmond M, Yoder AD, Zeigler DR, Zehr S, et al. The challenges faced by living stock collections in the USA. eLife. 2017;6:e24611.
235. Daley M, Knauf VC, Summerfelt KR, Turner JC. Co-transformation with one *Agrobacterium tumefaciens* strain containing two binary plasmids as a method for producing marker-free transgenic plants. Plant Cell Rep. 1998;17:489–96.
236. Padilla-Guerrero IE, Bidochka MJ. *Agrobacterium*-mediated co-transformation of multiple genes in *Metarhizium robertsii*. Mycobiology. 2017;45:84–9.
237. de Boer P, Bronkhof J, Dukik K, Kerkman R, Touw H, van den Berg M, Offringa R. Efficient gene targeting in *Penicillium chrysogenum* using novel *Agrobacterium*-mediated transformation approaches. Fungal Genet Biol. 2013;61:9–14.
238. Wang Y, DiGuistini S, Wang T-CT, Bohlmann J, Breuil C. *Agrobacterium*-meditated gene disruption using split-marker in *Grosmannia clavigera*, a mountain pine beetle associated pathogen. Curr Genet. 2010;56:297–307.
239. Krenek P, Samajova O, Luptovciak I, Doskocilova A, Komis G, Samaj J. Transient plant transformation mediated by *Agrobacterium tumefaciens*: principles, methods and applications. Biotechnol Adv. 2015;33:1024–42.
240. Nødvig CS, Nielsen JB, Kogle ME, Mortensen UH. A CRISPR-Cas9 system for genetic engineering of filamentous fungi. PLoS ONE. 2015;10:e0133085.
241. Pohl C, Kiel JAKW, Driessen AJM, Bovenberg RAL, Nygård Y. CRISPR/Cas9 based genome editing of *Penicillium chrysogenum*. ACS Synth Biol. 2016;5:754–64.
242. Schuster M, Schweizer G, Reissmann S, Kahmann R. Genome editing in *Ustilago maydis* using the CRISPR-Cas system. Fungal Genet Biol. 2016;89:3–9.
243. Prieto M, Wedin M. Dating the diversification of the major lineages of Ascomycota (Fungi). PLoS ONE. 2013;8:e65576.
244. Giesbert S, Schumacher J, Kupas V, Espino J, Segmüller N, Haeuser-Hahn I, Schreier PH, Tudzynski P. Identification of pathogenesis-associated genes by T-DNA-mediated insertional mutagenesis in *Botrytis cinerea*: a type 2A phosphoprotein phosphatase and an SPT3 transcription factor have significant impact on virulence. Mol Plant Microbe Interact. 2012;25:481–95.
245. Yemelin A, Brauchler A, Jacob S, Laufer J, Heck L, Foster AJ, Antelo L, Andresen K, Thines E. Identification of factors involved in dimorphism and pathogenicity of *Zymoseptoria tritici*. PLoS ONE. 2017;12:e0183065.

# 15

# A framework for an organelle-based mathematical modeling of hyphae

Rudibert King*

**Abstract**

**Background:** Although highly desirable, a mechanistic explanation for the outstanding protein secretion capabilities of fungi such as *Aspergilli* is missing. As a result, a rational and predictive design of strains as cell factories for protein production is still out of reach. The analysis of the secretion apparatus is not only hampered by open issues concerning molecular cell biological processes, but as well by their spatial fragmentation and highly dynamic features. Whereas the former issues are addressed by many groups, an account of the space- and time-dependent processes, which is best done by means of mathematical models, is lacking. Up to now, mathematical models for hyphal organisms mainly focus on one of two extremes. Either macroscopic morphology, such as pellet or mycelium growth, is addressed, or a microscopic picture is drawn predicting, for instance, the form of a hyphal tip. How intra-hyphal transport and organelle distribution works, however, has not been tackled so far mathematically.

**Results:** The main result of this contribution is a generic modeling framework to describe the space- and time-dependent evolution of intracellular substances and organelles. It takes intrahyphal, passive and active transport of substances into account and explains exponential and then linear length growth by tugor-driven uptake of water. Experimentally observed increasing concentration levels of organelles towards the tip can be well explained within the framework without resorting to complex biological regulations. It is shown that the accumulation can be partly explained by geometrical constraints, besides a necessary deceleration of the active transport velocity. The model is formulated such that more intricate intracellular processes can be included.

**Conclusions:** Results from steady-state experiments are easy to be interpreted. In a hyphal network, however, new branches are produced at an exponential rate. Moreover, passive and active transport processes give rise to a spatial distribution of organelles and other cytoplasmatic constituents inside hyphae. As a result, most of the data obtained in experiments will be from a non-steady and space dependent state. A quantitative and mechanistic explanation of the processes occurring will only be possible if these dependencies are taking into account while evaluating experimental findings.

**Keywords:** Morphological model, Hypha, Vesicles

## Background

The ecological and technical relevance of fungi is outstanding. They are integrated in most ecosystems, act as detrimental agents for plants and humans, decompose waste materials, and are exploited in the synthesis of valuable products [1, 2], to name just a few areas in which they play a major role. Their most striking feature is polarized growth and branching which leads to more or less dense mycelia or pellets [3, 4]. Concomitantly, *Aspergilli* such as *A. niger*, *A. oryzae* and *A. terreus* have astounding capabilities to secrete interesting enzymes, mainly through the apical region [5]. A rational design to obtain modified strains as optimized cell factories, however, is still limited by the incomplete picture of their growth, production and secretion machinery. By the very nature of living cells all occurring processes are highly dynamic and the behavior of a cell does not only depend on the actual stimuli but what had happened to the cell in the past. For fungal organisms, interpretation of physiological data is even more challenging. Besides

*Correspondence: rudibert.king@tu-berlin.de
Chair of Measurement and Control, Technische Universität Berlin, Berlin, Germany

the compartmentalization of biological functions in distinct organelles, space- and time-dependent distributions occur. This relates to organelles, other cytoplasmic compounds, and stimuli in and around a mycelium, and, therefore, impede the deduction of meaningful knowledge and hypotheses. All of which could be addressed in the context of mathematical models.

So far most mathematical models for hyphal organisms respecting morphological features focused on macroscopic processes in dense mycelia and pellets. Here, the interplay between nutrient transport by diffusion and space-dependent growth was addressed, see for example [1, 6–18] and references therein. A second group of models tries to predict the geometric appearance of rather small mycelia [19–25]. In these models, very little detailed biological information is needed, if used at all, to give rather realistic pictures. The results of simulations shown in Figure 1, as an example, just use the information that the apical growth rate in three dimensions is constant and that septa and branches are formed when a critical length of a hyphal compartment is obtained. Including a random growth direction in the simulations gives a visual impression which is hard to distinguish from real photographs. As for the first group of models, these models will not help in deciphering the secretion apparatus. This is true as well for so-called morphologically structured models, which do not even account for the space-dependency, see, e.g., [7, 21, 26].

A more detailed account of microscopic features of individual hyphae is given by a last group of models to describe, for instance, the shape of a tip, or the growth in length. A well-known example for the first class of problems, which will be used in what follows, is given by Bartnicki-Garcia et al. in [27]. Here, the geometrical form of an apex is predicted with a simple model. It is based on a set of hypotheses how vesicles are transported ballistically from the Spitzenkörper to the wall. The model has been refined in future works to better account for the three-dimensional shape of a tip, the way vesicles are transported to the wall by diffusion, by representing the cell wall as a flexible membrane, or by a better account of vesicle fusion with the cell wall [28–32].

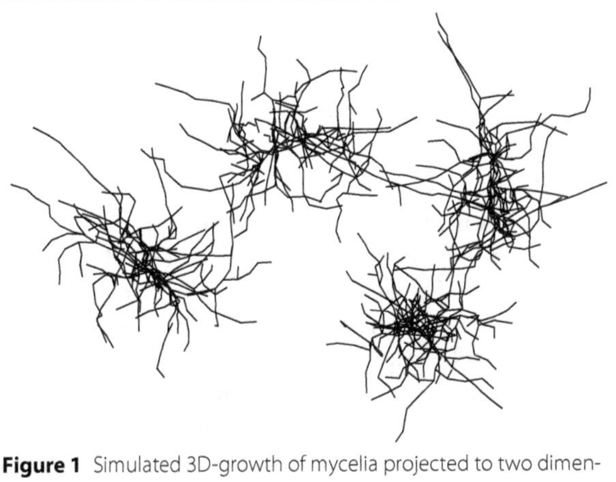

**Figure 1** Simulated 3D-growth of mycelia projected to two dimensions.

Mathematical descriptions of the growth in length, as the second class of microscopic features, were given recently with two different approaches. In [33], the long-range transport of material in hyphae is depicted by a particle transport along a single, hypothetical microtubule extending over the whole length of a hypha. The amount of material reaching the tip of the hypha determines length growth. A changing velocity, however, is neither considered nor a movement of the microtubules with the cytoplasmatic flow. In contrast, [34] explain length growth of *Phanerochaete velutina* mathematically by a turgor driven intra-hyphal flow towards the tip. In all these approaches a constant length growth rate is considered which is not true for the germ tube. Moreover, new branches of a mycelium very often show a lower initial velocity as well. As a result, and as a mycelium grows exponentially by an exponential production of new branches, a significant part of a mycelium will not be in a kind of quasi-steady state which is assumed above. In major parts of a mycelium, organelles and intracellular substances have not yet reached their quasi-steady state distribution, which might be important for a quantitative prediction of the growth of the mycelium. Likewise, if septa are closed and opened by Woronin bodies, intra-hyphal flow has to stop or will resume resulting in even more complex situations.

In our former works [11, 19], we explained the initially observed exponential and then linear growth with the limitation by a hypothetical intracellular compound. We had to resort to a hypothetical compound at that time as neither for fungi nor for actinomycetes details about the mechanism were known. Especially for fungi, this situation has changed drastically in recent years. Molecular methods, bioinformatics and image analysis have provided us with a whelm of information if not give rise to a 'Big Data Tsunami' [35]. More specifically, for the processes addressed here, which are responsible for length growth and (product) secretion, much more is known today. Excellent recent reviews about growth in length of fungal hyphae are given, e.g., by [36, 37], and about secretion in [32]. The importance of turgor driven length extension is stretched by [38, 39] in a series of papers.

We therefore think that the time is ripe to try to condense at least a small part of the available knowledge in a mathematical model. This can form the basis to discuss hypotheses and to account for the effects of space- and time-dependencies in the interpretation of experimental data. The model structure derived in this contribution therefore serves two main purposes.

- First of all, it represents a basic model structure, with which the initial exponential and then linear growth of a hypha can be described with a minimal amount of assumptions. This turgor driven evolution of the intra-hyphal flow forms the 'backbone' for all other processes occurring in a hyphae, and, therefore, has to be considered first. To be more specific, besides the postulation of some kinetics, no further biological regulations will be introduced to describe the experimentally observed growth evolution. If this is possible, already simple physical transport processes combined with implicitly formulated regulations through kinetic expressions can be used to explain the observed behavior without resorting to complex biological mechanisms. This, of course, does not rule out such regulations which additionally may occur. If experimental evidence is given, these processes can be included readily.
- Secondly, the model structure derived serves as a basis for future work when experimental data is interpreted and condensed in a mathematical framework. As an example, the distribution of vesicles in a hypha will be considered here which shows a distinct profile along the length of a hypha. Again simple physical arguments, mainly with respect to the active transport velocity and the geometry of the tip, will be enough to explain experimental data where a significant increase in concentration is observed toward the tip.

In the long run, such kind of models might help in answering questions raised in the endeavor toward a rational strain design. Examples are [5]: How many vesicles carrying proteins of interest can be used without interfering with vesicles for growth? Where are the bottlenecks in vesicle-mediated protein secretion? How many proteins can be channeled through the secretory pathway in order to provide each protein sufficient time to become correctly folded? Extending this list of questions will naturally occur when a model is at hand.

The rest of the paper is organized as follows. After a problem statement in the next section, the model of Bartnicki-Garcia et al. is revisited to determine the volume and surface area in the tip region. This will then be used to correct experimental data. The general model is formulated next. As a first application, length growth by turgor driven water uptake is described. Extending the model with vesicles allows for a comparison with experimental data in the last section before the paper finishes with some conclusions.

## Problem statement

What will be described below, will neither encompass the process of sporulation nor branching. An attempt to model germination can be found, e.g., in [40]. It is assumed that a very short part of a tip already exists, presumably from a mother hypha. In the simulations shown, depending on the boundary conditions, the mother hypha will either have no influence on the developing branch or it will supply material for growth. After some time, say $t = t_t$, the retrograd end of the new hypha will posses a final radius of $r_t(t_t) = R$ and the length of the tip will be $L_t(t_t) = L_{tmax}$. For $t > t_t$, the hypha will consist of this constant tip part with fixed length $L_{tmax}$ and a seemingly growing distal region of length $L_d(t)$, i.e., the overall length of the hypha will then be given by

$$L(t) = L_d(t) + L_{tmax}. \tag{1}$$

Consequently, it will be assumed that, after tip completion, its geometrical form and size stay constant. Both assumptions together simplify the mathematical treatment significantly as apparent growth is associated to a subapical region where the cell wall becomes rigidified, probably by the action of cross-linking enzymes [37]. With this assumption, the complex processes involved in plasma membrane and cell wall synthesis actually occurring in the tip region are approximated. A more detailed model is given by [32]. A constant form and size of the tip region, on the other side, necessitates a constant radius of the distal part. This is in contrast to the model introduced in [34] where the authors correlate high internal volume flows $Q$ with large diameters of hyphae. This will not be considered here for simplicity although the general model could account for it.

For a simpler numerical implementation, the tip and the distal part will be modeled separately with appropriate conditions to account for the connection between them, see Figure 2. In what follows, the variables related to the tip will be denoted by an index $t$ whereas those related to the distal part with a $d$.

The general model will be obtained by a formulation of balance equations applied to an infinitesimal intra-hyphal balancing volume of length $dx$, see the space between shaded areas in Figure 2. Terms of production, consumption and transport via the cell membrane, the consequences of an intra-hyphal flow, and, finally, due to active translocation will be considered. Space-dependent uptake of nutrients could be included readily, but is not done here. Most importantly, a constant physiological and functional state is assumed along a hypha. If this is not the case, the model developed in this contribution has to be combined with approaches proposed, e.g., by Nielsen and Villadsen [18]. As the transport from and to the environment will be proportional to the local surface area and production and

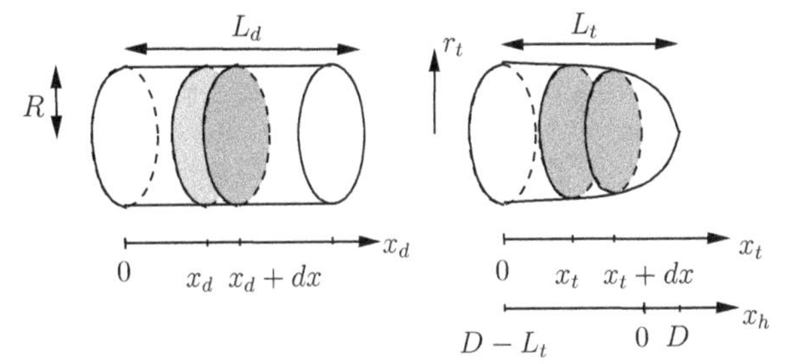


**Figure 2** Sketch of the distal part (*d*) with a constant radius *R*, and the tip (*t*) of a hypha.

consumption rates will be given based on the local volume, these quantities have be determined first.

## Geometrical model of the tip

For simplicity, the 2D-model proposed in [27] describing the form of a hyphal tip is used to derive expressions for the local surface area $A(x_t)$ and volume $V(x_t)$. Although the initial conjecture that the actual 3D-form of a hypha can be produced by a rotation of the solution of the 2D-model was corrected in [28], see as well [29], the simpler approach is used here. This is motivated by the fact that the actual differences in the forms obtained are small while the calculation of the 3D-form is rather involved.

In [27] it is proposed that: (1) the cell surface expands from materials discharged by wall-destined vesicles, (2) vesicles are released from a postulated vesicle supply center (VSC), (3) vesicles move from the VSC to the surface in any random direction. Based on these propositions, they derive the following model.

If the VSC is located in the origin of an $(x_h, y_h)$-coordinate system, as used by Bartnicki-Garcia et al. [27], where the cartesian coordinate $y_h$ equals the radius $r_t$ in Figure 2, i.e., $y_h = r_t$, the 2D-geometry reads

$$x_h = r_t \cot \frac{r_t}{D}. \tag{2}$$

In this coordinate system, the hypha extends to negative $x_h$-values. The foremost point of the tip is at $x_h = D$. With $x_t = x_h + L_t - D$, see Figure 2, the geometry of the tip is given by

$$x_t = r_t \cot \frac{r_t}{D} + L_t - D \tag{3}$$

in the coordinates used in this work.

For the calculation of the volume $V(x_t)$ and surface area $A(x_t)$, $r_t(x_t)$ is locally approximated by a straight line of length $l_t$ connecting $r_t(x_t)$ and $r_t(x_t + dx)$. Rotating this line defines the area and the enclosed volume.

Neglecting higher order terms, the infinitesimal surface area of such a truncated cone is given by

$$A(x_t) \approx 2\pi \sqrt{1 + \left(\frac{dr_t}{dx_t}\right)^2}\, r_t(x_t)dx, \tag{4}$$

and the infinitesimal balancing volume is approximated by

$$V(x_t) \approx \pi r_t^2(x_t)dx. \tag{5}$$

The derivative needed in the former expression can be obtained from Eq. 3 by implicit differentiation, see Appendix A.

For a stationary observer, see Figure 3, the balance volume increases when the tip grows out of a considered section $[x_t, x_t + dx]$.

As it is assumed that the form of the tip stays constant, every point of the surface moves with the extension rate $\dot{L}$ of the hypha, hence,

$$\dot{V}_t = 2\pi r_t \frac{dr_t}{dx_t}\frac{dx_t}{dt}dx = 2\pi r_t \frac{dr_t}{dx_t}(-\dot{L})dx. \tag{6}$$

A negative sign is included as for a stationary observer the volume increases while $dr_t/dx_t$ is negative.

The tip and the distal part must be connected without a step in radius, i.e., $r_t(x_t = 0) = R$. As a result, $D$ cannot be chosen independently when $L_{tmax}$ is fixed. From a biological point of view, fixing $D$ might be a better alternative. This, however, would lead to a very long tip region which is ruled out here for numerical reasons. From Eq. 3, when $L_{tmax}$ is prescribed, an implicit expression is given for $D$

$$0 = R \cot \frac{R}{D} + L_{tmax} - D. \tag{7}$$

For small $L_{tmax}$, this value of $D$ is an approximation of the real distance of the VSC from the apex. Figure 4 gives an impression of the obtained form of the tip, where some distal part is shown as well. The calculations were done with $R = 3.5\ \mu m$, $D = 1.2\ \mu m$, and $L_{tmax} = 15.6\ \mu m$.

## Experimental data

Although the main goal of this contribution lays in the derivation of a generic model structure, some comparisons with experimental data will be done.

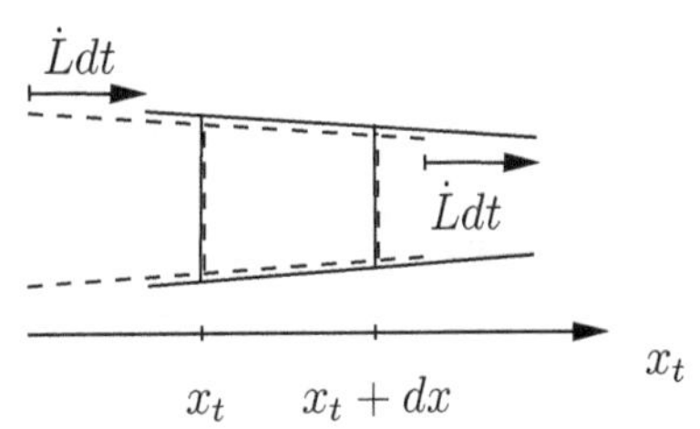


**Figure 3** The balance volume for time instant *t* is depicted by a *box* with *broken lines* and for *t* + *dt* with *solid lines*. As the tip moves to the *right*, the volume increases for a stationary observer.

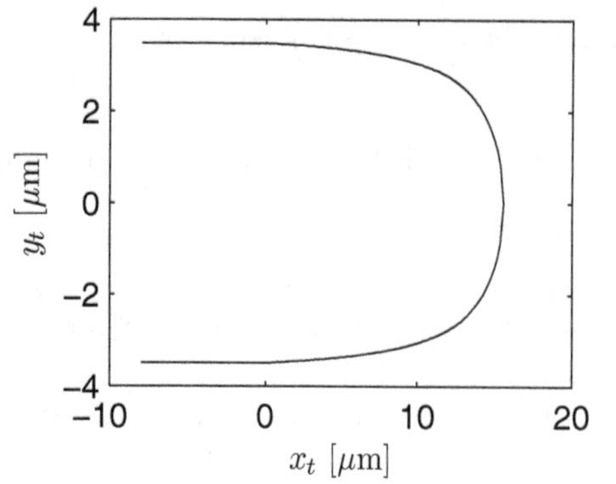


**Figure 4** Cross-section of a simulated tip of a hypha for $x_t \geq 0$. Some distal part with constant radius is shown as well for $x_t < 0$.

Unpublished experimental data with vesicle measurements from *Aspergillus niger* is kindly supplied by F. Spanhoff, A. Ram and V. Meyer. They visualized the secretory vesicle concentration of individual hyphae by the intensity of the fluorescent R-SNARE protein SynA using a Zeiss confocal microscope. Pictures were taken at an equidistant spacing of 0.2 μm along the length of a hypha for typically 10–12 layers (z-stacks) across the diameter. Hyphae were obtained from the periphery of a fungal colony. Uncalibrated fluorescence data is obtained by adding up all intensity values for each z-stack to obtain $I$. The mean value for 7 different hyphae is shown in Figure 5 as a function of the distance to the apex. The origin of the $x$-coordinate system used here is chosen $D = 1.2$ μm behind the apex, which coincides with the VSC used later. As the scanned volume in the apical dome is smaller than in subapical regions, the data is corrected here for this geometrical effect. Again the model of [27] is used to calculate a local volume, and, hence, from the intensity data, a local normalized measured concentration $V_m$. As absolute information about the number of vesicles is missing, the data is additionally normalized by an arbitrary scaling factor of 35 to compare against simulation results later. The scaling factor is chosen such that in the subapical region a normalized concentration of approximately 1 is obtained. The corrected data is shown in Figure 6. Observe the higher ratio of maximal to subapical values in the corrected data compared to the given intensity values in Figure 5.

In the experiments described above, hyphal length was not measured for *A. niger* as a function of time for newly developing branches. Typically, an initial exponential growth will be observed followed by a linear one. As the model will be able to describe this, another set of data is used here for comparison. Experimental results from [41] are exploited. This rather old set of data was already used by us in [11]. In this former work, a much simpler model was proposed to describe length evolution. Using the data again, both approaches can be compared. Fiddy and Trince [41] measured the evolution of a primary branch of *Geotrichum candidum* extending out of an intercalary compartment just behind a septum. They observed a correlation of the decreasing length extension rate of this branch with septation occurring after some time in this branch. However, the extension rate of the primary branch continued to increase, despite septation, until a length of about 700 μm was reached. From Figure 3b in [41], a maximal extension rate of 2.5 μm/min can be estimated. The data will be given later together with the results of a simulation.

## Generic model

To derive a generic model, a substance $\mathcal{S}_i$, $i = d, t$ is balanced in a segment extending from $x_i$ to $x_i + dx$, see Figure 2. In what follows, $\mathcal{S}_t$ represents vesicles ($\mathcal{V}_t$), Spitzenkörper ($\mathcal{K}_t$), osmolytes ($\mathcal{O}_t$), etc. in the tip, i.e., $\mathcal{S}_t = \{\mathcal{V}_t, \mathcal{K}_t, \mathcal{O}_t, \ldots\}$. Accordingly, $\mathcal{S}_d = \{\mathcal{V}_d, \mathcal{K}_d, \mathcal{O}_d, \ldots\}$ will denote variables in the distal part. A radial

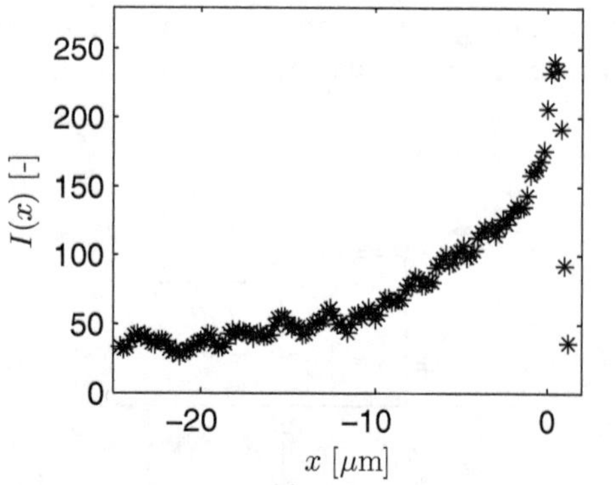


**Figure 5** Relative expression intensity $I$ of the fluorescent protein SynA::eGFP in *A. niger* as a function of the distance to the apex. Original unpublished data from F. Spanhoff et al.

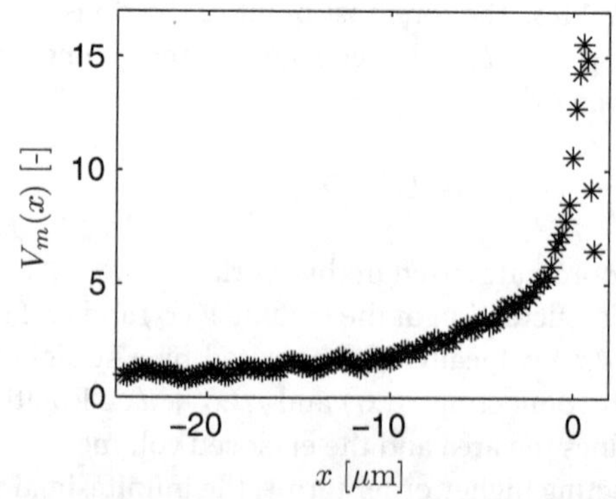


**Figure 6** Corrected and normalized vesicle concentration $V_m$.

distribution inside the hypha and diffusion in all directions are neglected. Diffusion in the $x$-direction could be included readily without complicating the numerical solution much. It would, however, make less sense for organelles. See [39] for a discussion of diffusion coefficients of different cellular components compared against the intra-hyphal flow velocity.

The volumetric mass concentration of a substance $\mathcal{S}_i$ is represented here by the very same symbol in the equations, i.e., $\mathcal{S}_i$ is used to denote the concentration of the generic compound. Hence,

$$\mathcal{S}_i(x_i,t) = \frac{m_{\mathcal{S}_i}(x_i,t)}{V(x_i)}. \tag{8}$$

Balancing the mass $m_{\mathcal{S}_i}$ of substance $\mathcal{S}_i$ in a segment of volume $V(x_i)$ with surface area $A(x_i)$ and infinitesimal length $dx$ leads to

$$\begin{aligned}\frac{\partial m_{\mathcal{S}_i}}{\partial t} &= \left(\mu_{\mathcal{S}_ip} - \mu_{\mathcal{S}_ic}\right)V(x_i) + \mu_{\mathcal{S}_it}A(x_i)\\ &\quad + Q_i(x_i,t)\mathcal{S}_i(x_i,t) - Q_i(x_i+dx,t)\mathcal{S}_i(x_i+dx,t)\end{aligned} \tag{9}$$

with production ($_{\dots p}$) and consumption ($_{\dots c}$) rates $\mu_{\mathcal{S}_ip}$ and $\mu_{\mathcal{S}_ic}$, respectively. Transport ($_{\dots t}$) from or to the surroundings is modeled by $\mu_{\mathcal{S}_it}$. Whereas the former reaction rates are given as a temporal change of mass per volume $V_i = V(x_i)$, the latter is based on the local external surface area $A_i = A(x_i)$ of the segment.

The last two terms in Eq. 9 represent intra-hyphal flow in and out of the balance volume, i.e., flow through the shaded areas in Figure 2. As due to turgor pressure hyphae take up water from the surroundings, and as only the apical region can extend in the real hypha, an intra-hyphal flow is set up. Hence, the volumetric flow rate $Q_i(x_i,t)$ is both a function of space $x_i$ and time $t$.

Intra-hyphal flow of a substance can be the result of the flow of the cytoplasm (*cyt*) transporting $\mathcal{S}_i$. For other substances, long-distance transport is realized via an active (*act*) dislocation along microtubules. In the latter case, microtubules can be transported as well with the flowing cytoplasma, see [42, 43], resulting in a superposition of flow and dislocation velocities for $\mathcal{S}_i$,

$$Q_i(x_i,t) = Q_{i,cyt}(x_i,t) + Q_{i,act}(x_i,t). \tag{10}$$

Alternatively, it can be assumed that microtubules stay fixed with respect to the cell wall and that all of $\mathcal{S}_i$ is attached to them. Then, $Q_{i,cyt} = 0$ for this specific substance.

The splitting up of $Q_i$ will now be used for a volume balance which is only affected by the cytoplasmic flow. As the cytoplasm is incompressible,

$$\frac{\partial V_i}{\partial t} = \dot{V}_i = Q_{i,cyt}(x_i,t) - Q_{i,cyt}(x_i+dx,t) + \mu_{V_ip}A(x_i) \tag{11}$$

where $\mu_{V_ip}$ represents the volume production, e.g., through turgor-driven uptake of water from the environment through the local surface $A_i = A(x_i)$.

To finally set up the generic model structure based on the balances given above, several steps are necessary which are detailed in the Appendix B:

1. The last term of Eq. 9 and the second term of the right-hand side of Eq. 11 are expanded in a Taylor series, neglecting all terms in $(dx)^n, n \geq 2$.
2. All equations are combined.

This leads to

$$\begin{aligned}\frac{\partial \mathcal{S}_i}{\partial t} &= \mu_{\mathcal{S}_ip} - \mu_{\mathcal{S}_ic} + \rho_{i1}\mu_{\mathcal{S}_it} - \rho_{i1}\mu_{V_ip}\mathcal{S}_i\\ &\quad - \rho_{i2}\mathcal{S}_i\frac{\partial Q_{i,act}}{\partial x_i} - \rho_{i2}Q_i\frac{\partial \mathcal{S}_i}{\partial x_i}\end{aligned} \tag{12}$$

with

$$\rho_{i1} = \frac{A_i}{V_i}, \quad \rho_{i2} = \frac{dx}{V_i}. \tag{13}$$

Initial and boundary conditions, and $Q_i$ will be specified below.

#### Generic model of the constant, distal part

For simplicity, it is assumed that the active translocation velocity along microtubules in the distal part, if it exists at all, is constant, i.e., for its gradient

$$\frac{\partial Q_{d,act}}{\partial x_d} = 0.$$

Only at the tip, a deceleration will be considered later. Furthermore, after setting $i = d$, and $A_d = 2\pi R dx$, $V_d = \pi R^2 dx$, Eq. 12 reads

$$\begin{aligned}\frac{\partial \mathcal{S}_d}{\partial t} &= \mu_{\mathcal{S}_dp} - \mu_{\mathcal{S}_dc} + \rho_{d1}\mu_{\mathcal{S}_dt} - \rho_{d1}\mu_{V_dp}\mathcal{S}_d\\ &\quad - \rho_{d2}Q_d\frac{\partial \mathcal{S}_d}{\partial x_d}\end{aligned} \tag{14}$$

with

$$\rho_{d1} = \frac{2}{R}, \quad \rho_{d2} = \frac{1}{\pi R^2}. \tag{15}$$

The fourth term on the right hand side can be interpreted as a dilution term due to intra-hyphal flow. From Eq. 31, see Appendix, with $\dot{V}_d = 0$, an expression can be given describing the spatial evolution of the intra-hyphal flow

$$\frac{\partial Q_{d,cyt}}{\partial x_d} = 2\pi R\mu_{V_dp}. \tag{16}$$

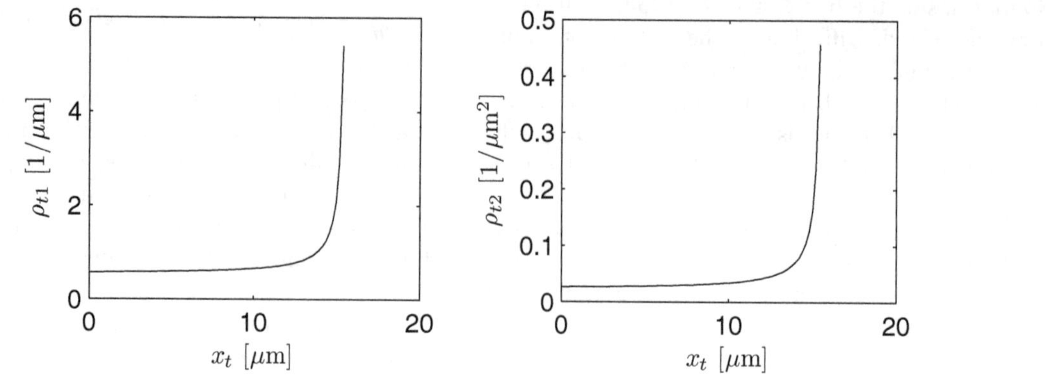


**Figure 7** $\rho_i$'s for $R = 3.5\,\mu$m, $D = 1.2\,\mu$m, and $L_t = 15.6\,\mu$m.

As long as the volume production $\mu_{V_{dp}} \neq 0$, Eq. 16 gives rise to a monotone increase of $Q_d$ with $x_d$, i.e., more and more fluid will be transported toward the tip.

The boundary conditions, $Q_d(0,t) = Q_{dx0}(t)$ and $S_d(0,t) = S_{dx0}(t)$, describe the information coming from a spore or from a branching site of a mother hypha. Whereas a mathematically 'convenient' boundary condition $Q_{dx0}(t) =$const. would make sense, as it describes an active transport, which can be zero as well, $S_d(0,t) = S_{dx0}(t) =$ const. would be more difficult to justify biologically. This would mean that the mother hypha or spore would not change its value of $S$ irrespective of what is going on in the new hypha. This could only be explained by a source of $S$ of infinite strength. If the spore or the mother hyphae are not determined by a separate model,

$$\left.\frac{\partial S_d(x_d,t)}{\partial x_d}\right|_{x_d=0} = 0$$

is a better choice, at least from a numerical point of view. An initial condition for $S_d$ must not be specified as the simulation will start without a distal region.

Before specifying the individual production and consumption rates this generic model equation will be adapted to the non-constant-area and non-constant-volume case seen in the tip.

### Generic model of the tip

Due to the non-constant surface area $A_t$ and volume $V_t$ of the balancing volume the expressions get more involved. With Eqs. 4, 6 and 31 the resulting cytoplasm flow reads

$$\frac{\partial Q_{t,cyt}}{\partial x_t} = 2\pi\sqrt{1+\left(\frac{dr_t}{dx_t}\right)^2}\, r_t(x_t)\mu_{V_tp} + 2\pi r_t \frac{dr_t}{dx_t}\dot{L}. \tag{17}$$

The first term on the right hand side increases the flow due to turgor driven volume production $\mu_{V_tp}$. The second term, however, as $dr_t/dx_t$ is negative, decreases the flow to account for the volume needed for length growth in the tip region.

Using Eq. 12,

$$\begin{aligned}\frac{\partial S_t}{\partial t} &= \mu_{S_tp} - \mu_{S_tc} + \rho_{t1}\mu_{S_tt} - \rho_{t1}\mu_{V_tp}S_t \\ &\quad - \rho_{t2}S_t\frac{\partial Q_{t,act}}{\partial x_t} - \rho_{t2}Q_t\frac{\partial S_t}{\partial x_t}\end{aligned} \tag{18}$$

with

$$\rho_{t1}(x_t) = \frac{2\sqrt{1+\left(\frac{dr_t}{dx_t}\right)^2}}{r_t(x_t)}, \quad \rho_{t2}(x_t) = \frac{1}{\pi r_t^2(x_t)} \tag{19}$$

results. A plot of these $\rho_t$-terms is given in Figure 7. In the distal part, the corresponding terms are constant, see Eq. 15.

Besides an initial condition $S_t(x_t,0) = S_{tt0}(x_t)$, a boundary condition has to be specified. If a distal part is not yet formed,

$$\left.\frac{\partial S_t}{\partial x_t}\right|_{x_t=0} = 0$$

is applied. If, on the other side, a distal compartment already exists,

$$S_t|_{x_t=0} = S_d|_{x_d=L_d}. \tag{20}$$

### Model of length growth

As the radius of the distal part and the form of the tip stay constant, $\dot{L}$ is given by the overall volume produced divided by the area of the growing end. If the tip is still developing, this area is $\pi r_t^2$. Hence,

$$\frac{dL(t)}{dt} = \frac{1}{\pi r_t^2} Q_{tp}(L_t, t). \tag{21}$$

When the tip is finished, with $L_t = L_{tmax}$, it is assumed that length growth is realized by extension of the right hand side of the distal part with area $\pi R^2$, see Figure 2. For a tip with constant geometry, this translates into a growth in length of the hypha of

$$\frac{dL(t)}{dt} = \frac{1}{\pi R^2} Q_{tp}(L_t, t). \tag{22}$$

Here, a hypothetical flow rate $Q_{tp}(L_t, t)$ at the tip $x_t = L_t$ has to be used. It is determined by the gross increase of the volume flow rate in the tip region

$$\frac{\partial Q_{tp}(x_t, t)}{\partial x_t} = 2\pi \sqrt{1 + \left(\frac{dr_t}{dx_t}\right)^2} r_t(x_t) \mu_{V_t p}, \tag{23}$$

with a proper boundary condition for $x_t = 0$. Actually, the real flow rate in the tip region has to decrease towards the tip as it is 'consumed' everywhere in the tip to fill up the new volume formed due to tip growth. In the model, however, length increase is attributed to an increase in the distal part or left hand end of the tip before it is fully developed. Therefore, as volume is taken up in the tip region as well, the gross increase of the volume flow rate has to be known to determine $\dot{L}(t)$. To distinguish this hypothetical from the real flow rate, it is denoted by $Q_{tp}$, instead of $Q_{t,cyt}$. The boundary condition at the left hand side of the tip, see Figure 2, is $Q_{tp}(0, t) = Q_d(L_d, t)$ or 0 when only the tip exists.

### Complete generic model

In summary, after specifying the individual production, consumption and transport rates, $\mu_j$, and the change in active transport in the tip, $\partial Q_{t,act}/\partial x_t$, the following equations have to be solved in the *generic model* assuming that the tip has already reached its maximal extension:

1. Integration of Eq. 16 determines the intra-hyphal flow rate at $x_d = L_d$ which sets the boundary condition for Eq. 23.
2. Integration of latter equation leads to the hypothetical flow rate at the tip, $Q_{tp}(L_t, t)$,
3. and, with Eq. 22, to the actual extension rate $\dot{L}$.
4. In a moving boundary framework, as $L(t)$ grows, Eqs. 14 and 18 are solved to determine $\mathcal{S}_i$, $i = d, t$.

Initially, only the tip region exists. Hence, Eqs. 16 and 18 are not needed.

## Modeling pressure regulation via osmolytes

We consider osmolytes which are responsible for maintaining a certain pressure and pressure gradient inside hyphae, see [38]. It is assumed that osmolytes are produced until a certain pressure is obtained for which intracellular sensors must exist. For the MAPK pathway, OS-1 is discussed as a sensor in [39]. For simplicity, an intracellular substance called osmolyte $\mathcal{O}_i$, $i = d, t$, is introduced, which represents both the osmolyte, and, indirectly, the pressure. To obtain a mass flow toward the apex, its concentration must be higher in subapical parts. Using the equations derived above, $\mathcal{S}_i$ is now replaced by $\mathcal{O}_i$.

To start with the most simple model, it is assumed that osmolytes are not transported actively, $Q_{i,act} = 0$, and are not consumed or degraded, i.e.,

$$\mu_{\mathcal{O}_i c} = 0.$$

Furthermore, they are not transported over the cell wall, hence,

$$\mu_{\mathcal{O}_i t} = 0.$$

For the production of $\mathcal{O}_i$, as a first approach, a logistic law-like expression is used

$$\mu_{\mathcal{O}_i p} = k_1(\mathcal{O}_{max} - \mathcal{O}_i), \quad i = d, t, \tag{24}$$

with a maximal osmolyte concentration $\mathcal{O}_{max}$. It has to be pointed out that these assumptions can be changed easily.

In the long run, for all osmolytes produced, water has to be taken up. If this process is fast, it can be assumed that, in a kind of quasi-steady-state point of view, water is taken up proportional to the synthesis rate of the osmolytes. As the first one is formulated based on the surface and the latter based on the volume,

$$\mu_{V_i p} A_i \sim \mu_{\mathcal{O}_i p} V_i$$

leads to an expression for the volume production rate $\mu_{V_i p}$ in part $i$ of the hypha

$$\mu_{V_i p} \sim \mu_{\mathcal{O}_i p} \frac{V_i}{A_i}.$$

With Eq. 24, for the distal part,

$$\mu_{V_d p} = k_2(\mathcal{O}_{max} - \mathcal{O}_d)\frac{V_d}{A_d} = k_2(\mathcal{O}_{max} - \mathcal{O}_d)\frac{1}{\rho_{d1}} \tag{25}$$

and for the tip

$$\mu_{V_t p} = k_3(\mathcal{O}_{max} - \mathcal{O}_t)\frac{V_t}{A_t} = k_3(\mathcal{O}_{max} - \mathcal{O}_t)\frac{1}{\rho_{t1}} \tag{26}$$

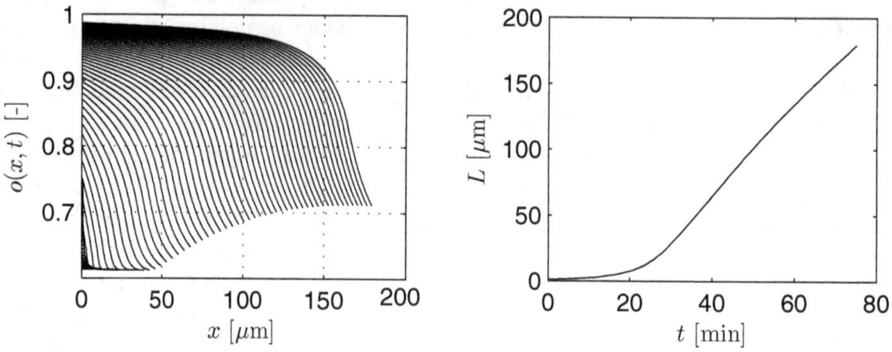


**Figure 8** *Left* evolution of the normalized osmolyte concentration for different time instants with $\Delta t = 1$ min. *Right* length of the hypha.

follows. Different constants, $k_2$ and $k_3$, are introduced to possibly account for the fact that most water is taken up in the tip region, as hypothesized by [39]. This difference in water uptake velocity might be the result of the plasticity of the wall in the tip region. As the maximal extend of the tip, $L_{tmax}$, considered here is fixed arbitrarily, this has to be observed during parameter identification and analysis of the simulation results.

With these kinetics and the generic equations derived in the last section, the models of the distal part and the tip can be formulated. They are omitted here for brevity. Furthermore, to reduce the number of kinetic parameters and to ease parameter identification, a normalization is done with $o_i = \mathcal{O}_i/\mathcal{O}_{max}$ and $q_i = Q_{i,cyt}/Q_{max}$, and $o_i, q_i \in [0,1]$ for $i = d, t$, $\dot{L}_{max} = Q_{max}/(\pi R^2) = Q_{max}\rho_{d2}$, $\theta_1 = k_1$, $\theta_2 = k_2\mathcal{O}_{max}$, $\theta_3 = k_3\mathcal{O}_{max}$.

For the numerical solution of the partial differential equations, the spatial coordinate is discretized equidistantly with a step size of $\Delta x = 0.2$ μm. The method of lines is applied for the equation describing the evolution of the osmolytes approximating the spatial derivatives by a first-order backward difference operator. In the beginning, when only the tip exists, the left most discretization segment of the tip, see Figure 2, is allowed to grow in length according to $\dot{L}$ until it exceeds a length of 0.3 μm. After the distal part is formed, its right most discretization segment takes over this task and grows accordingly until it exceeds a length of 0.3 μm. Then, this segment is split up into a segment of constant length (0.2 μm) and a growing one with an initial length <0.2 μm and the calculations are continued as before. The normalized flow rate $q_i$ is obtained accordingly from Eqs. 16 and 23 exploiting the trapezoidal rule. Parameters $\theta_i$ have to be chosen such that $q_i \leq 1$ is guaranteed.

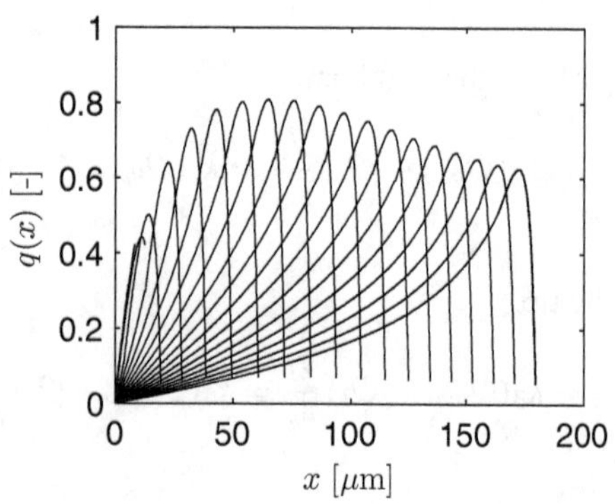


**Figure 9** Normalized intrahyphal flow with $\Delta t = 3$ min. Parameters as in Figure 8.

As this work concentrates on the formulation of the generic model and not on a parameter fit or selection of appropriate kinetic expressions to describe a very specific problem, a simple approach was chosen to find kinetic parameters for simulation studies. Measurements performed by Spanhoff et al. were done with hyphae growing with approximately $\dot{L}_{max} = 3\mu$m/min. The parameters $\theta_{1,2,3}$ are determined such that $L(t)$ shows an initial exponential increase followed, after a transition, by a phase of constant growth velocity of approximately $\dot{L}_{max} = 3\mu$m/min. To this end, an optimization problem was formulated. In lack of real data for this first study, a 'desired' evolution $L_{des}(t)$ was determined to allow for an adaption of the $\theta_i$'s.

Simulation results are shown in Figures 8 and 9. Parameters used are $\theta_1 = 0.3079$, $\theta_2 = 0.2619$, and $\theta_3 = 0.5032$ with vanishing flow from the mother hypha or spore. The initial tip starts with four discretization segments, i.e., a length of 0.8 μm, with exponentially decaying values of the initial, normalized osmolyte concentration $o_{t,k} = [0.750, 0.7351, 0.7206, 0.7063]$.

In the simulations given in Figure 8, growth is followed up to a length of about 170 μm, i.e., in the end more than 850 discretization segments are used. Due to production and consumption of the osmolyte, and due to length growth, the osmolyte profiles change dynamically over time. These osmolyte profiles, on the other hand, determine the overall volume production, see Figure 9, and, hence, length increase. As a result, all processes are highly interwoven.

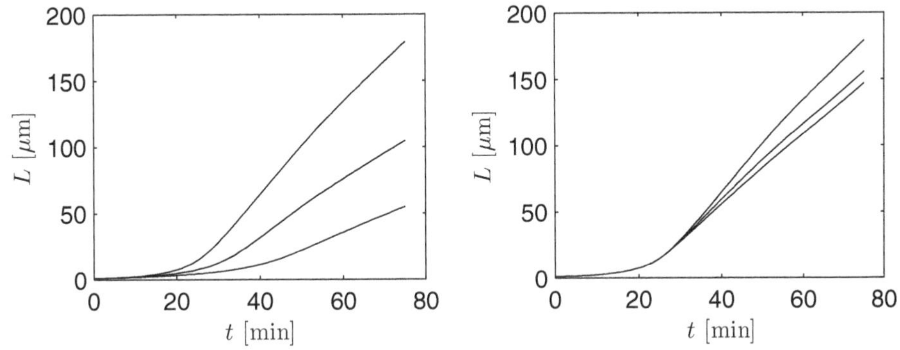


**Figure 10** *Left* length evolution for $\theta_1 = 0.3079$(top)/0.3579/0.4079. *Right* Length evolution for $\theta_2 = 0.2619$(top)/0.2119/0.1619.

When the rate of osmolyte production is increased by choosing higher values of $k_1 = \theta_1$, slower growth results, see left part of Figure 10. Here, as in the right part of this figure, the other parameters are as in Figure 8. Lowering $\theta_2 = k_2\mathcal{O}_{max}$ has the similar, though less pronounced effect. An explanation of this feature can be seen in Figure 11 where the evolutions of the osmolyte and the normalized flow rate are shown for an increased value of $\theta_1 = 0.3079 + 0.05$ compared to the case of Figure 8. A larger osmolyte production rate leads to higher values of $o(x, t)$, and, hence, to a lower volume production rate $k_i(\mathcal{O}_{max} - \mathcal{O}), i = 2, 3$.

Using the same volume production parameters $\theta_2 = \theta_3 = 0.2616$ in the distal and tip region results in a much too slow growth corroborating different uptake rates. These simple studies show that an intricate relation exists between parameters and the processes described by the model. Predicting the outcome of parameter or kinetic variations is difficult. Therefore, numerical studies are indispensable.

In the first simulation study shown above, which will be used again in the next section, linear growth already starts at a length of approximately $L = 30$ μm. By another choice of parameters, the exponential phase can be elongated. In Figure 12, as an example, parameters are chosen such that experimental data obtained by [41] can be described by the model. In this case, an elongated exponential phase can be observed. Truly linear growth does not start before 700 μm as pointed out by the authors. As the primary branch considered in [41] originates from a parent compartment, the normalized flow rate coming from this compartment was fitted as a parameter as well. For simplicity, a constant value $q_i(0, t)$ was assumed. A more detailed study should be performed, though, based on the general model introduced above, to better account for a variable influence of the mother compartment. Likewise, septation could be included. This would lead to a leveling out of the profiles inside the mother compartment and a variable supply of the new branch. Such a detailed study, however, is outside the scope of this contribution and has to be postponed to future work.

## Modeling of vesicle distribution

As a first example of an organelle, vesicles are considered which have been already used above to get an idea of the form of the tip. They are, beside other functions, responsible for material transport to the growing tip. According

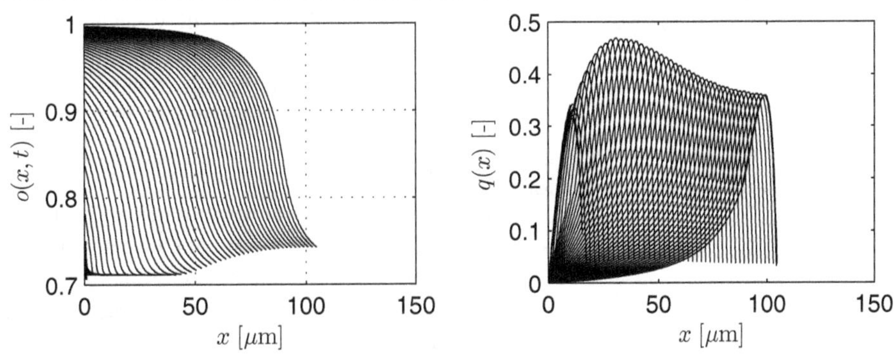


**Figure 11** Normalized osmolyte concentration and flow rate for $\theta_1 = 0.3579$. The remaining parameters are as in Figure 8.

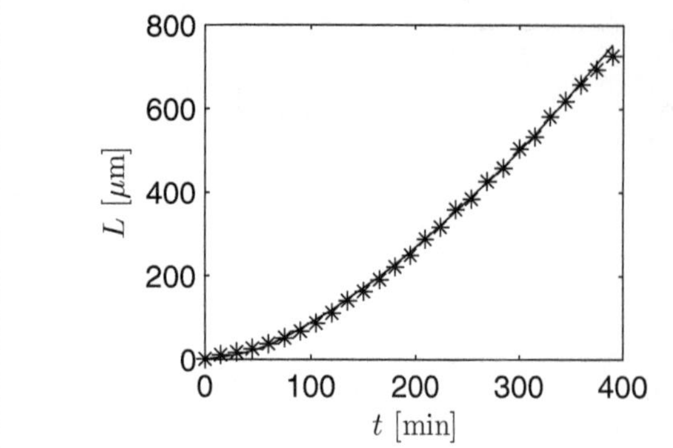


**Figure 12** Length of a hypha. Simulation (*solid line*) done with $\theta_1 = 0.4792, \theta_2 = 0.4811, \theta_3 = 0.5282$, and $q_i(0,t) = 0.0357$. Experimental data obtained by [41] are marked with *asterisk*.

to the model of [27], vesicles are used up in the apical part of a hypha. Therefore, the consumption rate in the tip will be $\mu_{\mathcal{V}_{tc}} \neq 0$, while $\mu_{\mathcal{V}_{dc}} = 0$ is assumed in the distal region.

As vesicles are produced inside hyphae, the transfer rate to the surroundings is zero in both parts, i.e., $\mu_{\mathcal{V}_{it}}$, $i = d, t$. For vesicle production $\mu_{\mathcal{V}_{ip}}$, again a logistic-like expression is chosen. However, as vesicles are accumulated in the tip region, the expression is modified describing a production only as along as $\mathcal{V}_i < \mathcal{V}_{max}$ and no production otherwise

$$\mu_{\mathcal{V}_{ip}} = \max\left\{k_4(\mathcal{V}_{max} - \mathcal{V}_i), 0\right\}. \qquad (27)$$

Active transport is modeled by microtubules. It is assumed that all vesicles are bound immediately to microtubules. Therefore, the vesicle concentration can be calculated as the product of the local microtubules concentration, $m_{MT}/V$, times the loading of the microtubules, $m_{\mathcal{V}_i}/m_{MT}$

$$\mathcal{V}_i = \frac{m_{MT}}{V}\frac{m_{\mathcal{V}_i}}{m_{MT}} = \frac{m_{\mathcal{V}_i}}{V}.$$

Due to the discrete nature of microtubules, this is a rough approximation. Still, if the local microtubules concentration stays constant up to the apex, the active volume flow is simply given by

$$Q_{i,act} = \pi r_i^2 u(x_i, t),$$

where $u(x_i, t)$ and $r_i$ represent the local transport velocity and hyphal radius, respectively. With this expression, the volume flow at the apex is zero which does make sense as $r_t(L_{tmax}) = 0$. The change in flow rate over the length of a hypha is given by

$$\frac{\partial Q_{i,act}}{\partial x_i} = 2\pi r_t u(x_i, t)\frac{dr_i}{dx_i} + \pi r_i^2 \frac{du}{dx_i}.$$

In the distal part, with $r_i = R = \text{const.}$,

$$\frac{\partial Q_{d,act}}{\partial x_d} = \pi R^2 \frac{du}{dx_d} = \frac{1}{\rho_{d2}}\frac{du}{dx_d}.$$

For a constant velocity $u$,

$$\frac{\partial Q_{d,act}}{\partial x_d} = 0,$$

as assumed above. For the tip,

$$\frac{\partial Q_{t,act}}{\partial x_t} = 2\pi r_t u(x_t, t)\frac{dr_t}{dx_t} + \pi r_t^2 \frac{du}{dx_t}.$$

When the second term in the last equation is neglected, an increase in vesicle concentration, as seen in Figure 6, can only be observed directly behind the tip due to the decelerated transport coming from the geometrical influence ($dr_t/dx_t < 0$). This does not coincide with experimental findings. For this reason, as a first trial, it is assumed that $u$ stays constant in most of the distal part at $u = u_{max}$. From $u_{max}$ it decreases linearly to zero in the foremost 15 μm. For the following simulation, $u_{max} = 35.62$ μm/min is chosen. Typical literature values for kinesin-1 are in the range of 24–54 μm/min. For kinesin 7 in *A. nidulans*, [44] report 10 μm/min compared to the conventional kinesin of this strain with 120 μm/min.

The equations are normalized again with $v_i = \mathcal{V}_i/\mathcal{V}_{max}$ and $\theta_4 = k_4$.

Consumption of vesicles is assumed to be proportional to the actual length growth rate for which it is used, and proportional to the local vesicle concentration $v_t$

$$\mu_{\mathcal{V}_{tc}} = \theta_5 \dot{L} v_t. \qquad (28)$$

As the volume corrected measurements by Spanhoff et al. indicate a falling vesicle concentration only for the last three segments with a width of $\Delta x = 0.2$ μm, for the rest of the hypha $\mu_{\mathcal{V}_{tc}} = 0$ is used.

A comparison of simulated data against the measurements introduced above is given in Figure 13. Additional parameters used are $\theta_4 = 0.0407$, and $\theta_5 = 2.2701$. The remaining parameters $\theta_i$, $i = 1, 2, 3$, are as in Figure 8.

With these parameters the experimentally observed accumulation profile of vesicles can be reproduced. A key-enabler is the model assumption that the transport velocity must decrease towards the tip. This could be tested in future experiments.

## Conclusions

As constantly new branches and septa are produced in a mycelium and a flow of cytoplasma towards the different apices occurs, quasi-steady state concentration profiles of intracellular substances and organelles hardly establish. Moreover, due to these processes and due to active and cytoplasmatic flows intracellular components are

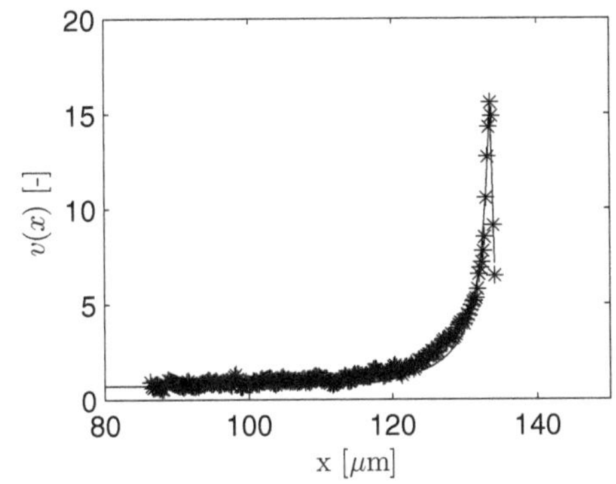


**Figure 13** Comparison of a measured (*asterisk*) and simulated (*solid line*) evolution of normalized vesicle concentration as a function of the coordinate $x$, given for $t = 60$ min. Only the foremost part of the hypha is given neglecting the retrograd part with constant $v$.

not only a function of time but as well a function of space inside a hypha. A comprehensive and quantitative interpretation of experimental data of individual hyphae will therefore only be possible when these space- and time-dependent processes are taken into account. To this end, a generic mathematical model is proposed here which first of all describes turgor driven length extension. By this, an initially accelerated and then linear growth can be predicted as seen in microscopic experiments. A much simpler model, see [11], could be used to predict the length evolution though. That model, however, cannot to be extended so easily to describe other constituents of a hypha as it is done here. The turgor driven length extensions forms the 'backbone' of a generic model to study, for instance, the time-dependent distribution of organelles and other compounds. These may be transported actively or passively towards the tip. Diffusion is not considered yet, but can be included readily. For a complete specification of the model, kinetic expressions have to be stated. In this contribution, very little effort is invested to study the effect of different kinetics and parameters, e.g., with respect to osmolyte or vesicle production and consumption. The emphasis is rather on the formulation of a generic model. Effects of different kinetics will be tackled in future works when more experimental data is available. However, even with simple kinetics chosen here it can be shown, for example, that the experimentally observed accumulation of vesicles near the tip can be explained. A crucial assumption to be able to do this is the postulation of a decreasing active transport velocity in the tip region. Without this, the experimentally observed apical accumulation of vesicles cannot be described in the chosen setting. The model structure can be readily extended to study the effect of different organelles and cytoplasmatic compounds. Before doing so, however, the numerical solution of the partial differential equations with a moving boundary should be revisited to hopefully decrease the computational burden. This was not done yet, as septation and branching have to be included in future works.

**Acknowledgements**
The author would like to thank Frederik Spanhoff, Arthur Ram and Vera Meyer for supplying unpublished vesicle measurements and VM who inspired this work.

**Competing interests**
The author declares that he has no competing interests.

## Appendix A: Gradient of the radius

From Eq. 3,

$$F(x_t, r_t(x_t)) = r_t \cot \frac{r_t}{D} + L_t - D - x_t = 0.$$

Implicit differentiation leads to

$$\frac{dr_t}{dx_t} = -\frac{\dfrac{\partial F}{\partial x_t}}{\dfrac{\partial F}{\partial r_t}} = \frac{D \sin^2 \dfrac{r_t}{D}}{D \cos \dfrac{r_t}{D} \sin \dfrac{r_t}{D} - r_t}.$$

## Appendix B: Derivation of the generic model

Expanding the last term in Eq. 9 into a Taylor series results in

$$\begin{aligned} &Q_i(x_i + dx, t)\mathcal{S}_i(x_i + dx, t) \\ &\quad = Q_i(x,t)\mathcal{S}_i(x,t) + \mathcal{S}_i \frac{\partial Q_i}{\partial x_i} dx + Q_i \frac{\partial \mathcal{S}_i}{\partial x_i} dx + \cdots \end{aligned}$$

This yields upon substitution and neglect of terms in $(dx)^n, n \geq 2$

$$\begin{aligned} \frac{\partial m_{\mathcal{S}_i}}{\partial t} &= \left(\mu_{\mathcal{S}_i p} - \mu_{\mathcal{S}_i c}\right) V_i + \mu_{\mathcal{S}_i t} A_i \\ &\quad - \mathcal{S}_i \frac{\partial Q_i}{\partial x_i} dx - Q_i \frac{\partial \mathcal{S}_i}{\partial x_i} dx. \end{aligned}$$

In the distal part, the balancing volume $V(x_d) = V_d$ stays constant, i.e., $\dot{V}(x_d) = \dot{V}_d = 0$. In the tip region, however, a stationary observer sees a growing volume $\dot{V}(x_t) = \dot{V}_t$ as the apex moves away and the local radius increases. Hence, generally, with $m_{\mathcal{S}_i} = V_i \mathcal{S}_i$

$$\begin{aligned} V_i \frac{\partial \mathcal{S}_i}{\partial t} + \mathcal{S}_i \dot{V}_i &= \left(\mu_{\mathcal{S}_i p} - \mu_{\mathcal{S}_i c}\right) V_i + \mu_{\mathcal{S}_i t} A_i \\ &\quad - \mathcal{S}_i \frac{\partial Q_i}{\partial x_i} dx - Q_i \frac{\partial \mathcal{S}_i}{\partial x_i} dx, \end{aligned}$$

or

$$\begin{aligned} \frac{\partial \mathcal{S}_i}{\partial t} &= \mu_{\mathcal{S}_i p} - \mu_{\mathcal{S}_i c} + \mu_{\mathcal{S}_i t} \frac{A_i}{V_i} - \mathcal{S}_i \frac{\partial Q_i}{\partial x_i} \frac{dx}{V_i} - Q_i \frac{\partial \mathcal{S}_i}{\partial x_i} \frac{dx}{V_i} \\ &\quad - \mathcal{S}_i \frac{\dot{V}_i}{V_i}. \end{aligned} \tag{29}$$

The surface area $A_i$ and volume $V_i$ will differ depending on the part considered.

Taylor series of the volume balance, see Eq. 11, which is only affected by the cytoplasmic flow, leads to

$$\begin{aligned}\frac{\partial V_i}{\partial t} &= \dot{V}_i \\ &= Q_{i,cyt}(x_i, t) - Q_{i,cyt}(x_i + dx, t) + \mu_{V_ip} A(x_i) \\ &\approx -\frac{\partial Q_{i,cyt}}{\partial x_i} dx + \mu_{V_ip} A_i. \end{aligned} \tag{30}$$

Rearrangement gives

$$\frac{\partial Q_{i,cyt}}{\partial x_i} dx = \mu_{V_ip} A_i - \dot{V}_i, \tag{31}$$

and

$$\begin{aligned}\frac{\partial Q_i}{\partial x_i} dx &= \frac{\partial Q_{i,cyt}}{\partial x_i} dx + \frac{\partial Q_{i,act}}{\partial x_i} dx \\ &= \mu_{V_ip} A_i - \dot{V}_i + \frac{\partial Q_{i,act}}{\partial x_i} dx.\end{aligned}$$

Combining it with Eq. 29 leads to the *generic model*

$$\begin{aligned}\frac{\partial S_i}{\partial t} &= \mu_{S_ip} - \mu_{S_ic} + \mu_{S_it} \frac{A_i}{V_i} - \mu_{V_ip} S_i \frac{A_i}{V_i} \\ &\quad - S_i \frac{\partial Q_{i,act}}{\partial x_i} \frac{dx}{V_i} - Q_i \frac{\partial S_i}{\partial x_i} \frac{dx}{V_i}.\end{aligned} \tag{32}$$

## References


1. Papagianni M (2004) Fungal morphology and metabolite production in submerged mycelial processes. Biotechnol Adv 22(3):189–259
2. El-Enshasy HA (2007) Filamentous fungal cultures—process characteristics, products, and applications. Bioprocess Value Added Prod Renew Resour New Technol Appl 9:225–261
3. Metz B (1976) From pulp to pellet. PhD thesis, Technische Universität Delft, Delft
4. Kossen NWF (2000) The morphology of filamentous fungi. Adv Biochem Eng Biotechnol 70:1–33
5. Meyer V, Fiedler M, Mitsche BH, King R (2014) The cell factory *Aspergillus* enters the Big Data era: opportunities and challenges for optimizing product formation. Adv Biochem Eng Biotechnol. doi:10.1007/10_2014_297
6. Cohen D (1967) Computer simulation of biological pattern generation processes. Nature 216:246–248
7. Megee RD, Kinoshita S, Fredrickson AG, Tsuchiya HM (1970) Differentiation and product formation in molds. Biotechnol Bioeng 12(5):771–801
8. Edelstein L (1982) The propagation of fungal colonies: a model for tissue growth. J Theor Biol 98:679–701
9. Prosser JI (1995) Mathematical modelling of fungal growth. In: The Growing Fungus. Springer, pp 319–335
10. Tough AJ, Prosser JI (1996) Experimental verification of a mathematical model for pelleted growth of *Streptomyces coelicolor* a3 (2) in submerged batch culture. Microbiology 142(3):639–648
11. King R (1998) Mathematical modelling of the morphology of *streptomyces* species. Adv Biochem Eng Biotechnol 60:95–119
12. Lejeune R, Baron GV (1997) Simulation of growth of a filamentous fungus in 3 dimensions. Biotechnol Bioeng 53(2):139–150
13. Davidson FA (2007) Mathematical modelling of mycelia: a question of scale. Fungal Biol Rev 21(1):30–41
14. Krull R, Cordes C, Horn H, Kampen I, Kwade A, Neu TR et al (2010) Morphology of filamentous fungi: Linking cellular biology to process engineering using *Aspergillus niger*. Adv Biochem Eng Biotechnol 121:1–21
15. Celler K, Picioreanu C, van Loosdrecht MCM, van Wezel GP (2012) Structured morphological modeling as a framework for rational strain design of *Streptomyces* species. Anton Leeuw Int J G 102(3):409–423
16. Davidson FA, Boswell GP, Fischer MWF, Heaton L, Hofstadler D, Roper M (2011) Mathematical modelling of fungal growth and function. IMA Fungus 2(1):33
17. Hopkins SM (2012) A hybrid mathematical model of fungal mycelia: tropisms, polarised growth and application to colony competition. PhD thesis, University of Glamorgan
18. Nielsen J, Villadsen J (1994) Morphologically Structured Models. In: Bioreaction Engineering Principles. Springer, pp 229–269
19. Yang H, King R, Reichl U, Gilles ED (1992) Mathematical model for apical growth, septation, and branching of mycelial microorganisms. Biotechnol Bioeng 39:49–58
20. Meškauskas A, Fricker MD, Moore D (2004) Simulating colonial growth of fungi with the neighbour-sensing model of hyphal growth. Mycol Res 108(11):1241–1256
21. Boswell GP, Jacobs H, Ritz K, Gadd GM, Davidson FA (2007) The development of fungal networks in complex environments. B Math Biol 69(2):605–634
22. Coradin JH, Braun A, Viccini G, Krieger N, Mitchell DA (2011) A three-dimensional discrete lattice-based system for modeling the growth of aerial hyphae of filamentous fungi on solid surfaces: a tool for investigating micro-scale phenomena in solid-state fermentation. Biochem Eng J 54(3):164–171
23. Boswell GP, Davidson FA (2012) Modelling hyphal networks. Fungal Biol Rev 26(1):30–38
24. Heaton L, Obara B, Grau V, Jones N, Nakagaki T, Boddy L et al (2012) Analysis of fungal networks. Fungal Biol Rev 26(1):12–29
25. Nieminen L, Webb S, Smith MCM, Hoskisson PA (2013) A flexible mathematical model platform for studying branching networks: experimentally validated using the model actinomycete. *Streptomyces coelicolor*. PloS one 8(2):54316
26. Nielsen J (1992) Modelling the growth of filamentous fungi. In: Fiechter A (ed) Adv Biochem Eng Biotechnol, vol 46. Springer, pp 187–223
27. Bartnicki-Garcia S, Hergert F, Gierz G (1989) Computer simulation of fungal morphogenesis and the mathematical basis for hyphal (tip) growth. Protoplasma 153(1–2):46–57
28. Gierz G, Bartnicki-Garcia S (2001) A three-dimensional model of fungal morphogenesis based on the vesicle supply center concept. J Theor Biol 208(2):151–164
29. Tindemans SH, Kern N, Mulder BM (2006) The diffusive vesicle supply center model for tip growth in fungal hyphae. J Theor Biol 238(4):937–948
30. Goriely A, Tabor M (2003) Biomechanical models of hyphal growth in actinomycetes. J Theor Biol 222(2):211–218
31. Boudaoud A (2003) Growth of walled cells: From shells to vesicles. Phys Rev Lett 91:018104
32. Caballero-Lima D, Kaneva IN, Watton SP, Sudbery PE, Craven CJ (2013) The spatial distribution of the exocyst and actin cortical patches is sufficient to organize hyphal tip growth. Eukaryot cell 12(7):998–1008
33. Sugden KEP, Evans MR, Poon WCK, Read ND (2007) Model of hyphal tip growth involving microtubule-based transport. Phys Rev E 75(3):031909
34. Heaton LM, López E, Maini PK, Fricker MD, Jones NS (2010) Growth-induced mass flows in fungal networks. Proc R Soc Lond Ser B 277:3265–3274
35. Agrawal D, Bernstein P, Bertino E, Davidson S, Dayal U, Franklin M et al (2012) Challenges and opportunities with big data. CRA white paper
36. Harris SD, Read ND, Roberson RW, Shaw B, Seiler S, Plamann M et al (2005) Polarisome meets spitzenkörper: microscopy, genetics, and genomics converge. Eukaryot Cell 4(2):225–229
37. Riquelme M (2013) Tip growth in filamentous fungi: a road trip to the apex. Ann Rev Microbiol 67:587–609
38. Lew RR (2005) Mass flow and pressure-driven hyphal extension in *Neurospora crassa*. Microbiology 151(8):2685–2692

39. Lew RR (2011) How does a hypha grow? The biophysics of pressurized growth in fungi. Nat Rev Microbiol 9(7):509–518
40. Bizukojc M, Ledakowicz S (2006) A kinetic model to predict biomass content for Aspergillus niger germinating spores in the submerged culture. Process Biochem 41(5):1063–1071
41. Fiddy C, Trinci APJ (1976) Nuclei, septation, branching and growth of *Geotrichum candidum*. J Gen Microbiol 97:185–192
42. Mouriño-Pérez RR, Roberson RW, Bartnicki-García S (2006) Microtubule dynamics and organization during hyphal growth and branching in *Neurospora crassa*. Fungal Genet Biol 43(6):389–400
43. Ramos-García SL, Roberson RW, Freitag M, Bartnicki-García S, Mouriño-Pérez RR (2009) Cytoplasmic bulk flow propels nuclei in mature hyphae of *Neurospora crassa*. Eukaryot Cell 8(12):1880–1890
44. Schunck T, Herrero S, Fischer R (2011) The *Aspergillus nidulans* cenp-e kinesin kipa is able to dimerize and to move processively along microtubules. Curr Genet 57(5):335–341

16

# Production of cyathane type secondary metabolites by submerged cultures of *Hericium erinaceus* and evaluation of their antibacterial activity by direct bioautography

T. Shen[1], G. Morlock[2] and H. Zorn[1*]

**Abstract**

**Background:** Fungi of the phylum Basidiomycota are well-known to form a broad spectrum of biologically active secondary metabolites, especially low molecular weight compounds such as terpenoids. *Hericium erinaceus* produces various cyathane type diterpenoids including erinacines. However, no quantitative data and production kinetics have been reported on the biosynthesis of the erinacines C and P in submerged cultures. In the present study, the production of erinacine C was optimized, and the product formation kinetics as well as the antimicrobial activity were studied by high-performance liquid chromatography (HPLC), high-performance thin-layer chromatography (HPTLC) and direct bioautography.

**Results:** Oatmeal and Edamin® K were identified to be crucial media components for an efficient production of erinacine C. The highest concentrations of erinacine C were obtained in the optimized culture medium on the 9th culture day (approximately 260 mg $L^{-1}$). The production of erinacine P was strongly time dependent. The maximum concentration of erinacine P of 184 mg $L^{-1}$ was observed on the third culture day. Afterwards, the concentrations of erinacine P decreased while the concentrations of erinacine C steadily increased. Comparable results were obtained by HPTLC with UV detection and HPLC with diode-array detection (DAD) analyses. Direct bioautography allowed for an additional analysis of the antimicrobial activity of the secondary metabolites.

**Conclusions:** The C and N sources oatmeal and Edamin® K induced the formation of erinacine C. Detailed product formation kinetics of the erinacines C and P have been reported for the first time. HPTLC combined with the *Aliivibrio fischeri* bioassay allowed for an instant detection of cyathane diterpenoids in crude extracts and for an evaluation of the antimicrobial activity of the secondary metabolites directly on the plate.

**Keywords:** *Hericium erinaceus*, Cyathane type diterpenoids, Erinacine C, Erinacine P, HPTLC, TLC-MS interface, *Aliivibrio fischeri*

## Background

Basidiomycetes are the highest developed fungi. Many of them, if not all, synthesize biologically active compounds with medicinal properties in their fruiting bodies and mycelia. *Hericium erinaceus* is an edible mushroom that belongs to the family *Hericiaceae* and has been used for the treatment of, e.g., digestive diseases in traditional Chinese medicine for more than 1000 years. Various compounds isolated from *H. erinaceus*, high molecular weight compounds as well as small molecules, have shown a variety of beneficial functions, such as anticancer, anti-inflammatory, and immunomodulatory properties [1–5]. Hericenones and erinacines are the major bioactive low molecular weight compounds formed by fruiting bodies and submerged cultures of *H. erinaceus*, respectively. Several of these compounds have been

*Correspondence: holger.zorn@uni-giessen.de
[1] Institute of Food Chemistry and Food Biotechnology, Justus Liebig University Giessen, Heinrich-Buff-Ring 17, 35392 Giessen, Germany

shown to significantly induce nerve growth factor (NGF) synthesis and to protect neuronal cells against endoplasmic reticulum (ER) stress- or oxidative stress-induced cell death. Therefore, the consumption of *H. erinaceus,* and of dietary supplements derived thereof, has been suggested for the prevention and treatment of dementia, and especially of Alzheimer's disease [1, 6–12].

The erinacines A to I, P, and Q are diterpenoid compounds with a cyathane skeleton consisting of five-, six-, and seven-membered rings. Among the erinacines A to G, erinacine C (Fig. 1) showed the strongest induction of NGF synthesis [1]. Various strategies have been reported for the chemical synthesis of cyathane type diterpenoids, e.g., the total synthesis of allocyathin $B_2$, erinacine A, and erinacine E [13–17]. However, all of the chemical syntheses are complex multistep processes and suffer from low yields. Watanabe and Nakada [15] described the total synthesis of erinacine E in 39 steps with a total yield of 0.9 %. Snider [17] reported the total synthesis of erinacine A in 19 steps with a yield of 1 %. Therefore, the biotechnological production of erinacine C was investigated in submerged cultures of *H. erinaceus* in the present study. The culture substrates were optimized and the production kinetics of erinacine C and of its supposed precursor erinacine P (Fig. 1) were recorded by HPLC–DAD analysis. Additionally, the combination of high-performance thin-layer chromatography (HPTLC) with a bioassay allowed for a fast evaluation of the antimicrobial activity of the erinacines.

## Results

### Optimization of erinacine C production

The concentrations of erinacine C formed in the mycelia and supernatants of the cultures grown in media 1–7 were different (Fig. 2). The presence of oatmeal was crucial for the biosynthesis of erinacine C, as erinacine C was not detected in medium 3 (without oatmeal). Besides, the use of Edamin® K increased the formation of erinacine C substantially. The concentration of erinacine C in medium 7, in which Edamin® K was used instead of Edamin® S, was about 8.5 times higher than in the reference medium 1.

Based on these results, the effect of oatmeal on erinacine C production was rechecked in medium 7 (with oatmeal) *versus* medium 8 (without oatmeal). Erinacine C was produced in medium 8 in spite of the missing oatmeal, but the production of erinacine C was higher in medium 7 (Fig. 3). Approximately 257 mg $L^{-1}$ erinacine C was detected in the supernatants in medium 7 on the 9th culture day, compared to ~85 mg $L^{-1}$ in the supernatants of medium 8. Therefore, medium 7 was chosen for the following experiments.

The concentrations of erinacine P and C were both found to be time-dependent (Fig. 4). The concentrations of erinacine P increased during the first three culture days and decreased thereafter. With the decreasing concentrations of erinacine P, the concentrations of erinacine C increased steadily. A maximum concentration of erinacine P of ~184 mg $L^{-1}$ was observed on the third culture day, while the maximum concentration of erinacine C was 257 mg $L^{-1}$ on the 9th culture day.

### HPTLC-UV/vis/FLD-*Aliivibrio fischeri* analysis

HPLC has been mainly applied for the analysis of natural products from complex matrices. However, the use of HPTLC for natural product analysis provides a number of advantages. It allows, e.g., for a screening for bioactivity directly on the plate, for quantification, and by coupling to mass spectrometers also for compound confirmation or structure suggestions.

The compound zones at $hR_F$ 34 (erinacine P) and $hR_F$ 49 clearly showed antibacterial activity (Fig. 5). Erinacine P was also detectable by inspection of the chromatogram at UV 254 nm, while the substance at $hR_F$ 49 was fluorescent at UV 366 nm. The results obtained by HPLC–DAD were clearly confirmed by HPTLC-UV/FLD analysis followed by bioautography. Also here, the concentrations of erinacine P increased during the first 3 cultivation days and decreased thereafter. It was evident that the unknown bioactive compound zones at $hR_F$ 49 clearly increased, similarly to erinacine C as discussed before.

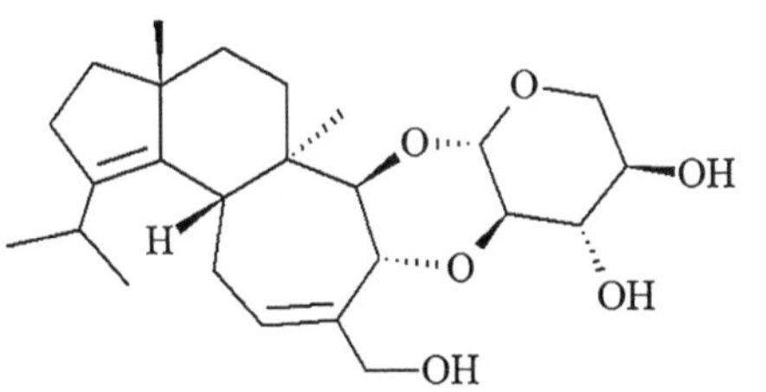

erinacine C

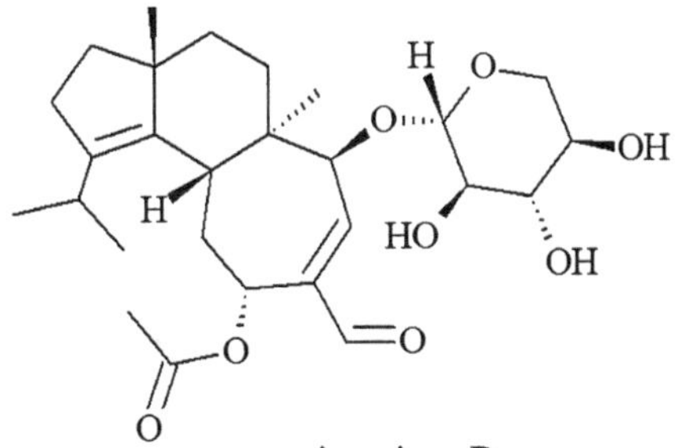

erinacine P

**Fig. 1** Structures of erinacine C and P

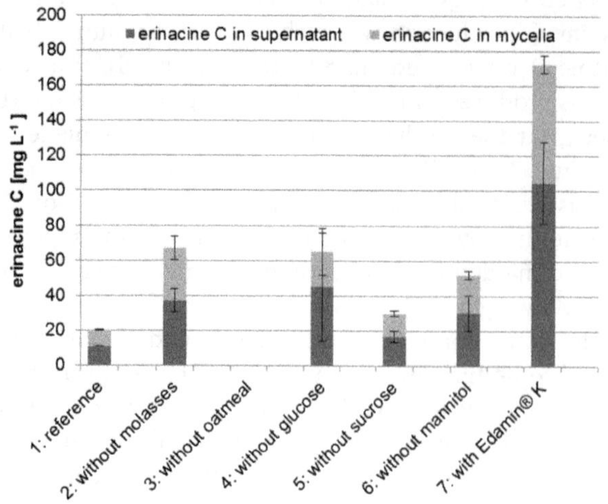


**Fig. 2** Effect of medium components on the production of erinacine C (means and standard deviations). The cultivation was stopped on day 27

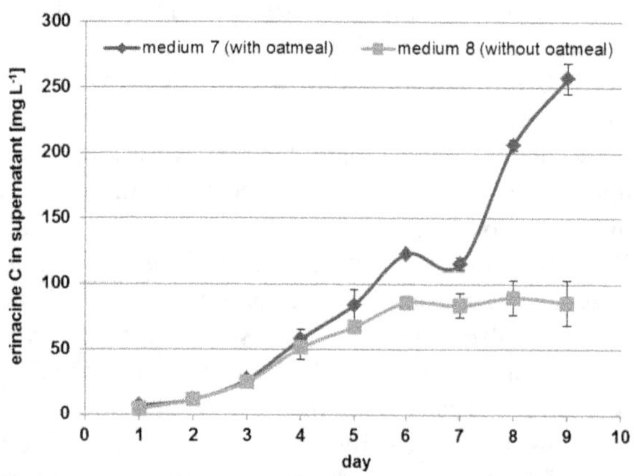


**Fig. 3** Production of erinacine C (means and standard deviations) in medium 7 (with oatmeal) *versus* medium 8 (without oatmeal)

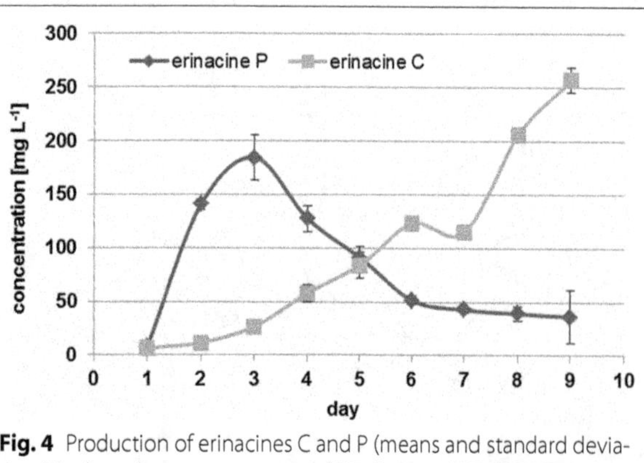


**Fig. 4** Production of erinacines C and P (means and standard deviations) in the culture supernatants of *H. erinaceus* (HPLC–DAD analysis at 210 nm, duplicate cultures)

### HPTLC-mass spectrometry

Extracts of mycelia from submerged cultures of *H. erinaceus* on day 4 and the reference compound erinacine P were both applied on the same HPTLC plate. The MS spectra of the substance zone at $hR_F$ 34 showed $m/z$ values of 515.2 $[M+Na]^+$ and 455.1 $[M+Na-CH_3COOH]^+$ in the $ESI^+$ mode, which were identical with those of the reference compound erinacine P (Fig. 6).

## Discussion

The production of erinacine C strongly depended on the medium composition. Oatmeal and Edamin® K supported the formation of erinacine C. Oatmeal as a complex substrate provided protein, fat, carbohydrates, dietary fiber, and mineral nutrients. It might act as a carbon source as well as a nitrogen source. Edamin® K represents a high quality source of peptides and amino acids produced by enzymatic digestion of lactalbumin, which acted in the medium as a nitrogen source.

The comparison of medium 7 to medium 1 proved the essential role of Edamin® K for the production of erinacine C. The maximum concentration of erinacine P of 184 mg $L^{-1}$ was observed after 3 culture days (Fig. 4). Afterwards, the concentrations of erinacine P decreased while the concentrations of erinacine C steadily increased. These data are in good agreement with the results of Kenmoku et al. [12] who suggested erinacine P as a precursor of erinacine C. Erinacine P is supposed to be an early metabolite of the cyathane xylosides family as, e.g., the erinacines A and B were formed from erinacine P. The biosynthetic relationship between erinacine Q, erinacine P, and erinacine C was elucidated by a $1'$-$^{13}$C-D-glucose labeling experiment. Based on NMR data, $1'$-$^{13}$C-erinacine Q was converted into $1'$-$^{13}$C-erinacine C via $1'$-$^{13}$C-erinacine P [11]. The erinacines Q and P may be the common biosynthetic intermediates of cyathane xylosides such as the erinacines A, B, C and H in *H. erinaceus*, the erinacines E, F and G in *H. romosum*, and the striatals and striatins in *Cyathus* spp. However, detailed product formation kinetics of erinacine P and erinacine C have been reported in the present study for the first time. The product formation kinetics suggested an optimum harvest time of 3 days for erinacine P, and of 9 days for erinacine C. The product concentrations of >250 mg $L^{-1}$ for erinacine C could form a basis for a profitable industrial fermentation process.

The production of erinacine C could be further optimized by, e.g., variation of the concentration of Edamin® K, regulation of the culture pH, adjustment of an ideal C/N ratio or addition of certain mineral elements. Krzycykowski et al. [18] reported on the effects of the carbon and nitrogen sources, mineral elements, and of the initial pH of the culture medium on the growth of *H. erinaceus*

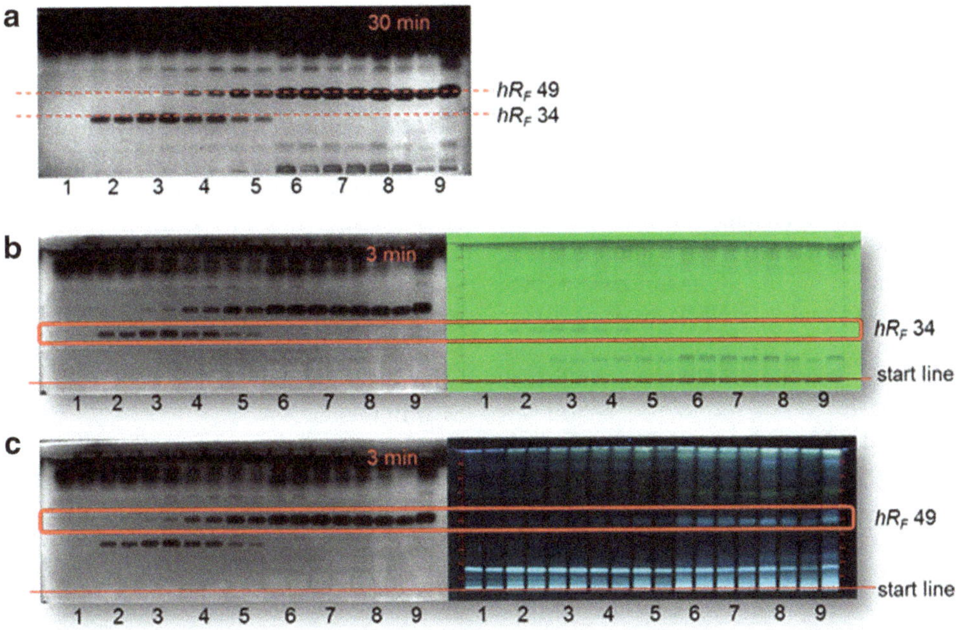


**Fig. 5** Direct bioautography (HPTLC-*Aliivibrio fischeri*) of secondary metabolites from submerged cultures of *H. erinaceus* from culture days 1–9 (applied in duplicate): **a** bioluminescent bioautogram documented as *grey scale* image after 30 min and after 3 min compared to detection at **b** UV 254 nm and **c** UV 366 nm

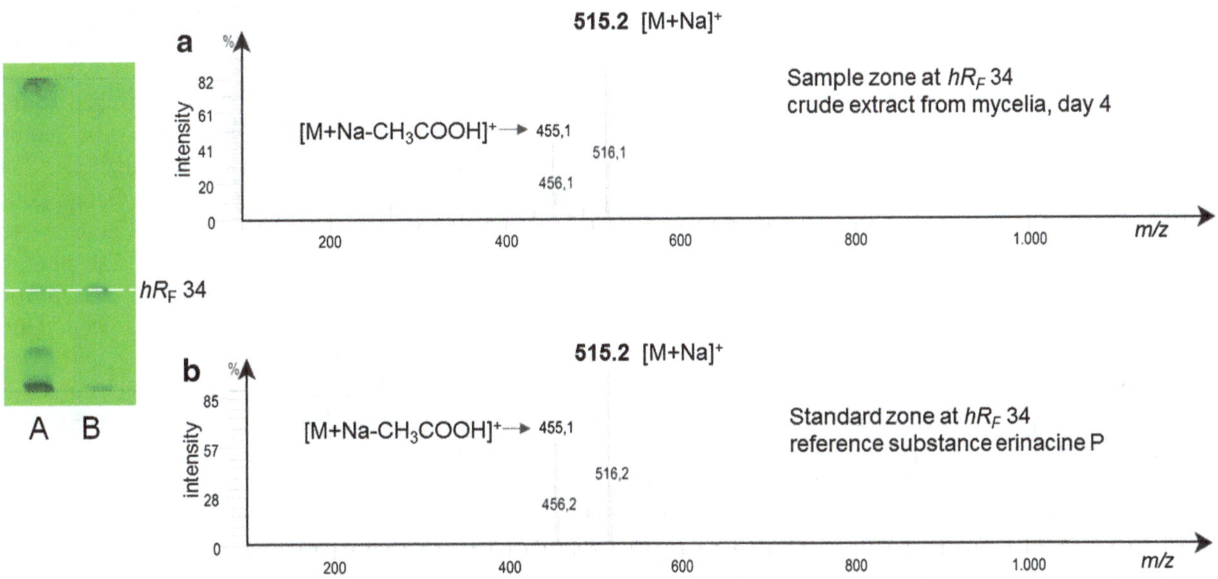


**Fig. 6** HPTLC-MS analysis. **a** Chromatogram and full scan MS spectra of the substance zone at $hR_F$ 34 from crude mycelial extracts of culture day 4 in comparison to and **b** those of the reference erinacine P

and the production of erinacine A. A more acidic pH of the culture medium (i.e. pH 4.5) positively affected both mycelial growth and erinacine A production. Likewise, the addition of $ZnSO_4$ and NaCl improved the synthesis of erinacine A. Considering the role of erinacine P as a precursor of cyathane xylosides, it may be necessary to optimize the yields of erinacine P to achieve higher product concentration of erinacine C.

HPTLC is commonly used in food analysis to detect, e.g., dyes, sugars, antioxidants, or artificial sweeteners. However, HPTLC is also broadly applied for the analysis of natural products. Recently, isolated sesquiterpenoids from *Cyperus rotundus* have been screened for their antioxidative potential [19]. The authors described a validated HPTLC method for the identification and quantification of solavetivone, aristolone and nootkatone, which showed no significant variation to the common HPLC method. The combination of HPTLC and bioassays (bioautography) or HPTLC and MS/NMR indicated that HPTLC might be considered as a fast and highly reliable method for compound identification or zone confirmation [20–24].

In the present study, HPTLC-UV/vis/FLD-bioassay was used in addition to HPLC–DAD for the analysis of secondary metabolites from submerged cultures of *H. erinaceus*. While erinacine C remained at the start line under the chromatographic conditions applied, the substance at $hR_F$ 34 was confirmed to be erinacine P. This was substantiated by mass spectrometric data, UV/vis spectra, and comparison to the authentic reference compound. Comparable product formation kinetics for erinacine P were obtained by HPLC–DAD and HPTLC analyses (Additional file 1: Figures S1, S2). Different from HPLC–DAD, HPTLC allowed for visualization of antibacterial compounds directly on the plate. The antimicrobial activity of erinacine P has been described previously in [26]. Apart from erinacine P, a second compound with strong activity against *Aliivibrio fischeri* bacteria ($hR_F$ 49) was detected in the supernatants of *H. erinaceus*. HPTLC-MS analyses did not allow for an unambiguous identification of this compound and of at least two further metabolites with antibacterial activity. Further HPTLC-high-resolution MS studies will be necessary to elucidate the structures of these metabolites.

## Conclusions

Oatmeal and Edamin® K were found to be essential medium components for the production of erinacine C. The maximum concentration of erinacine P was observed after 3 culture days, while thereafter the concentration of erinacine C steadily increased. The proposed HPTLC-UV/vis/FLD-bioassay method allowed for an efficient detection and also quantification of erinacine P. Bioactive secondary metabolites from submerged cultures of *H. erinaceus* were observed instantly via direct bioautography (HPTLC-*Aliivibrio fischeri*). The direct link to antimicrobial compounds and confirmation by mass spectrometry was considered as streamlined tool for natural product analysis.

## Methods

### Fungal strain and reference compounds

*Hericium erinaceus* (FU70034, isolated from basidiocarp tissue) was obtained from InterMed Discovery (IMD) Natural Solutions, Dortmund, Germany. The fungus was maintained on a solid medium containing 20 g $L^{-1}$ malt extract (Fluka, Neu-Ulm, Germany) and 15 g $L^{-1}$ agar–agar (Roth, Karlsruhe, Germany). The identity of the strain was confirmed by comparison of its ITS nrDNA sequences with reference data in NCBI BLAST® (for sequence data cf. Additional file 1: Figure S3). The strain FU70034 showed 95 % identity to *Hericium erinaceus* from different geographic origins (Additional file 1: Table S1 with references).

Reference substances of erinacine C and erinacine P were isolated from extracts of *H. erinaceus* cultures by preparative HPLC and identified by MS and NMR analyses as described previously [25].

### Medium composition and cultivation

Yeast malt medium consisted of D-(+)-glucose monohydrate 4.0 g $L^{-1}$, malt extract 10.0 g $L^{-1}$, and yeast extract 4.0 g $L^{-1}$ (all Roth). The pH was adjusted to 6.3, and the medium was autoclaved at 121 °C for 20 min prior to use. Various main culture media were tested (data not shown), and the production of erinacine C was confirmed only in sugar molasses medium (medium 1). The optimization of erinacine C production was thus based on medium 1. One of the carbon sources was omitted in the media 2–7, respectively, and the nitrogen source Edamin® S was changed to Edamin® K in media 7 and 8 (Table 1). Molasses was obtained from Südzucker, Offstein, Germany and oatmeal was from Dr. Oetker, Düsseldorf, Germany. Sucrose and ammonium sulfate were purchased from Roth, and D-Mannitol, Edamin® S, Edamin® K, and calcium carbonate were purchased from Sigma, Taufkirchen, Germany. The media were autoclaved at 121 °C for 20 min prior to use.

Pre-cultures of *H. erinaceus* were grown submerged in 250 mL Erlenmeyer flasks containing 100 mL yeast malt medium at 24 °C and 150 rpm for 7 days. Afterwards, 40 mL homogenized mycelia were inoculated into 400 mL main culture medium in 1000 mL Erlenmeyer flasks and incubated at 24 °C and 150 rpm for further 9 days. All samples for HPTLC analysis were obtained from cultures in medium 7.

### Sample preparation

Twenty mL culture supernatant were extracted with 20 mL ethyl acetate (≥99.5 %, Ph. Eur., Roth). The ethyl acetate phase was dried over sodium sulfate (water free, >99 %, Roth) and 15 mL of the extract was

**Table 1 Main culture media 1–8 for optimization of erinacine C production**

| Medium concentration (g $L^{-1}$) | 1 | 2 | 3 | 4 | 5 | 6 | 7 | 8 |
|---|---|---|---|---|---|---|---|---|
| Molasses | 5 | – | 5 | 5 | 5 | 5 | 5 | 5 |
| Oatmeal | 5 | 5 | – | 5 | 5 | 5 | 5 | – |
| D-(+)-glucose monohydrate | 1.5 | 1.5 | 1.5 | – | 1.5 | 1.5 | 1.5 | 1.5 |
| Sucrose | 4 | 4 | 4 | 4 | – | 4 | 4 | 4 |
| D-mannitol | 4 | 4 | 4 | 4 | 4 | – | 4 | 4 |
| Edamin® S or K | 0.5 S | 0.5 S | 0.5 S | 0.5 S | 0.5 S | 0.5 S | 0.5 K | 0.5 K |
| Ammonium sulfate | 0.5 | 0.5 | 0.5 | 0.5 | 0.5 | 0.5 | 0.5 | 0.5 |
| Calcium carbonate | 1.5 | 1.5 | 1.5 | 1.5 | 1.5 | 1.5 | 1.5 | 1.5 |

evaporated to dryness. The residue was dissolved in 1 mL acetonitrile (Chromasolv, gradient grade, Sigma) for HPLC and HPTLC analyses.

Additionally, the concentrations of erinacine C in the mycelia were analyzed by HPLC–DAD. Therefore, the mycelia were separated from the culture supernatants by centrifugation (2880×*g*, 4 °C, and 10 min) and washed 3 times with water. Afterwards, the mycelia were extracted with 20 mL ethyl acetate twice. The combined ethyl acetate phases were dried over sodium sulfate, and 30 mL of the extract was evaporated to dryness. The residue was dissolved in 1 mL acetonitrile for analysis.

Mycelia from culture day 4 were used for HPTLC-MS analyses. The mycelia were prepared as described before, except that 20 mL of the extract was evaporated to dryness. All experiments were performed in duplicate.

### HPLC–DAD analysis

The HPLC system consisted of pump L-7100, autosampler L-7200, interface D-7000, and diode array detector L-7455 (Merck Hitachi, Darmstadt, Germany). A reversed phase Nucleosil® 100-5 $C_{18}$, CC 125/3 mm with a respective guard column (CC 8/3 mm) was used (Macherey–Nagel, Düren, Germany). A mixture of acetonitrile (A) (HPLC gradient grade, Sigma) and bidistilled water (B) served as eluent. Gradient: 30 % A (0 min)—50 % A (15 min)—50 % A (16 min)—100 % A (23 min)—100 % A (38 min)—30 % A (43 min)—30 % A (47 min). The flow rate was 0.6 mL $min^{-1}$.

### HPTLC-UV/vis/FLD analysis

HPTLC plates silica gel 60 $F_{254}$, 20 cm × 10 cm, were obtained from Merck, Darmstadt, Germany. Prior to use, the plates were pre-washed with methanol—water (4:1, v/v) and dried (110 °C, 15 min). The sample extracts were applied as 8 mm bands onto the HPTLC plate using the Automatic TLC Sampler ATS 4 (CAMAG, Muttenz, Switzerland). The track distance was 10 mm. In the Twin Trough Chamber 20 × 10 cm (CAMAG), the plate was developed with 9 mL *n*-hexane–ethyl acetate–methanol (6:11:1, v/v/v). For an acidic plate conditioning, 1 mL acetic acid was filled in the opposite chamber trough. After development, the plate was dried at room temperature and documented at UV 254 nm, UV 366 nm, and white light illumination using the TLC Visualizer (CAMAG). The recording of spectra (200–700 nm) and densitometric absorbance measurement at 210 nm were performed by TLC Scanner 4 (CAMAG). All data were processed and evaluated by winCATS version 1.4.7.2018 (CAMAG).

### Bioautography

For bioautography, the HPTLC plate was automatically immersed into the bioluminescent *Aliivibrio fischeri* bacteria suspension prepared according to DIN EN ISO 11348-1, Sect. "Methods". The immersion speed was 4 cm/s and the immersion time 1 s using the Chromatogram Immersion Device (CAMAG). The bioautogram was documented as greyscale image with an image accumulation time of 50 s using the BioLuminizer (CAMAG). Changes were monitored by capturing an image every 3 min over a 30-min period. *Aliivibrio fischeri* bioactive zones were instantly visible as darkened or enlightened zones on the luminescent background.

### HPTLC-mass spectrometry

On the chromatogram, the target zone erinacine P ($hR_F$ 34) was marked at UV 254 nm with a soft pencil. The TLC-MS Interface equipped with the oval elution head of 4 mm × 2 mm (CAMAG) was coupled to the electrospray interface of a single quadrupole mass spectrometer (CMS, Advion, Ithaca, NY, USA). The target zone was eluted with methanol at a flow rate of 0.1 mL $min^{-1}$. The total ion current (TIC) full scan mass spectra (*m/z* 100–1000) were recorded in the positive ionization mode. A plate/system background spectrum was recorded at a comparable migration distance and subtracted from the analyte spectrum. The elution head was flushed with eluent and dried after each measurement.

**Authors' contributions**

TS performed the experiments and drafted the manuscript. GM was responsible for HPTLC analyses and data interpretation and helped to draft the manuscript. HZ designed the study and helped to draft the manuscript. All authors read and approved the final manuscript.

**Author details**

[1] Institute of Food Chemistry and Food Biotechnology, Justus Liebig University Giessen, Heinrich-Buff-Ring 17, 35392 Giessen, Germany. [2] Institute of Nutritional Science, and Interdisciplinary Research Center for Biosystems, Land Use and Nutrition, Justus Liebig University Giessen, Heinrich-Buff-Ring 26-32, 35392 Giessen, Germany.

**Acknowledgements**

Financial support by the excellence initiative of the Hessian Ministry of Science and Art for the LOEWE Center for Insect Biotechnology and Bioresources is gratefully acknowledged.

**Competing interests**

The authors declare that they have no competing interests.

**References**

1. Kawagishi H, Zhuang C. Compounds for dementia from *Hericium erinaceum*. Drug Future. 2008;33:149–55.
2. Li G, Yu K, Li F, Xu K, Li J, He S, et al. Anticancer potential of *Hericium erinaceus* extracts against human gastrointestinal cancers. J Ethnopharmacol. 2014;153:521–30.
3. Jiang S, Wang S, Sun Y, Zhang Q. Medicinal properties of *Hericium erinaceus* and its potential to formulate novel mushroom-based pharmaceuticals. Appl Microbiol Biotechnol. 2014;98:7661–70.
4. Lee KF, Chen JH, Teng CC, Shen CH, Hsieh MC, Lu CC, et al. Protective effects of *Hericium erinaceus* mycelium and its isolated erinacine A against ischemia-injury-induced neuronal cell death via the inhibition of iNOS/p38 MAPK and nitrotyrosine. Int J Mol Sci. 2014;15:15073–89.
5. Wang M, Gao Y, Xu D, Konishi T, Gao Q. *Hericium erinaceus* (Yamabushitake): a unique resource for developing functional foods and medicines. Food Funct. 2014;5:3055–64.
6. Ma BJ, Shen JW, Yu HJ, Ruan Y, Wu TT, Zhao X. Hericenones and erinacines: stimulators of nerve growth factor (NGF) biosynthesis in *Hericium erinaceus*. Mycology. 2010;1:92–8.
7. Kawagishi H, Shimada A, Hosokawa S, Mori H, Sakamoto H, Ishiguro Y, et al. Erinacines E, F, and G, stimulators of nerve growth factor (NGF)-synthesis, from the mycelia of *Hericium erinaceum*. Tetrahedron Lett. 1996;37:7399–402.
8. Kawagishi H, Shimada A, Shirai R, Okamoto K, Ojima F, Sakamoto H, et al. Erinacines A, B and C, strong stimulators of nerve growth factor (NGF)-synthesis, from the mycelia of *Hericium erinaceum*. Tetrahedron Lett. 1994;35:1569–72.
9. Kawagishi H, Simada A, Shizuki K, Ojima F, Mori H, Okamoto K, et al. Erinacine D, a stimulator of NGF-synthesis, from the mycelia of *Hericium erinaceum*. Heterocycl Commun. 1996;2:51–4.
10. Lee EW, Shizuki K, Hosokawa S, Suzuki M, Suganuma H, Inakuma T, et al. Two novel diterpenoids, erinacines H and I from the mycelia of *Hericium erinaceum*. Biosci Biotechnol Biochem. 2000;64:2402–5.
11. Kenmoku H, Shimai T, Toyomasu T, Kato N, Sassa T. Erinacine Q, a new erinacine from *Hericium erinaceum*, and its biosynthetic route to erinacine C in the basidiomycete. Biosci Biotechnol Biochem. 2002;66:571–5.
12. Kenmoku H, Sassa T, Kato N. Isolation of erinacine P, a new parental metabolite of cyathane-xylosides, from *Hericium erinaceum* and its biomimetic conversion into erinacines A and B. Tetrahedron Lett. 2000;41:4389–93.
13. Takano M, Umino A, Nakada M. Synthetic studies on cyathins: enantioselective total synthesis of (+)-allocyathin $B_2$. Org Lett. 2004;6:4897–900.
14. Trost BM, Dong L, Schroeder GM. Total synthesis of (+)-allocyathin $B_2$. J Am Chem Soc. 2005;127:2844–5.
15. Watanabe H, Nakada M. Biomimetic total synthesis of (−)-erinacine E. J Am Chem Soc. 2008;130:1150–1.
16. Wender PA, Bi FC, Brodney MA, Gosselin F. Asymmetric synthesis of the tricyclic core of NGF-inducing cyathane diterpenes via a transition-metal-catalyzed [5+2] cycloaddition. Org Lett. 2001;3:2105–8.
17. Snider BB, Vo NH, O'Neil SV. Synthesis of (±)-allocyathin $B_2$ and (+)-erinacine A. J Org Chem. 1998;63:4732–40.
18. Krzyczkowski W, Malinowska E, Herold F. Erinacine A biosynthesis in submerged cultivation of *Hericium erinaceum*: quantification and improved cultivation. Eng Life Sci. 2010;10:446–57.
19. Priya Rani M, Padmakumari KP. HPTLC and reverse phase HPLC methods for the simultaneous quantification and in vitro screening of antioxidant potential of isolated sesquiterpenoids from the rhizomes of *Cyperus rotundus*. J Chromatogr B. 2012;904:22–8.
20. Adhami HR, Scherer U, Kaehlig H, Hettich T, Schlotterbeck G, Reich E, et al. Combination of bioautography with HPTLC-MS/NMR: a fast identification of acetylcholinesterase inhibitors from galbanum. Phytochem Anal. 2013;24:395–400.
21. Morlock GE, Brett N. Correct assignment of lipophilic dye mixtures? A case study for high-performance thin-layer chromatography–mass spectrometry and performance data for the TLC–MS Interface. J Chromatogr A. 2015;1390:103–11.
22. Meier B, Spriano D. Modern HPTLC—a perfect tool for quality control of herbals and their preparations. J AOAC Int. 2010;93:1399–409.
23. Gössi A, Scherer U, Schlotterbeck G. Thin-layer chromatography-nuclear magnetic resonance spectroscopy—a versatile tool for pharmaceutical and natural products analysis. Chimia. 2012;66:347–9.
24. Naumoska K, Vovk I. Analysis of triterpenoids and phytosterols in vegetables by thin-layer chromatography coupled to tandem mass spectrometry. J Chromatogr A. 2015;1381:229–38.
25. Shen T, Hof LM, Hausmann H, Stadler M, Zorn H. Development of an enzyme linked immunosorbent assay for detection of cyathane diterpenoids. BMC Biotechnol. 2014;14:98.
26. Anke T, Rabe U, Schu P, Eizenhöfer T, Schrage M, Steglich W. Studies on the biosynthesis of striatal-type diterpenoids and the biological activity of herical. Z Naturforsch. 2002;57c:263–71.

# 17

# Multiple genotypes within aecial clusters in *Puccinia graminis* and *Puccinia coronata*: improved understanding of the biology of cereal rust fungi

Anna Berlin*, Berit Samils and Björn Andersson

**Abstract**

**Background:** Cereal rust fungi (*Puccinia* spp.) are among the most economically important plant pathogens. These fungi have a complex life cycle, including five spore stages and two hosts. They infect one grass host on which they reproduce clonally and cause the cereal rust diseases, while the alternate host is required for sexual reproduction. Although previous studies clearly demonstrate the importance of the alternate host in creating genetic diversity in cereal rust fungi, little is known about the amount of novel genotypes created in each successful completion of a sexual reproduction event.

**Results:** In this study, single sequence repeat markers were used to study the genotypic diversity within aecial clusters by genotyping individual aecial cups. Two common cereal rusts, *Puccinia graminis* causing stem rust and *Puccinia coronata* the causal agent of crown rust were investigated. We showed that under natural conditions, a single aecial cluster usually include several genotypes, either because a single pycnial cluster is fertilized by several different pycniospores, or because aecia within the cluster are derived from more than one fertilized adjoining pycnial cluster, or a combination of both.

**Conclusion:** Our results imply that although sexual events in cereal rust fungi in most regions of the world are relatively rare, the events that occur may still significantly contribute to the genetic variation within the pathogen populations.

**Keywords:** Stem rust, Crown rust, Cereal rusts, Life cycle

## Background

Cereal rust fungi (*Puccinia* spp.) are among the most studied plant disease-causing agents, as they affect cereals and grasses in all parts of the world, potentially causing devastating yield losses. Some of the most important cereal rust diseases are stem rust (caused by *P. graminis*), stripe rust (*P. striiformis*), leaf rust on wheat (*P. triticina*) leaf rust on rye (*P. recondita*), barley leaf rust (*P. hordei*) and crown rust on oats (*P. coronata*) [1, 2]. Cereal rust species can be subdivided based on their host specificity [3], and all cereal rusts may infect a wide variety of wild grass species [4]. The different specializations are referred to as *formae speciales* (f. sp.), i.e. the wheat-infecting *P. graminis* is referred to as *P. graminis* f. sp. *tritici* and the oat-infecting type as *P. graminis* f.sp. *avenae* [3].

Cereal rust fungi are obligate biotrophs that have macrocyclic and heteroecious life cycles, including five spore stages and two hosts. The gramineous hosts enable efficient clonal reproduction while the alternate hosts are necessary for sexual reproduction, which constitutes an important source of genetic diversity [5]. The spore stage causing disease on cereals and other grasses is the uredinial stage, where clonally propagated dikaryotic urediniospores re-infect the gramineous hosts in several

*Correspondence: anna.berlin@slu.se
Department of Forest Mycology and Plant Pathology, Swedish University of Agricultural Sciences, Box 7026, 750 07 Uppsala, Sweden

infection cycles. The urediniospores are thick-walled and withstand solar irradiation [6], which enable rusts to potentially spread with winds over large distances [7]. As the gramineous host matures, the fungus initate teliospore production. The teliospores of *P. graminis* appear in the same pustules (or sori) as the urediniospores, while teliospores of *P. coronata* create a characteristic blackish ring around the pustule of urediniospores [8]. In temperate climates, the teliospores are the overwintering spores. *Puccinia graminis* and *P. coronata* undergo karyogamy in autumn, whereupon meiosis starts and is completed in spring when the teliospores germinate and haploid basidiospores develop [9, 10]. The basidiospores can only disperse in the range of hundreds of meters [11] before infecting young tissue of the alternate hosts, where pycnia and pycniospores (syn. spermogonia and spermatia) develop. One infection originating from a single basidiospore usually gives rise to multiple pycnia in a tight group (pycnial cluster) on the upper side of the host leaf [12]. The pycniospores are produced in a sugary nectar that attracts insects that act as vectors and transfer the spores between pycnia [10, 13]. Pycniospores can also disperse by splashing raindrops. When a pycniospore from one pycnial cluster successfully fertilizes the receptive hyphae of a pycnial cluster of the opposite mating type, a dikaryotic mycelium is formed which grows towards the lower side of the leaf where an aecium develops, and dikaryotic aeciospores are formed. The aecium has a cup-like form and usually appears close together with other aecia in an aecial cluster (or cup cluster), opposite the pycnial cluster [13]. It has been suggested that spores within an aecium are genetically identical [8, 14]. The aeciospores are released from the aecia and infect the gramineous hosts, where new uredinia and urediniospores are produced.

The role of the alternate hosts of rust fungi on cereals has regained attention after the emergence of the virulent *P. graminis* race Ug99 [15] and the discovery of barberry as the aecial host of the stripe rust fungus *P. striiformis* [16]. The life cycle of *P. graminis* has been known for centuries [17], and both *P. striiformis* and *P. graminis* have the shrub barberry (*Berberis* spp.) and Mahonia (*Mahonia* spp.) as their alternate hosts [16, 17]. The alternate hosts for *P. coronata*, the fungus causing crown rust on oats and other grasses, are *Rhamnus* spp. and *Frangula alnus*. However, the literature reports that the *forma specialis* infecting oats only produce aecia on *Rhamnus cathartica* [4, 8].

Jin et al. [18] showed that the presence of the alternate hosts *Berberis vulgaris*, *Mahonia repens* and *M. aquifolium* in North Western United States of America maintained the diversity within the *P. graminis* population, whereas it has declined to a single clonal linage on the west side of the Rocky Mountains where barberry has been successfully eradicated. Other studies also show that the genetic diversity within the pathogen populations on the gramineous hosts are high in regions where the growing seasons are clearly separated and the alternate host are needed for the rusts survival [19–21]. In areas where the fungus can survive year-round in the clonal, uredinial, stage on its gramineous host, the genetic diversity originating from sexual reproduction is suggested to be limited. One example of this is the Ug99 race lineage of *P. graminis*. This race was first detected in Uganda in 1999 [15] and fifteen years later it dominated the *P. graminis* population in East Africa [22].

Traditionally, the studies of variation within cereal rust fungi have been based on race analysis where virulence against known resistance genes is tested. Many rust resistance genes have been identified and characterized [23]. The race identification is performed by inoculating urediniospore isolates on a set of differential host lines, each carrying different resistance genes. Specific virulence phenotypes (races or pathotypes) are defined by the phenotypic reactions on the differential host lines, giving a specific resistance pattern for each fungal isolate [24, 25]. Earlier studies, using race analysis of uredinial offspring, show that phenotypically different individuals can be produced from a single aecial cluster [26, 27]. Although virulence phenotyping is useful for its purpose, it will not identify overall genetic variation or separate among individual genotypes efficiently. By applying molecular methods and DNA-based markers, genetic relationships among fungal isolates can be determined based on neutral markers and with higher resolution [28].

In our study, we aimed to determine the genotypic diversity after a single completion of the sexual part of the life cycle of cereal rust fungi under natural conditions. Aecial clusters from natural infections of *P. graminis*, collected on *B. vulgaris*, and *P. coronata*, collected on *R. cathartica* and *F. aluns*, were analyzed. The genotypic diversity among cups within single aecial clusters on the alternate hosts of these fungi was investigated using single sequence repeat (SSR) markers.

## Results

In total, the eight SSR markers used to analyze the aecial collections of *P. graminis* identified 46 alleles, and the ten SSR markers used in the aecial collections of *P. coronata* identified 40 alleles (Table 1). For *P. coronata*, the allele distribution showed a clear population differentiation between the two aecial hosts *F. alnus* and *R. cathartica*. Only one of the identified alleles (locus PcaSSR B02, allele 167) was shared between samples collected from the two hosts. One SSR marker (PcaSSR A66) completely failed to amplify in the samples collected from *F. alnus*, and

**Table 1 Allele sizes of SSR loci for *Puccinia coronata* and *Puccinia graminis***

| *Puccinia coronata* | *Rhamnus cathartica* | *Frangula alnus* | *Berberis vulgaris* |
|---|---|---|---|
| PcaSSRA59[a] | 137, 140, 143 | 134 | |
| PcaSSRA66[a] | 175, 185, 187 | | |
| PcaSSRA67[a] | 189, 191 | 172 | |
| PcaSSRA73[a] | 150 | 156, 160 | |
| PcaSSRB02[a] | 159, 161, 164, 167* | 167*, 172 | |
| PcaSSRB09[a] | 142, 151 | 125 | |
| PcaSSRB25[a] | 197, 201, 203, 205 | | |
| PcaSSRB33[a] | 182, 186, 188, 196 | 166 | |
| PcaSSRC52[a] | 192, 200, 202 | 190 | |
| PcaSSRC76[a] | 143 | 137, 139, 141, 152 | |
| ***Puccinia graminis*** | | | |
| Pgestssr021[b] | | | 231, 237, 240, 243, 246, 249, 252 |
| PgtSSR21[c] | | | 167, 169, 171, 173, 175 |
| Pgestssr024[b] | | | 121, 130, 133, 156, 159 |
| Pgestssr109[d] | | | 159, 162, 168, 171, 174 |
| Pgestssr255[b] | | | 228, 231, 234, 237, 240, 243, 246 |
| Pgestssr279[d] | | | 169, 175, 178, 181, 184 |
| Pgestssr368[d] | | | 229, 232, 235, 238, 241, 245, 248 |
| PgtCAA53[e] | | | 176, 185, 202, 211, 214 |

* Allele shared between hosts

Source of SSR markers: [a] Dambroski et al. [29], [b] Berlin et al. [20], [c] Szabo [30], [d] Zhong et al. [31], [e] Jin et al. [32]

one (PcaSSR B25) only amplified two out of 18 samples (Additional file 1: Table S1).

In total 22 aecial clusters were analyzed; 12 representing *P. coronata,* where 7 were excised from *R. cathartica* and 5 from *F. alnus*, and 10 representing *P. graminis* which were all excised from *B. vulgaris* (Table 2, Additional file 1: Table S1 and Additional file 2: Table S2). The number of aecial cups analyzed for each cluster reflects the size of the particular cluster, and ranged between 2 and 21 cups genotyped (N), with an overall average of 8.4 cups genotyped per cluster.

The multilocus genotype (MLG) for each individual aecial cup (aecium) was determined by the combination of the alleles from all individual SSR markers for that particular sample. Multiple genotypes were detected in all but one aecial cluster, with an average of 4.1 MLGs for *P. coronata* and 4.9 MLGs per cluster for *P. graminis* (Table 2). It should be noted that in particular for *P. coronata*, the high number of non-detected alleles for samples collected from *F. alnus* could lead to an underestimation of the number of MLGs. Within each aecial cluster, in most cases, one of the two alleles for each SSR locus is invariant among aecial cups whereas the other allele may differ (Figs. 1, 2). A few exceptions to this pattern were detected in three of the 10 clusters in *P. graminis*, where one or a few aecial cups did not share the common allele with other cups in the same cluster (cluster cups 1603 D-H; 1613 A-D; 1617 A, Additional file 2: Table S2).

## Discussion

The present study aimed at determining the fine-scale genotypic diversity in the aecial stage of cereal rust fungi. Our results showed that under natural conditions, different genotypes are usually produced within each aecial cluster on the alternate host of both *P. graminis* and *P. coronata*. This implicates that a single mating event will contribute substantially to the genetic variation in a cereal rust fungal population. Multiple genotypes within aecial clusters have previously been indicated by virulence phenotyping [26, 27]. After artificial fertilization of one pycnial cluster, aeciospores from a single aecial cup (or aecium) produced one virulence phenotype, while other cups within the same aecial cluster produced offspring with a different phenotype. Here we used molecular markers to demonstrate that multiple genotypes within aecial clusters are common in nature, and as many as ten different multilocus genotypes were detected within a single cluster.

Multiple genotypes within an aecial cluster suggest multiple fertilization events within a pycnial cluster. We showed that the aecial cups within a cluster usually shared one allele for each SSR loci, while the second allele in many cases differed between the cups. This pattern

**Table 2 Number of aecial cups genotyped (N) and number of identified multilocus genotypes (MLGs) within each aecial cluster of *Puccinia coronata* and *Puccinia graminis* respectively, collected on *Rhamnus cathartica, Frangula alnus* and *Berberis vulgaris***

| | Host | N | MLG |
|---|---|---|---|
| *P. coronata* | | | |
| 1604 | *Rhamnus cathartica* | 5 | 4 |
| 1605 | *Rhamnus cathartica* | 21 | 10 |
| 1606 | *Rhamnus cathartica* | 6 | 6 |
| 1607 | *Rhamnus cathartica* | 6 | 4 |
| 1608 | *Rhamnus cathartica* | 4 | 3 |
| 1609 A–G | *Rhamnus cathartica* | 7 | 6 |
| 1609 H–J | *Rhamnus cathartica* | 3 | 2 |
| 1611 A–C | *Fragula aluns* | 3 | 1 |
| 1611 D–F | *Fragula aluns* | 3 | 2 |
| 1611 G–M | *Fragula aluns* | 7 | 6 |
| 1611 N–O | *Fragula aluns* | 2 | 2 |
| 1611 P–R | *Fragula aluns* | 3 | 3 |
| Total | | 70 | 49 |
| *P. graminis* | | | |
| 1601 | *Berberis vulgaris* | 13 | 4 |
| 1603 | *Berberis vulgaris* | 16 | 9 |
| 1612 A–F | *Berberis vulgaris* | 6 | 3 |
| 1612 G–I | *Berberis vulgaris* | 3 | 2 |
| 1612 J–O | *Berberis vulgaris* | 6 | 2 |
| 1613 | *Berberis vulgaris* | 14 | 7 |
| 1614 | *Berberis vulgaris* | 17 | 5 |
| 1615 | *Berberis vulgaris* | 15 | 6 |
| 1616 | *Berberis vulgaris* | 7 | 4 |
| 1617 | *Berberis vulgaris* | 18 | 8 |
| Total | | 115 | 49 |

indicates that one pycnial cluster (consisting of pycnia with identical genotypes) are fertilized by genotypically different pycniospores, i.e. coming from different pycnial clusters. The shared allele would thus originate from the resident pycnial cluster, and the second allele would originate from the fertilizing pycniospores. This pattern was observed for both *P. graminis* and *P. coronata*, indicating that it may be consistent across different cereal rust species. A similar allelic pattern has been observed in the blister rust fungus of Scots pine, *Cronatium flaccidum*, where one allele was shared within aecial lesions, indicating multiple mating's between the spermogonium (syn. pycnium) and fertilizing spermatia (syn. pycniospores) [33]. In nature, multiple fertilizations are supported by insects who are attracted by the pycnispore-containing nectar in the pycnia and travel from leaf to leaf and transfer pycniospores among pycnia. It is also possible that closely located pycnial clusters may contribute to the same aecial cluster, and in such case both alleles of the SSR loci might differ between some aecial cups. Such a pattern was not common in the present study, but was detected in a few cases for *P. graminis* (Additional file 2: Table S2).

As a result of meiosis taking place in the teliospore, the offspring produced from a selfing of a clonal linage of a rust fungus will be genetically diverse. This was demonstrated by two recent studies [14, 34]. For example, Tian et al. [34] selfed a *P. striiformis* isolate, resulting in 118 viable offspring's determined as 82 multilocus genotypes and 24 virulence phenotypes. When diversity is described by virulence phenotyping, the true genotypic diversity will be underestimated, since only variation caused by a limited number of virulence genes is assessed, and additionally, virulence genes are usually recessive und can thus be hidden in the avirulent phenotypes. The study by Tian et al. [34] clearly shows this discrepancy and it also illustrates the level of overall genetic variation originating from sexual recombination within a single clonal linage. These results, in combination with our findings, imply that there will be large amounts of genetic and genotypic variation in the aeciospore inoculum reaching and infecting the grass hosts.

Our results on host specific allele sizes of SSR loci confirm the usefulness of SSR markers to distinguish between *formae speciales* of *P. coronata*, as they agree with the previously described DNA sequence based genetic differentiation of the two *formae speciales* [35] and their ability to only infect specific alternate hosts [4].

A practical implication of our results for population genetic studies as well as virulence phenotyping is the scale of sampling required and interpretations of virulence tests. Algorithms in population genetic analyses are often based on single isolates (i.e. single individuals) and thus sampling in the aecial stage of cereal rust fungi should be done on individual aecial cups, rather than entire aecial clusters. If sampling and analyses are done on genotypic mixtures it may give confusing and erroneous results. The same is also applicable for virulence phenotyping. If mixtures of aeciospore genotypes are tested, the results could imply complex virulence races where in reality simple virulence races were mixed together. Additionally, there is some inconsistency in terminology of aecial morphology, which could lead to misinterpretations. According to many textbooks and other literature [14, 36, 37], a single cup in an aecial cluster of *P. graminis* represents an aecium. However in some studies, the whole aecial cluster (with several aecial cups) is denoted as an aecium [38, 39]. In other studies, it may be unclear whether an aecium refers to a single aecial cup or a cup cluster.

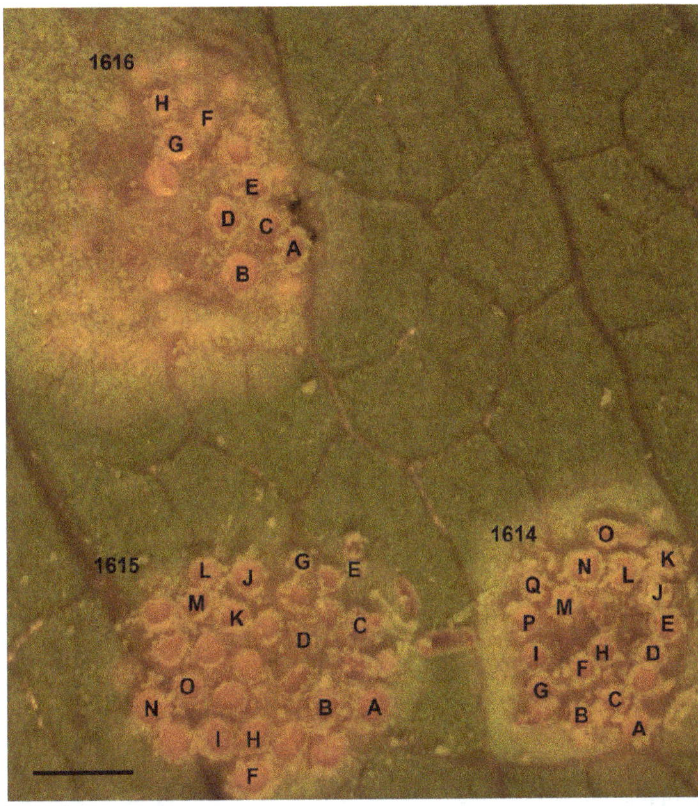


| Sample ID | PgtSSR21 | | Pgestssr024 | | Pgestssr279 | | Pgestssr368 | |
|---|---|---|---|---|---|---|---|---|
| 1614 A | 169 | 171 | 133 | 133 | 175 | 175 | 229 | 235 |
| 1614 B | 169 | 169 | 133 | 133 | 175 | 175 | 229 | 235 |
| 1614 C | 169 | 169 | 133 | 133 | 175 | 175 | 229 | 235 |
| 1614 D | 169 | 169 | 133 | 133 | 175 | 175 | 229 | 235 |
| 1614 E | 169 | 169 | 133 | 133 | 175 | 175 | 229 | 235 |
| 1614 F | 169 | 169 | 133 | 133 | 175 | 175 | 229 | 235 |
| 1614 G | 169 | 169 | 133 | 133 | 175 | 175 | 229 | 235 |
| 1614 H | 169 | 169 | 133 | 133 | 175 | 175 | 229 | 235 |
| 1614 I | 169 | 169 | 133 | 133 | 175 | 175 | 229 | 235 |
| 1614 J | 169 | 169 | 133 | 130 | 175 | 175 | 229 | 235 |
| 1614 K | 169 | 169 | 133 | 133 | 175 | 175 | 229 | 235 |
| 1614 L | 169 | 169 | 133 | 130 | 175 | 175 | 229 | 235 |
| 1614 M | 169 | 169 | 133 | 133 | 175 | 175 | 229 | 235 |
| 1614 N | 169 | 169 | 133 | 133 | 175 | 175 | 229 | 235 |
| 1614 O | 169 | 169 | 133 | 133 | 175 | 175 | 229 | 235 |
| 1614 P | 169 | 169 | 133 | 133 | 175 | 175 | 229 | 235 |
| 1614 Q | 169 | 169 | 133 | 133 | 175 | 175 | 229 | 235 |
| 1615 A | 171 | 171 | n. d. | n. d. | 175 | 178 | 241 | 232 |
| 1615 B | 171 | 171 | 133 | 133 | 175 | 178 | 241 | 232 |
| 1615 C | 171 | 171 | 133 | 133 | 175 | 178 | 241 | 232 |
| 1615 D | 171 | 171 | 133 | 133 | 175 | 178 | 241 | 232 |
| 1615 E | 171 | 171 | 133 | 133 | 175 | 175 | 241 | 241 |
| 1615 F | 171 | 171 | 133 | 133 | 175 | 178 | 241 | 232 |
| 1615 G | 171 | 171 | 133 | 133 | 175 | 178 | 241 | 232 |
| 1615 H | 171 | 171 | 133 | 133 | 175 | 178 | 241 | 232 |
| 1615 I | 171 | 171 | 133 | 133 | 175 | 178 | 241 | 232 |
| 1615 J | 171 | 171 | 133 | 133 | 175 | 178 | 241 | 232 |
| 1615 K | 171 | 171 | 133 | 130 | 175 | 178 | 241 | 232 |
| 1615 L | 171 | 171 | 133 | 133 | 175 | 178 | 241 | 232 |
| 1615 M | 171 | 171 | 133 | 133 | 175 | 178 | 241 | 232 |
| 1615 N | 171 | 171 | 133 | 133 | 175 | 178 | 241 | 232 |
| 1615 O | 171 | 171 | 133 | 133 | 175 | 178 | 241 | 232 |
| 1616 A | 171 | 171 | 133 | 133 | 175 | 175 | 235 | 229 |
| 1616 B | 171 | 171 | 133 | 130 | 175 | 175 | 235 | 232 |
| 1616 C | 171 | 171 | 133 | 133 | 175 | 175 | 235 | 229 |
| 1616 D | 171 | 171 | 133 | 133 | 175 | 175 | 235 | 229 |
| 1616 E | n. d. | n. d. | 133 | 133 | 175 | 175 | 235 | 229 |
| 1616 F | 171 | 171 | 133 | 133 | 175 | 175 | 235 | 229 |
| 1616 G | 171 | 171 | 133 | 130 | 175 | 175 | 235 | 229 |

**Fig. 1** *Puccinia graminis* aecial clusters and SSR allele sizes of selected markers. *Bar* 1 mm

## Conclusions

The genetic diversity of the uredinial, clonal stage of both *P. graminis* and *P. coronata* on the cereal hosts, has been extensively studied in several areas of the world. In the present study, we emphasise the importance of the aecial stage on the alternate host in creating genotypic diversity, where one successful completion of the sexual cycle in natural condition may result in aecial clusters containing several novel genotypes. This means that genetic diversity is introduced to the cereal rust fungal populations more efficiently than previously acknowledged. The main conclusion of our result is that although sexual events of cereal rust fungi in most regions of the world are relatively rare, the events that do occur may still contribute significantly to the genetic variation within the pathogen populations.

## Methods

Collections of natural infected leaves showing clear symptoms of cluster cup rust were performed at two locations (Hågadalen N59°49′16″ E17°36′4″ and Skerike N59°37′40″ E16°30′13″) in late June. Both locations were located approximately 500 m from the closest cereal fields. Samples of *Berberis vulgaris* and *F. alunus* leaves were collected in Hågadalen and *R. cathartica* leaves in Skerike. The leaf samples were put in a plant press before returning to the laboratory.

The aecia were photographed and single aecial cups were carefully cut to enable genotyping of the individual cups within an aecial cluster. Sample identities were indicated on the photo for later correlation between sample position within an aecial cluster and genotype. Aecia on *B. vulgaris* and *R. cathartica* were commonly found and were excised from both the same and different leaves. However, aecial clusters on *F. alnus* were scarce and all were excised from the same leaf. For *P. coronata*, 7 aecial cluster were sampled from *R. cathartica* and 5 from *F. alnus*, and 3–21 and 2–7 individual cups per cluster, respectively, were successfully genotyped. For *P. graminis*, 10 aecial clusters were sampled, and 3–18 aecial cups were successfully genotyped. An average of 7 cups per cluster was sampled, the differences in number of samples largely reflect the actual number of aecial cups within the sampled cluster.

For each of the sampled aecial cups, DNA was extracted using the OmniPrep Kit (G-Biosciences) for fungal tissues, with the adjustment of the amounts of reagents to half to adjust for small sample sizes. Eight

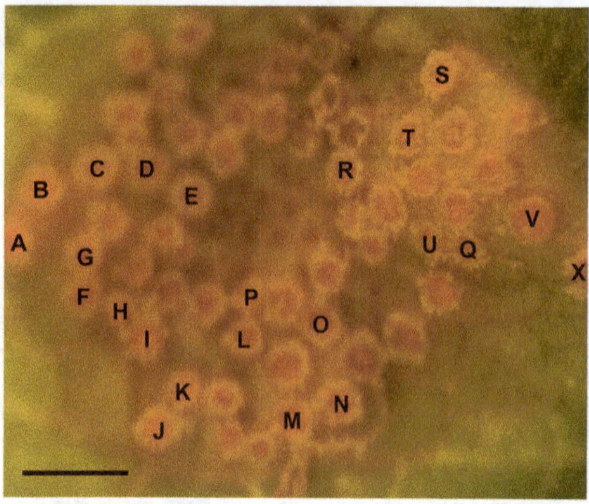


| Sample | PcaSSRA67 | | PcaSSRB33 | | PcaSSRA59 | | PcaSSRC52 | | PcaSSRB02 | |
|---|---|---|---|---|---|---|---|---|---|---|
| 1605 A | 191 | 191 | 182 | 182 | 143 | 143 | 200 | 200 | 161 | 159 |
| 1605 C | 191 | 191 | 182 | 182 | 143 | 143 | 200 | 200 | 161 | 161 |
| 1605 D | 191 | 191 | 182 | 182 | 143 | 143 | 200 | 200 | 161 | 161 |
| 1605 E | 191 | 191 | 182 | 188 | 143 | 143 | 200 | 200 | 161 | 161 |
| 1605 F | 191 | 191 | 182 | 182 | 143 | 143 | 200 | 200 | 161 | 161 |
| 1605 G | 191 | 191 | 182 | 182 | 143 | 143 | 200 | 200 | 161 | 164 |
| 1605 H | 191 | 191 | 182 | 182 | 143 | 143 | 200 | 200 | 161 | 161 |
| 1605 I | 191 | 191 | 182 | 182 | 143 | 143 | 200 | 200 | 161 | 161 |
| 1605 J | 191 | 191 | 182 | 182 | 143 | 143 | 200 | 200 | 161 | 161 |
| 1605 K | 191 | 191 | 182 | 182 | 143 | 143 | 200 | 200 | 161 | 161 |
| 1605 L | 191 | 191 | 182 | 182 | 143 | 143 | 200 | 200 | 161 | 161 |
| 1605 M | 191 | 191 | 182 | 182 | 143 | 143 | 200 | 200 | 161 | 161 |
| 1605 N | 191 | 191 | 182 | 182 | 143 | 143 | 200 | 200 | 161 | 167 |
| 1605 O | 191 | 191 | 182 | 182 | 143 | 143 | 200 | 200 | 161 | 161 |
| 1605 P | 191 | 191 | 182 | 182 | 143 | 143 | 200 | 200 | 161 | 161 |
| 1605 Q | 191 | 191 | 182 | 182 | 143 | 143 | 200 | 200 | 161 | 161 |
| 1605 S | 191 | 191 | 182 | 182 | 143 | 143 | 200 | 200 | 161 | 161 |
| 1605 T | 191 | 189 | 182 | 188 | 143 | 140 | 200 | 200 | 161 | 167 |
| 1605 U | 191 | 189 | 182 | 182 | 143 | 143 | 200 | 200 | 161 | 161 |
| 1605 V | 191 | 189 | 182 | 188 | 143 | 140 | 200 | 192 | 161 | 159 |
| 1605 X | 191 | 191 | 182 | 188 | 143 | 140 | 200 | 200 | 161 | 159 |

**Fig. 2** *Puccinia coronata* aecial cluster and SSR allele sizes of selected markers. *Bar* 1 mm

single sequence repeat (SSR) markers developed for *P. graminis* [20, 30–32] and 10 SSR markers developed for *P. coronata* [29] were applied on the different fungal samples to study the genetic diversity within the aecial clusters (Table 1). The PCR reaction was performed in a 14 μl reaction with the final concentration of 14 ng DNA, 0.4 μM of the each forward and reverse primers respectively, 3.5 mM $MgCl_2$, 0.02 mM of each of the dNTP, 0.05 U/μl of DreamTaq DNA polymerase and Dream Taq buffer according to manufacturer's recommendation (Thermo Fischer Scientific). PCR conditions were initial denaturation at 94 °C, followed by 35 cycles of denaturation at 94 °C for 30 s, annealing at 55 °C for 30 s and extension at 72 °C for 30 s, and a final extension of 72 °C for 10 min. The length of the amplicons was determined using ABI 3730xl DNA Analyzer (SciLifeLab, Uppsala, Sweden) and was scored using the software GeneMarker (Softgenetics). To calculate the number of genotypes within each aecial cluster, the R add-in software PoppR was used [40].

#### Authors' contributions

A. Berlin collected and prepared the samples, performed the data analysis, and all authors equally contributed in the data interpretation and writing of the manuscript. All authors read and approved the final manuscript.

#### Acknowledgements

This work was funded by the Carl Tryggers foundation for scientific research.

#### Competing interests

The authors declare that they have no competing interests.

#### Consent for publication

The authors all agree on publish this manuscript.

#### Funding

This work was funded by the Carl Tryggers foundation for scientific research.

#### References

1. Savile DBO. Taxonomy of the cereal rust fungi. Orlando: Academic Press; 1984.
2. Bolton MD, Kolmer JA, Garvin DF. Wheat leaf rust caused by *Puccinia triticina*. Mol Plant Pathol. 2008;9(5):563–75.
3. Eriksson J, Henning E. Die Getrideroste ihre Geschichte und Natur sowie Massregeln gegen dieselben. Stockholm: P. A. Nordstedt & Söner; 1896.
4. Gäumann E. Die Rostpilze Mitteleuropas. Bern: Buchdruckeri Büchler & Co; 1959.
5. Roelfs AP, Groth JVA. Comparison of virulence phenotypes in wheat stem rust populations reproducing sexually and asexually. Phytopathology. 1980;70(9):855–62.
6. Maddison AC, Manners JG. Sunlight and viability of cereal rust uredospores. Trans Br Mycol Soc. 1972;59(3):429–43.
7. Nagarajan S, Singh DV. Long-distance dispersion of rust pathogens. Annu Rev Phytopathol. 1990;28:139–53.
8. Simons MD. Crown rust. In: Roelfs AP, Bushnell WR, editors. The cereal rusts. St. Paul: Academic Press Inc.; 1985. p. 131–72.
9. Hacquard S, Delaruelle C, Frey P, et al. Transcriptome analysis of poplar rust telia reveals overwintering adaptation and tightly coordinated karyogamy and meiosis processes. Front Plant Sci. 2013;4:456. doi:10.3389/fpls.2013.00456.
10. Leonard KJ, Szabo LJ. Stem rust of small grains and grasses caused by *Puccinia graminis*. Mol Plant Pathol. 2005;6(2):99–111.
11. Roelfs AP. Wheat and rye stem rust. Orlando: Academic Press; 1985.
12. Anikster Y, Eilam T, Mittelman L, et al. Pycnial nectar of rust fungi induces cap formation on pycniospores of opposite mating type. Mycologia. 1999;91(5):858–70.
13. Craigie JH. Disovery of the function of the pycnia of the rust fungi. Nature. 1927;120(3030):765–7.
14. Rodriguez-Algaba J, Walter S, Sørensen CK, et al. Sexual structures and recombination of the wheat rust fungus *Puccinia striiformis* on *Berberis vulgaris*. Fungal Genet Biol. 2014;70:77–85.
15. Pretorius ZA, Singh PP, Wagorie WW, Payne TS. Detection of virulence to wheat stem rust resistance gene *Sr31* in *Puccinia graminis*. f. sp. *tritici* in Uganda. Plant Dis. 2000;84(2):203.
16. Jin Y, Szabo LJ, Carson M. Century-old mystery of *Puccinia striiformis* life history solved with the identification of *Berberis* as an alternate host. Phytopathology. 2010;100:432–5.

17. Large EC. The advance of the fungi. London: Henderson and Spalding; 1946.
18. Jin Y, Rouse M, Groth J. Population diversity of *Puccinia graminis* is sustained through sexual cycle on alternate hosts. J Integr Agric. 2014;13(2):262–4.
19. Ali S, Gladieux P, Rahman H, Saqib MS, Fiaz M, Ahmad H, Leconte M, Gautier A, Justesen AF, Hovmøller MS, Enjalbert J, De Vallavieille-PopeC. Inferring the contribution of sexual reproduction, migration and off-season survival to the temporal maintenance of microbial populations: a case study on the wheat fungal pathogen *Puccinia striiformis* f.sp. *tritici*. Mol Ecol. 2014;23:603–17.
20. Berlin A, Djurle A, Samils B, Yuen J. Genetic variation in *Puccinia graminis* collected from oat, rye and barberry. Phytopathology. 2012;102:1006–12.
21. Berlin A, Rahmatov M, Muminjanov H, et al. Sexual reproduction contributes to genotypic variation in the population of *Puccinia graminis* in Tajikistan. Eur J Plant Pathol. 2015;141(1):159–68.
22. Newcomb M, Olivera PD, Rouse MN, et al. Kenyan isolates of *Puccinia graminis* f. sp. *tritici* from 2008 to 2014: virulence to SrTmp in the Ug99 race group and implications for breeding programs. Phytopathology. 2016;106(7):729–36.
23. McIntosh RA, Wellings CR, Park RF. Wheat rusts: an atlas of resistance genes. Netherlands: Springer; 1995.
24. Roelfs AP, Martens JW. An international system of nomenclature for *Puccinina graminis* f.sp. *tritici*. Phytopathology. 1988;78:526–33.
25. Chong J, Leonard KJ, Salmeron JJ. A North American system of nomenclature for *Puccinia coronata* f. sp. *avenae*. Plant Dis. 2000;84:580–5.
26. Wang Z, Zhao J, Chen X, et al. Virulence variations of *Puccinia striiformis* f. sp. *tritici* isolates collected from *Berberis* spp. in China. Plant Dis. 2015;100(1):131–8.
27. Ingram DS, Williams PH, Sidhu GS. Advances in plant pathology. Vol. 6, genetics of plant pathogenic fungi. London: Academic Press; 1988.
28. Zhan G, Chen X, Kang Z, et al. Comparative virulence phenotypes and molecular genotypes of *Puccinia striiformis* f. sp. *tritici*, the wheat stripe rust pathogen in China and the United States. Fungal Biol. 2012;116(6):643–53.
29. Dambroski HR, Carson ML. Development and characterization of novel, polymorphic microsatellite markers for oat crown rust, *Puccinia coronata*. Mol Ecol Resour. 2008;8:1395–8.
30. Szabo LJ. Development of simple sequence repeat markers for the plant pathogenic rust fungus, *Puccinia graminis*. Mol Ecol Notes. 2007;7:92–4.
31. Zhong S, Leng Y, Friesen TL, Faris JD, Szabo LJ. Development and characterization of expressed sequence tag-derived microsatellite markers for the wheat stem rust fungus *Puccinia graminis* f. sp. *tritici*. Phytopathology. 2009;99(3):282–9.
32. Jin Y, Szabo LJ, Rouse MN, Fetch T, Pretorius ZA, Wanyera R, Njau P. Detection of virulence to resistance gene Sr36 within the TTKS race lineage of *Puccinia graminis* f. sp. *tritici*. Plant Dis. 2009;93:367–70.
33. Samils B, Ihrmark K, Kaitera J, Stenlid J, Barklund P. New genetic markers for identifying *Cronartium flaccidum* and *Peridermium pini* and examining genetic variation within and between lesions of scots pine blister rust in Sweden. Fungal Biol. 2011;115:1303–11.
34. Tian Y, Zhan G, Chen X, et al. Virulence and simple sequence repeat marker segregation in a *Puccinia striiformis* f. sp. *tritici* population produced by selfing a Chinese isolate on *Berberis shensiana*. Phytopathology. 2015;106(2):185–91.
35. Liu M, Hambleton S. Laying the foundation for a taxonomic review of *Puccinia coronata* s.l. in a phylogenetic context. Mycol Prog. 2013;12(1):63–89.
36. Schumann GL, Leonard KL. Stem rust of wheat (black rust). Plant Health Instr. 2000; doi:10.1094/PHI-I-2000-0721-01.
37. Agrios GN. Chapter eleven—plant diseases caused by fungi. Plant pathology. 5th ed. San Diego: Academic Press; 2005. p. 385–614.
38. Zhao J, Wang L, Wang Z, Chen X, Zhang H, Yao J, Zhan G, Chen W, Huang L, Kang Z. Identification of eighteen *Berberis* species as alternate hosts of *Puccinia striiformis* f. sp. *tritici* and virulence variation in the pathogen isolates from natural infection of barberry plants in China. Phytopathology. 2013;103:927–34.
39. Knott DR. The wheat rusts—breeding for resistance. Berlin: Springer; 2012.
40. Kamvar ZN, Tabima JF, Grünwald NJ. Poppr: an R package for genetic analysis of populations with clonal, partially clonal, and/or sexual reproduction. PeerJ. 2014;2:e281.

# PERMISSIONS

All chapters in this book were first published in FB&B, by BioMed Central; hereby published with permission under the Creative Commons Attribution License or equivalent. Every chapter published in this book has been scrutinized by our experts. Their significance has been extensively debated. The topics covered herein carry significant findings which will fuel the growth of the discipline. They may even be implemented as practical applications or may be referred to as a beginning point for another development.

The contributors of this book come from diverse backgrounds, making this book a truly international effort. This book will bring forth new frontiers with its revolutionizing research information and detailed analysis of the nascent developments around the world.

We would like to thank all the contributing authors for lending their expertise to make the book truly unique. They have played a crucial role in the development of this book. Without their invaluable contributions this book wouldn't have been possible. They have made vital efforts to compile up to date information on the varied aspects of this subject to make this book a valuable addition to the collection of many professionals and students.

This book was conceptualized with the vision of imparting up-to-date information and advanced data in this field. To ensure the same, a matchless editorial board was set up. Every individual on the board went through rigorous rounds of assessment to prove their worth. After which they invested a large part of their time researching and compiling the most relevant data for our readers.

The editorial board has been involved in producing this book since its inception. They have spent rigorous hours researching and exploring the diverse topics which have resulted in the successful publishing of this book. They have passed on their knowledge of decades through this book. To expedite this challenging task, the publisher supported the team at every step. A small team of assistant editors was also appointed to further simplify the editing procedure and attain best results for the readers.

Apart from the editorial board, the designing team has also invested a significant amount of their time in understanding the subject and creating the most relevant covers. They scrutinized every image to scout for the most suitable representation of the subject and create an appropriate cover for the book.

The publishing team has been an ardent support to the editorial, designing and production team. Their endless efforts to recruit the best for this project, has resulted in the accomplishment of this book. They are a veteran in the field of academics and their pool of knowledge is as vast as their experience in printing. Their expertise and guidance has proved useful at every step. Their uncompromising quality standards have made this book an exceptional effort. Their encouragement from time to time has been an inspiration for everyone.

The publisher and the editorial board hope that this book will prove to be a valuable piece of knowledge for researchers, students, practitioners and scholars across the globe.

# LIST OF CONTRIBUTORS

**Sandrine Pigné, Mohammed Kerkoud, Roxane Raulo, Nelly Bataillé-Simoneau, Muriel Marchi, Anthony Kwasiborski, Guillaume N'Guyen, Philippe Simoneau and Thomas Guillemette**
IRHS, Agrocampus-Ouest, INRA, Université d'Angers, SFR 4207 QuaSaV, 49071 Beaucouzé, France

**Etienne Janod and Stéphane Cuenot**
UMR 6502, Institut des Matériaux Jean Rouxel 2, Rue de la Houssinière, BP 32229, 44322 Nantes Cedex 3, France

**Agata Zykwinska**
UMR 6502, Institut des Matériaux Jean Rouxel 2, Rue de la Houssinière, BP 32229, 44322 Nantes Cedex 3, France
Laboratoire Ecosystèmes Microbiens et Molécules Marines pour les Biotechnologies, IFREMER, Rue de l'île d'Yeu, BP 21105, 44311 Nantes Cedex 3, France

**Guillaume Mabilleau**
Plateforme SCIAM, Institut de Biologie en Santé, CHU, Université d'Angers, 4, Rue Larrey, 49933 Angers Cedex, France

**Thiemo Zambanini, Lisa M. Schmitz, Hamed Hosseinpour Tehrani, Elena Geiser, Melanie Beudels, Dominik Venc, Lars M. Blank and Nick Wierckx**
Institute of Applied Microbiology - iAMB, Aachen Biology and Biotechnology - ABBt, RWTH Aachen University, Worringerweg 1, Aachen 52074, Germany

**Sandra K. Hartmann**
Institute of Applied Microbiology - iAMB, Aachen Biology and Biotechnology - ABBt, RWTH Aachen University, Worringerweg 1, Aachen 52074, Germany
BioSC, c/o Forschungszentrum Jülich, 52425 Jülich, Germany

**Markus Schwarzländer**
BioSC, c/o Forschungszentrum Jülich, 52425 Jülich, Germany
Institute of Crop Science and Resource Conservation (INRES), Rheinische Friedrich-Wilhelms-Universität Bonn, 53113 Bonn, Germany
Institute of Plant Biology and Biotechnology (IBBP), Westfälische Wilhelms- Universität Münster, 48143 Münster, Germany

**Linda Büttner**
Department of Biology, Philipps-University Marburg, Karl-von-Frisch-Straße 8, 35032 Marburg, Germany

**Georg Wandrey and Jochen Büchs**
AVT-Aachener Verfahrenstechnik, Biochemical Engineering, RWTH Aachen University, Forckenbeckstraße 51, 52074 Aachen, Germany

**Martin Weinhold**
Menschenfotograf, Am Goldmannpark 21, 12587 Berlin, Germany

**Edeltraud Mast-Gerlach and Vera Meyer**
Department of Applied and Molecular Microbiology, Institute of Biotechnology, Technische Universität Berlin, Gustav-Meyer-Allee 25, 13355 Berlin, Germany

**Alexander Idnurm, Andrew S. Urquhart, Dinesh R. Vummadi and Angela P. Van de Wouw**
School of BioSciences, University of Melbourne, Building 122, Parkville, VIC 3010, Australia

**Steven Chang and Francisco J. López-Ruiz**
Department of Environment and Agriculture, Centre for Crop and Disease Management, Curtin University, Bentley, WA 6102, Australia

**Zinnat Shahina and Taranum Sultana**
Department of Chemistry and Biochemistry, University of Regina, 3737 Wascana Parkway, Regina, SK, Canada

**Tanya E. S. Dahms**
Department of Chemistry and Biochemistry, University of Regina, 3737 Wascana Parkway, Regina, SK, Canada
Regina Qu'Appelle Health Region, Regina, SK, Canada

**Amira M. El-Ganiny**
Microbiology and Immunology Department, Faculty of Pharmacy, Zagazig University, Zagazig, Egypt

**Jessica Minion**
Regina Qu'Appelle Health Region, Regina, SK, Canada

**Malcolm Whiteway**
Centre for Structural and Functional Genomics, Concordia University, Montreal, QC, Canada

**Pedro F. Souza Filho, Patrik R. Lennartsson and Mohammad J. Taherzadeh**
Swedish Centre for Resource Recovery, University of Borås, 50190 Borås, Sweden

**Ramkumar B. Nair**
Mycorena AB, Stena Center 1A, 411 92 Gothenburg, Sweden

**Dan Andersson**
Faculty of Caring Science, Work Life and Social Welfare, University of Borås, 50190 Borås, Sweden

**Timothy C. Cairns, Corrado Nai and Vera Meyer**
Department of Applied and Molecular Microbiology, Institute of Biotechnology, Technische Universität Berlin, Gustav-Meyer-Allee 25, 13355 Berlin, Germany

**Sonia Chadha, Sayaji T. Mehetre, Ravindra Bansal and Prasun K. Mukherjee**
Nuclear Agriculture and Biotechnology Division, Bhabha Atomic Research Centre, Trombay, Mumbai 400085, India

**Alan Kuo, Andrea Aerts and Igor V. Grigoriev**
U.S. Department of Energy Joint Genome Institute, Walnut Creek, CA 94598, USA

**Irina S. Druzhinina**
Research Area Biochemical Technology, Institute of Chemical and Biological Engineering, TU Wien, 1060 Vienna, Austria

**Qusai Al Abdallah, Ana Camila Oliveira Souza, Adela Martin-Vicente, Wenbo Ge and Jarrod R. Fortwendel**
Department of Clinical Pharmacy and Translational Science, University of Tennessee Health Science Center, Memphis, TN, USA

**Alice Rassinger, Robert L. Mach and Astrid R. Mach-Aigner**
Institute of Chemical, Environmental and Bioscience Engineering, TU Wien, Gumpendorfer Str. 1a, 1060 Vienna, Austria

**Joseph Strauss**
Fungal Genetics and Genomics Lab, Department of Applied Genetics and Cell Biology, BOKU-University of Natural Resources and Life Sciences, Konrad Lorenz Str. 24, 3430 Tulln/Donau, Austria

**Agnieszka Gacek-Matthews**
Fungal Genetics and Genomics Lab, Department of Applied Genetics and Cell Biology, BOKU-University of Natural Resources and Life Sciences, Konrad Lorenz Str. 24, 3430 Tulln/Donau, Austria
Institute of Microbiology, University of Veterinary Medicine Vienna, Veterinärplatz 1, 1210 Vienna, Austria

**Sietske Grijseels, Zahida Wasil, Jens C. Frisvad, Kristian Fog Nielsen, Mhairi Workman, Thomas Ostenfeld Larsen and Rasmus John Normand Frandsen**
Department of Biotechnology and Biomedicine, Technical University of Denmark, 2800 Kgs. Lyngby, Denmark

**Carsten Pohl, Yvonne Nygård and Arnold J. M. Driessen**
Molecular Microbiology, Groningen Biomolecular Sciences and Biotechnology Institute, University of Groningen, 9747 AG Groningen, The Netherlands

**Jens Christian Nielsen**
Department of Biology and Biological Engineering, Chalmers University of Technology, 412 96 Gothenburg, Sweden

**Jens Nielsen**
Department of Biology and Biological Engineering, Chalmers University of Technology, 412 96 Gothenburg, Sweden
Novo Nordisk Foundation Center for Biosustainability, Technical University of Denmark, 2800 Kgs. Lyngby, Denmark

**Julian Brandl, Anders Noerregaard and Mikael R. Andersen**
Technical University of Denmark, Soeltofts Plads, Building 223, 2800 Kongens Lyngby, Denmark

**Maria Victoria Aguilar-Pontes and Ronald P. de Vries**
Fungal Physiology, Westerdijk Fungal Biodiversity Institute and Fungal Molecular Physiology, Utrecht University, Uppsalalaan 8, 3584 CT Utrecht, The Netherlands

**Mikko Arvas**
VTT Technical Research Centre of Finland, Tietotie 2, 02044 Espoo, Finland
Finnish Red Cross Blood Service, Helsinki, Finland

**Adrian Tsang**
Concordia University, 7141 Sherbrooke Street West, H4B1R6 Montreal, Québec, Canada

**Arthur F. J. Ram**
Leiden University, Sylviusweg 72, 2333 BE Leiden, The Netherlands

**Paul Schäpe and Vera Meyer**
Berlin University of Technology, Gustav-Meyer-Allee 25, 13355 Berlin, Germany

**Michel Flipphi, Norbert Ág, Levente Karaffa, Napsugár Kavalecz and Erzsébet Fekete**
Department of Biochemical Engineering, University of Debrecen, Debrecen 4032, Hungary

**Gustavo Cerqueira**
Broad Institute of MIT and Harvard, Cambridge, MA, USA

**Claudio Scazzocchio**
Department of Microbiology, Imperial College London, London, UK
Institut de Biologie Intégrative de la Cellule, CEA/CNRS, Université Paris-Saclay UMR, 9198 Orsay, France

**Alexander Idnurm and Candace E. Elliott**
School of BioSciences, University of Melbourne, Melbourne, VIC 3010, Australia

**Andy M. Bailey and Gary D. Foster**
School of Biological Sciences, University of Bristol, Bristol, UK

**Timothy C. Cairns**
Department of Applied and Molecular Microbiology, Technische Universität Berlin, Berlin, Germany

**Giuseppe Ianiri**
Department of Molecular Genetics and Microbiology, Duke University Medical Center, Durham, USA

**Junhyun Jeon**
College of Life and Applied Sciences, Yeungnam University, Gyeongsan, South Korea

**Rudibert King**
Chair of Measurement and Control, Technische Universität Berlin, Berlin, German

**T. Shen and H. Zorn**
Institute of Food Chemistry and Food Biotechnology, Justus Liebig University Giessen, Heinrich-Buff-Ring 17, 35392 Giessen, Germany

**G. Morlock**
Institute of Nutritional Science, and Interdisciplinary Research Center for Biosystems, Land Use and Nutrition, Justus Liebig University Giessen, Heinrich-Buff-Ring 26-32, 35392 Giessen, Germany

**Anna Berlin, Berit Samils and Björn Andersson**
Department of Forest Mycology and Plant Pathology, Swedish University of Agricultural Sciences, 750 07 Uppsala, Sweden

# Index

www.ingramcontent.com/pod-product-compliance
Lightning Source LLC
LaVergne TN
LVHW080717210926
848485LV00003B/482